APPLIED SPECTROSCOPY
A Compact Reference for Practitioners

APPLIED SPECTROSCOPY
A Compact Reference for Practitioners

Edited by

Jerry Workman, Jr.
KIMBERLY-CLARK CORPORATION
NEENAH, WI

Art W. Springsteen
LABSPHERE, INC.
NORTH SUTTON, NH

ACADEMIC PRESS

San Diego London Boston
New York Sydney Tokyo Toronto

This book is printed on acid-free paper. ∞

Academic Press
1300 Boylston Street, Chestnut Hill, MA 02167, USA
525 B Street, Suite 1900, San Diego, CA 92101, USA
http://www.apnet.com

United Kingdom Edition published by
Academic Press Limited
24–28 Oval Road, London NW1 7DX
http://www.hbuk.co.uk/ap/

Library of Congress Cataloging-in-Publication Data

Workman, Jerry.
Applied spectroscopy : a compact reference for practitioners / Jerry Workman, Art W. Springsteen.
p. cm.
Includes bibliographical references and index.
ISBN 0-12-764070-3 (alk. paper)
1. Spectrum analysis—Handbooks, manuals, etc. 2. Spectrum analysis—Industrial applications—Handbooks, manuals, etc. 3. Solids—Spectra—Handbooks, manuals, etc. I. Springsteen. Art W. II. Title.
QC453.W67 1997
621.36′1—dc21 97–41175
CIP

PRINTED IN THE UNITED STATES OF AMERICA
98 99 00 01 02 IC 9 8 7 6 5 4 3 2 1

Contents

4. Spectroscopic Quantitative Analysis

James H. Duckworth

5. Spectroscopic Qualitative Analysis

James H. Duckworth

Section 2 Reflectance Measurements of Solids

6. Reflectance Spectroscopy: An Overview of Classification and Techniques

Art Springsteen

7. Spectroscopy of Solids

Jerry Workman, Jr.

8. Standards for Reflectance Measurements

Art Springsteen

9. Reflectance Measurements of Diffusing Surfaces Using Conic Mirror Reflectometers

Keith A. Snail and Leonard M. Hanssen

10. Bidirectional Reflectance Distribution Function as a Measure of Optical Scatter

S. H. C. P. McCall

14. Spectroscopy of Ceramics

James F. Cordaro

15. Spectroscopy Using Flowing Systems for Chemical, Pharmaceutical, and Biological Applications

Emil W. Ciurczak

16. Textile Application of Molecular Spectroscopy

Subhas Ghosh

17. Solar Measurements

J. Murray Stewart

Contributors

Numbers in parentheses indicate the pages on which the authors' contributions begin.

EMIL W. CIURCZAK (329, 423), Consultant, 13110 Oriole Drive, Calverton, MD 20705

JOHN COATES (49), Coates Consulting, P.O. Box 3176, Newtown, CT 06470

JAMES F. CORDARO (399), QivaStar, Inc., 667 Unit B, 50 Richmond Rd. Ridgecrest, CA 93555

JAMES H. DUCKWORTH (93,165), Galactic Industries Corporation, 395 Main Street, Salem, NH 03079

SUBHAS GHOSH (437), Director, Institute of Textile Technology, 2551 Ivy Road, Charlottesville, VA 33903

LEONARD M. HANSSEN (269), Optical Technology Division, National Institute of Standards & Technology, Gaithersburg, MD 20899

S.H.C.P. MCCALL (299), Stellar Optics Research International Corporation (SORIC),78 Normark Drive, Thornhill, Ontario L3T 3R1, Canada

KEITH A. SNAIL (269), Optical Sciences Division, Naval Research Laboratory, Washington, DC 20375

ART SPRINGSTEEN (193, 247), Labsphere, Inc., Shaker St., North Sutton, NH 03260

J.MURRAY STEWART (343, 459), Pilkington-Libbey Owens Ford Company, Technical Center, 1701 East Broadway Avenue, Toledo, OH 43606

SUSAN WHITE (389), Thermal Protection Materials and Systems, NASA Ames Research Center, Moffet Field, CA 94035

JERRY WORKMAN, JR. (3, 29, 225), Kimberly-Clark Corporation, Analytical Science and Technology, 2100 Winchester Road, Neenah, WI 54956

Preface

The scope of this book started as a grandiloquent treatise on the endless subjects pertaining to molecular spectroscopy. The original definition was due in part to the encouragement of Jane Ellis, then an editor at Academic Press. Since its inception, Art and I redefined the book, due to practical limitations, to a more focused manuscript.

The refined scope of this book is to delineate practical, tested general spectroscopic methods (for ultraviolet, visible, and infrared spectrometry) in clear language for novice users and to provide a reference resource for advanced spectroscopists. We have sought to make the text practical in the sense of containing important information and equations that will be referred to on a continuing basis by practicing spectroscopists.

Book topics include practical aspects of spectrometers and spectrometry, sample preparation, chemometrics, and calibration practices; reflectance measurements; and standard materials measurements. We have placed an emphasis on reflectance and color measurements due to their common usage in today's spectroscopic laboratories. Methods for selecting a measurement technique are included, as well as solar and color measurements. Spectrometry of new materials, ceramics, and textiles are covered by respective experts in their fields. An appendix of practical reference data for UV-VIS, NIR, IR bench top, IR microspectroscopy, and UV Sun Protection Factor (SPF) measurements is included.

Jerry Workman, Jr.
Neenah, Wisconsin, 1997

Acknowledgments

As the lead editor on this volume I would like to extend our thanks to the authors of the individual chapters, who in most cases maintained enthusiasm and vigor during the reviewing and updating process. Many thanks to Dr. John Coates, Jamie Duckworth, Dr. Keith Snail, Dr. Leonard Hanssen, Dr. Susan McCall, Dr. Emil Ciurczak, Murray Stewart, Dr. Susan White, Dr. John Cordaro, and Dr. Subhas Ghosh—your efforts in particular are greatly appreciated by the editors.

I would also like to extend my appreciation to my wife, Rebecca, who with her patient friendship continually encourages my work; and to my children, Cristina, Stephannie, Daniel, Sara, and Michael, who have grown up wondering when this primer would be completed. My thanks certainly go out to Art who has acted with me as an editorial "tag-team." Art is a dedicated scientist and a reflectance expert in the truest sense, an eclectic conversationalist on topics from abaci to zymurgy, and a true friend. It is our hope that this book will be well-worn over the years from frequent use.

Jerry Workman, Jr.
Neenah, Wisconsin, 1997

It has been a long and interesting road since Jerry and I agreed to do this book over beignets at the Cafe du Monde in New Orleans. Jerry is not only a fine spectroscopist but a great friend with whom to trade Civil War stories. I dare say there is not another scientist around who could provide you, on a moment's notice, with George Armstrong Custer's demerit list at West Point.

Thanks certainly go out to my wife, Dr. Kathryn Springsteen, and daughter Anne (a seventh grader when this project started, a freshman at Wellesley now). I would also like to thank my coworkers in the Reflectance Research group, Jim Leland (my theorist whom I count on to challenge my empirical thinking . . .), Trudy Ricker, Dr. Shannon Storm, Ramona Yurek, and especially Michelle Bennett, who helps me keep my sanity.

The real dedication, though, must go to Phillip Lape, the founder of Labsphere. Phil's support, encouragement, and total confidence in my abilities have allowed us both to thrive and Labsphere to grow. Phil's retirement last year was a joy to him and a loss to me—he's what kept me intellectually stimulated and inventing over the past ten years. In these days of musical jobs, I am forever grateful to have been doing what I love to do over the past ten years. So Phil, enjoy your retirement—this book is for you!

Art Springsteen
North Sutton, New Hampshire, 1997

Section 1

THE PRACTICAL BASICS OF SPECTROMETRY

OPTICAL SPECTROMETERS

JERRY WORKMAN, JR.

Kimberly-Clark Corporation,
Analytical Science and Technology

ISBN 0-12-764070-3/$25.00.

I. Introduction

The first assumption in spectroscopic measurements is that Beer's law relationship applies between a change in spectrometer response and the concentration of analyte material present in a sample specimen. *The Bouguer, Lambert, and Beer relationship* assumes that the transmission of a sample within an incident beam is equivalent to 10 exponent the negative product of the molar extinction coefficient (in $L \cdot mol^{-1}\ cm^{-1}$), multiplied by the concentration of a molecule in solution (in $mol \cdot L^{-1}$) times the pathlength (in cm) of the sample in solution. There are some obvious (and not so obvious) problems with this assumption. The main difficulty in the assumed relationship is that the molecules often interact, and the extinction coefficient (absorptivity) may vary due to changes in the molecular configuration of the sample. The obvious temperature, pressure, and interference issues also create a less than ideal situation for the analyst. However, for many (if not most) analytical problems the relationship holds well enough.

The following are properties of the Bouguer, Lambert, and Beer (Beer's law) relationship:

$$T = \frac{I}{I_0} = 10^{-\varepsilon cl},$$

where T is the transmittance, I_0 is the intensity of incident energy, I is the intensity of transmitted light, ε is the molar extinction coefficient (in $L \cdot mol^{-1}\ cm^{-1}$), c is the concentration (in $mol \cdot L^{-1}$), and l is the pathlength (in cm).

To simplify the previous equation into its more standard form showing absorbance as a logarithmic term, used to linearize the relationship between spectrophotometer response and concentration, gives the following expression as the relationship between absorbance and concentration:

$$\text{Abs.} = A = -\log\left(\frac{I}{I_0}\right) = -\log(T) = \varepsilon cl.$$

The following statements hold true for what is most often termed Beer's law: (1) The relationship between transmittance and concentration is nonlinear, and (2) the relationship between absorbance and concentration is linear. Beer's law is the common basis for *quantitative analysis*. Knowledge of Beer's law allows us to calculate the maximum theoretical dynamic range for an instrument using a few simple mathematical relationships.

The goal in the design of an optical spectrometer is to maximize the energy (or radiant power) from a light source through the spectrometer to the detector. The optical throughput for a spectrometer is dependent on multiple factors, such as the light source area, the apertures present within the light path, lens transmittance and mirror reflectance losses, the exit aperture, and the detector efficiency. A simplified model to describe such a system is given below (Note: the reader is referred to Section 4.7 for a definition of throughput). From *paraxial optical theory* it is known that

$$A_s \Omega_{Ent} = A_d \Omega_{Exit}.$$

This relationship applies when A_s is the illuminated area of the light source, A_d is the illuminated detector area, Ω_{Ent} is the solid angle subtended by the source at the entrance aperture of the spectrometer, and Ω_{Exit} is the solid angle subtended at the detector by the exit aperture. This relationship is essential to determining the distances of the source and detector from the main spectrometer optical system as well as the size of the entrance and exit apertures (or slits).

Several terms are useful in any discussion of spectrometry; these include the following: *selectivity* is the specific sensor response to the *analyte of interest*; *sensitivity* is the *quantifiable level of response* from a sensor with respect to the concentration of a specified analyte; and *detection limit* is the smallest concentration difference that can be detected above the background noise level of a sensor. This is often estimated using two or three times the background root mean square (RMS) noise as a signal and estimating the concentration for this signal level using a calibration curve. Additional details on the physics of spectrometers can be found in Blaker (1970), Bracey (1960), Braun (1987), Ditchburn (1965), Fogiel (1981), and Wist and Meiksin (1986).

II. Types of Spectrometers

The reader is referred to Bracey (1960), Braun (1987), and Wist and Meiksin (1986) for details on conventional spectrometer design.

A. Discrete Photometers

Discrete photometers consist of an irradiance source, discrete interference filters, a sample compartment, and detector, along with appropriate electronics for signal amplification and stabilization of detector signals. The optical configuration for a generic photometer is shown in Fig. 1.

B. Single Beam

Single-beam spectrometers have a single optical channel that is configured to measure either sample or reference channel but not both simultaneously. The resultant spectrum is the ratio of the transmission spectra from sample and reference measurements. In practice the response of the detector mea-

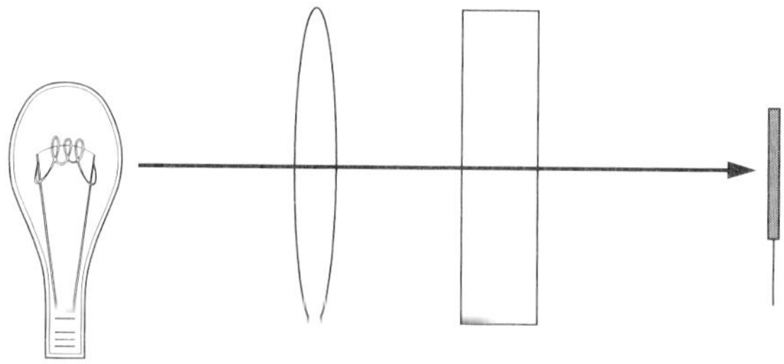

FIG. 1. Interference filter photometer optical configuration.

sured with zero photon energy flux is measured as the *dark current* (Dark). The final transmission (in T units) spectrum from this device is given as

$$\text{Spectrum} = \left(\frac{\text{Sample}_T - \text{Dark}}{\text{Reference}_T - \text{Dark}}\right).$$

If the spectrometer has any instabilities (optical, mechanical, or electronic), the time constant between sample and reference measurements becomes critical. Single-beam instruments must either be more inherently stable or alternate between sample and reference measurements at a frequency so as to make negligible the rate of change of the spectrometer. Designs for single- and double-monochromator spectrophotometers, as well as a diode array design, are shown in Figs. 2A–2C. These designs can be configured as single or double beam.

C. Double Beam

A double-beam spectrometer consists of both sample and reference channels and measures both channels simultaneously. The separation of the light beam from the source is accomplished using a beam splitter so that approximately 50% of the source-emitted energy is divided to both sample and reference channels. The use of the dual-beam concept compensates for instrument instabilities inherent to all spectrometers. The resultant spectrum is the ratio of the sample and reference channels in transmission. As in the case of single-beam instruments, in practice the final spectrum from this device is given as

$$\text{Spectrum} = \left(\frac{\text{Sample}_T - \text{Dark}}{\text{Reference}_T - \text{Dark}}\right).$$

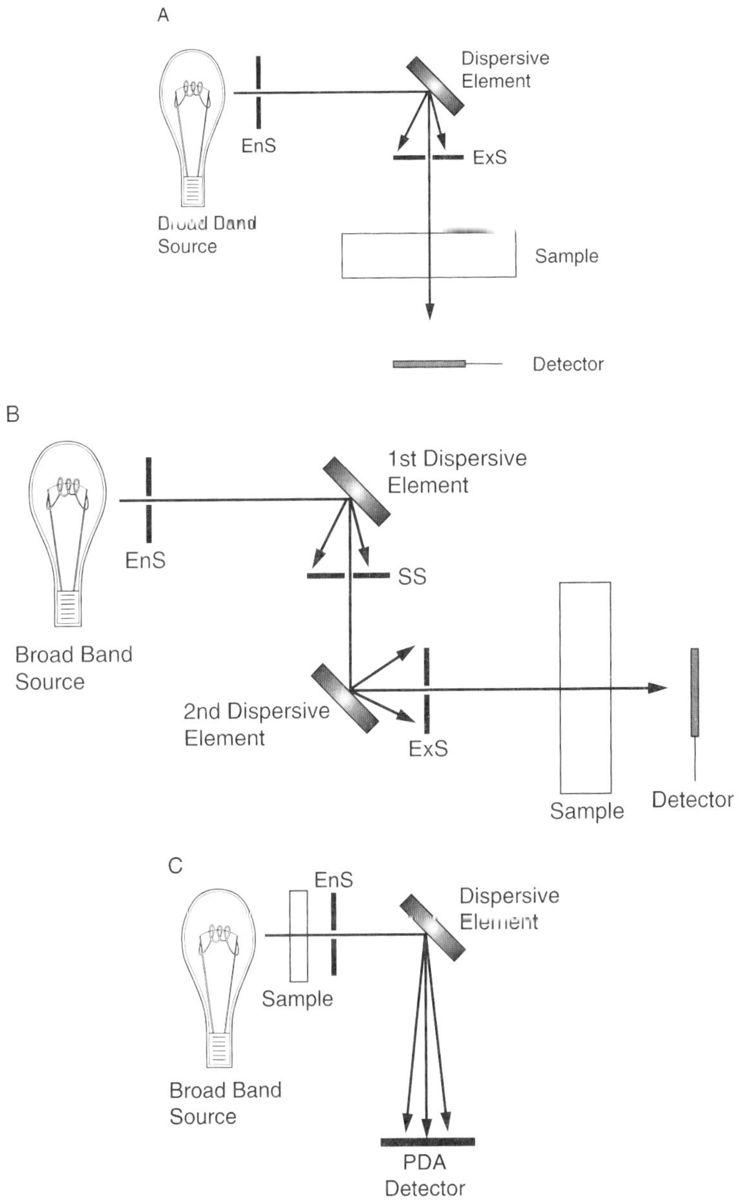

FIG. 2. (A) Single monochromator system (dispersive) optical configuration. EnS and ExS designate entrance slit and exit slit, respectively. (B) Double monochromator system (dispersive) optical configuration. EnS, SS, and ExS designate entrance slit, second slit, and exit slit, respectively. (C) Diode array spectrophotometer (dispersive) optical configuration. EnS designates the entrance slit.

D. Double Beam/Dual Wavelength

Double-beam/dual-wavelength spectrometers are capable of making measurements at two nominal frequencies simultaneously. They possess the capability of illuminating a specimen with two wavelengths (λ_1 and λ_2) while measuring the spectrum at both wavelengths. This is accomplished by using two dispersive elements (e.g., diffraction gratings) to disperse the incident energy from the source onto the sample specimen at the two different wavelengths. A shutter interrupts one of the beams, whereas the other is incident to the detector. The signals from (λ_1 and λ_2 are processed in such a manner that the displayed signal is a differential absorbance of ($A_{\lambda_1} - A_{\lambda_2}$); this differential absorbance is proportional to concentration. Dual-wavelength scanning is used to cancel the effects of background when measuring turbid samples, for quantitative determination of a single component in multicomponent mixtures, or for quantitative determination of high-speed reactions.

E. Interferometer Based

Interferometry is often used to measure wavelengths; this type of measurement is conducive to spectroscopic measurements. In spectroscopy, the accuracy of wavelength measurements can be critical to from 6 to 10 significant figures. Wavelength measurements from interferometer-type devices are much more accurate than those generally made using dispersive-type instruments. The most common type of interferometer is the Michelson type shown in Fig. 3. A parallel beam of light from the spectrometer source is directed into a beam splitter at an angle of 45° from normal. The beam splitter consists of a 50% reflective surface. The reflected portion of light from the beam splitter is reflected to a highly reflective movable mirror (MM), whereas the light transmitted through the beam splitter strikes a fixed mirror (FM). From the perspective of the detector (D) the following phenomena are "observed." If the pathlengths to FM and MM are the same, an observer at the D position will see a bright fringe. If the pathlengths are not the same the phase difference will determine whether the fringe observed at the detector is light or dark. The fixed mirror is only adjusted for tilt to keep it normal to the incident beam. As the MM is moved, the path becomes different and the phase changes. A movement of MM (d) is related to the number of fringes in the interference pattern ($\#f$) observed at D by the relationship:

$$\frac{d}{\#f} = \frac{\lambda}{4},$$

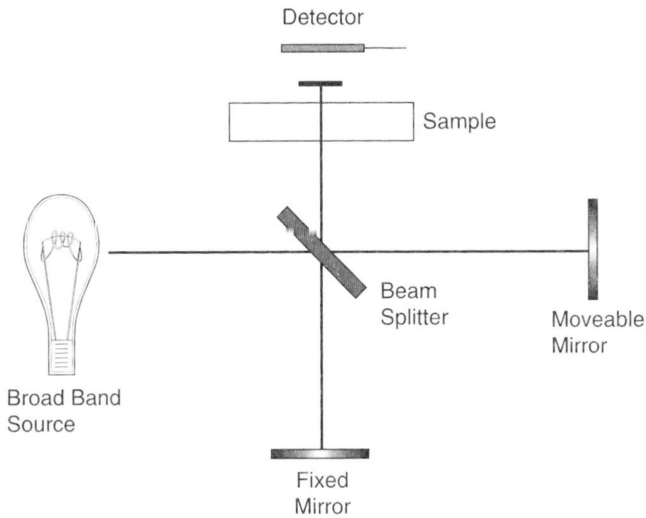

FIG. 3. Fourier transform spectrophotometer (interferometer) optical configuration.

where λ is the wavelength of parallel incident light entering the interferometer, and d is expressed in units identical to wavelength units. The optical design configuration for an interferometer-based instrument is shown in Fig. 3.

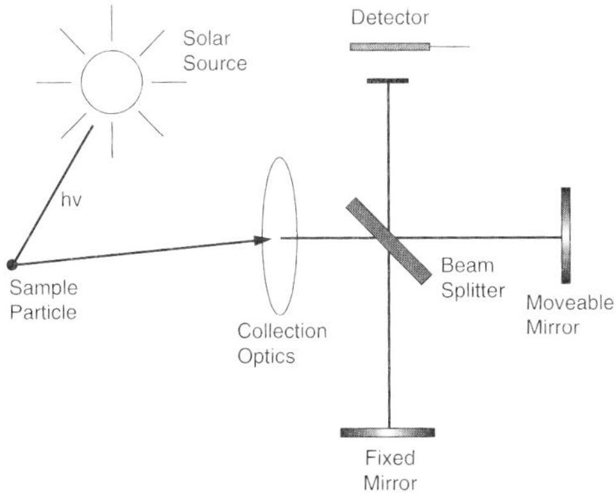

FIG. 4. Open-path sampling configuration using an interferometer.

F. Open-Path and Emission Spectrometers

An open-path spectrometer is designed according to the features of dispersive and interferometer-based instruments with the exception that the sample is located remotely from the instrument and the light source is either sunlight or laser power. The basic configuration for sampling in the open path design is shown in Fig. 4.

III. Details of Spectrometer Components

A. Light Sources

All materials emit electromagnetic radiation, the nature of which depends on the material and its temperature. Gases or vapors of atoms at less than atmospheric pressure emit radiation at discrete wavelengths when an electric current is passed through them. Excited gases containing molecules emit spectra consisting of multiple lines very close together that comprise emission bands. Solid materials emit or radiate continuous spectra across all frequencies. The following relationships define the physics of light sources.

Kirchoff's law states that for a thermal radiator at constant temperature, and in thermal equilibrium, the emissivity (e) at any given frequency is equal to the absorptance (a) for radiation from the same direction so that $e = a$, where *absorptance* is defined as the ratio of energy absorbed by the surface to that of incident energy striking the surface.

An ideal surface absorbing all the energy that strikes its surface could be termed a *black body* (no radiation is reflected or emitted, thus it appears perfectly black at all frequencies). It would also stand from Kirchoff's law that this black body would also be the ideal emitter of radiation.

The *emissivity* (ε) for any black body source is defined as the ratio of the emitted radiance (ρ) by the source to the radiance of a black body (BB) at the same temperature and frequency (wavelength). The following illustrates this:

$$\varepsilon = \frac{\rho}{\rho_{\mathrm{BB}}}.$$

The spectral radiance (ρ) is defined as the radiant power (or photon flux density) emitted per unit source area per unit solid angle (conventionally expressed in $\mathrm{W/m^2/sr}$).

Max Planck derived the principles of quantum mechanics and expressed this using the concept of light quanta as discrete energy "packets" that are transferred with an energy (E) proportional to the frequency of the electromagnetic radiation involved. Thus, *Planck's law* has become one of the most widely known concepts in the physics of electromagnetic radiation. The expression is given as

$$E = h\nu,$$

where E is the energy content of each light quantum, and Planck's constant $(h) = 6.63\ldots \times 10^{-34}$ J · sec; and ν = frequency in $\sec^{-1}$.

From Planck's work, the *Planck's radiation formula* is used to calculate the radiance emitted by from a black body at a given wavelength (λ), emissivity (e), and temperature (T) as

$$\rho_{BB} = 3.745 \times 10^4 \lambda^{-5}\left(\frac{1.4388 \times 10^4}{e^{\lambda T} - 1}\right),$$

where the units for ρ_{BB} are in W/cm^2/μm.

From Planck's radiation formula, other relationships for a black body spectrum can be derived. The *Stefan Boltzmann law* gives the total radiance emitted (ρ_{BB}) from a black body as a function of the black body temperature:

$$\rho_{BB} = \sigma \times T^4,$$

where σ is the Stefan's constant $= 5.67 \times 10^{-8}$ W/m^2 °K^4. The Wien's displacement law demonstrates the shift (to shorter wavelengths) of the maximum peak position for a black body spectrum as a function of temperature:

$$\lambda_{max} = \frac{2898\ \mu\text{m} \times {}^\circ K}{T},$$

where λ_{max} is the peak maximum position from the black body spectrum in micrometers (μm), and T is in °K. [Note: T(°K) = T(°C) + 273.]

B. Detectors

There are two basic types of photon detectors: photoemissive and solid state. The photoemissive type is generally represented by the photomultiplier tube detectors, whereas the solid-state type detectors are represented by photodiode detectors, pyroelectric detectors, and infrared detectors.

In defining the physics of detector devices, several terms deserve explanation. First, the term *specific detectivity* (D*) is essential. This D* value is defined as the detectivity of a radiation detector as a function of the square root of the product of the active detector element area (A) and the bandwidth (ω, in cycles per second), divided by the noise equivalent power

(NEP, in Watts) of the detector element:

$$D^* = \frac{\sqrt{A \times \omega}}{\text{NEP}}.$$

The D^* is reported in cm $\text{Hz}^{\frac{1}{2}}$ W^{-1} units.

C. Filters

Two basic types of interference filters exist: *bandpass filters* and *edge filters.* Bandpass filters transmit light only for a defined spectral band. The transmitted spectral bands may be from less than 1 nm full bandwidth at half maximum (FWHM) transmission band height to 50 nm or more FWHM. Edge filters transmit light either above or below a certain wavelength region; these are referred to as "cut-on" or "cut-off" types, respectively. These filters transmit efficiently throughout a broad region until the transmission limit of the filter substrate material is reached.

Interference filters consist of a solid Fabry–Perot cavity. This is a device made of a sandwich of two partially reflective metallic layers separated by a transparent dielectric spacer layer. The partially reflective layers are made of higher refractive index than the dielectric spacer layer and are $\lambda/4$ in thickness, where λ is the peak wavelength (wavelength of maximum transmission) for the filter. The lower refractive index spacer layer is made to $\lambda/2$ thickness. The thickness of the dielectric spacer layer determines the actual peak transmission wavelength for the filter. Only the $\lambda/2$ light transmits with high efficiency; the other wavelengths experience constructive interference between the multiple-order reflections from the two partially reflective layers.

The wavelength position of the transmittance peak (λ_t) through either a Fabry–Perot interferometer or bandpass interference filter is given as

$$\lambda_t = \frac{2 \times n_\sigma t_\sigma \cos \alpha}{o},$$

where λ_t is the wavelength of maximum transmittance for the filter, n_σ is the refractive index of the dielectric spacer, t_σ is the thickness of the dielectric spacer in micrometers, α is the angle of incidence of the light impinging onto the dielectric spacer, and o is the order number for the interference (a nonzero integer as 0, 1, 2 ...).

The assumption for all interference filters is that incident energy striking the filter is collimated and at a normal incidence. The wavelength of peak transmittance for the filter can be moved by varying the angle of incidence (α) of light impinging on the surface of the filter. The relationship defining the peak position of the maximum transmission is given by

$$\lambda_{\text{New}} = \lambda_{\text{Max}} \sqrt{1 - \left(\frac{n_\varepsilon}{n_\sigma}\right) \sin^2 \alpha},$$

where λ_{New} is the wavelength of the new peak transmission position at incident angle α (when the incident angle is nonnormal; that is, $\geq 0°$), λ_{Max} is the wavelength of the current peak maximum for the interference filter with impinging light at an incident angle ($\alpha = 0°$), n_ε is the refractive index of the surrounding medium (air = 1.0003), n_σ is the refractive index of the dielectric spacer (sometimes referred to as the *effective refractive index*), and α is the angle of incidence of the light impinging onto the filter surface.

D. Gratings

When incident light strikes a diffraction grating, the light is separated into its component wavelengths of light with each wavelength scattered at a different angle. To calculate the particular angle (θ_d) at which each wavelength of light (λ_a) is scattered from a diffraction grating, the following expression is used:

$$\theta_d = \sin^{-1}\left(\frac{o \times \lambda_a}{s}\right),$$

where θ_d is the angle of diffracted light from the normal angle, o is the order number (integers such as 1, 2, 3), s is the spacing of the lines on the grating (in the same units as wavelength), and λ_a is the wavelength of the incident light in air (for white light it represents the wavelength of interest for calculation of the diffraction angle).

The dispersion of a grating refers to how broadly the monochromator disperses (or spreads) the light spectrum at the sample specimen position. Dispersion is generally expressed in units of nm per millimeter. Dispersion depends on the groove density (number of grooves per millimeter) of the grating.

The intensity distribution of light (I) from the surface of a diffraction grating is given by

$$I = \left[\left(\frac{I_0}{S^2}\right)\left(\frac{\sin b}{b}\right)\left(\frac{\sin S \times a}{\sin a}\right)\right],$$

where I is the intensity from the surface; I_0 is the incident energy intensity; S is the number of slits passed by the light following diffraction; $b = \pi \times s/\lambda_a$, where s is the spacing of the lines on the grating (in the same units as wavelength) and λ_a is the wavelength of interest for calculation of the diffraction angle; and $a = \pi \times \sigma/\lambda_a$, where σ is the spacing of the slits (in the same units as wavelength) and λ_a is the wavelength of the incident light in air (for white light it represents the wavelength of interest for calculation of the diffraction angle).

The resolution (P) of a diffraction grating is given by the following expression:

$$P = oN = \frac{\lambda}{\Delta\lambda},$$

where o is the order of the interference pattern, N is the total number of lines on the diffraction grating surface, λ is the wavelength at which the resolution is determined, and $\Delta\lambda$ is the distance in wavelength between the lines or optical phenomena to be resolved.

E. Beam Splitters

Beam splitters are optical devices used to divide and recombine an optical beam (or beams) of light. They can be produced using half-silvered mirrors that reflect approximately 50% of the energy incident to them; the remaining 50% is transmitted through the beam splitter.

F. Prisms

Prisms can be used either to disperse light into its spectral components or as right-angle prisms with reflective coatings on the hypotenuse side to bend light at a 90° angle. The dispersive properties of prisms have been known since the late 17th century. The deflection or dispersion of a prism is described using *Snell's law* at each optical surface of the prism, taking into account the refractive index of the prism at each wavelength. Snell's law describes the change in the path of light crossing an interface between two different materials (with different refractive indices) when the incident angle of the first surface is other than 90°. The wavefronts or wave propagation angle through the interface must move toward the normal angle as shown in Fig. 5.

The numerical explanation of Snell's law is given by

$$n_1 \sin \alpha_1 = n_2 \sin \alpha_2,$$

where n_1 is the refractive index of the first medium at the interface of two materials of varying refractive indices, n_2 is the refractive index of the second medium at the interface of two materials of varying refractive indices, α_1 is the angle of incidence and reflection of the impinging light onto the surface of the second medium, and α_2 is the angle of refraction of

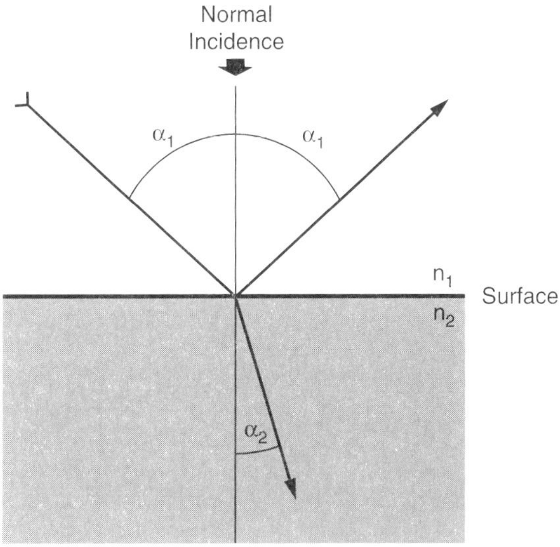

FIG. 5. A demonstration of Snell's Law showing the light wave propagation angles of reflection and refraction at the interface between materials of differing refractive indices as n_1 and n_2.

the light passing through the interface of two materials of varying refractive indices.

For an equilateral (dispersive) prism, the *wave propagation angle* (δ) (shown in Fig. 6) through the prism is given by

$$\delta = \alpha_1 + \sin^{-1}[(\sin\theta)(n^2 - \sin^2\alpha_1)^{1/2} - (\sin\alpha_1 \cos\theta] - \theta,$$

where α_1 is the angle incident to the surface of the prism, θ is the wedge angle of the prism (60° for an equilateral triangle), n is the refractive index of the prism at the frequency (wavelength) of incident energy, and δ is the wave propagation angle (angle of refraction) through the prism.

G. INTERFEROMETER ASSEMBLIES

Multiple interferometer types exist in modern spectrometers; the reader is referred to Candler (1951), Jamieson *et al.* (1963), Steel (1983), and Strobel and Heineman (1989) for a more exhaustive description of the optical configurations for these devices. The classical interferometer design is represented by the Michelson interferometer as shown in Fig. 3. A MM is displaced linearly at minute distances to yield an interference pattern (or interferogram) as a series of sine waves when the interference pattern is observed from a specific field of view (Fig. 7A).

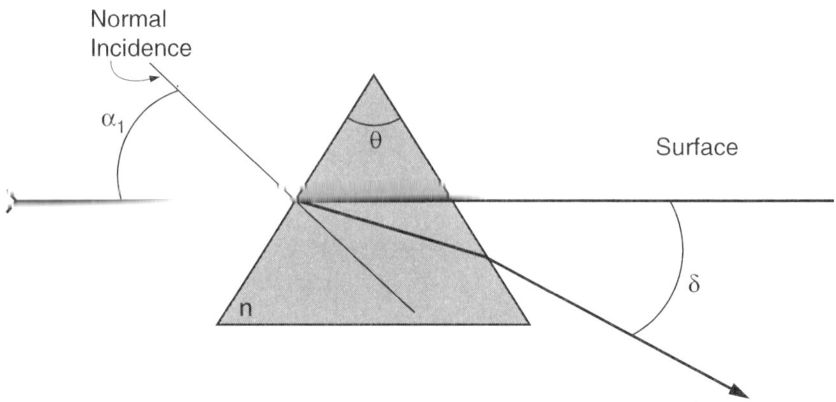

FIG. 6. The wave propagation angles through an equilateral prism for light dispersion.

The interferogram is plotted as the intensity of light (y axis) versus the mirror position (x axis); thus, the signal from the interferometer is a function of time [as the mirror is in motion at a constant velocity (Fig. 7B)]. The raw interferogram, as it is sometimes termed, is converted to a spectrum using the Fourier transform, and a spectrum is determined by ratioing a spectrum determined with a sample in the beam (as the sample spectrum) to a spectrum determined with no sample in the beam (as the background spectrum).

A laser beam of known frequency is used to signal a separate sensor to provide near perfect sampling of the mirror position. The resolution of the interferometer depends on the distance of motion in the movable mirror. The precise distance traversed by the movable mirror can be determined using the number of fringes observed to pass a given field of view in time (t), given the wavelength of light from the laser. The distance traversed by the mirror is given as

$$d = \frac{f\lambda}{2},$$

where d is the distance traveled, f is the number of fringes in an interference pattern passing a specified field of view, and λ is the wavelength of laser light used.

High resolution involves moving the mirror at integral cycles over greater distances per unit time compared to low-resolution measurements. A complete explanation of the physics of interferometry is beyond the scope of this chapter. The reader is referred to Griffiths and de Haseth (1986).

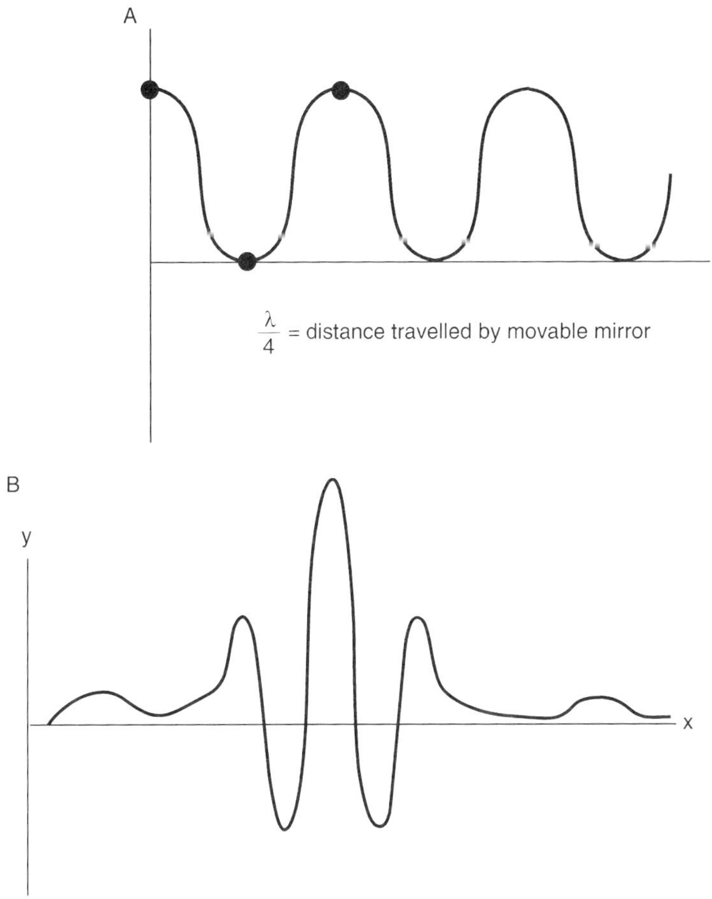

FIG. 7. (A) Interferometric output as a sine wave function. (B) The interferogram is plotted as a function of the light intensity (y axis) versus the mirror position (x axis), thus the signal is a function of time (as the mirror is moved at a constant rate). The raw interferogram is subjected to the multiple-step Fourier transformation and a spectrum results.

H. POLARIZERS

A diversity of polarizing elements exist for the purpose of rotating or selecting light of a specific electronic vector orientation. When the electronic direction vector of light incident to a surface is parallel to the electronic field vector of the surface, increased interaction of the incident light (absorption) occurs. This principle is important in characterizing the surface chemistry for optical components, thin films, metal surfaces, and semiconductor interfaces. The use of a quartz plate can act as a polarizer that will rotate the plane of linearly polarized light. This rotation (P) can be described using the relationship:

$$P = [\pi t(|n_{\Lambda} - n_{P}|)] \times \frac{1}{\lambda_0},$$

where t is the thickness of a quartz plate cut perpendicularly to the optical axis, n_{Λ} is the refractive index for right circularly polarized light, n_P is the index of refraction for left circularly polarized light, and λ_0 is the vacuum wavelength of light entering the quartz plate. Note: for quartz, $n_{\Lambda} = 1.5582$ and $n_P = 1.5581$.

I. Electronic Components Used in Spectrometry

For references describing the electronic components of optical spectrometers, see Braun (1987), Jamieson *et al.* (1963), and Strobel and Heineman (1989). Two special categories of electronic devices deserve representation when discussing spectrometers. The first includes detectors and detector electronics that produce a current or voltage signal proportional to the photon flux striking the detector. Detector stability provided by correct electronic circuitry allows the detector signal to be selectively amplified with the minimum introduction of noise; thus, electronic circuitry enhances the signal-to-noise ratio of the detector signal.

Digital microcomputers comprise the second essential electronic element for modern spectrometers. With the addition of appropriate software, sophisticated instrument control and data processing can enhance the usefulness and user friendliness of spectrometers. These issues are described in detail in Jamieson *et al.* (1963) and Strobel and Heineman (1989).

IV. Properties of Spectrometers

A. Aperture Diameter

To calculate the aperture (a) required by an optical system to resolve two objects with known linear separation, the Rayleigh criterion for resolution is used as

$$a = \frac{1.22 \times \lambda}{\alpha_R},$$

where α_R is the angle of separation from the measuring device exit aperture to the objects to be resolved (this is calculated as tan α = opp./adj., where α_R is expressed in radians), λ is the wavelength of light observed from the objects, and a is the aperture of the optical system.

B. Entrance and Exit Pupils

The *entrance pupil* refers to the size and location of the entrance aperture between the light source and the remainder of an optical system. The *exit pupil* refers to the size and location of the exit aperture within an optical system (or train) just prior to the detector.

C. Bandpass and Resolution

The terms bandpass and resolution are used to express the capability of a spectrometer to separate spectral bands or lines that are separated by some finite distance. For an instrument that disperses energy over a prespecified spectral region of the electromagnetic spectrum, the bandpass of a spectrometer is used to describe which portion of the spectrum can actually be isolated by the spectrometer in a "pure" wavelength form. The spectrometer bandpass is dependent on the dispersion of the grating (see Section 3.4) and the entrance and exit slit width. An example is often used to illustrate the

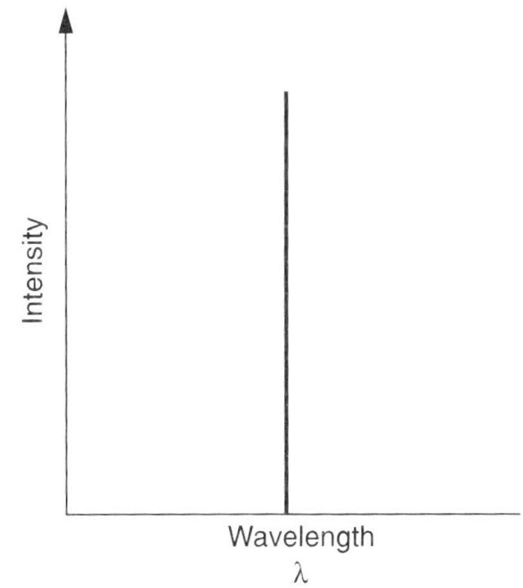

FIG. 8. The bright line emission spectrum at a single wavelength as it would appear in an ideal spectrophotometer.

problem associated with measuring monochromatic light using conventional spectrometers. If the ideal spectrometer was used to measure a bright line emission spectrum at a single wavelength (λ_1), the spectrum would appear as a single line (Fig. 8). What really occurs when such a spectrum is measured using a conventional spectrometer is a broadband spectrum as shown in Fig. 9. The spectrum assumes a Gaussian-like (or bell-shaped) curve. This characteristic broadening of a line spectrum through the spectrometer is an illustration of the spectrometer bandpass. The actual bandpass for any instrument is assigned a value by determining the FWHM height of the bell-shaped spectrum. Thus, for the band in Fig. 9, the FWHM could be empirically determined by finding the wavelength at which maximum intensity occurs and measuring the peak height at this position. This height measurement is divided in half and the bandwidth is measured at this half-height on the band as illustrated in Fig. 10.

The actual shape of a band is the result of several instrumental characteristics including the overall quality of the optics and detector systems, as well as the width and positions of the entrance and exit slits. Every dispersive spectrometer consists of a dispersive element (e.g., diffraction grating) in combination with an entrance and an exit slit. The image of the entrance slit and exit slit determines the spectrometer bandpass, which is sometimes referred to as the slit function. Actually, the slit function is the result of the

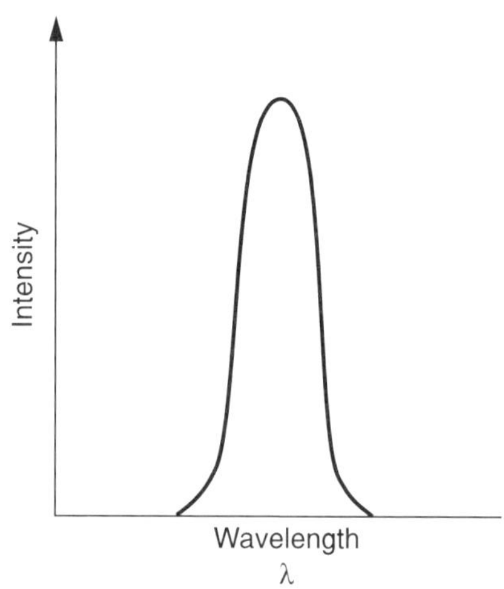

FIG. 9. Spectrum of a bright line emission source (e.g., deuterium lamp). The characteristic broadening is an illustration of the bandpass of a spectrophotometer.

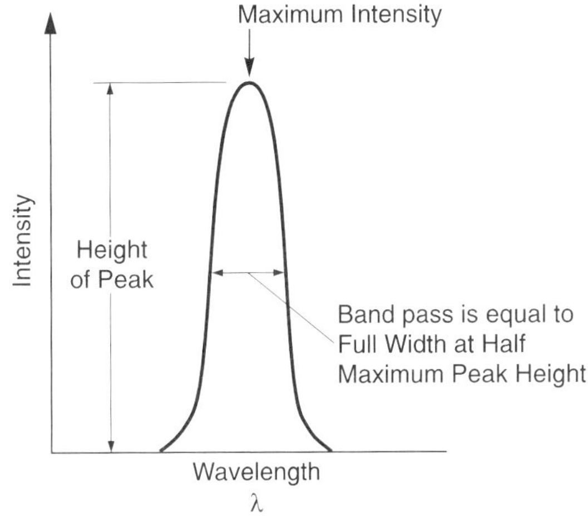

FIG. 10. Illustration of the determination of bandpass using the bell-shaped peak obtained by using a bright line source projected through a monochromator optical system.

convolution (combination) of the images of these two slits. The bandshape of a dispersive spectrometer is shown in Figs. 9 and 10. Other factors associated with optical and electronic quality cause a rounded overall shape. The bandpass of a spectrometer is equal to the FWHM. Often, texts dealing with instrumentation will state that the bandpass of a spectrometer is approximated by the product of the linear dispersion of the monochromator and the entrance or exit slit width (whichever is larger).

The resolution of a spectrometer can be defined as the minimum distance between two peaks that can be detected by the spectrometer under designated operational performance settings. Resolution is calculated by multiplying the slit width (generally expressed in mm) by the dispersion of the monochromator (in nm per millimeter). Due to practical issues and nonideal optics, the actual resolution of a spectrometer must be slightly greater (poorer) than the theoretical value.

To summarize, bandpass and resolution are identical in practice. Only for the resolution specification of a spectrometer is the expression of bandpass under the specified measuring conditions of an instrument dependent on the slit width settings.

The empirical resolution of a spectrometer is determined by measuring the FWHM in mm for two narrow bands that are completely resolved (to the baseline) using the spectrometer. The spatial difference between the maximum absorbance (λ_{max}) is determined between the bands (in mm), simultaneously noting the difference between the λ_{max} points in nm. The

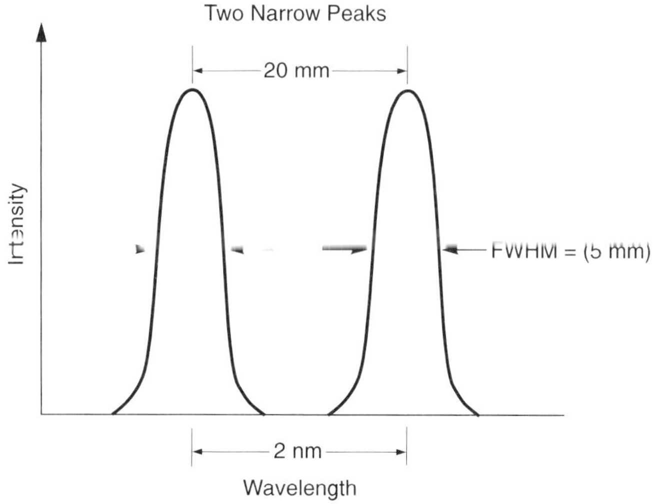

FIG. 11. The method to determine maximum resolution of a spectrometer under specific measurement conditions. In the example shown, bandpass = resolution = FWHM × dispersion = (2 nm/20 mm) × 5 mm = 0.50 nm resolution.

various measurements required for this calculation are shown in Fig. 11 and illustrated by the following relationship:

$$\text{The bandpass} = \text{resolution} = \frac{\text{band difference in nm}}{\text{band difference in mm}} \times \text{FWHM in mm}.$$

D. Numerical Aperture

The numerical aperture (NA) is a measure of how much light can be collected by any optical system. The NA is expressed as the product of the refractive index of the incident material (n_i) times the sine of the ray angle maximum (θ_{max}) from normal incidence; the function is given by

$$\text{NA} = n_i \times \sin\theta_{max}.$$

E. Attenuation (Light Transmittance Losses in Optical Systems)

Losses in transmitted light through spectrometers are due to absorption, reflection, scattering, and optical misalignment; the losses can vary with temperature and wavelength. The quantity of optical loss is expressed as an attenuation rate in decibels (dB) of optical power per unit distance (cm).

Typical losses result from launch optics, temperature variations, optical couplings within the optical path, aging of mirrored surfaces, and soiled optical surfaces. The losses in energy transmitted through a spectrometer can be calculated by using Beer's Law. Beer's Law states that the irradiance of energy through an absorbing medium falls exponentially with the distance of transmission following the relationship:

$$I_d = I_0 \times 10^{(\alpha d/10)},$$

where I_d is the irradiance at distance (d) from the source, I_0 is the source irradiance at $d = 0$, α is the attenuation (absorption) coefficient in units of dB/cm, and d = distance in cm.

Attenuation losses are wavelength dependent, thus the value for α is a function of the incident wavelength (λ).

F. Etendue

The etendue (or relative throughput advantage) for an optical system is the product of the potential illuminated surface area (A) times the solid angle of the optic. Traditionally, this is represented by the following two equations, where ε' represents the etendue, and Ω_S represents the solid angle. Thus, the etendue is represented as

$$\varepsilon' = A \times \Omega_S,$$

and the solid angle is given by

$$\Omega_S = 2\pi(1 - \sqrt{1 - (\mathrm{NA})^2}).$$

Therefore, the previous equations allow us to calculate the relative improvement for an optical spectrometer. As can be seen, the NA and aperture diameter are preeminent factors for throughput in optical systems as shown in Table 1.

G. Throughput

The relative throughput (T) represents the overall effectiveness of an optical system to transmit light relative to the amount of light introduced into the system (I_0) from the light source. It is defined as the ratio of light energy passing into an optical system to the light energy passing out of the optical system. For dispersive spectrometers, this relationship is defined by

$$T = \frac{\pi D w_s}{4f^2(R_1 \times R_2 \times \cdots \times R_k)(\Sigma_g)},$$

TABLE 1
THE ETENDUE AND RELATIVE THROUGHPUT AS A FUNCTION OF NUMERICAL APERTURE

Numerical aperture	Relative etendue (ε') for 1-mm diameter aperture	Relative throughput
0.20	0.10	1
0.40	0.40	4
0.60	1.00	10

where D is the dispersion constant in mm/nm, w_s is the slit width in mm (or exit slit), f is the f/number of the optical system [f/number = 1/2(NA)], R_k is the reflectivity of mirrors or other optical surfaces, and Σ_g is the spectral efficiency of the grating (approx 0.80 at the blaze wavelength).

H. SIGNAL-TO-NOISE RATIO

The theoretical total signal (S) from an optical system is proportional to

$$S = R_S B_\lambda \varepsilon' \tau q,$$

where R_S is the light source spectral radiance, B_λ is the spectral bandwidth, ε' is the etendue of spectrometer optical system, τ is transmission losses (and emissivity), and q is the quantum efficiency.

The measured signal to noise (s/n) from an optical system can be calculated as the full transmitted signal divided by the RMS noise (in transmittance units). Thus, for a 100% line with RMS noise as 0.001 %T, the s/n = 100/0.01 = 10,000 : 1. This applies when RMS noise is calculated as

$$\text{RMS} = \sqrt{\frac{1}{n} \sum_{i=1}^{n} (T_i - \bar{T})^2},$$

where T_i is the individual transmission value at each data channel i, and $\bar{T}$ is the mean transmission value for all data channels.

I. DYNAMIC RANGE

The range of a specified analyte concentration over which a sensor response is directly proportional to a change in concentration is the dynamic range of a spectrometer. Dynamic range is stray light and noise limited. Knowledge of Beer's law (previously discussed) allows us to calculate the maximum theoretical dynamic range for an instrument using a few simple mathematical relationships and a calculation of the relationship to stray light and maximum observable absorbance value, given by the following relationship (note: The error in a measurement due to stray light can be computed using this expression):

$$A_i = \log\left(\frac{1 + \dfrac{S_l}{I_0}}{\dfrac{I}{I_0} + \dfrac{S_l}{I_0}}\right) = \log\left(\frac{100 + I_s}{T + I_s}\right),$$

where I_0 is the incident light intensity, I is the transmitted light intensity, S_l is the stray light intensity, I_s is the stray light as percentage of I_0, and T is the percentage transmittance of the measurement under test.

Error in a measurement is also attributable to the combined noise of the measurement system given as

$$\text{Noise (as \%)} = \frac{\text{RMS} \times 100}{A}.$$

The relative dynamic range of a spectrometer is written as

$$A_{\text{DR}} = \log\left(\frac{100 + I_s}{I_s}\right) - k(\text{RMS}_A),$$

where k is the multiplier for desired confidence level, and RMS_A is the root mean square noise measurement (in A)

Note: to simplify calculations, the $-k(\text{RMS}_A)$ term can be dropped, yielding an estimated value for dynamic range.

J. Stray Radiant Energy

The relationship between absorbance and stray light is given by

$$A_i = \log\left(\frac{1 + \dfrac{S_l}{I_0}}{\dfrac{I}{I_0} + \dfrac{S_l}{I_0}}\right) = \log\left(\frac{100 + I_s}{T + I_s}\right).$$

This stray light calculation relationship in absorbance units applies where I_0 is the incident light intensity, I is the transmitted light intensity, S_l is the

TABLE 2
RELATIONSHIPS BETWEEN PERCENTAGE TRANSMITTANCE, TRANSMITTANCE, AND ABSORBANCE

% Transmittance	Transmittance	Absorbance
100.0	1.0	0.0
10.0	0.1	1.0
1.0	0.01	2.0
0.1	0.001	3.0
0.01	0.0001	4.0
0.001	0.00001	5.0
0.0001	0.000001	6.0

stray light intensity (as a fraction of I_0), I_s is the stray light (as a fraction of I_0), and T is the percentage transmittance. The relationship between transmittance (T) and absorbance (A) is given in Table 2.

The calculation of percentage error in a measurement due to stray radiant energy is given by

$$E(\%) = 100\left[1 - \left(\frac{\log\left(\dfrac{100 + I_s}{T + I_s}\right)}{A_t}\right)\right],$$

where I_s is the stray light as percentage of I_0, T is the percentage transmittance of measurement, and A_t is the true absorbance level of sample specimen measured.

K. WAVELENGTH ACCURACY

The accuracy in wavelength measurements is determined by taking a standard reference material (or emission line spectrum) of known wavelength position (λ_k) and making measurements of these known positions using the spectrometer. The difference between the known position(s) and the positions as measured using the spectrometer (λ_S) is reported as the wavelength accuracy of the spectrometer expressed as $\lambda_k - \lambda_S$.

L. WAVELENGTH REPEATABILITY

Wavelength repeatability (λ_r) is the precision with which a spectrometer can make repeated measurements at the same nominal wavelength over temporal and environmental changes. This specification is calculated as

$$\lambda_{\mathrm{r}} = \sqrt{\frac{1}{n} \sum_{i=1}^{n} (\lambda_i - \bar{\lambda})^2},$$

where λ_i is the wavelength determined at each of multiple measurements taken n times, and $\bar{\lambda}$ is the mean wavelength determined using each of the multiple measurements taken n times.

M. Photometric Accuracy

The accuracy in photometric measurements is determined by taking a standard reference material of known transmission (T_{k}) values and making measurements at specific wavelengths of these known photometric values using the spectrometer. The difference between the known transmission and the transmission as measured using the spectrometer (T_{S}) is reported as the photometric accuracy of the spectrometer expressed as $T_{\mathrm{k}} - T_{\mathrm{S}}$.

N. Photometric Repeatability

Photometric repeatability (T_{r}) is the precision with which a spectrometer can make repeated measurements at the same nominal transmission value over temporal and environmental changes. This specification is calculated as

$$T_{\mathrm{r}} = \sqrt{\frac{1}{n} \sum_{i=1}^{n} (T_i - \bar{T})^2},$$

where T_i is the transmission determined at each of multiple measurements taken n times, and $\bar{T}$ is the mean transmission determined using each of the multiple measurements taken n times.

References

Blaker, J. W. (1970). *Optics II—Physical and Quantum Optics*, pp. 91. Barnes & Noble, New York.

Bracey, R. J. (1960). *The Techniques of Optical Instrument Design*, pp. 316. English Universities Press, London.

Braun, R. D. (1987). *Introduction to Instrumental Analysis*, pp. 1004. McGraw-Hill, New York.

Candler C. (1951). *Modern Interferometers*, pp. 502. Hilger & Watts, Glasgow.

Ditchburn, R. W. (1965). *Light*, 2nd ed., pp. 833. Blackie, London.

Fogiel, M. (Ed.) (1981). *The Optics Problem Solver*, pp. 817. Research & Education Association, New York.

Griffiths, P. R., and de Haseth, J. A. (1986). *Fourier Transform Infrared Spectrometry*, pp. 1–55. Wiley, New York.

Jamieson, J. A., McFee, R. H., Plass, G. N., Grube, R. H., and Richards, R. G. (1963). *Infrared Physics and Engineering*, pp. 673. McGraw-Hill, New York.

Steel, W. H. (1983). *Interferometry*, 2nd ed., pp. 308. Cambridge Univ. Press, Cambridge, UK.

Strobel, H. A., and Heineman, W. R. (1989). *Chemical Instrumentation: A Systematic Approach*, 3rd ed., pp. 1210. Wiley, New York.

Wist, A. O., and Meiksin, Z. H. (1986). *Electronic Design of Microprocessor Based Instruments and Control Systems*, pp. [illegible]. Prentice Hall, New York.

Further Reading

Ball, C. J. (1971). *An Introduction to the Theory of Diffraction*, pp. 134. Permagon, Oxford, UK.

Clark, G. L. (Ed.) (1960). *The Encyclopedia of Spectroscopy*, pp. 787. Reinhold, New York.

Cohen, J. B. (1966). *Diffraction Methods in Material Science*, pp. 357. Macmillan, New York.

Guild, J. (1960). *Diffraction Gratings as Measurement Scales*, pp. 211. Oxford Univ. Press, London.

Jenkins, F. A., and White, H. E. (1957). *Fundamentals of Optics*, pp. 637. McGraw-Hill, New York.

Johnson, C. S., Jr., and Pedersen, L. G. (1986). *Problems and Solutions in Quantum Chemistry and Physics*, pp. 429. Dover, New York.

Knowles, A., and Burgess, C. (Eds.) (1984). *Practical Absorption Spectrometry*, pp. 234. Chapman & Hall, New York.

Mach, E. (1926). *The Principles of Physical Optics*, pp. 324. Dover, New York.

Smith, R. J. (1980). *Electronics: Circuits and Devices*, 2nd ed., pp. 494. Wiley, New York.

ULTRAVIOLET, VISIBLE, AND NEAR-INFRARED SPECTROMETRY

JERRY WORKMAN, JR.
Kimberly-Clark Corporation
Analytical Science and Technology

I. Introduction

Ultraviolet (UV; 190–380 nm), visible ((VIS; 380–750 nm), shortwave near-infrared (SW-NIR; 750–1100 nm), longwave near-infrared (LW-NIR; 1100–2500 nm), infrared (2500 nm–400 cm^{-1}), and Raman spectroscopy comprise the bulk of the electronic and vibrational mode measurement techniques.

ISBN 0-12-764070-3/$25.00.

The measurement modes for UV–VIS–NIR spectroscopy are given as follows:

Basic UV–VIS–NIR Measurement Modes

Instrument measurement mode	Description of measurement
Scan	Absorbance vs wavelength
Timedrive	Absorbance vs time at a specific wavelength
Individual wavelength(s)	Individual absorbance(s) at selected wavelength(s)
Chemometrics and quantitative methods	Concentration of analyte vs absorbance
Kinetics	Kinetic rates of reaction

Basic spectroscopic measurements involve the instrumental concepts of bandpass and resolution, signal-to-noise ratio, dynamic range, stray light, wavelength accuracy and precision, and photometric accuracy and precision. These concepts were described in Chapter 1.

II. Instrumentation

A. Optical Configurations

The basic optical configuration for UV–VIS–NIR instrumentation is shown in (Figs. 1–5). Double-monochromator (dispersive; Fig. 1) instruments

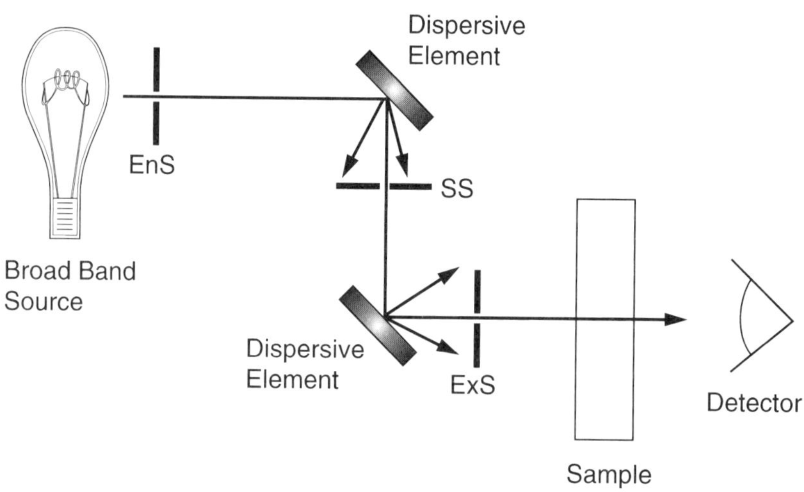

Fig. 1. Double monochromator system (dispersive) optical configuration.

provide a traditional means for reducing the stray light of an instrument at the point of measurement. Double-monochromator instruments are generally used for applications in which a high degree of photometric accuracy and repeatability is required. Optical components with strict specifications are often tested using double-monochromator systems. These systems demonstrate extremely low stray light specifications on the order of 0.0001% T or better.

Single monochromators (dispersive; Fig. 2) are lower cost than double-monochromator systems and are generally used as "work horses" within the laboratory. These systems typically meet the basic requirements for routine quantitative and qualitative work for relatively nondemanding applications. The dynamic range of these systems is stray light limited.

Diode array detection (dispersive; Fig. 3) offers the advantage of the absence of moving parts, extending the longevity and reliability of such systems compared to more traditional spectrophotometers. The main advantage of these instruments is the rapidity with which data can be collected (e.g., on the millisecond scale versus the traditional scanning instrument that makes spectral measurements on the seconds to minutes scale).

Interferometers (Fig. 4) provide the main optical element for Fourier-transform infrared spectrophotometry. Interferometer-based Fourier-transform spectrophotometers are extremely accurate in the frequency scale

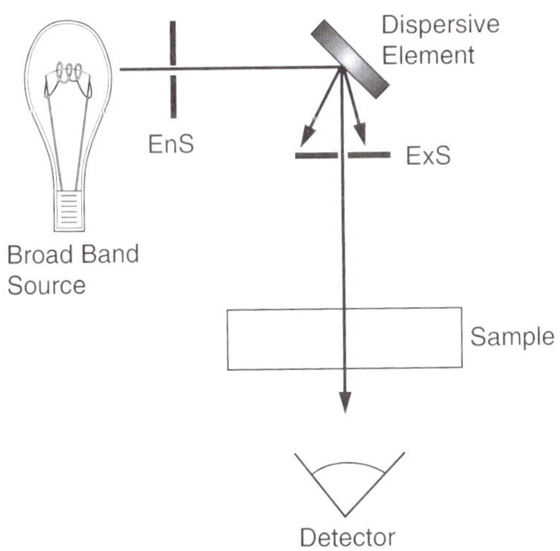

FIG. 2. Single monochromator system (dispersive) optical configuration.

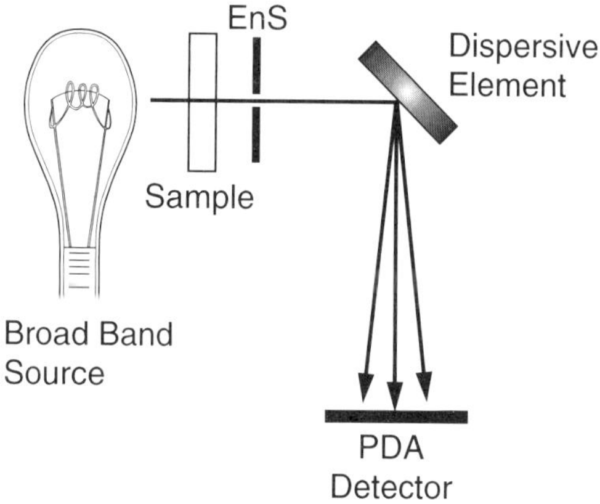

FIG. 3. Diode array spectrophotometer (dispersive) optical configuration.

but have sometimes lacked precision in the photometric domain when compared to their older counterparts, the dispersive-based instruments. However, interferometry is an extremely accurate means for measuring spectra with respect to frequency-dependent measurements.

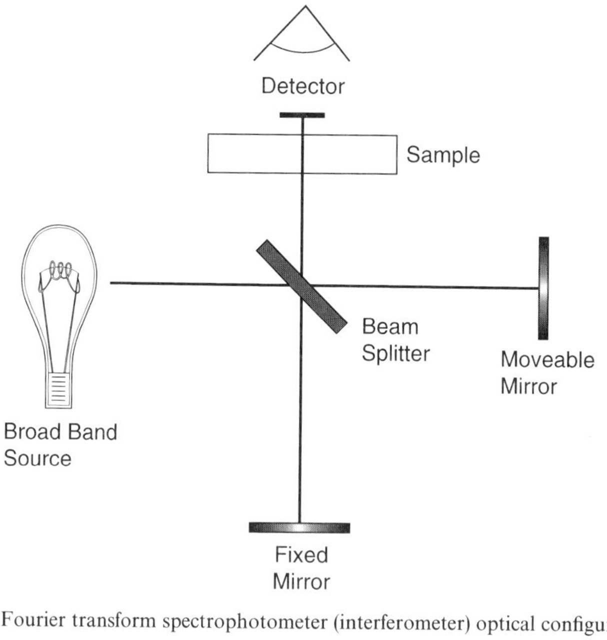

FIG. 4. Fourier transform spectrophotometer (interferometer) optical configuration.

Interference filter photometers (dispersive; Fig. 5) provide a low-cost, rugged alternative to grating or interferometer-based instrumentation. These instruments typically contain from 5 to 40 interference filters that select the proper wavelengths for quantitative analysis based on previous work with scanning instruments or based on theoretical positions for absorption.

Light-emitting diodes provide emission of prespecified wavelengths of light. Interference filters are sometimes used to further reduce the band width of the light used for analysis.

Basic instrument configurations are shown in Figs. 1–5. Abbreviations used in the figures are: BBS, broad band source; EnS, entrance slit; DE, dispersive element (grating or prism type); SS, second slit; ExS, exit slit; S, sample; DET, detector; PDA DET, photodiode array detector; BS, beam splitter; FM, fixed mirror; MM, movable mirror; NB-IF, narrow bandpass interference filter.

B. Typical Lamp Sources (Useful Working Ranges)

Most instrumentation used to measure the region from 190 to 2500 nm utilizes a combination of the quartz Tungsten–halogen lamp for the visible and near-infrared regions (approximately 350–2500 nm) and the DC deuterium lamp for the ultraviolet region (from 190 to 350 nm). The useful working ranges for the most common sources are:

Quartz Tungsten–halogen filament lamp (220–2700 nm)
DC deuterium lamp for UV (185–375 nm)
Pulsed xenon arc lamp (180–2500 nm)
DC arc lamp (200–2500 nm)

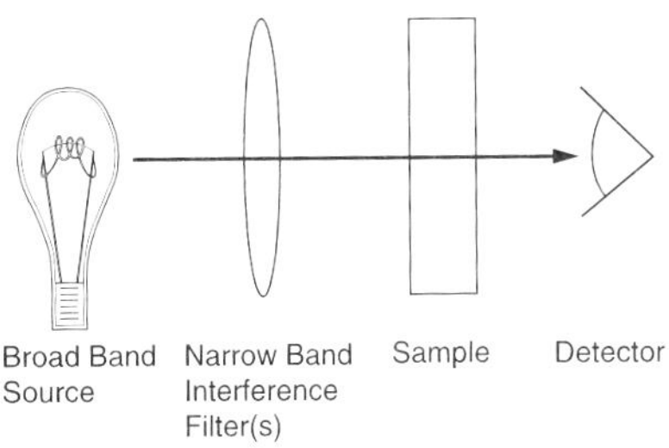

FIG. 5. Interference filter photometer optical configuration.

Unusual sources can be found when making measurements within the ultraviolet or near-infrared region. These can include lasers such as the nitrogen laser (337.1 nm) and a variety of dye lasers (350–750 nm).

1. *Calibration Lamps (with Emission Line Locations in Nanometers)*

Calibration lamps are used to check the wavelength accuracy for ultraviolet and visible spectrophotometers. The main lamps used include

Mercury (argon) lamp (253.7, 302.2, 312.6, 334.0, 365.0, 404.7, 435.8, 546.1, 577.0, 579.0 nm)
Mercury (neon) lamp (339.3, 585.2, 793.7, 812.9, 826.7, 837.8, 864.7, 877.2, 914.9, 932.7, 953.4 nm)

C. Detectors (Useful Working Ranges)

A variety of detectors are available for UV–VIS–NIR measurements. High-performance instruments utilize photomultiplier tube technology from the ultraviolet into the visible region. Lead sulfide is the detector of choice for near-infrared measurements. The more common detectors are given below with their useful operating ranges indicated. More information on detector performance is given in Chapter 1.

Silicon photodiode (350–1100 nm)
Photomultiplier tubes (160–1100 nm: total detection range using PMT technology)
PbS (lead sulfide) (1000–3000 nm)
CCDs (charge-coupled devices) (180–1100 nm)
Photodiode arrays (silicon-based PDAs) (180–1100 nm)
InGaAs (indium gallium arsenide) (800–1700 nm; 1300–2200 nm; and 1500 to 2500 nm)

D. Window and Cuvet Materials

A variety of window materials are available for sample cells and optical elements within spectrophotometers. These materials are selected for their optical clarity for use in specific wavelength regions as well as their strength and cost characteristics. The following table lists data for the most common materials used for sample cells and optical components in UV–VIS–NIR instrumentation.

Characteristics of Window/Cuvet Materials

Optical material	Transmittance range (nm)	Refractive index at 600 nm	Relative rupture strength (sapphire = 100)
Methacrylate	250–1100	—	—
UV-grade fused silica	200–2500	1.4580	10.9
Synthetic fused silica	230–2500	1.4580	10.9
Crystalline quartz (SiO_2)	240–2500	1.5437	2.3
Quartz, extremely low O-H	190–2500	1.5437	2.3
Flint glass (SF 10)	380–2350	1.7268	3.8
Flint glass (SF 8)	355–2350	1.6878	3.8
BK 7 glass	315–2350	1.5165	3.7
Optical crown glass	320–2300	1.5226	3.7
Borosilicate crown glass	360–2350	1.4736	3.7
Pyrex	360–2350	1.4736	3.8
Tempax	360–2350	1.4736	3.8
Sapphire (Al_2O_3)	150–5000	1.7677	100
Sodium chloride	250 nm–16 μm	1.5400	0.5
Suprasil 300	190–3600	1.54	3.8
Diamond	220–4000	2.40	83.7

E. Fiber Optic Materials (Useful Working Range)

New core and cladding materials are becoming available for use in fiber optic construction. The most common materials in current use include those listed here. The useful working range for typical samples of these materials is listed in parentheses:

- SiO_2 (pure fused silica or quartz) (0.2–1.25 μm)
- Anhydrous quartz (0.4–2.64 μm)
- ZrF (zirconium fluoride) (0.9–4.76 μm)
- Chalcogenide (2.22–11.1 μm) for use in NIR–IR measurements

F. Methods for Testing UV–VIS Instrumentation

Linearity checks are performed by using three neutral-density glass filters available from NIST as SRM 930D. These glasses have nominal percentage transmittance at 10, 20, and 30%. Solutions of nickel and cobalt in nitric and perchloric acids are available as SRM 931. Metal on quartz transmittance standards are available as SRM 2031 with nominal percentage transmittance at 10, 30, and 90%.

Photometric accuracy is measured for UV using SRM 935, consisting of a solution of potassium dichromate in perchloric acid. An additional material consisting of potassium acid phthalate in perchloric acid is available as SRM 84.

Wavelength accuracy is measured using ASTM Practice E 275-83, "Practice for Describing and Measuring the Performance of Ultraviolet, Visible, and Near Infrared Spectrophotometers." A second method is used as E 958-83, "Practice for Measuring Practical Spectral Bandwidth of Ultraviolet-Visible Spectrophotometers." Potassium dichromate in perchloric acid at pH 2.9 exhibits known maxima at 257 and 350 nm, with minima at 235 and 313 nm. Samarium perchlorate can be used for the wavelength region 225–520 nm with excellent results. Holmium oxide glass filters exhibit bands at 279.3, 287.6, 333.8, 360.8, 385.8, 418.5, 446.0, 453.4, 536.4, and 637.5 nm. In addition the holmium glass exhibits bands within the 750- to 1200-nm region. Didymium oxide glass filters are available for use from 250 to 2000 nm.

Stray light measurements are made using a sharp cutoff filter. Examples of these filter materials include saturated solutions of such compounds as potassium ferromanganate or lithium carbonate. Other solutions exhibiting abrupt cutoff wavelengths include KBr, KCl, NaI, $NaNO_3$ solutions, and acetone. Refer to ASTM E 169-87, "Practice for General Techniques of Ultraviolet-Visible Quantitative Analysis."

III. Sampling Considerations

A. Sample Presentation Geometry

A variety of sample presentation methods are available to the analyst. These include transmission (straight and diffuse), reflectance (specular and diffuse), transflectance (reflectance and transmission), and interactance (a combination of reflectance and transmission). These methods are discussed in greater detail in Chapter 7.

B. Typical Sampling Accessories

Sample cells	Outer Dimensions (mm)	Pathlength (mm)	Capacity (ml)
Standard transmittance	45 (H) × 12.5 (W) × 3.5 (L)	1.0	0.3
	45 (H) × 12.5 (W) × 7.5 (L)	5	1.5
	45 (H) × 12.5 (W) × 12.5 (L)	10	3.0
Semimicro	45 (H) × 12.5 (W) × 12.5 (L)	10	1.0 or 1.5
Micro cell	25 (H) × 12.5 (W) × 12.5 (L)	10	0.5
Cylindrical cells	10 (L) × 22 (Diameter)	10	3.1
	20 (L) × 22 (D)	20	6.3
	50 (L) × 22 (D)	50	16
	100 (L) × 22 (D)	100	31
Micro flow cell	50 (H) × 12.5 (W) × 12.5 (L)	10	0.4 or 0.6
round	75 (H) × 12 (D)	~10	5.9
	105 (H) × 19 (D)	~17	23.8
	150 (H) × 19 (D)	~17	34.0

C. Cuvet Cleaning

Light cleaning: Detergent wash, followed by multiple pure water rinses
Heavy cleaning: Repeat the above followed by cleaning with a chromic-sulfuric acid solution wash and multiple pure water rinses.

D. Selection of Pathlength

UV, 190–350 nm, 1 mm–10 cm
SW-NIR, 800–1100 nm, 5–10 cm
LW-NIR, 1050–3000 nm, 0.1–2 cm

E. Matrix/Measurement Techniques

Clear solids (optical materials): transmittance
Translucent or opaque solids: diffuse reflectance or diffuse transmittance (for turbid samples)
Reflecting optical surfaces: specular reflectance
Clear liquids: transmittance
Translucent or opaque liquids: reflectance or diffuse transmittance
High optical density (highly absorbing): tiny pathlengths in transmittance

F. Sample Optical Properties

Clear samples are measured using transmission spectroscopy.

Colored samples are generally measured using transmission spectroscopy unless the optical density exceeds the linear range of the measuring instrument. At this point, either dilution or narrowing the pathlength is the technique of choice.

Fine scattering particulates within a sample are measured by diffuse transmission or diffuse reflectance methods. The scattering produces a pseudo-pathlength effect that must be compensated for by using scatter correction data processing methods when making quantitative measurements on scattering materials (see description in Appendix B).

Large scattering particulates present a challenge for measurements because the particles intercept the optical path at random intervals. Signal averaging can be employed to compensate for random signal fluctuations as well as careful monitoring and tracking of signal changes. Reflectance spectroscopy can be used to measure the size, velocity, and concentration of scattering particulates within a flowing stream.

High-absorptivity (*optically dense*) materials with absorbances above 4–6 OD are difficult to measure accurately without the use of double-monochromator instruments with stray light specifications below 0.0001% T. Measurements can be made with extremely slow scanning speeds and by opening the slits during measurement. These measurements should be avoided by the novice unless high-performance instrumentation and technical support are available.

IV. Applications

A. Typical UV Solvents and Approximate UV Cutoff Wavelengths

Solvent	UV Cutoff (nm)
Acetonitrile	190
Water	190
Cyclohexane	195
Isooctane	195
n-Hexane	201
Ethanol (95 vol.%)	205
Methanol	205
Trimethyl phosphate	210
Acetone	220
Chloroform	240
Xylene	280

The grouping of atoms producing a characteristic absorption is called a chromophore (i.e., chromo = color + phore = producer). A specific grouping of atoms produces a characteristic absorption band at specific wavelengths. The intensity and location of these absorption bands will change with structural changes in the group of atoms and with solvent changes. The location of bands associated with UV chromophores is shown in the following section.

B. Absorptions of UV Chromophores (160–210 nm)

Chromophore	Absorption band (nm)
Nitriles (R–C ≡ N)	160[a]
Acetylenes (–C ≡ C–)	170[a]
Alkenes (>C=C<)	175[a]
Alcohols (R–OH)	180 (175–200)[a]
Ethers (R–O–R)	180[a]
Ketones (R–C=O–R′)	180,[a] 280
Amines, primary (R–NH_2)	190[a] (200–220)
Aldehydes (R–C=O–H)	190,[a] 290
Carboxylic acids (R–C=O–OH)	205
Esters (R–C=O–OR′)	205
Amides, primary (R–C=O–NH_2)	210
Thiols (R–SH)	210
Nitrites (R–NO_2)	271
Azo group (R–N=N–R)	340

[a] Absorptions below the cutoff for common solvents would not be observed in solvent solution measurements.

C. Use of UV–VIS for Life Sciences

1. *Enzymatic Methods*

The table below gives a summary of the various enzymatic methods of analysis using spectroscopic technique.

Spectroscopic Assay Measurements for Enzymes

Enzyme name	Reaction type and assay wavelength (nm)
Lactate dehydrogenase	Direct absorbance at 350
Alkaline phosphatase	Direct absorbance at 550
NADH and NADPH	Direct absorbance at 340
Alcohol dehydrogenase	Increased abs. NADH at 340
Aldolase	Increased abs. at 240; decreased abs. at 340

continued

continued

D-Amino acid oxidase	Decreased abs. NADH at 340
L-Amino acid oxidase	Change in abs. at 436
α-Amylase	Color reaction at 540
L-Arginase	Color reaction at 490
Arylsulfatase	Hydrolysis reaction at 405
Catalase	Liberation of H_2O_2 abs. at 240
Cholinesterases	Color reaction at 340
α-Chymotrypsin	Hydrolysis products at 280
Creatine kinase	Increased abs. NADH at 340
Deoxyribonuclease I	Depolymerization product at 260
Diamine oxidase	Decreased abs. NADH at 340
β-Galactosidase	Hydrolysis product at 405 or 436
Glucose oxidase	Color reaction at 436
Glucose-6-phosphate dehydrogenase	Increased abs. NADH at 340
Glucose phosphate isomerase	Increased abs. NADPH at 340
β-Glucosidase	Increased abs. NADPH at 340
β-Glucuronidase	Color reaction at 540
Glutamate-oxaloacetate transaminase	Decreased abs. NADH at 340
Glutamate pyruvate transaminase	Decreased abs. NADH at 340
γ-Glutamyl transferase	*p*-Nitroaniline at 400
Hexokinase	Increase in abs. NADH at 340
α-Hydroxybutyrate dehydrogenase	Decreased abs. NADH at 340
Isocitrate dehydrogenase	Increase in abs. NADH at 340
Lactate dehydrogenase	Decrease in abs. NADH at 340
Leucine amino peptidase	Presence of *p*-nitroalinine at 405
Lipase	Color reaction at 540
Monoamine oxidase	Color reaction at 456
Pepsin	Color reactions at 578, 691, and 750
Peroxidase	Color reaction at 460
Acid phosphatase	Increased abs. at 300
Alkaline Phosphatase	Color reaction at 530
Pyruvate kinase	Decreased abs. at 340
Sorbitol dehydrogenase	Decrease in abs. NADH at 340
Trypsin	Increased abs. at 280
Urease	Color reaction at 580, 630
Xanthine oxidase	Color reaction at 530–580

D. Use of SW-NIR for Organic Composition Analysis

The table following demonstrates the basic functional group measurements that have useful signal in the SW-NIR region (800–1100 nm) of the electromagnetic spectrum. As can be seen from the table, the SW-NIR region is used to measure molecular vibrations as combination bands for C–H groups and for second overtones of O–H and N–H groups and for third-overtone C–H group measurements. All NIR spectroscopy is used to measure these basic organic functional groups resulting from molecular vibrations. The advantages of SW-NIR include high signal-to-noise ratio from readily available technologies (typically 25,000 : 1) as well as high throughput using fiber-optic cabling. An additional advantage of SW-NIR over other IR regions includes the use of flow cell pathlengths sufficiently large for industrial use (most often 5–10 cm). This range of pathlengths is useful in obtaining representative sample size measurements and in preventing fouling of internal cell optics. Figure 6 demonstrates the positions of third-overtone C–H stretch vibrations.

C–H, N–H, and O–H Stretch Absorption Bands for Specific SW-NIR (800–1100 nm) functional groups (Second or Third Overtones).

Structure	Bond vibration	Approx band location (nm)
ArCH (aromatics)	C–H str. 3rd over.	857–890
CH=CH (methylene)	C–H str. 3rd over.	930–935
CH_3 (methyl)	C–H str. 3rd over.	912–915
	C–H combination	1010–1025
R–OH (alcohols)	O–H str. 2nd over.	940–970
ArOH (phenols)	O–H str. 2nd over.	947–980
HOH (water)	O–H str. 2nd over.	960–990
Starch	O–H str. 2nd over.	967
Urea	Sym. N–H str. 2nd over.	973
$CONH_2$ (primary amides)	N–H str. 2nd over.	975–989
CONHR (secondary amides)	N–H str. 2nd over.	981
Cellulose	O–H str. 2nd over	993
Urea	Sym. N–H str. 2nd over.	993
$ArNH_2$ (aromatic amines)	N–H str. 2nd over.	995
NH (amines, general)	N–H str. 2nd over.	1000
Protein	N–H str. 2nd over.	1007
Urea	N–H str. 2nd over.	1013
RNH_2 (primary amines)	N–H str. combination	1020
Starch	O–H str. combination	1027
CONH (primary amides)	N–H str. combination	1047
$=CH_2$ (methylene)	C–H str. combination	1080

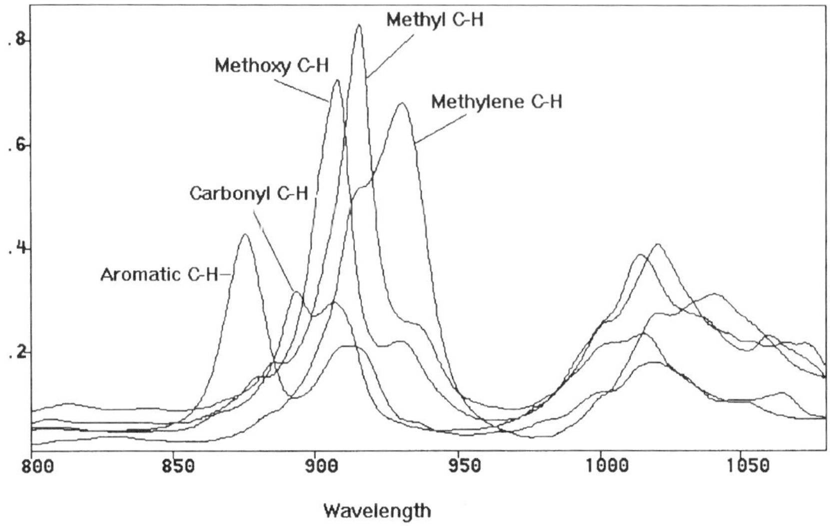

FIG. 6. Predominant SW-NIR hydrocarbon spectral features (absorbance as a function of wavelength).

E. TYPICAL UV–VIS–NIR METHODS

Analyte or analytical method	Wavelength(s) used for measurement (nm)
Aromatics, amines, aldehydes, naphthalenes, phenols, ketones (in ppm levels)	254 or 313
Violet[a]	400
ASTM D1209[b]	430
International color (blue)	440
Blue[a]	450
ASTM D1209[b]	455
ASTM D1209[b]	480
Green[a]	500
ASTM D1209[b]	510
International color (green)	520
Yellow-green	550
Orange[a]	600
International color (red)	620
Red[a]	650
Dark red[a]	700
Cu(II) ion	820
Reference wavelength	830
Ar C–H 3rd over.	875

continued

continued

Carbonyl attached C–H 3rd over.	895
Turbidity	900
Methoxy C–H str. 3rd over.	905–909
Methyl C–H str. 3rd over.	915
Methylene C–H str. 3rd over.	930
Alcohol O–H 2nd over.	960–970
Water O–H str. 2nd over.	958–960

[a] Used for tristimulus, chromaticity coordinates, color distance, and CIE coordinates.

[b] Used for ASTM E-346-78 (color in methanol), D2108-71 (color in halogenated solvents), and E 202-67 (color in glycols).

The following table illustrates the potential for SW-NIR measurements. The approximate band locations shown are based on spectra of pure, undiluted materials measured with a low-resolution dispersive instrument. Most of the data were extracted from spectra shown in the interpretive spectroscopy references at the conclusion of this chapter (see also Fig. 7).

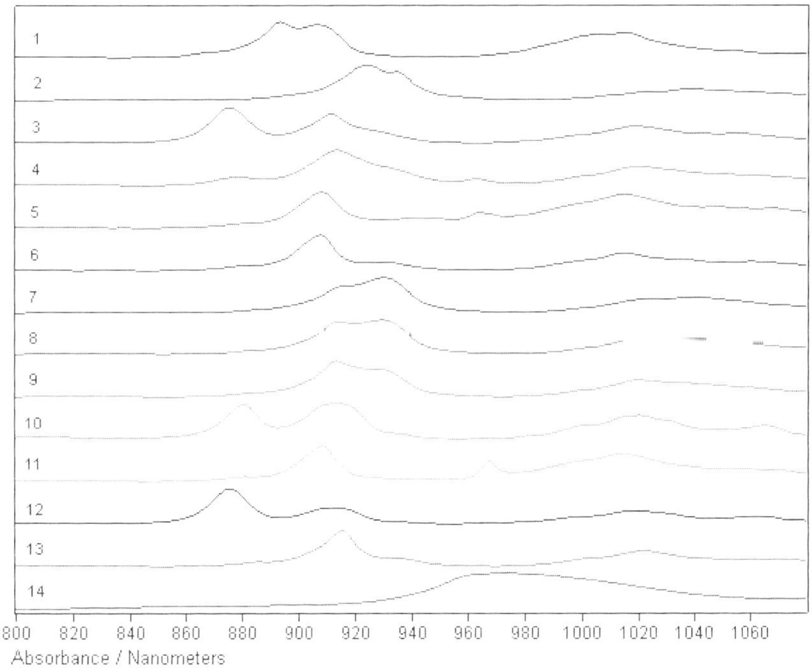

FIG. 7. Short-wavelength near-infrared spectra for selected hydrocarbons and water. 1, Acetone; 2, cyclohexane; 3, ethylbenzene; 4, gasoline with added ethanol; 5, isopropanol; 6, *tert*-butyl methyl ether; 7, *n*-decane; 8, *n*-heptane; 9, pentane; 10, *p*-xylene; 11, *tert*-butanol; 12, toluene; 13, trimethyl pentane; 14, water.

F. Approximate SW-NIR Band Locations for Common Organic Compounds

Undiluted, pure compound	Approx. SW-NIR band locations (nm)
Acetic acid, ethyl ester	895, 995
Acetic anhydride	890, 990
Acetone	893, 906, 1002, 1015
Acetonitrile	875, 980, 1030
Benzene	865, 1005
Benzoic acid	855
Benzl alcohol	875, 925, 975, 1020
Benzoyl chloride	866, 1010
Butanol	912, 960, 1015
Butyl ether	900, 1010
Butyric acid	895, 1005
Chlorobenzene	868, 1015, 1050
Chlorocyclohexane	895, 1010
2-Chloroethanol	885, 970
1-Chloropropane	895, 1005, 1040
α-Chlorotoluene	870, 895, 1015
p-Chlorotoluene	870, 910, 1010
p-Cresol	855, 890, 970
Cumene	870, 890, 1000
Cyclohexane	923, 934, 1020, 1037
Cyclohexanol	917, 960, 1030
Cyclohexanone	905, 1020
Cyclohexene	880, 910, 1025
Cyclopentanol	910, 960, 1030
Cyclopentanone	890, 910, 1020
n-Decane	915, 930, 1020, 1040
1-Decanol	930, 960, 1040
1,2-Dichloroethane	880, 1010
p-Dichlorobenzene	850
N,*N*-Dimethylaniline	855, 885, 930, 995, 1040
Diphenylamine	1015
Ethanol	963
Ethylbenzene	875, 911, 1019, 1059
Formic acid	880, 1020
Glycerol	920, 990
n-Heptane	914, 929, 1019, 1032
Heptanoic acid	915, 1030
2-Heptanol	925, 960
3-Heptanol	915, 930, 960, 1030
Hexane	900, 1000
Hexanoic acid	942, 1010, 1060
Hexylamine	805, 930, 1010, 1020, 1040

continued

continued

Hexadecanol	930, 960, 1040
Isobutanol	915, 960, 1015
Isooctane	885, 916, 935, 1020
Isopropanol	910, 963, 1014
Lauric acid	925, 1035
Methyl *tert*-butyl ether (MTBE)	879, 908, 930, 1015
Methyl cyclohexane	925, 933, 1020, 1033, 1066
2-Methylfuran	840, 880, 900, 990, 1090
Naphthalene	860
Nitrobenzene	860, 1005
Nitroethane	885, 995
2-Nitropropane	860, 980
Nonane	930, 1005, 1030
2-Nonanone	830, 910, 925, 1020, 1035
Octane	915, 1000, 1015
1-Octanol	930, 960, 1040
n-Pentane	913, 928, 1019
Pentanol	915, 950, 1020
Propanol	915, 960, 1030
Propionic acid	905, 990, 1010, 1045
Pyridine	865, 990
Styrene	860, 1000
tert-Butyl alcohol (-butanol)	880, 908, 967, 1014
Toluene	875, 912, 999, 1019, 1065
Trichloroethylene	890, 1033
α-3,4-Trichlorotoluene	860, 890, 1030
Trimethylpentane	885, 915, 936, 1022
Water	955–960
m-Xylene	880, 913, 1002, 1019, 1065
o-Xylene	880, 913, 1002, 1019, 1065
p-Xylene	880, 913, 1002, 1019, 1065

G. Approximate LW-NIR Band Locations for Common Organic Compounds

Traditional LW-NIR spectroscopy included the use of filter instruments to measure the major constituents of nutritional interest in grain and forage materials. Selected wavelengths commonly chosen for filter instrument measurements of these components included the following wavelengths with the corresponding nutritional parameters (see also Fig. 8).

2270 nm, lignin
2310 nm, oil
2230 nm, reference region

2336 nm, cellulose
2180 nm, protein
2100 nm, carbohydrate
1940 nm, moisture
1680 nm, reference region

Dominant near-infrared spectral features include the measurement of the overtones and combination absorption bands from the following functional groups. The table and composite spectra further define the locations of these groups in typical NIR spectra.

Methyl C–H
Methylene C–H
Methoxy C–H
Carbonyl C–H
Aromatic C–H
Hydroxyl O–H
N–H from primary amides, secondary amides; both alkyl and aryl group associations
N–H from primary, secondary, and tertiary amines
N–H from amine salts

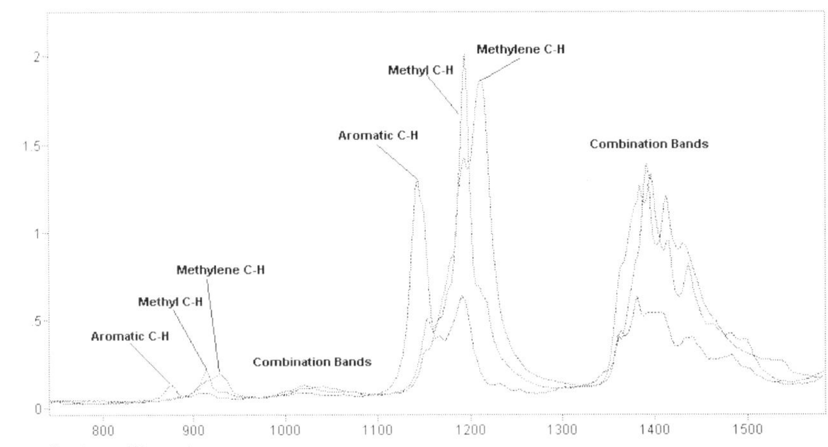

FIG. 8. Predominant long-wavelength near-infrared (LW-NIR) hydrocarbon spectral features (showing absorbance as a function of wavelength).

C–H, N–H, and O–H Stretch Absorption Bands for Specific LW-NIR (1100–2500 nm) Functional Groups (First and/or Second) Overtones)

Structure	Bond vibration	Location (nm) of 2nd overtone	Location (nm) of 1st overtone
ArCH (aromatics)	C–H str.	1143–1187	1714–1780
CH=CH (methylene)	C–H str.	1240–1247	1860–1870
CH3 (methyl)	C–H str.	1216–1220	1824–1830
	C–H combination	1347–1367	2020–2050
R–OH (alcohols)	O–H str.	—	1410–1455
ArOH (phenols)	O–H str.	—	1421–1470
HOH (water)	O–H str.	—	1440–1485
Starch	O–H str.	—	1451
Urea	Sym. N–H str.	—	1460
$CONH_2$ (primary amides)	N–H str.	—	1463–1484
CONHR (Secondary amides)	N–H str.	—	1472
Cellulose	O–H str.	—	1490
Urea	Sym. N–H	—	1490
$ArNH_2$ (aroamtic amines)	N–H str.	—	1493
NH (amines, general)	N–H str.	—	1500
Protein	N–H str.	—	1511
Urea	N–H str.	—	1520

Further Reading

Classics

J. R. Edisbury (1967). *Practical Hints on Absorption Spectrometry* (*Ultra-Violet and Visible*). Plenum, New York.

R. A. Sawyer, (1963) Experimental Spectroscopy. Dover, New York.

General

A. Knowles and C. Burgess (Eds.) (1984). *Practical Absorption Spectrometry* (*Techniques in Visible and Ultraviolet Spectrometry*, Vol. 3). Chapman & Hall, London.

G. A. Vanasse (ed.) (1981). *Spectrometric Techniques—Vol. II*. Academic Press, Boston.

Color

The Committee on Colorimetry: Optical Society America (1973). *The Science of Color.* Washington, DC.

Life Sciences

G. G. Guilbault (1976). *Handbook of Enzymatic Methods of Analysis.* Dekker, New York.

Interpretive Spectroscopy

A. Bonanno and P. Griffiths (1993). *J. Near Infrared Spectrosc.* **1**, 13–23.

T. Hirschfeld and A. Zeev Hed (1981). *The Atlas of Near Infrared Spectra.* Sadtler Research Laboratories, Philadelphia.

J. J. Kelly and J. B. Callis (1990). *Anal. Chem.* **62**, 1444.

J. J. Kelly, C. H. Barlow, T. M. Jinguji, and J. B. Callis (1989). *Anal. Chem.* **61**, 313.

D. M. Mayes and J. B. Callis (1989). *Appl. Spectrosc.* **43**, 27.

A REVIEW OF SAMPLING METHODS FOR INFRARED SPECTROSCOPY

JOHN COATES
Coates Consulting

ISBN 0-12-764070-3/$25.00.

I. Introduction

The quality of an infrared spectrum is strongly dependent on the method of sample preparation and the optical interface between the sample and the infrared instrument—usually defined by a sampling accessory. It must be appreciated that the physical state and how the sample is treated will influence the appearance of the spectrum. Furthermore, different sampling accessories that use different principles of optical measurement produce variations in the appearance of the final spectrum for the same sample. Taking care and time at the moment of sample preparation, and understanding the impact of different measurement techniques, can save time and much anguish later on.

This chapter is intended to serve as a basic guide to provide direction for the optimum method of sampling for a given type of sample. It is designed differently than traditional texts in that it is focused on the practical rather than the theoretical aspects of the subject. To that end, the text is divided into four functional parts. The first part deals with the sample itself, where the emphasis is placed on the physical state of the sample, the potential impact of the chemical properties of the sample, and the sampling approaches that are amenable to that type of sample. Having established the available options—that is, the methods and procedures for sample handling—the individual techniques are reviewed briefly in the second section. Most times, it is preferred to use the sample in its natural, unmodified state. However, sometimes this is not practical because the sample is not in a convenient form for sampling. In order to provide options for such situations, a section dedicated to procedures for sample modification is also provided. The final section deals with the actual acquisition of the spectrum and the options available for handling the spectral data from its raw form to the final presentation.

There are no guarantees that a specific recommended procedure will work because individual samples can and often do behave differently from the norm. In addition to the specified procedure, it is important to document all known facts about the sample, as well as any abnormalities observed during sample preparation. The cross-referenced sections dealing with sample type versus available procedures should be used to provide a definition of the sample characteristics and the method to be used. Often there is more than one method available for a given sample type. The best advice is keep it simple; often, a straightforward transmission measurement will provide the best quality data. The actual choice may well depend on the final application—that is, the purpose for recording the spectrum in the first place—such as a qualitative identification of an "unknown," a quality comparison for raw material screening, or a quantitative measurement for chemical composition.

For many applications, the basic, unprocessed spectral data should be adequate. If any form of spectral processing is used, such as baseline correction, digital smoothing, or spectral subtraction, it is important to appreciate the impact and the consequences of the specific data manipulation. Sometimes, irreversible distortions are introduced into the spectrum that can lead to erroneous quantitative calculations or incorrect conclusions in spectral interpretation. Also, when sharing spectral data, or storing or archiving spectra, it is essential to note the form of processing used and any numerical terms or factors that are applied. If spectral subtraction is used, it is strongly recommended that the original sample spectrum, and the spectrum used for subtraction, are also made available. Refer to Section 6 for recommendations for spectral acquisition, data and presentation formats, and comments regarding spectral data manipulations.

II. General Considerations

Infrared is a form of electromagnetic radiation nominally extending from the visible (approximately 700 nm) to just before the microwave region—practically defined as approximately 200 μm. In this text, it will be described as radiation or light, and although "light" is normally used to describe visible radiation, there is sufficient overlap in optical characteristics of materials to justify the use of this as a generic term. Most infrared interactions with matter are recorded as absorptions of energy, either as a direct or indirect measurement. Emission measurements may be performed, but these are less frequent for analytical chemistry applications. Measurements may be made via light transmission through the sample or by reflection directly from the sample surface or from an interface. The absorption cross section of molecules in the mid-infrared can be extremely high, and this often dictates which method should be used for recording the spectrum.

It is assumed that the reader has a basic knowledge of optical spectroscopy and the molecular theory of chemistry. Infrared (IR) spectroscopy is one of the analytical techniques available for the measurement of absorption or emission of radiation as a consequence of vibrations within a molecule between the component atoms. For an understanding of underlying theory, the reader is directed to a standard reference text on organic chemistry or spectroscopy (1, 2) or to texts focused on the structure/spectra relationships of molecular compounds (3–5).

The term infrared spectroscopy can be generally applied to measurements in at least three spectral regions: the near infrared, the mid-infrared, and the far infrared. Commonly, when the term infrared is applied, this implies the mid–infrared spectral region, today defined by the frequency range 4000–400 cm^{-1} (2.5–25 μm: 2500–25,000 nm in units of wavelength). This range tends to be arbitrarily defined, but the upper spectral limit does define the extent of all the fundamental molecular vibrations, with the exception of that of hydrogen fluoride.

The lower limit is normally defined by the sampling method or matrix or by the optics of the instrument. In the latter case, the standard potassium bromide (KBr) beam splitter substrate of a Fourier transform IR (FTIR) instrument provides an optical cutoff at approximately 400 cm^{-1}. Fundamental vibrations do extend below this cutoff, and these are normally associated with either low-energy lattice vibrations or the vibrations of chemical bonds involving the heavier elements. The region below 400 cm^{-1} (this limit is subjective) is referred to as the far infrared.

The spectral region above 4000 cm^{-1} (2500 nm) and extending out to the visible region of the electromagnetic spectrum (approximately 700 nm) is known as the near infrared (NIR). Commercially, the instrumentation for this region has been differentiated from instrumentation covering the fundamental regions (mid- and far-IR), and so NIR as a technique has been treated separately from traditional IR spectroscopy (6). Today, these artificial barriers are disappearing, and it is more appropriate to discuss all spectral regions under the generic term IR spectroscopy. From a sampling perspective, there are differences between NIR and the mid-infrared spectral region. These are due to various factors, such as large differences in absorptivities and the efficiency of coupling of the radiation with the material as a function of wavelength.

The NIR spectral region is composed of absorptions related to vibrational overtones and combination absorptions. The absorptivity associated with these is less, by one or more orders of magnitude, than the fundamental vibrational absorptions. As noted earlier, the magnitude of the absorption may be used to define the best procedure for sampling. It is commonly accepted that NIR is easier for sample handling than traditional IR spectroscopy. However, this is a subjective statement, and with modern sampling technologies it can be challenged for a broad range of sample types, dependent on the final application. In this text, the focus will be on sampling technology as developed for mid-IR spectral region: In some cases, the concepts can be extrapolated to the NIR region. For the most part, sampling for NIR analysis involves transmission or reflectance (normally diffuse reflectance) or a combination of both. For a general overview of the technique and its related sampling procedures, the reader is referred to a standard text on the subject (6).

In many cases, factors such as price, convenience of operation, optical performance/efficiency, and final application may be the final arbitor in the selection of a sampling method. The actual application can be an important deciding factor. A host of techniques are available for simple qualitative analysis for material identification, whereas for a high-precision/high-accuracy quantitative method the field may be narrowed to only one or two techniques. Likewise, if the application is very specific, such as a microscale analysis or the characterization of a chromatographic fraction directly from a chromatograph, then a specialized approach is dictated. Infrared spectroscopy, both near- and mid-IR, are both used for factory-based measurements during manufacturing. These may be manual or automated, off-line batch-oriented or on-line for continuous monitoring. In all scenarios, the need for a rugged, simple to implement (and use), and reproducible technique will limit the available technologies. For background reading, the reader is referred to Refs. (7)–(13).

III. Sample Type versus Sampling Method

One of the most important practical aspects of infrared spectroscopy is that it may be applied to almost any type of sample in any physical state, form, or modification. In this section, sampling will be discussed in terms of physical state and/or physical nature of the sample. One point that will be emphasized throughout is the importance of documentation of all experimental procedures, sampling methods and conditions, and any abnormalities observed during the sample preparation procedure.

If the sample (solution or mixture) is known to contain water, then not only should this information be recorded but also extreme care must be taken to ensure that the sampling method is compatible with water being present. Comments are provided in Section 5 regarding the removal of water and other volatile materials that can interfere with an analysis. Also, if a sample is seen to be made up of more than one phase, it is helpful to obtain spectra from each separated phase. Techniques such as filtration, decanting, and centrifuging should be used if practical to ensure representative sampling of a particular phase. Record the nature of the phase being sampled (liquid or solid), and note whether it is an upper or lower layer (this obviously provides some idea of relative densities). Additional procedures that can aid in the separation of mixed phases are highlighted in Section 5.

A. Gases and Vapors

The sample handling for gases and vapors is relatively straightforward, and the main issues in selecting the best sampling conditions are pressure (concentration), temperature, and the relative concentrations of the analytes. For concentrated samples it is normal to use a transmission cell with a short path length, typically in the range of 1–10 cm, the final choice being dependent on the concentration of the most important analytes. For gases, it is important to record the pressure or partial pressures that are used, especially when performing a quantitative measurement. This also applies if a nonabsorbing makeup gas, such as nitrogen or air, is used; the type of makeup gas, if used, should always be noted.

Another important application of infrared gas analysis is for trace analysis, such as for the analysis of a dilute mixture (in the ppm range) or an environmental specimen. In such cases, individual analytes are measured from the high ppb levels to the 10's or 100's of ppm. For such analyses, extended path lengths are required, and typically multipass gas cells from 1 to 20 m in path length are used. One very specific application is an open-path measurement for ambient air monitoring in manufacturing plants or in toxic waste sites in which no cell is used. Instead, a source and interferometer combination are focused on a remote detection system with the aid of special telescope optics. In such cases, several hundreds of meters of effective path length are used.

Two special forms of gas or vapor analysis involve the interfacing of an infrared instrument to a gas chromatograph (GC) or a thermogravimetric analysis (TGA) instrument. In both cases, relatively high concentrations of gas or vapor are produced within an extremely small volume. In such applications, the key is to remove dead volume between the two instruments (GC-FTIR or TGA-FTIR) and to have an optimum optical coupling with the sample at minimal sample volume. In the case of gas chromatography, the separated fractions can be measured either with a heated lightpipe or by a cooled trapping method, such as a cryo-focusing method, the Cryolect (Mattson) or a subambient method, or the Tracer (BioRad/Digilab). It is important to appreciate that these last two are very different methods of measurement: Both are measured in the solid phase compared to the gas phase used with the lightpipe.

B. Liquids

1. *General*

Traditionally, liquids have been considered to be relatively easy except for low-molecular-weight compounds, which often suffer from volatility problems and extremely high absorptivity that typically require extremely small optical path lengths (often 10 μm or less). The main methods for liquids are transmission in a sealed or semipermanent cell or by attenuated total Reflectance (ATR; internal reflectance). Today, ATR-based methods are becoming a preferred method because of the relative simiplicity, ease of cleaning, and a good match to the required sample path length. For volatile liquids, there is the option to handle the sample in vapor form. A heated gas cell is required for this approach.

For liquids in general, it is very helpful to note if the sample is known to be a single material, a mixture, or a solution. If the sample is a solution, the major solvents, if known, should also be recorded, preferably during the same time frame. In this way, it is possible to differentiate the solute and solvent, especially with the aid of computer subtraction techniques. Other issues relevant to knowledge of the sample include the following: It is important to know if the sample is "wet", especially if salt optics (KBr, NaCl, or CsI) are used, and it is beneficial to know if there are volatile components present. Sometimes a special treatment is required for volatile materials (see Section 3.4).

2. *Mobile*

If the sample is a nonvolatile or semivolatile liquid, then it is possible to use a short path-length transmission cell with either a sealed or semipermanent design. Path length is an important issue, and it is important to select a path length that provides adequate intensity to minor absorptions, while ensuring that information is not lost in the major bands due to a "bottoming out" at 0% T. A new class of ATR accessories for handling liquids has become available during the past decade. These include ATR liquid cells, based on traditional ATR designs; cylindrical ATR cells; horizontal ATR trough cells; and dip probes. For routine sample handling, the horizontal ATR format has become the most popular. This format is preferred for ease of use and ease of cleaning. In the case of an enclosed or sealed cell, make sure that the cleaning solvent is effective for removing the sample (test this fact outside of the cell) and is compatible with the window materials. Note that with cylindrical ATR cells, there is a tendency for sample and/or solvent to become trapped between the sealing O-ring and the surface of the internal reflectance element (IRE).

One of the most important factors to consider is the compatability of the cell optics with the sample, especially with an ATR sensing element (the IRE), because of their relative high cost. Accessories featuring diamond-based optics are now available, and these provide corrosion and chemical resistance for virtually any type of sample.

3. *Volatile*

There are several important issues to consider with volatile liquids, such as the need to use a sealed cell, potential loss of sample during analysis, protection of the instrument (in particular sensitive optics), and general fire and safety precautions. The traditional approach for volatile liquids and solutions (based on volatile solvents) is to use a sealed, fixed path-length transmission cell. If the sample is extremely volatile, then thermal control of the cell (cooling) may be advised to prevent leakage or damage to the seals of the cell. Various configurations of liquid ATR cell are available, as previously noted. One of the best options for samples of high-volatility is to use the high-pressure flowthrough style of cells, with thermal control (cooling, if available).

As commented earlier, it is possible to analyze the sample in the vapor phase by placing it within a heated gas cell. The latter suggestion is useful when the total material is volatile or when there is the need to obtain information about volatile components only. If a vapor-phase approach is used, the sample temperature in the gas cell must ideally be known because this will impact the appearance of the final spectrum. Note that the appearance of a vapor-phase spectrum may be significantly different than the corresponding spectrum in the condensed phase. Several commercial spectral collections of common organic compounds in the vapor phase are available in both hardcopy and digital formats (14, 15).

4. *Viscous*

A simple and convenient method for viscous, nonvolatile materials is to produce a capillary film between a pair of transmission windows or, if the material is very viscous, to form a thin smear film on a single window. It is recommended that barium or calcium fluoride, or comparable window materials, be used for aqueous-based materials. It is very difficult to fill traditional sealed cells with viscous liquids, attempts to force-fill the cell can result in permanent damage, particularly to the seals. Also, once filled, the cells can be extremely difficult to clean. If the sample is extremely viscous, such as a tar or a resineous material, it is possible to dissolve the sample in a suitable volatile solvent and evaporate a film of the material from solution on the surface of a single infrared window. The main caution here is to ensure that all traces of solvent are removed, noting that there will be a tendency for solvent residues to be retained. Heating under an infrared lamp or in a vacuum oven will help this process.

As previously discussed, liquids ATR cells may be used and, in general, the horizontal ATR format is the most appropriate and convenient—in use and for cleaning. Alternatively, either a boat cell cylindrical ATR accessory or a ATR dipping probe may be used. In all cases, pay careful attention to cleaning. Trace contamination of the surface of the ATR element will result in a major interference to the sample spectrum.

C. Pastes, Emulsions, and Slurries

Dependent on the constituents, pastes and slurries generally may be handled in a similar manner as proposed for viscous liquids. Always be aware of potential homogeneity problems. Also be cautious of the effect of any abrasive materials, retained in suspension, on optical window or crystal surfaces. This is particularly so for any system designed for continuous flow. Constant abrasion will result in light losses (due to light scattering) and eventually permanent damage to the optical surfaces. Emulsions containing water should be handled with either barium or calcium fluoride (or a comparable nonhygroscopic material) windows or with a ATR cell. Make sure that separation does not occur at the window or the IRE interface. Sometimes the polar nature of the optical element will cause segregation to occur. Common examples are emulsions, such as mayonaise, and protein-containing materials; in this latter case, the protein may actually form a coating on the optical surface. Also, some emulsions are stabilized by amino or ammonium compounds, and these will attack many of the window materials, especially those containing heavy metals (such as BaF_2 and ZnSe).

D. Solids

1. *General*

Of the three physical phases, solids pose the greatest challenge for infrared spectroscopy. The main issues are how to get effective interaction between the infrared radiation and the sample; how to minimize artifacts and distortions caused by refractive index, scattering, and polarization effects; and how to sample reproducibly and in a manner that is representative of the sample. A very wide range of sampling techniques are available for solids, and their suitablity is dependent on the nature and physical form of the sample. These include transmission, reflectance (various forms), emission, and absorption (photoacoustic) measurements.

2. *Powders*

A standard approach for grindable powders is the preparation of a compressed potassium halide (KBr or KCl) pellet or a mineral oil mull. In both situations, the objective is to grind the sample, ideally, down to a submicrometer particle size and to disperse the finely ground material in a matrix with a closely matched refractive index. This approach is intended to minimize or eliminate the impact of light scattering caused by the particulate nature of the sample. If the particle size is similar or greater in magnitude to the wavelength of the infrared radiation, then a significant portion of the light is lost (not transmitted) due to light scattering. Reduction of the particle size below this level, by up to an order of magnitude, will significantly reduce the light losses, and this is further aided by dispersing the material in an index-matching matrix (the alkali halide or the mulling agent).

In the case of compressed halide pellets, the sample is first ground to the desired nominal particle size and is then mixed (with partial grinding) with the halide salt (usually KBr but sometimes KCl). The mixture is then compressed in a special high-pressure die, with a nominal pressure of approximately 10 tons (for a 13-mm-diameter pellet). It is important to ensure that there is no ionic interaction between the sample and the potassium halide. Also, be aware of materials that undergo phase changes when subjected to the extreme pressures of the compressed pellet technique. Make sure that the potassium halide is kept in a clean, dry area, and if there is any doubt about its quality, run a blank pellet without sample. For materials that react with alkali halides, or for spectra to be recorded below 400 cm^{-1}, polyethylene powder may be used as an alternative compressed pellet medium.

In the preparation of an oil mull, the sample is ground in the same way as described previously, and a small quantity of an oil, usually a white mineral oil (such as a medicinal white oil), is added and mixed with the sample to form a stiff paste—to a consistency of petroleum jelly. The paste is distributed between a pair of infrared windows (usually KBr or NaCl) to form a thin translucent film. If the film appears to be totally opaque, then there is a probability that the particle size is still too large. The mull spectrum will obviously contain the hydrocarbon (C–H) absorptions of the mulling agent. A second mulling agent, known as fluorolube (a perfluorinated hydrocarbon oil), which does not create such an interference, may be used to provide transparency in these regions.

Coarse or hard powders are not well served by either the compressed pellet or mull techniques mainly because of difficulties associated with grinding these materials. In such situations, the best approaches require the use of an accessory, such as diffuse reflectance or photoacoustic. Be aware that diffuse reflectance is also particle-size dependent. Mixing the sample with a potassium halide may be beneficial to reduce anomalous optical behavior at the surface of the sample. This is especially helpful with inorganic compounds or other materials that have high absorptivities or have a high index of refraction. As a rule of thumb, organic compounds can be measured in about a 10% mixture and inorganics with approximately 1 or 2% mixture (in KBr or KCl).

Note that both diffuse reflectance and photoacoustic may be applied to most forms of powdered solids. As a rule, photoacoustic measurements, which are the only form of true absorption measurement, are not significantly influenced by sample morphology.

An alternative procedure for powders is ATR, either with a traditional accessory or with a flat plate-style horizontal accessory equipped with a pressure applicator. Note that the use of pressure is recommended to ensure intimate contact between the sample and the IRE surface. Normally, the sample must conform to the surface of the IRE, and because the strength of the IRE is typically limited the procedure is recommended only for soft powders. However, with the recent introduction of high-pressure horizontal ATR devices equipped with small-area diamond IREs, it is possible to handle most types of powdered material

3. *Amorphous Materials (Organic), Continuous Sheets, and Polymeric Films*

Moldable materials may be prepared as self-supporting films for transmission measurements. One must be aware of the probability for the occurrence of interference fringes in the spectrum caused by internal reflections of the front and back surfaces of the film. Materials not pre-existing as films, such as polymer pellets, may be molded or hot-pressed; accessories are available for producing films with well-defined path lengths.

As an alternative to the hot-press film method, materials that are soluble in a volatile solvent may be cast into a film on a transmission window (or on the surface of a IRE of an ATR accessory) from solution. If this approach is used, ensure that all residual solvent is removed by evaporation usually under an infrared lamp or in a vacuum oven. Care must be exercised with the rate of evaporation: If this is carried out too fast, small bubbles will form in the film. If this occurs, photometric distortions can occur in the recorded spectrum. Always record the solvent used in the preparation of a cast film in case residual solvent remains.

Certain materials already exist as a sheet or film, including plastics, metals (with coatings), paper, etc. If the material is transparent (to IR radiation) then is may be analyzed directly as a self-supported film by transmission. Alternatively, one of the reflectance methods may be used. For polymers and materials such as paper, the ATR approach may be used—in particular, the horizontal ATR equipped with a pressure applicator. This will work for coated or uncoated samples.

When dealing with a coating on a metal surface—a metal sheet or a foil—either the ATR or an external reflectance/specular reflectance device may be used. Under these conditions the measurement is defined as a reflection/absorption or a transflectance measurement. This method works best for thin to medium film thicknesses. If the coating is very thin, a grazing angle reflectance measuring device may be used. Very thick films may cause problems by transflectance due to anomalous optical effects at the surface, resulting in spectral distortions (a function of the refractive index of the material). If distorted spectra are obtained, one should consider using an alternative sampling method, such as ATR. If an alternative is not available, it is possible to apply a function known as the Kramers–Kronig transformation (16) to separate the refractive index and the absorption components of the spectrum. This will provide a more conventional form of the absorbance spectrum, relatively free from distortion.

Less common situations to consider are when studies of surface contamination are required and when the surface cannot be separated and brought to the instrument. In the first example, a practical solution is to grind a small amount of alkali halide powder (KBr or KCl) at the point of the contamination. This will result in the transfer of some of the contaminant to the alkali halide, and this in turn may be examined either by the compressed pellet technique or by diffuse reflectance. A way to assist this surface extraction is to add a few drops of a solvent, such as methylene chloride, in the area of the contamination. In the second situation, sampling may be performed by gently abrading the surface with silicon carbide paper. This action will transfer a small amount of the surface to the abrasive paper, and this in turn may be examined with the aid of a diffuse reflectance accessory.

Finally, one technique that can work for a limited number of applications, in which thin films are to be measured on a surface, is emission spectroscopy. This can be a difficult measurement to control, although some accessory companies do offer devices to retain the sample and to provide controlled heating of the sample (for emission to be effective, it is necessary for the sample temperature to exceed that of the instrument's detector) (17). Generally, this is not a simple measurement, and factors such as self-absorption from the film and absorption from the substrate can constitute significant interferences.

4. *Lumps, Granules, or Pellets*

This class of solids is an extension of the sample types already discussed, and many of the procedures already highlighted may be used here. If the material dissolves in a suitable solvent, then a cast film may be prepared on an IR transmitting window or on the surface of an ATR element. Moldable materials, such as polymer pellets, may be prepared as hot-pressed films, with care taken to ensure that material does not thermally degrade. Grindable materials can be handled as previously discussed for powders using the compressed halide pellet, mineral oil mull, or diffuse reflectance methods to acquire the spectrum.

Elastomeric or rubber-like materials are typically difficult to use with transmission-based techniques; however, dependent on the ingredients, they are ideal for ATR-based measurements. Most elastomers conform well with the IRE surface, providing good intimate contact. Elastomers with a high filler content, and in particular with dispersed carbon black, will cause problems because of the absorption characteristics of the carbon. In such cases, either ATR with a high refractive index IRE (such as germanium) or a photoacoustic measurement may be employed. In the event that such an approach is not available, then a destructive method such as pyrolysis can be used (see Section 5).

Fibrous materials may be analyzed by a number of different methods, including the standard method for sample preparation—compressed halide pellets. In this latter case, normally the fiber structure is destroyed during sample preparation. Alternatively, one can consider using diffuse reflectance, photoacoustic, or infrared microscopy. With due consideration to the preparation procedures, the fibers may be retained in their original form with these methods. In some cases, such as with IR microscopy, it is possible to study the orientation characterisitics of the fiber material and to correlate this information to certain mechanical properties of the fiber. The use of IR polarizers is implied with this type of measurement. Diamond compression cells work well for single-filament fibers by both transmission (diamond anvil cell) and ATR.

5. *Intractable Materials*

As noted previously, elastomeric materials filled with inorganics or carbon black can be problematic, and they often are considered to be intractable. If all accessory-based methods fail, then pyrolysis for characterization of the base polymer may suffice. Note that if the pyrolysis procedure is taken to completion (such as in a TGA apparatus) then an analysis on the residue or ash can lead to information on any inorganics present (excluding carbon). If only a small fragment of material exists then a diamond compression cell may provide adequate information, as long as the carbon content is not too high. A good alternative for filled polymers is to use photoacoustic detection, even for high-carbon-containing compounds. With this procedure there is a size constraint—it must be possible to place a representative sample within the accessory and this is typically constrained to a few millimeters.

There is a broad range of materials classified as insoluble and intractable; for example, cured resins, thermosetting polymers, wood and other cellulosic materials, certain inorganics, etc. These materials may be handled by one of the techniques already mentioned, such as photoacoustic, external reflectance, IR microscopy, or even pyrolysis (or TGA). Potential problems may be experienced with reflectance methods (as discussed earlier) in terms of spectral distortions. In such cases, it may be necessary to apply the Kramers–Kronig transformation procedure.

One other technique, previously described, is the abrasion of the surface of the sample with a silicon carbide paper. The abraded material may be collected as a dust and analyzed as a compressed alkali halide pellet or by diffuse reflectance. Alternatively, the spectrum of the sample from the surface of the abrasive paper may be obtained with a diffuse reflectance accessory.

IV. Sampling Method versus Sample Type

This section provides similar information and cross-referenced information to that of the previous section, but in this case it is more procedure oriented. Hints and tips are provided for the actual method rather than focusing on the sample and its properties.

A. Transmission Windows

The most basic sampling device is the IR transmission window; see Table 1 for a summary of common IR transmitting materials. This is primarily used

TABLE 1
COMMON IR TRANSMITTING WINDOW MATERIALS

Material	Useful range (cm^{-1}; transmission)	Refractive Index at 1000 cm^{-1}	Water Solubility (g/100 ml, H_20)
Sodium chloride (NaCl)	40,000–590	1.49	35.7
Potassium bromide (KBr)	40,000–340	1.52	65.2
Cesium iodide (CsI)	40,000–200	1.74	88.4
Calcium fluoride (CaF_2)	50,000–1140	1.39	Insoluble
Barium fluoride (BaF_2)	50,000–840	1.42	Insoluble
Silver bromide (AgBr)	20,000–300	2.2	Insoluble
Zinc sulfide (ZnS)	17,000–833	2.2	Insoluble
Zinc selenide (ZnSe)	20,000–460	2.4	Insoluble

for cast films, capillary sandwich films, or smear films. These are summarized as follows:

- Smear films: best suited for viscous and grease-like materials (including solid samples prepared as a mull)
- Cast films: formed by evaporation of a material from solution; can be applied to any nonvolatile material; typically applied to viscous samples, soluble polymers, nonvolatile liquids, and soluble amorphous materials
- Capillary films: used for viscous or semiviscous, nonvolatile liquids

The following are words of caution about the preparation of evaporated films from solvent-based solutions: If KBr or NaCl windows are used, ensure that any solvents or solutions used are water free. When the film is cast, allow the solvent to evaporate slowly (to prevent moisture condensation) and finish evaporation under a heat lamp to remove any residual solvent. If a solid is cast from solution, make sure that crystallization has not occurred. Note that for some crystalline solids a preferred orientation can occur (such as along the crystal lattice lines of the salt window), causing intensity variations as a function of position of the evaporated film. Generally, in the preparation of any film—cast, smear, or capillary—ensure that the film is regular and free from voids, air bubbles, or pinholes. Failure to do this will result in photometric errors in the recording of the spectrum.

B. Transmission Cells (Liquids)

Several forms of transmission cells exist, and these are a combination of some form of framework or assembly (cell body) and a pair of transmission windows. These may be defined in terms of demountable cells, semipermanent cells, fixed path-length cells, and variable (continuously) path-length cells. Demountable cells can be used for neat (undiluted) liquids or solutions and the benefit of this style of cell is that it is easily disassembled for cleaning. Fixed path-length or semipermanent cells are typically used for solutions or nonviscous liquids. The path length, defined by a spacer (Fig. 1), may be calibrated by the interference fringe method. When filling an assembled cell via the filling ports, it is important to ensure that the sample is drawn through with gentle suction. Attempts to push the sample through can result in cracking of the windows due to excess pressure buildup, especially with viscous liquids.

Be aware that cells made from nontraditional window materials, such as zinc selenide, zinc sulfide, germanium, or other medium to high refractive index materials, will usually result in the production of interference fringes, even in the presence of the sample.

A product produced by 3M, known as the IR card, uses a thin polymer film (polyethylene or PTFE) as a substitute for a traditional window. This is a relatively inexpensive method of sampling. However, if used, it is normally

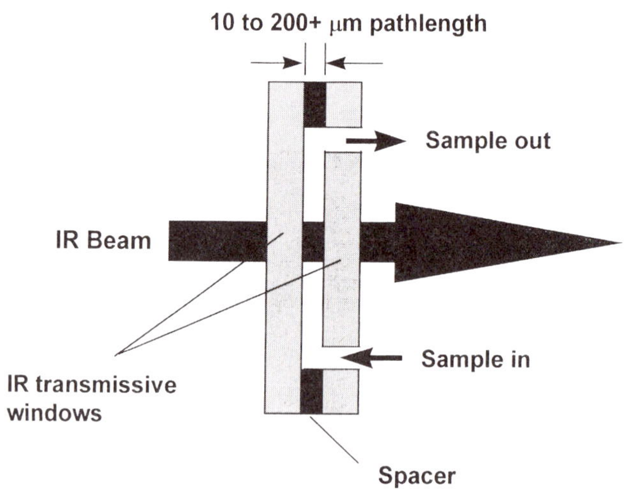

Fig. 1. A cross-sectional schematic of an infrared transmission cell.

necessary to record a reference spectrum of the card substrate material in the absence of the sample; the sample spectrum is then obtained following spectral subtraction of the substrate spectrum. A novel alternative, produced by Janos Technology, is the ECRAN screen card. This card uses a fine mesh, with essentially no infrared signature, to support the sample.

C. Gas Cells

1. *Short Pathlength*

Gas cells may be considered a special form of transmission cell. The difference is primarily in the construction of the body of the cell and its extended pathlength. In essence, a traditional gas cell (short path length) is a tube (with filling ports) with windows mounted and sealed at both ends. Like liquid cells, gas cells are available in demountable (removable windows) and permanent formats (bonded windows). Materials of construction can be important, especially if corrosive gases or vapors are studied. The most inexpensive cells feature glass tubular bodies, and these are ideal for many sample types. For a more rigid construction, cell bodies made from stainless steel, Monel, and Hastelloy C are available.

Short pathlength gas cells, with path lengths in the range of 1–20 cm, are used for the analysis of gases at high concentrations. The cells are usually filled by evacuating with a vacuum pump and backfilling with the sample to the desired pressure. Remember that pressure and concentration for a gas are equivalent, so pressure regulation is important. Note that if the sample pressure is less than atmospheric pressure, and ambient pressure is required, the cell may be "topped up" with a nonabsorbing gas such as dry nitrogen. Air may be used as a makeup gas, but it must be remembered that untreated air contains water vapor and carbon dioxide, both of which can constitute an interference. Also, note that some pressure broadening and intensification of the sample absorbance bands may occur with the introduction of makeup gas. If analyzing gas from a dynamic system, the cell may be placed within a sampling loop and the gas sampled by continuous flow through the cell. Once the cell is flushed by at least four cell volumes, the cell may be isolated and placed into the instrument.

Temperature is also an important parameter, and the heating of cells may be required for samples that condense at ambient temperatures. Various forms of heated gas cells are available for such applications. If the cell is heated, it is necessary to record the temperature because this will impact the quantitative aspects of any gas-phase measurement (because of the P–V–T relationship of gases). Measurements at high temperatures will also result in a broadening of the spectral lines.

2. *Long Pathlength*

Long pathlength gas cells, with pathlengths in the range of 1–20 m, are used for the analysis of gases at very low concentrations—typically in the range of 1–1000 ppm. Such cells differ in design to the conventional short pathlength versions. The extended pathlength is generated inside the cell by internal optics that provide a folded optical path with multiple reflections of the infrared beam. Some cell designs use adjustable optics that may be tuned to different pathlengths. Because the optics are enclosed within the cell and are in contact with the sample, it is necessary to provide some degree of protection to the optical surfaces. Normally, these surfaces are gold coated, which provides a high degree of both reflectivity and protection.

Filling procedures for long path cells are similar to those described for short pathlength cells. However, it should be realized that most long pathlength cells have an internal cell volume of 1 liter or more. Control of pressure is very important. At ambient pressures or below, many gaseous species provide a linear response to changes in concentration, up to approximately 1000 ppm, assuming that no gas-phase interactions occur between component species. Above this concentration, pressure effects can result in nonlinear response, which can be accommodated by suitable calibration procedures (accounting for temperature and pressure effects).

It is essential to ensure that no condensed material is formed on the internal optical surfaces prior to use. For some samples, it may be necessary to use the cell at elevated temperature to prevent condensation from occurring. If this is the case, it is important to be certain that the cell is constructed for operation at higher than ambient temperatures. This includes the materials used for construction (such as bonding materials for the internal optics) as well as the ability of the cell to retain optical throughput (due to dimensional changes).

D. Compressed Pellets

Compressed pellets or "KBr disks" are a standard procedure for the preparation of powdered or friable solids. The correct procedure requires careful attention to the experimental technique and, if possible, some understanding of the sample characteristics. The following are the guidelines for this technique:

- The sample is ground to a submicrometer particle size (less than 0.1 μm) in either a pestle and mortar (preferably made from agate) or a mechanized grinder.
- The ground sample is mixed uniformly with dry ground KBr or KCl powder.
- The mixture is placed in a clean, dry pressure die and compressed at a nominal 10-ton pressure for 5–10 min. A vacuum may be applied to the die to remove traces of moisture.
- Once pressed, the resultant pellet is placed carefully into a holder, which in turn may be placed in the instrument. Care must be taken not to touch the surface of the freshly pressed pellet. Be aware that the fresh pellet surfaces are very sensitive to moisture.

Note the following:

1. *The need to grind the sample down to particle sizes of below 1 μm is important to reduce or eliminate the effects of light scattering* (*Fig. 2*). *Time taken on the preparation of a pellet is usually time well spent. A poorly prepared pellet will usually provide low transmission and even a distorted spectrum* (*if the sample is not adequately ground*).
2. *make sure that the KBr or KCl powder used is pure and is stored in a clean, dry container. Make sure that it does not come into contact with any plasticized polymer surfaces.*
3. *be aware that some samples will react with KBr via a mechanism of halogen ion exchange. Also, note that some oxidizing agents may react*

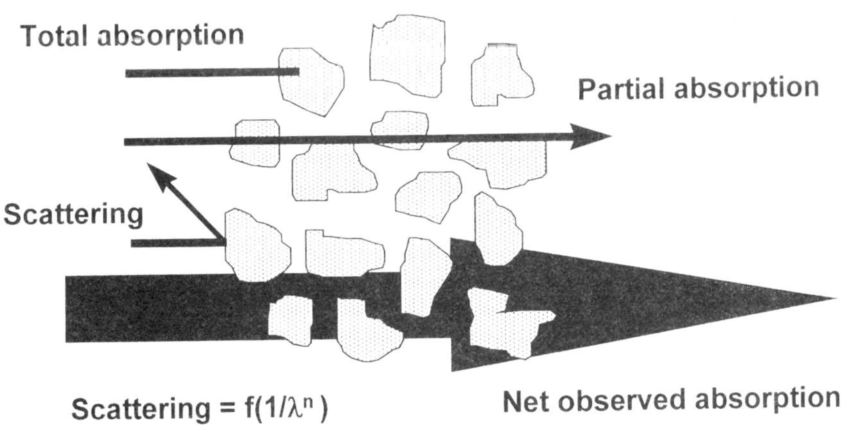

FIG. 2. The effect of scattering of light transmitted in solid sampling applications.

with KBr, liberating free bromine—indicated by an orange appearance to the pellet.

4. *a normal ratio of KBr to sample is approximately 300 : 1—a good pellet is usually obtained from a mixture of 1 mg of sample with 250–350 mg of KBr.*
5. *trace quantities of sample can be handled from solution, where the solution is dropped on the KBr powder and the solvent is allowed to dry. Take care to prevent condensation from forming and ensure all traces of residual solvent are removed by heating (possibly under a heat lamp or in a vacuum oven).*

E. Oil-Based Mulls

The oil mull technique is a companion method to the alkali halide compressed pellet. In this procedure, the sample is ground, as previously described, to a submicrometer particle size (the previous issues regarding light scattering apply here). A few drops of mulling agent, a medium-viscosity oil, are added and the sample is reground/mixed to produce a paste with a consistency of petroleum jelly (Vaseline). This resultant paste is applied to a KBr window as a smear on a single window or as a capillary film between two windows. The normal mulling agent is a hydrocarbon white oil (light paraffin or a medicinal white oil), also known by the trade name of Nujol. Another common mulling agent is Fluorolube, a perfluorinated hydrocarbon oil.

F. Self-Supporting Films

Some samples exist in the form of a film. Such samples can be analyzed directly by transmission, assuming that the film is not too thick (10–100 μm, dependent on the material) and that the material naturally transmits IR radiation. Regular, well-formed films can produce interference fringes, which are formed by internal reflections between the surfaces of the film. This effect may be reduced by roughening the surface of the film or by placing the film between a pair of infrared windows. Note that roughening will cause light scattering to occur with a subsequent loss in sample transmission.

Other samples, such as polymers in powder or pellet form, can be converted into a film. The normal methods of film production are either casting from solution or by hot-pressing between heated metal plattens. Commercial accessories are available for producing hot-pressed films with a well-defined thickness. The thickness may be checked after preparation by use of a micrometer gauge. Special card mounts can be purchased from accessory manufacturers for permanent mounting of prepared films. This is an ideal approach for sample documentation.

G. Internal Reflectance or Attenuated Total Reflectance

There are many types of ATR accessory. The old-style versions are the fully demountable type, with the ATR element (IRE) mounted vertically (traditional). The more modern variants include the horizontal ATR and the cylindrical ATR (such as the Circle and Tunnel cells). The "traditional" accessory is suitable for studying continuous surfaces, such as sheets or small blocks of compliant materials, and today it has relatively limited application. The horizontal accessories are ideal for liquids and pastes (trough versions) as well as soft powders and sheet films (flat plate versions). If a powder or film is to be analyzed with a horizontal ATR, pressure must be applied. The accessories are normally supplied with an optional pressure applicator for this purpose—often equipped with a micrometer gauge for reproducible setting of the applied pressure. The cylindrical ATR accessories are normally limited to mobile liquids.

The internal reflectance technique has been available in several different forms for many years (18, 19). It is also known as ATR and multiple internal reflectance or frustrated MIR (MIR/FMIR).

The measurement principle is based on the transmission of light through an optical element made from a high refractive index material. The angle of incidence used in the accessory is set up to generate internal reflection. The geometry of the element (known as an IRE) is normally designed to provide multiple bounces (reflections) on the internal faces (Fig. 3). If an infrared-absorbing material is placed on the air–IRE interface at the point(s) of internal reflectance, an interaction occurs between the infrared light and the sample. This results in an attenuation of the internally reflected light as a function of the absorption of the sample, hence the name ATR. A spectrum similar in appearance to a transmission spectrum, except for intensity variations as a function of wavelength (λ), is produced.

The ATR phenomenon occurs within the first micrometer or two from the IRE surface. The *depth of penetration* (D_p) into the sample is defined by the relative refractive indices or the sample and the IRE, the wavelength of the radiation, and the angle of incidence, as indicated by the following equation:

$$D_p = \frac{\lambda}{2\pi n_1 \sqrt{(\sin^2\theta - n_{12}^2)}},$$

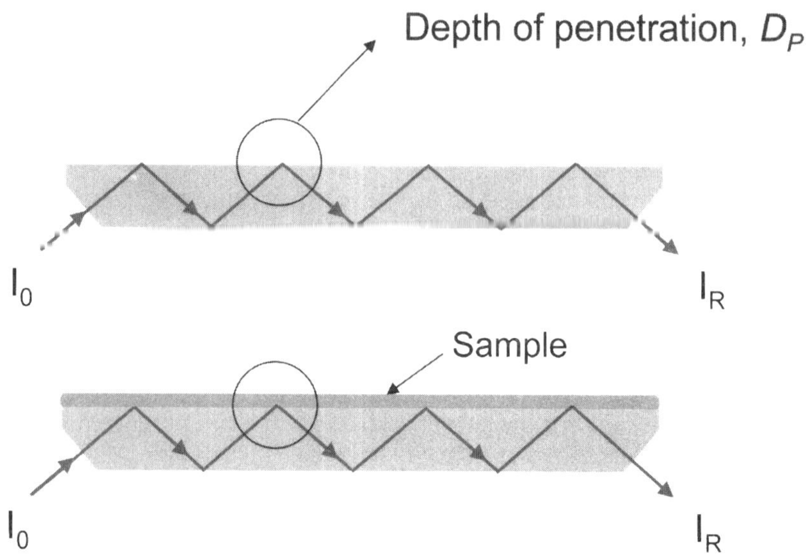

FIG. 3. A conceptual cross-sectional drawing of a multireflection IRE.

where D_p is the depth of penetration (usually in μm), λ is the wavelength (usually in μm), n_1 is the refractive index of the IRE, n_{21} is the ratio of the refractive indices of sample and IRE, and θ is the angle of incidence.

Note: *In a multiple internal reflection configuration, the effective path length of the sampling surface is the product of the number of reflections at the measurement surface and the depth of penetration. Also, refer to Table 2 for common IRE materials.*

It is essential that the sample is placed in intimate contact with the IRE. For compliant samples, pressure may be applied to ensure adequate contact. Air voids between the sample and IRE will reduce the intensity of the spectrum. Make sure that even pressure is applied across the sample; excess or uneven pressure can permanently damage the IRE.

Samples with hard, granular, or textured surfaces may not provide adequate contact with the IRE. Examples are certain fibrous materials, some fabrics, or noncompliant foams such as polyurethane foams. Such materials will often provide a poor spectrum by the ATR method. One solution here is to utilize one of the diamond-based ATR systems. These accessories permit the use of significantly higher pressures, and under these conditions even relatively rigid samples will deform to provide surface contact with the diamond IRE.

Note: *It must be appreciated that the ATR technique is a surface phenomenon. This requires that good cleaning practices are used, and it is essential that no surface contaminants are present prior to use.*

TABLE 2

COMMON IR TRANSMITTING MATERIALS USED FOR ATR INTERNAL REFLECTANCE ELEMENTS

Material	Useful range (cm^{-1}; transmission)[a]	Refractive Index at 1000 cm^{-1}	Water solubility (g/100 ml, H_2O)
Zinc sulfide (ZnS)	17,000–833	2.2	Insoluble
Zinc selenide (ZnSe)	20,000–460	2.4	Insoluble
Cadmium telluride (CdTe)	20,000–320	2.67	Insoluble
AMTIR[b]	11,000–625	2.5	Insoluble
KRS-5[c]	20,000–250	2.37	0.05
Germanium (Ge)	5,500–600	4.0	Insoluble
Silicon (Si)	8,300–660	3.4	Insoluble
Cubic zirconia (ZrO_2)	25,000–~1600	2.15	Insoluble
Diamond (C)	45,000–2,500 1,650–<200	2.4	Insoluble
Sapphire	55,000–~1800	1.74	Insoluble

Note. Materials such as cubic zirconia, diamond, and sapphire may be used for transmission windows for special applications.

[a] The usable range is normally less than the transmission range because of the higher optical attenuation that results from the extended optical path length of an IRE.

[b] AMTIR: Infrared glass made from germanium, arsenic, and selenium.

[c] Eutectic mixture of thalium iodide/bromide.

H. EXTERNAL OR SPECULAR REFLECTANCE

The external reflectance method provides a mirror-style reflection from the surface of the sample. It is used only for samples that have flat, reflective surfaces. The most common applications of external reflectance are for the direct measurement for films, coatings, or surface contaminants on metal surfaces. These measurements are sometimes called transflectance or reflection/absorption (Fig. 4).

When studying a surface film by transflectance, the light passes through the material twice—to and from the reflective surface; the resultant spectrum is therefore at least twice the normal intensity of the normal transmission through the film (coating). Both the angle of incidence and the reflected angle from the surface are important primarily because they impact the intensity of the final spectrum. For thicker films, one typically uses an angle close to that of normal incidence. Conversely, for very thin films or surface contaminants, a grazing angle of incidence (as high as 80°–85° from normal incidence) should be used.

External reflection can be applied to samples that have a reflective surface but are not coated on a metal surface. Such samples usually provide a distorted (mixed-mode) spectrum—a convolution of the refractive index

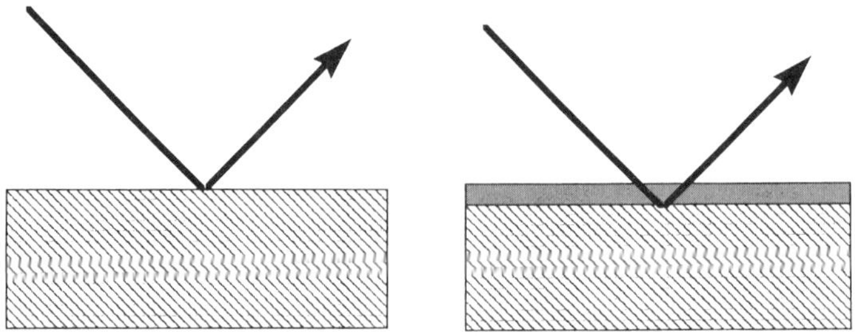

FIG. 4. External reflectance regimes.

and absorption spectra. It is very difficult to provide a meaningful interpretation from such spectral data. A numerical operation, known as the Kramers–Kronig transformation (16), can be performed on such data to provide a separation of the refractive index and absorbance components to yield a spectral representation equivalent to the absorbance form of the data. Note, however, that this is not always a simple operation, and complete separation may not be achievable.

I. Diffuse Reflectance

Diffuse reflectance (also known as DRIFTS) in some ways can be considered to be a parallel technique to external reflectance. However, in general it is more versatile (20). The method is used when a light-diffusing surface, such as a powder or a roughened surface, is to be studied. For the study of many types of powdered materials, it is possible for the sample to be analyzed directly. Although the technique has a wide range of applications in the mid-IR, it is more widely used in NIR. Optical effects which primarily occur close to the surface of the sample (Fig. 5), can impact the quality of the final spectrum. These tend to limit some of the mid-IR applications, especially for measurements involving the sample as received. In general, such limitations do not exist with NIR measurements.

Note that particle size is important. Reducing the particle size by grinding will often improve the quality of the final spectrum. Also, a uniform particle size or a narrow particle size distribution is beneficial. For some organic solids, and most inorganic materials, it is necessary to dilute the sample with a nonabsorbing matrix, such as KBr or KCl powder. This helps to reduce spectral distortions that originate from anomalous surface

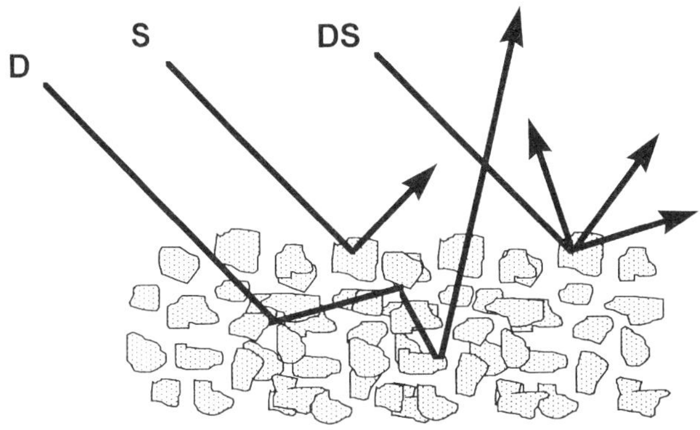

FIG. 5. Reflection modes encountered with diffuse reflectance.

reflections (Fig. 5). For organics, the dilution ratio should be approximately 10:1 KBr to sample, and for inorganics dilution ratios as high as 100:1 may be necessary. Mixing with a nonabsorbing matrix is recommended if a distorted or a poor-contrast spectrum is obtained from the sample. It is often beneficial to experiment with the measurement with or without dilution to ascertain the best method of sample preparation.

Nonpowdered samples can be studied directly from the sample surface as long as the surface is irregular (granular) and not polished. In some cases, a thin film of KBr powder placed on the surface of the sample may help to improve the final spectrum quality; this works well with fibers, fabrics, and paper samples. In this case, the KBr film reduces the incidence of unwanted surface reflections and helps to ensure better coupling between the sample and the infrared beam by promoting penetration beyond the surface.

The spectrum produced from a diffuse reflectance measurement does not bear a direct numerical relationship between peak intensity and composition, in contrast to traditional transmission measurements in which the absorbance form of the data is directly proportional to concentration. With diffuse reflectance, spectral distortions are generated by a constantly varying effective path length, which is defined by the depth of penetration of the beam into the sample. This is dependent on the absorptivity of the sample at a particular wavelength (frequency) and the measured intensity throughout the spectrum varies accordingly. A correction that can be applied to the diffuse reflectance spectrum, which can help linearize the data, is the Kubelka–Munk function:

$$f(R_\infty) = \frac{(1 - R_\infty)^2}{2R_\infty}$$

Where $f(R_\infty)$ is the corrected spectrum and R_∞ is the ratio of the sample diffuse reflectance spectrum and a nonabsorbent reference (normally KBr or KCl), both measured at infinite depth. In the mid-IR region, infinite depth corresponds to approximately 3 mm.

The consequence of the $1 - R_\infty$ is to invert the spectrum and provide an absorbance-like format. It will be noted from the previous expressions that this corrected function can be correlated to concentration. Note that this function is normally only applied to undiluted (or concentrated) powdered samples.

Diffuse reflectance can be used as a trace analysis method. In this case, the material to be studied is applied as a small quantity of a solution in a volatile solvent to a powdered KBr matrix. The solvent is evaporated and the sample is allowed to coat the KBr crystals. The normal sampling cup dimensions are approximately 13 mm in diameter × 3 mm deep; however, smaller diameter cups are available with diameters of 1.5–2 mm. These are closely matched to the typical instrument beam diameter after imaging with the diffuse reflectance accessory primary focusing mirror (typically a 6:1 off-axis ellipsoid). With this microsampling extension, it is possible to obtain good spectra from sample loadings as low as a few 100 ng.

One novel approach to the measurement of difficult solid samples is to use a silicon carbide abrasive paper (as mentioned in Section 3). The silicon carbide surface provides an irregular reflective surface that scatters all incident radiation. If the sample to be analyzed is rubbed against the abrasive surface, a small amount of sample is transferred to surface. In reality, the rubbing action distributes the sample as a thin film across the irregular, reflective surface. The sample, in turn, is measured directly on the surface by the use of a diffuse reflectance accessory; note that this is really a transflectance measurement (measurement of a coating on a roughened surface) and not a classical diffuse reflectance measurement. Any spectral contributions of the silicon carbide can be removed by a ratio of the spectrum of a clean portion of the abrasive paper.

J. Photoacoustic

Analysis with photoacoustic detection is the only method that involves the direct measurement of the absorption of the infrared radiation. The technique involves the conversion of the infrared energy into thermal energy (within the sample), followed by thermal energy transfer from the sample phase into a gas phase (immediately above the sample). In the case of gaseous samples the transfer of infrared to thermal energy occurs directly. The efficiency of the process is linked to the absorptivity of the sample and to the thermal conductivities of the sample and the gas phases. High sample absorption combined with good thermal conductivities (both phases) provide a situation conducive to effective photoacoustic measurements. Note that because absorption is the primary process, carbon black, a universal absorber, is used as a reference material.

The technique is sometimes referred to as the method of last resort. In practice, it is a very versatile technique (21, 22) and is ideal for difficult samples, such as heavily filled polymers, coatings on metal wires or metal powders, or samples generally described as "intractable." The sample is placed in a small chamber, which is sealed at the time of the measurement. Prior to sealing, the sample and chamber are purged with a clean dry gas—dry nitrogen or, preferably, helium or argon (both gases have better thermal conductivities than nitrogen).

It is always advisable to adhere closely to the manufacturer's instructions, especially in regard to the use of a dry purge gas. Failure to do this will result in data dominated by the spectrum of water vapor. Photoacoustic measurements are based on thermal processes, which are intrinsically slow, and in order to maximize sensitivity, slow interferometer scan rates are recommended for best performance. Note that the sample size is normally limited to a few millimeters. Always document the purge gas that is used and the scan speed of the instrument because these impact the intensity and quality of the final spectrum.

K. Microtransmission: Cells/Pellets

In cases in which insufficient sample is available for the traditional sampling procedures, a micro approach may be adopted. Microversions of compressed halide pellets (down to 0.5 mm), transmission cells, and reflectance accessories can be used. Dependent on the method and the sample type, the sampling size can be reduced to a few micrograms or even a few hundred nanograms, for solids and approximately 1 μl (approx 1 mg) for liquids. Reducing the physical sample size reduces the sample weight required for a given spectrum intensity; the ground rules are defined by the constraints of the Beer–Lambert law, which establishes the basic relationship in terms of units of weight/area.

In concept, the smaller the sample size, the lower the amount of material required (a logical conclusion); however, the trade is optical thoughput. Traditional instruments are limited by the image size of the infrared beam at the sampling point, which is typically between 6 and 8 mm in diameter.

An improved matching between sample size and beam size can be achieved by the use of a beam condenser accessory, providing beam size reductions on the order of 4:1 or 6:1 times. With such accessories it is possible to handle physical sample sizes in the range of 0.5–0.25 mm in diameter. For sample sizes smaller than this, it is often necessary to use an infrared microscope.

When working with any of the microsampling methods, cleanliness is of utmost importance. Traces of impurities picked up during sample preparation can have higher concentrations, and therefore greater impact on the final spectrum, than the sample itself. This attention to cleanliness must extend to the purity of any solvent used during sample preparation and also to preparation materials, such as KBr, if micropellets are to be prepared.

L. Diamond-Based Accessories

Diamond is considered to be an ideal infrared sampling optic for several reasons: its high mechanical strength (resistance to abrasion), its chemical inertness, and its reasonable optical transmission over most of the mid-IR spectral region (with the exception of a small area in the region of 2000 cm^{-1}). Several commercial accessories are available featuring diamond optics; these include diamond compression cells (for high-pressure transmission measurements) and diamond ATR sampling devices in both horizontal ATR and cylindrical ATR formats.

1. *Compression Cell: Diamond Anvil Cell*

Large-scale diamond compression (high-pressure) cells have been available for some time. However, during recent years, a microversion of the cell, also known as the diamond anvil cell, has been produced. This small diamond anvil cell is ideal for the study of physically hard (but compliant) samples, certain intractable samples, and samples that are optically too thick for normal transmission measurements. Note that this small form of the accessory is intended for samples 1 mm or less in size. The aperture through the diamonds is small, and therefore the use of a beam condensor accessory is recommended for optimal throughput. Alternatively, the cell may be used in combination with an IR microscope.

2. *Diamond ATR*

Several diamond ATR-based accessories are available. A small cylindrical ATR, featuring a diamond IRE, used in flow cell or immersion probe geometries, provides an ideal tool for the continuous monitoring of chemically reactive media. An alternate configuration, based on a small diamond sensing surface combined with a zinc selenide support optic, is available in both insertion probe and horizontal ATR formats. The probe formats are ideal for industrial process monitoring applications. The horizontal ATR format provides a more or less "universal" sampling platform for both liquids and solids. Its benefits over traditional horizontal ATR accessories include the strength of the diamond, for handling hard noncompliant surfaces, and the chemical inertness, for handling corrosive media such as strong acid and strong alkaline solutions.

M. Microscopes

Infrared microscopes have become popular accessories during the past decade. These accessories are extremely versatile and can be used for a wide range of samples. Their applications extend far beyond the discussion in this chapter, and the reader is directed to reference texts dedicated to the subject (23, 24). In terms of optimum use and operation, users are advised to follow the manufacturer's recommended procedures. It is important to always use matching apertures for sample and background scans and always document the aperture size used. Also, most commercial microscopes may be used in transmission and reflection modes and with or without infrared (and visible) polarizers. A broad range of new applications have been developed around the imaging ability of the modern microscopes in combination with computer video graphics and computer controlled sampling stages.

N. Gas Chromatography-IR

The combination of gas chromatography and infrared is ideal for the analysis of complex mixtures (25, 26). In one standard configuration, the eluted components are transferred, down heated transfer lines, to a long path-length heated lightpipe. The lightpipe, which may be approximately between 10 and 20 cm long, has a small internal diameter (1 or 2 mm) and is internally coated with gold. Issues to consider are the need to minimize the opportunity for thermal degradation of the eluted components while maintaining a continuously heated path through the interface between the GC and the IR instrument and ensuring that there is minimal dead volume in the transfer lines and in the lightpipe (and its connections). Things to remember and document are the type of column used for the separation, the temperatures used throughout (injector block, column, and lightpipe cell), the number of spectra coadded around a separated GC peak, and the solvent used for injection.

Note the following:

1. *Make sure that there is no column bleed or that any bleed is compensated.*
2. *Remember that injectors and columns are usually very hot. For some thermally unstable compounds these can appear as highly reactive surfaces. As a result, what goes on the column may be different from what elutes; be prepared for surprises—pyrolysis and thermal decomposition can occur.*
3. *Coelution can occur, and therefore the final coadded spectrum may still be a mixture. Evaluate possible differences between the spectra obtained from the leading and trailing edges of the eluted GC peak.*

O. Liquid Chromatography-IR

In recent years, a new focus has been applied to the combination of liquid chromatography (LC) and FTIR. This is of particular interest for the analysis of polymeric materials (for molecular weight studies), pharmaceutical products, and natural and biological products. The main problem has always been how to record the spectrum of the eluted component, which is at relatively low concentration, in the presence of a high concentration of solvent. This is particularly a problem with reversed-phase separations involving binary (e.g., water–methanol) and ternary (e.g., water–methanol–acetonitrile) solvent systems. New accessories have been developed for off-line measurement of eluted fractions in which the components are deposited on the surface of a rotating reflective plate. The deposited components are measured in an automated specular reflectance unit to produce a series of spectra, which are processed to generate a reconstructed chromatogram.

One variant, close to LC, that has been successfully interfaced to infrared spectroscopy, is supercritical fluid chromatography (SFC). In this case, one of the key mobile phases, super critical carbon dioxide, is virtually transparent for most of the mid-IR region. This may be operated directly on-line with a high-pressure liquid flow cell.

P. TGA-IR

Thermogravimetric analysis is a standard laboratory procedure for studying the thermal degradation characterisitics of materials in oxidative, inert, and reducing environments. The technique may be coupled on-line to an infrared instrument to permit monitoring of the gaseous degradation products. This is an important combination for evaluating thermal degradation, oxidation mechanisms, and pyrolysis mechanisms for polymers. After a thermal cycle, there is always the opportunity to analyze the residue (ash) to determine the presence of inorganic compounds in composite materials.

Q. Other Techniques

There are a plethora of sampling methods, and this chapter has presented by no means an exhaustive listing. If a unique sampling procedure is used, make sure that it has been thoroughly documented and commented on. Note, in particular, any unique sample treatments and always record the instrument operating parameters.

V. Sample Modification and/or Purification Methods

Where possible, use a sampling method that does not involve any sample pretreatment or modification; that is, analyze the sample "as is." It is recognized that this is not always possible. For routine sampling, some of the newer diamond-based accessories allow many different sample types to be analyzed in an original, native form. However, it is recognized that not all laboratories have this facility, therefore, sample pretreatment may be the only answer. A few of the common sample pretreatments are noted in the following Sections. If any of these procedures are used, it is important to ensure that a comment is provided relative to the use of the method within the sample documentation. Failure to do this can lead to erroneous results or misleading answers.

A. Solutions

In the past, solution-based measurements were popular mainly because sampling was limited to the simple procedures involving transmission measurements. One possible benefit is that both solids and liquids may be studied in solution. In this case, the optical effects caused by differences between these two types of sample phase are removed. The problem has always been the selection of an appropriate solvent. All liquids have an infrared spectrum, and almost all liquids have relatively intense and complex spectra. There are few materials that have simple spectra that have clear "windows" for broad-range transmission measurements. As a general comment, the materials that are the best solvents, particularly for polar compounds, by nature are strong infrared absorbers. Therefore, it is very difficult to find convenient solvents that have good solvent characterisitics, that are relatively involatile, and that are convenient to handle (e.g., are nontoxic).

Favorite materials in the past were carbon tetrachloride, chloroform, and carbon disulfide. Carbon tetrachloride and carbon disulfide are only modest solvents, in terms of solvent power, but have very simple spectra. Also, conveniently, the regions of major absorption of these two compounds are very different, and by use of a two-solvent strategy it is possible to construct a single-solution spectrum from two fragment spectra, recorded individually from each solvent. Unfortunately, these materials are banned from laboratory use in many countries, and so their use as IR solvents has almost ceased.

Chloroform is a good solvent, and its use has been constrained but not necessarily banned in all countries. It is important to be aware that commercial grades of chloroform contain stabilizers (to inhibit reactive carbene formation): Low-molecular-weight olefins or ethanol are typically used for this purpose. Other halogenated solvents, such as methylene chloride (dichloromethane), perchloroethylene, and certain halothanes [Freons—(perchlorofluorocarbons)], can be used as infrared solvents (27). However, although the Freons are relatively innocuous and do not pose a safety hazard, their use is regulated in North America and some other countries on environmental grounds. Even so, standard methods, particularly environmental procedures for extraction and enrichment, are still in use based on these solvents. When using volatile halogen-based solvents, it is important to ensure that the cells do not leak and that the instrument is protected either by an active purge or by environmental seals. Halogenated solvents decompose in the region of the hot infrared source, and failure to protect the instrument can result in excessive corrosion of exposed metal components within the instrument, including aluminum-coated optical surfaces.

Hydrocarbons, such as hexane, heptane, and isooctane, may be used, but they have reduced solvent powers and are typically limited in use for the dissolution of nonpolar materials (the old rule of like-dissolves-like applies here). More polar solvent, such as alcohols (methanol, ethanol, isopropyl alcohol, etc.), dioxane, tetrahydrofuran, sulfolane, and dimethyl formamide, may be used for special applications, but it must be realized that the spectra of these solvents are very intense and often quite complex. As a result, there are limited windows of transparency at which measurements may be made. A number of excellent spectral collections have been published that feature solvent spectra (27–31), and it is recommended that the reader consult one of these texts before proceeding with a solvent-based method.

One important solvent that is often overlooked is water. In the past, this has been considered an inappropriate sampling medium. This was primarily because of the high intensity of the water absorptions and a lack of suitable sampling devices. However, with modern ATR cells, designed for liquid applications, it is practical to perform analyses on aqueous solutions. With careful control of temperature, it is possible to determine materials in water solutions at fractional percentage concentrations.

Two important issues for the measurement of materials in solution that must be taken into consideration are temperature and solvent–solute interactions. The impact of temperature was implied previously. One of the main reasons for considering thermal control when measuring solutions is that it is common practice to subtract (in absorbance) or ratio out the spectral components of the solvent from the solution spectrum. This exercise will be successful only if the temperatures of both the solution and the solvent are identical, or very similar, when the spectra are recorded. Wide differences in the temperatures will result in significant artifacts in the solvent-subtracted spectra due to spectral mismatch of the solvent components.

Solvent–solute interactions are also important, and it is important to realize that the spectrum of a solute in solution may appear to be very different from the spectrum of the solute in its free state. This especially applies to the spectra of solids recorded from solutions. Not only can the spectra change but also bandwidths and band positions can change as a result of mutual interaction between solute and solvent. This will apply to certain absorptions of the solvent spectra, especially with polar compounds, and in particular at high solute concentrations. Be aware that solutes such as certain inorganic salts may not have a significant infrared spectrum (such as sodium chloride in brine or saline solutions) but the presence of the salt will impact the spectrum. This is the result of ion solvation, which perturbs the spectrum of the solvent (water in the case of salt), and this is quite significant at high concentrations.

B. Drying or Solvent Removal

For some applications in which samples are in solution, or are "wet," it is desirable to remove the volatile solvent components prior to analysis. In the case of organic solvents, it is common to evaporate the solvent by heating either on a steam bath or in a rotary film evaporator. In such situations, the final traces of volatile material are typically removed in a vacuum oven. Caution: Make sure that oil traps are used on all vacuum lines; pump oil vapors from a vacuum pump will contaminate samples in a vacuum oven. In cases in which a thermally sensitive material must be removed from an aqueous solution, freeze-drying is a practical alternative for isolating the sample if suitable equipment is available.

A special case of drying is the removal of water from a wet sample. This may be important for two reasons: (i) to remove a major spectral interference caused by the presence of the water and (ii) to protect any moisture-sensitive optics within the instrument that may be used in the sampling accessory (the cell) or in extreme conditions. For organic liquids, it is recommended that a solid inorganic drying agent such as anhydrous sodium sulfate be used. A practical alternative is to use 3A molecular sieve, but this normally requires that the liquid be stored over the drying agent for a long period.

Powdered solids may be dried in a desiccator or a vacuum oven. There are a wide variety of desiccants in common use. Materials such as phosporus pentoxide may be used in cases in which an aggressive drying agent is required. Note that samples removed from a vacuum oven must ideally be allowed to cool in a desiccator prior to analysis; otherwise, moisture can be absorbed while the sample is cooling to room temperature.

C. Solvent Extraction

If a sample, solid or liquid, is known to be a mixture, solvent extraction may be used to selectively remove or extract a material. Solvent extraction is also a convenient method to enrich a particular component—either in the extracted phase or in the residual phase (what is left behind). Many of the comments provided earlier (Section 5.1) apply here. Obviously, it is necessary to find a solvent that has a high affinity for the component(s) to be extracted and a low affinity for the remaining material. It is not a good idea to select a solvent that is midway in its ability to extract an ingredient. If affinity exists for both phases then there is a risk of establishing a stable emulsion.

Obviously, it is not always possible to overcome this problem, especially if the system contains surfactant materials. In such cases, in which aqueous systems are involved, the addition of a polar ionic substance, such as salt or sodium sulfate, may be necessary to help break (salt-out) an emulsion. Once a material has been extracted, it may be measured directly either in the extraction medium (see comments about solutions in Section 5.1) or on the isolated material after removing the extraction solvent (see comments in Section 5.2). If the solvent is removed, make sure that all traces of solvent are eliminated prior to analysis, and always document what solvents have been used.

Examples of the practical use of solvent extraction are the use of water to remove water-soluble components from organic mixtures (the remaining organic phase is analyzed), the use of methanol with mineral oils or polymers to remove polar additives, the use of compound-selective solvents on powdered mixtures, and the use of Freon for the extraction and measurement of hydrocarbons and organics from soil and environmental water samples. In this last example, standard methods exist for the quantitative estimation of the extracted materials.

D. Partial Evaporation or Distillation

If an organic solution or mixture of solvents is known to contain components of different volatility, partial evaporation or selective distillation can provide preferential enrichment of one or more components. The collection of spectra at different stages of the evaporation process can provide an insight into the nature of the key components. Careful use of spectral subtraction can help in the numerical extraction of chemical components from the different fractions. Successive subtractions may be used, with moderate success, in multicomponent situations. The main obstacle in this case is spectral variations due to mutual interaction of components in solution. Obviously, if nonvolatile components are present, this approach provides a practical separation of some of the main components—by complete evaporation.

Normally, the analyses mentioned previously are performed on the residual materials. However, there is nothing to prevent measurements being made on the separated fractions or on condensate materials. In the case of samples containing water, be aware that the condensate will be enriched with water. An interesting approach to the controlled evaporation of materials is to use the combined techniques of thermogravimetric analysis and infrared. Although this technique is typically used for the determination of thermal decomposition products, it may also be used for controlled and reproducible selective removal of volatile components, with on-line monitoring of the volatile components.

E. Dialysis

Dialysis is not a technique widely used in analytical laboratories. However, it is a very convenient and useful method for separating mixtures of low- and high-molecular-weight (e.g., polymers) compounds. It is used in the oil industry for the separation of polymeric additives from base oil components and low-molecular-weight additives in blended lubricants. For this separation, the blended lubricant is placed within a latex rubber membrane. The membrane is placed within a Soxhlet extractor, and hexane or a light petroleum distillate is used as the extraction solvent. After several hours of reflux, the higher-molecular-weight materials are retained in the membrane as a hexane solution. The application is not limited to lubricating oils and may be applied to any polymer-based system in which materials may be extracted by a hydrocarbon-based solvent on the basis of molecular weight.

Note that the use of a hydrocarbon solvent is required to swell the membrane and to generate the required molecular porosity. Other membranes may be available for other chemical systems and for the use of other solvents.

F. Chromatography: Adsorption and Ion Exchange

The use of instrumental chromatographic methods was described earlier in the role of hyphenated or hybridized techniques when combined with infrared spectroscopy. In this case, the chromatographic front-end acts as a sophisticated sample preparation system for isolation of specific chemical species in a mixture. In cases in which a particular component needs to be removed or separated, it is not necessary to resort to an instrumental method. In such cases, the sample may be passed through a simple column containing the solid-separation phase. A convenient approach for liquids is to prepare small columns of adsorbent or ion-exchange resin in a Pasteur or dropper pipette. This is ideal as a method for sample cleanup or for selectively removing contaminants or specific chemical components.

In such cases, the eluted material can be analyzed directly and can be compared to the sample prior to elution. If a specific additive or chemical compound is removed by the solid-phase material, then its spectrum may be generated by subtraction of the spectra recorded before and after the column separation. Alternatively, the separated component may be eluted from the column by the use of an appropriate solvent.

This approach is not limited to liquids, and solid-phase adsorbents may be used for selective removal of components from gas streams. In cases in which molecular sieve compounds are used, the separated components may be studied following thermal desorption of the adsorbed component(s). A variant of this approach can be used for the analysis of particulates or dust in gas streams (or in the environment). In this case, the sample is drawn through a suitable membrane. Membranes made from materials such as PVC have sufficiently good infrared transmission so that separated components may be measured directly on the filter material. Membrane separations may also be applied to liquid systems. One convenient approach is to use a porous silver membrane and to measure any separated components (particulates or other insoluble matter) retained on the surface of the membrane directly via a reflection measurement.

G. Pyrolysis

The use of TGA as a controlled technique for monitoring the thermal breakdown of materials was mentioned earlier. A crude version of this approach is pyrolysis, in which the sample is heated in a Pyrex tube and the volatile thermal degradation products are collected as a condensate (also known as the pyrolysate) and analyzed. This is ideal for studying polymers, especially physically hard or heavily filled polymers, which are difficult to handle by the normal sample preparation procedures. It can be helpful to study both the pyrolysate (the condensate) and the pyrolysis residue. In this case, the residue can provide information on filler materials. Note that not all materials will form a condensate, and so it is also beneficial to analyze all gaseous components from the thermal degradation. Examples of materials that do not appear in the condensate are hydrogen chloride (from PVC), hydrogen cyanide (from urethanes and other nitrogenous polymers), and hydrogen sulfide (from sulfur-based polymers).

H. Abrasion

Very hard surfaces can be studied by the abrasion of the surface with a sheet of abrasive material, such as silicon carbide or carborundum paper. At this point a number of different methods may be used to analyze the abraded material. In essence, any solid sampling technique that is capable of handling fine powders—KBr pellet, diffuse reflectance, ATR, photoacoustic, etc.—may be used to study the material. An interesting variant is to use diffuse reflectance to study the abrasive (see the reference to the silicon carbide method in Section 4) for the residual material.

VI. Recommended Spectrum Acquisition, Presentation Format, and Data Manipulation Procedures

A. Recommended Spectrum Acquisition Parameters

It is difficult to be specific about the actual instrumental parameters for spectrum acquisition because these are dependent on both the instrument type and the method of sample preparation. The following are guidelines for average situations:

Normal sample, modern FTIR spectrometer: 1–4 minute scan time, with spectral resolution of 2–4 cm^{-1}, providing a signal-to-noise ratio (S/N) of between 1000:1 and 10,000:1. If a spectrum is to be used for qualitative purposes only, then a S/N of approximately 1000:1 to 2000:1 is normally adequate and is typically obtained in well under 1 min often within a few seconds. For quantitative applications, or applications involving spectral subtraction or resolution enhancement techniques, higher signal-to-noise performance is required. Note that scanning for more than 256 scans may approach a point of diminishing returns in terms of spectral quality and S/N improvement (limited by the $\sqrt{n}$ law, and the performance of the interferometer). Note that for quantitative work, 8 cm^{-1} resolution is often sufficient, and this is accompanied by a two-fold gain in S/N (from 4 cm^{-1}).

Normal sample, older-style dispersive spectrometer: 3- to 10-min scan, with a slit program (integrated scan mode) equivalent to 3–5 cm^{-1} (average value), providing a S/N of 500 : 1 or better.

Low transmission sample (microsample or optically opaque material): 2 to 8-min scan, with a spectral resolution of 4–8 cm^{-1}, providing a S/N of 500 : 1 or better. The use of a MCT detector is suggested for samples with an overall transmission of 5%T or less, especially in the absence of an optimizing accessory (such as a beam condensor for microsamples). When using a MCT detector, it is important that the potential for nonlinearity is appreciated. Also, it is important to be aware of the occurrence of detector saturation, which will lead to extreme nonlinearity, and spectral distortion.

Irrespective of the final selection of data acquisition parameters, full documentation of the experimental details should be recorded with all spectra, especially if regulated procedures are followed. Also, as a general rule, it is recommended that good laboratory practice guidelines are adopted throughout.

B. Recommended Spectral Range

Unless special circumstances dictate otherwise, all mid-infrared spectra are normally recorded within the traditional mid-infrared spectral range, nominally defined as the region extending from 4000 cm^{-1} (2.5 μm, 2500 nm) to a minimum of 400 cm^{-1} (25 μm, 25,000 nm). It is expected that if special windows are used, or a restricted range detector is employed, the spectral range may be restricted to a region less than the nominal 4000–400 cm^{-1}. With a modern FTIR instrument, it is normal that the entire spectral range is obtained, and for some instruments the upper limit may be set beyond the traditional limit. The exceptions would be if a detector or a component optic constrains the spectral range or if special filtration (optical or digital) is imposed to improve the S/N performance for a specific application. If the range is too restricted, it must be appreciated that important, diagnostic spectral information may be lost, relative to either qualitative or quantitative applications.

C. Recommended Presentation Formats

The normal intensity scale obtained for a raw measurement is %T, usually scaled from 0 to 100%. If an abbreviated or expanded scale is used, it is important to ensure that the scale limits are presented, preferrably with intermediate scale markings shown. Note that for all numerical operations on spectral data, it is necessary to perform the operations on the absorbance form of the spectrum, including baseline corrections, spectral subtractions, and any numerical scaling of data. An important exception is the numerical and/or digital treatment of noise. Noise is essentially linear over the entire scale in transmittance (or %T) and is logarithmic in absorbance. All processing that assumes a linear distribution of noise, such as digital smoothing, must be applied to the transmittance form of the data.

The standard linear wavenumber scale is normally used for mid-IR data: Fundamental vibration group frequencies are naturally associated with the wavenumber form of the scale. Wavelength-based scales are not normally used, except for the NIR. For mid-IR instruments, options to provide a linear display in wavelength are usually available as an output function for the graphics: The raw data are seldom transformed and are usually retained linear in the frequency scale (cm^{-1}).

D. Spectral Data Manipulation

At all times, it is strongly recommended that raw spectral data are retained. If practical, the storage of single-beam data, or raw interferometric data, may be preferred, especially if the original digital significance of the data is to be retained. For appearance purposes, some limited digital smoothing and nominal baseline corrections are acceptable if they improve overall readability of the final spectrum. If there are any doubts in terms of acceptability, it is recommended that original raw data are kept at hand to support any data that have been numerically manipulated (irrespective of the reasons for the manipulation). All documentation concerning data manipulation should be retained with the spectra. If spectral subtraction is used, it is important to retain all original raw spectral data including the component being subtracted.

E. Standardization

In the production of infrared spectral data, one of the greatest problems is a lack of standardization. This lack of standardization relates to sampling, the format of the data, the actual recording of the spectrum, and the final digital representation used for storage and archiving. For spectral archiving, there have been attempts in recent years to gain consensus between the instrument vendors for the production of a standard for data transfer. The result of this has been the JCAMP.DX spectrum data format under the auspices of the IUPAC (32). This documented ASCII format has been implemented for most modern instruments, although variation in the interpretation of the standard has led to some incompatibilities. Recently, a group involving the American Instrument Association and the FTIR vendors worked on a new standard for spectral data transfer based on a *NetCDF* protocol (33). This is not expected to replace JCAMP-DX, but it is expected to become an important standard for data transfer via the Internet and through various laboratory information management systems architectures.

However, the "physical" transfer of spectra between instruments is only one step in the complex chain of the standardization in spectra. The ideal is that a given sample provides a constant spectrum for a given physical state and a defined set of recording and sampling conditions. In the past, it was considered adequate to run a simple calibration standard, such as polystyrene. This is often sufficient as a simple validation of an instrument's performance relative to a prerecorded norm. However, it is not adequate for, and does not constitute, instrument standardization. Standardization implies a unified control of parameters, such as spectral resolution and band shape, actual spectral line position (wavelength calibration), and photometric recording accuracy, and all things that can impact these parameters in a practical measurement.

Currently, there is a focus on instrument standardization relative to the use of a fixed and reproducible method of sampling. Initially, the first implementations were for near-infrared analysis, in which the analysis was primarily quantitative and in which there was a need to preserve the integrity of calibrations between different instruments (34) (usually of the same design). The technology and concepts exist for extending the approach to other spectral regions such as the mid-infrared.

The next, and weakest, link in the chain is the sample itself and the method selected for sampling. A homogeneous sample, such as a gas/vapor or a clear liquid, can be relatively straightforward as long as factors such as pressure and temperature are addressed. Solids present a different problem because morphology and form become an overriding issue. Some sampling techniques, such as photoacoustic detection/sampling, come close to providing spectra that are independent of sample morphology. However, the more readily accessible techniques, especially the common KBr pellet, oil mull, and diffuse reflectance techniques, for powders and grindable solids, which are very dependent on particle size, are particularly problematic.

There are surface-oriented sampling techniques such as internal reflectance (ATR) and external (specular) reflectance, both providing data that are influenced in one way or another by the sample's refractive index (at the measurement wavelength), the refractive index of the sampling medium (in particular, ATR), and the polarization of the infrared beam (in particular, external reflectance). For ideal cases, the mathematical relationship between the recorded spectrum and an idealized absorption (or transmission) spectrum is understood. In these situations, a correction algorithm can be applied to generate a pseudo-absorbance spectrum. This assumes that the sampling technique is well understood, that the nature of the differences between the recorded spectrum and the "absorbance" spectrum are appreciated, and that a standardized approach to the sample preparation is adopted.

In general, there has not been a universal adoption of standardization regimes for infrared spectroscopic data. In some ways, this has limited the potential of the technique for generalized methods of analysis. For example, it has been stated by the US Food and Drug Administration that the reluctance to accept infrared spectroscopic methods (both near- and mid-infrared) stems from the inability to produce standardized spectral data. The goal is the ability to produce an exact and very reproducible standard spectrum for a given material by a prescribed method of sample preparation, independent of operator or instrument (model or vendor). Many of the factors that impact meeting this goal are now understood, and it is anticipated that future generations of instrumentation, with integrated sampling and data manipulation, may lead to a resolution of this problem for the majority of samples. In the meantime, readers are encouraged to utitlize the standard procedures for sample preparation as defined in texts, such as the ASTM's *Annual Book of Standards* (35).

VII. Summary

This chapter has focused on the issues of sampling as they pertain to routine sampling for qualitative and quantitative analysis. There are many specialized approaches to sampling that may be adopted for research applications. These are typically experiment dependent, and their success may be somewhat limited to the application for which they are customized. For this reason, such examples have been generally excluded from this text. The number of accessory companies has grown in the past 10–15 years, and the number of new sampling devices is growing steadily. It is beneficial to contact these companies to obtain the latest literature, and in this way stay current with the latest developments in infrared sampling.

References

1. Hollas, J. M. (1996). *Modern Spectroscopy*. Wiley, Chichester, UK.
2. Smith, A. L. (1991). *Infrared Spectroscopy, Practical Handbook of Spectroscopy* (J. W. Robinson, Ed.), pp. 481–535, CRC Press, Boca Raton, FL.
3. Colthup, N. B., Daly, L. H., and Wiberley, S. E. (1990). *Introduction to Infrared and Raman Spectroscopy* 3rd ed., Chap. 1. Academic Press, New York.
4. Feinstein, K. (1995). *Guide to Spectroscopic Identification of Organic Compounds*. CRC Press, Boca Raton, FL.

5. Silverstein, R. M., Bassler, G. C., and Morrill, T. C. (1991). *Spectrometric Identification of Organic Compounds*, 5th ed. Wiley, New York.
6. Burns, D. A., and Ciurczak, E. W. (Ed.) (1992). *Handbook of Near-Infrared Analysis*, Practical Spectroscopy Series, Vol. 13. Dekker, New York.
7. Smith, A. L. (1979). *Applied Infrared Spectroscopy: Fundamentals, Techniques, and Problem Solving*, Vol. 54 in Chemical Analysis. Wiley-Interscience, New York.
8. Schrader, B. (Ed.) (1995). *Infrared and Raman Spectroscopy: Methods and Applications*, VCH Publishers, New York.
9. Ferraro, J. R., and Krisnan, K. (Eds.) (1989). *Practical Fourier Transform Infrared Spectroscopy: Industrial and Chemical Analysis*. Academic Press, San Diego.
10. Griffiths, P. R., and de Haseth, J. A. (1986). *Fourier Transform Infrared Spectrometry*, Vol. 83 of Chemical Analysis. Wiley-Interscience, New York.
11. Coleman, P. B. (Ed.) (1993). *Practical Sampling Techniques Techniques for Infrared Analysis*, CRC Press, Boca Raton, FL.
12. Porro. T. J., and Pattacini, S. C. (1993). *Sample handling for mid-infrared spectroscopy, Part I: Solid and liquid sampling. Spectroscopy* **8**(7), 40–47.
13. Porro, T. J., and Pattacini, S. C. (1993). *Sample handling for mid-infrared spectroscopy, Part II: Specialized techniques. Spectroscopy* **8**(8), 39–44.
14. Pouchert, C. (1989). *Aldrich Library of FTIR Spectra, Vol. 3: Vapors*. Aldrich Chemical Company, Milwaukee, WI.
15. *Solvents Condensed Phase, Gas Phase and Mass Spectra, Book IV: Sprouse Collection of Infrared Spectra* (1988). Sprouse Scientific Systems, Charlotte, NC.
16. Ohta, K., and Ishida, I. (1988). *Comparison among several numerical integration methods for Kramers–Kronig transformation. Appl. Spectrosc.* **42**, 952–957.
17. Compton, S. V., Compton, D. A. C., and Messerschmidt, R. G. (1991). *Analysis of samples using infrared emission spectroscopy. Spectroscopy* **6**(6), 35–39.
18. Harrick, N. J. (1987). *Internal Reflection Spectroscopy*. Harrick Scientific, Ossining, NY (Original Publication by Wiley, New York, 1967).
19. Mirabella, Jr., F. M. (Ed.) (1993) *Internal Reflection Spectroscopy: Theory and Applications, Practical Spectroscopy Series*, Vol. 15. Dekker, New York.
20. Culler, S. R. (1993) *Diffuse reflectance spectroscopy: Sampling techniques for qualitative/* quantitative analysis of solids. In *Practical Sampling Techniques: Techniques for Infrared Analysis* (P. B. Coleman, Ed.), pp. 93–105. CRC Press, Boca Raton, FL.
21. McClelland, J. F. (1983). *Photoacoustic spectroscopy. Anal. Chem.* **55**(1), 89A–105A.
22. McClelland, J. F., Jones, R. W., Luo S., and Seaverson, L. M., (1993). *A practical guide to FTIR photoacoustic spectroscopy*. In *Practical Sampling Techniques: Techniques for Infrared Analysis* (P. B. Coleman, Ed.), pp. 107–144. CRC Press, Boca Raton, FL.
23. Harthcock, M. A., and Messerschmidt, R. G. (Eds.) (1988). *Infrared Microspectroscopy Theory and Applications*. Dekker, New York.
24. Humecki, H. J. (Ed.) (1995). *Practical Guide to Infrared Microspectroscopy*, Practical Spectroscopy Series, Vol. 19. Dekker, New York.
25. White, R. (1991). *Chromatography*/Fourier Transform Infrared Spectroscopy and its Applications. Dekker, New York.
26. Herres, W. (1987). *Capillary Gas Chromatography–Fourier Transform Infrared Spectroscopy, Theory and Applications*. Huthig, New York.
27. Craver, C. D. (1977). *Halogenated Hydrocarbons*, A Coblentz Society special collection of infrared spectra. Coblentz Society, Kirkwood, MO.
28. Craver, C. D. (1983). *Infrared Spectra of Regulated and Major Industrial Chemicals*, Special collection of infrared spectra from the Coblentz Society. Coblentz Society, Kirkwood, MO.

29. Pouchert, C. (1981). *Aldrich Library of Infrared spectra, 3rd Ed.* Aldrich Chemical Company, Milwaukee, WI.
30. *Solvents by Cylindrical Internal Reflectance*, Book II: *Sprouse Collection of Infrared Spectra* (1987). Sprouse Scientific Systems, Charlotte, NC.
31. *Infrared Spectra Handbook of Common Organic Solvents*, Infrared Spectra Handbooks (1983). Bio-Rad, Sadtler Research Laboratories Division, Philadelphia.
32. McDonald R. S., and Wilks, Jr., P. A. (1988), *J-CAMP-DX: A Standard form for exchange of infrared spectra in computer readable form. Appl. Spectrosc.* **42**(1), 151–162.
33. Lysakowski, R. S. (1994). *NetCDF—A defacto standard framework for analytical data exchange, storage and retrieval, Computerized Data Standards: Databases, Data Exchange, and Information Systems* (R. S. Lysakowski, and C.E. Gragg, Eds.), ASTM STP-1214, pp. 57–74. American Society for Testing and Materials, Philadelphia.
34. Workman, Jr. J., and Coates, J. (1993). Multivariate calibration transfer: The importance of standardizing instrumentation. *Spectroscopy* **8**(9), 36–42.
35. Molecular Spectroscopy (1996). In *ASTM Annual Book of Standards*, Vol. 03.06, pp. 589–785. West Conshohocken, PA 19428-2959.

SPECTROSCOPIC QUANTITATIVE ANALYSIS

JAMES H. DUCKWORTH
Galactic Industries Corporation

I. What Is "Chemometrics"?

Just the mere mention of the word chemometrics invokes images of incomprehensible statistics and mathematics that are more black magic than a useful tool. However, there is generally no reason for this type of reaction.

Literally translated, the word chemometrics means performing calculations on measurements of chemical data. This can be anything from calculating pH from a measurement of hydrogen ion activity to computing a Fourier transform interpolation of a spectrum. Recently, the common usage of the word refers to using linear algebra calculation methods to make either quantitative or qualitative measurements of chemical data, primarily spectra. Nearly all trained spectroscopists have the basic understanding of the concepts necessary to apply these methods. Unfortunately, like all specialty areas of science, chemometrics has a language all its own that makes it difficult for the beginner to understand.

The science of chemometrics gives spectroscopists many different ways to solve the calibration problem for analysis of spectral data. Some are very

ISBN 0-12-764070-3.

simple to understand, whereas others require a strong background in linear algebra. However, they all have one thing in common: They each solve an individual problem but do not address all possible problems. Some methods have the advantage of being simple to understand but may not be very robust for all possible samples. Others are very complex to understand and implement but give solutions that are very stable and can handle a large variety of "unknowns." The key to understanding chemometrics is not necessarily understanding the mathematics of all the different methods; it is knowing which model to use for a given analytical problem and properly applying it.

This chapter and Chapter 5 will attempt to "demystify" some of the more basic tenets of chemometric spectral analysis methods. The intent is not to completely explain all the complex mathematics behind the techniques (although the equations are provided in almost all cases). Instead, they are designed to give a basic understanding of how and why the mathematics work in order to properly apply them to solving analytical problems. These chapters may also serve as a foundation to learning more about chemometrics and, hopefully, to lead the reader to other sources of information.

II. The Beer–Lambert Law

One of the keys to quantitative analysis in any scientific field is the assumption that the amounts (concentrations) of the constituents of interest in the samples are somehow related to the data from a measurement technique used to analyze them. The ultimate goal is to create a calibration equation (or series of equations) that, when applied to data of "unknown" samples measured in the same manner, will accurately predict the quantities of the constituents of interest.

In order to calculate these equations, a set of "standard" samples are made that reflect the composition of the "unknowns" as closely as possible. These standards are designed to span the expected range of concentrations and compositions in the unknowns and are measured under the same conditions as the unknowns. The standards are then measured by an instrument. Together, this collection of known data (the composition of each standard) and the measured data from the instrument form what is known as a training set or calibration set. The calibration equations that describe the relationships between these two sets of information are calculated from these data. The exact equation or set of equations that make up the calibration is also known as a model. Thus, this process is often called "solving the calibration model."

Once the model equations have been selected and solved, they can be used to calculate the same quantities or properties in unknown samples. However, in order for this sample(s) to be predicted accurately, it must be measured under exactly the same conditions on the same instrument as the calibration set.

For many current applications, a spectrophotometer is increasingly becoming the measurement device of choice. Unlike other methods that give "single-point" measurements for each calibration and unknown sample (i.e., pH or single-element atomic absorption), the spectrum of a sample contains many data points. Every response value in a spectrum has some relation to the properties or constituent(s) that make up the measured sample. Using a spectrum of a sample that has many data points has some distinct advantages over single-point measurement techniques. One of the most important factors is that there are many more measurements per sample (spectral data points) to use in generating the calibration equations. As anyone who has performed quantitative analysis knows, the more measurements per sample, the more accurate the results. The problem for the analyst is to discover what those relationships are and use a calibration model that reflects them accurately.

One advantage of using spectroscopy as a measurement technique is that the Beer–Lambert law (also know as Beer's law) defines a simple linear relationship between the spectrum and the composition of a sample. This law, which should be familiar to all spectroscopists, forms the basis of nearly all other chemometric methods for spectroscopic data. Simply stated, the law claims that when a sample is placed in the beam of a spectrometer, there is a direct and linear relationship between the amount (concentration) of its constituent(s) and the amount of energy it absorbs. In mathematical terms:

$$A_\lambda = \varepsilon_\lambda bC,$$

where A_λ is the sample's absorbance value at specific wavelength (or frequency) λ, ε_λ is the absorptivity coefficient of the material (constituent) at that wavelength, b is the path length through the sample, and C is the concentration. The absorptivity coefficient for every material is different, but for a given compound at a selected wavelength this value is a constant. The only problem that remains is to discover the "real" value of that constant.

III. Classical Quantitation Methods

Classical methods of quantitative analysis are relatively straightforward in that they use well-understood relationships between the independent data (spectral absorbances) and the dependent data (concentrations). Typically, the calibration equations are solutions of equations directly relating these two pieces of data. Some methods are univariate (solve only one equation based on one measured value and have one calibration value per sample), whereas others are multivariate (solve a series of equations using many measurements per sample for one calibration value). Multivariate models have a distinct advantage over univariate ones in that they allow inclusion of more of the available data (spectral absorbances) and thus the solutions are generally more stable because of the averaging effect.

A. Least Squares Regression Model

One of the simplest models to understand and apply is simple linear regression. In this model, either height or area of a selected peak in the spectrum is assumed to be related to the constituent quantities by a polynomial equation. In this case, the model equations can appear as

$$C = B_1(\text{Area}) + B_0$$

$$C = B_2(\text{Height})^2 + B_1(\text{Height}) + B_0,$$

where C is the concentration of desired constituent, and the B's are the calibration coefficients. Notice that because only one measurement per sample (either peak height or area) is used to solve the equations, this a univariate type of model.

Calibrating the model requires that the analyst identify an isolated peak in the spectra of the calibration set that is directly related to the constituent of interest. This is critically important to the predictive ability of the final model. If the peak selected does not reflect the concentration of the desired constituents, then the model will not work. In addition, the spectral band must be isolated; in other words, it must be a single isolated band with no overlap with bands of any other constituent in the sample. Knowledge of the spectrum of the pure constituent is extremely useful in band selection. Once the peak is selected, the area or height of the band is measured in every training sample.

A technique called least squares regression is used to solve for the model equation using the peak height/area data and the known constituent quantities. This mathematical technique calculates the coefficients of a given equation such that the differences between the known spectral responses (peak areas or heights) and the predicted spectral responses are minimized. (The predicted spectral responses are calculated by reading the measurements off the calibration line at the known concentrations; see Fig. 1.) As

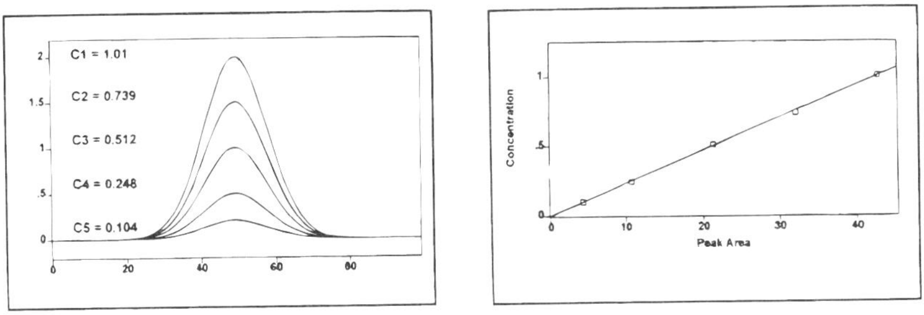

FIG. 1. Least squares (linear) regression: The areas of the constituent peak and the constituent concentrations (left) were used to compute the coefficients of the linear calibration equation (right).

can be seen from the equations, there can be more than one coefficient. The minimum number of calibration samples required to solve the equation is equal to the number of coefficients in the model. Thus, the equations shown previously would require a minimum of two and three calibration samples, respectively. However, more are usually measured to improve the accuracy of the calibration coefficients.

Once the coefficients of the model equation have been solved, the quantities of the same constituent in unknown samples can be calculated by substituting the peak areas or peak heights into the equation and solving for concentration.

In addition to being very simple to calculate and understand, solving a least squares regression equation is very fast on today's personal computers. Therefore, most of the time spent to properly apply this model will be in making the calibration samples, measuring the spectra, and selecting the constituent peak for calibration.

Unfortunately, there are a few problems with this type of model. First, if there is more than one desired constituent in the calibration samples, each individual constituent requires solving a separate model equation. More important, a separate peak in the spectrum must be identified for each constituent. Remember that each peak must be completely isolated from interference from other constituents. This is not always practical, and in mixture samples this can be nearly impossible. Therefore, the use of least squares regression models tends to be restricted to pure or simple mixture samples.

Another related problem is that of sample purity. If the unknowns contain a constituent that was not present in the original calibration set, and the contaminant has spectral bands in the region of the selected calibration peak, the accuracy of the predicted quantity will be significantly diminished.

Least squares regression is known as a univariate method. Univariate methods assume that there is a direct relationship between two sets of values that can be described by a single equation. In univariate methods, there is generally one independent variable (in this case, spectral response) and one dependent variable (concentration). By knowing either one of these values, a solution can be calculated for the other.

Finally there is the problem of selecting the correct polynomial. Most spectroscopic systems tend to follow linear relationships, but this is by no means true 100% of the time. Some systems are quadratic (parabolic) or even exponential in nature. Thus, the form of the model equation is another variable (along with the spectral band) to be carefully considered when using least squares regression models.

Least Squares Regression

Advantages	Disadvantages
• Easy to understand and calculate • Calculations are very fast • Used primarily for simple samples: pure compounds, binary mixtures, etc.	• Requires isolated spectral bands that are solely related to the constituent(s) of interest • Cannot be used for comlex mixture samples in which the individual constituents have overlapping spectral bands • Band selection can be difficult or impossible if spectrum of property of interest is not known • Selecting correct polynomial can be difficult; most spectroscopic systems can be solved with a simple straight line, but this is not 100% true • Large prediction errors will result from constituents with bands in same region of spectrum as calibration bands(s)

B. Classical Least Squares Model

This method is founded in using Beer's law to extend the calculation of the absorptivity coefficients across a much larger portion of the spectrum than the least squares regression model. Referring back to the description of Beer's law, notice that it defines a relationship between four different variables; the spectral response (A_λ), the constituent absorptivity constant (ε_λ), the path length of light (b), and the constituent concentration (C). The goal

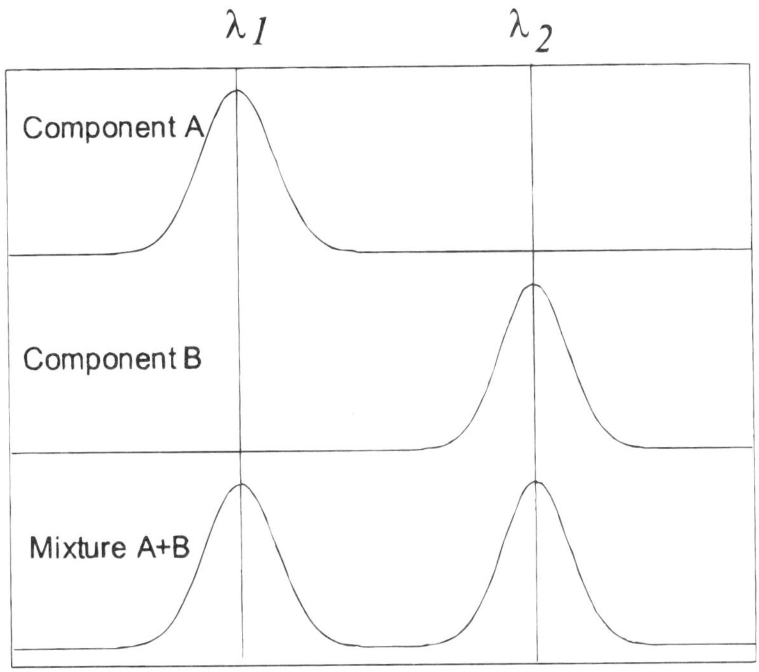

FIG. 2. Classical least squares: Hypothetical spectra of two different pure constituents A and B and a mixture of the two. Because the constituent bands in the spectra do not overlap, the selected wavelengths could be used to solve separate equations for both A and B.

of calibrating a spectroscopic quantitative method is solving for the absorptivity constants. However, if the path length of the samples is also kept constant (as it is for most quantitative experiments), Beer's law can be rewritten as

$$A_\lambda = K_\lambda C,$$

where the absorptivity coefficient and path length are combined into a single constant, K. This equation can be easily solved by measuring the absorbance of a single sample of known concentration and using these values to solve for K_λ. Predicting the concentration of an unknown sample is as simple as measuring the absorbance at the same wavelength and then applying the following:

$$C = \frac{A_\lambda}{K_\lambda}.$$

However, basing an entire calibration on a single sample is generally not a good idea. Due to limitations of noise, instrument error, sample handling error, and many other possible variations, it is best to measure the absorbances of a series of different concentrations and calculate the slope of the best fit line through all the data points. Just as in the case of least squares regression, this can be solved with a simple regression line of absorbance versus concentration.

However, the problem becomes more complex if the sample contains two constituents. In any algebraic solution, it is necessary to have as many equations as unknowns. In this case, it is necessary to set up two equations:

$$A_{\lambda 1} = K_{a,\lambda 1} C_a$$

$$A_{\lambda 2} = K_{b,\lambda 2} C_b,$$

where $A_{\lambda 1}$ and $A_{\lambda 2}$ are the absorbances at two different wavelengths, C_a and C_b are the concentrations of the two constituents ("A" and "B") in the mixtures, and $K_{a,\lambda 1}$ and $K_{b,\lambda 1}$ are the absortivity constants for the two constituents at those wavelengths. Again, it is possible to solve each equation independently provided that the spectrum of one constituent does not interfere with the spectrum of the other (i.e., the bands are well resolved) (Fig. 2).

Unfortunately, the previous equations make the assumption that the absorbance at wavelength 1 is entirely due to constituent A and the absorbance at wavelength 2 is entirely due only to constituent B (Fig. 3). As with the least squares regression model, this requires finding two wavelengths in the training set of spectra that exclusively represent constituents A and B. With complex mixtures, or even simple mixtures of very similar materials, this is a difficult if not impossible task.

However, it is possible to get around this by taking advantage of another part of Beer's law; the absorbances of multiple constituents at the same wavelength are additive. Thus, the two constituent equations for a single spectrum should really be

$$A_{\lambda 1} = K_{a,\lambda 1} C_a + K_{b,\lambda 1} C_b$$

$$A_{\lambda 2} = K_{a,\lambda 2} C_a + K_{b,\lambda 2} C_b.$$

All the equations presented so far assume that the calculated least squares line(s) that best fits the calibration samples is perfect. In other words, it has been assumed that there is no error in the measurements or in the predictive ability of the model when the equations are used to predict unknowns. Once again, this never happens in the real world; there is always some amount of error. The existence of error is what requires running more than two samples in the first place.

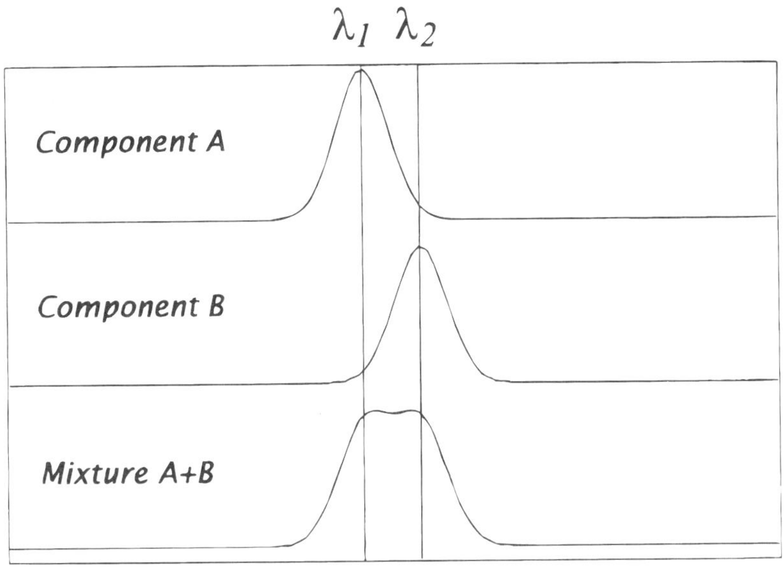

FIG. 3. Classical least squares: Hypothetical spectra of two alternative pure constituents, A and B, and a mixture of the two. In this case, the bands of the constituent spectra overlap, and the equations must be solved simultaneously for both A and B.

It is necessary to amend the equations one more time to add a variable to compensate for the errors in the calculation of the absorbance:

$$A_{\lambda 1} = K_{a,\lambda 1}C_a + K_{b,\lambda 1}C_b + E_{\lambda 1}$$
$$A_{\lambda 2} = K_{a,\lambda 2}C_a + K_{b,\lambda 2}C_b + E_{\lambda 2},$$

where $E_{\lambda 1}$ and $E_{\lambda 2}$ are the residual errors between the least squares fit line and the actual absorbances. When performing least squares regression, the "offset" coefficient (a.k.a., intercept and bias) performs the same function. In these terms, the E values can be thought of as the calibration offset or bias. It is obvious to see that this will always be zero when fitting only two points (i.e., only two calibration mixture samples). However, as with most calibration models, classical least squares usually requires many more training samples to build an accurate calibration. As long as the same number (or more) of wavelengths are used as there are constituents, it is possible to calibrate for all constituents simultaneously.

The next problem is how to solve all these equations. If you have ever tried to solve simultaneous equations by hand, you know this is a very tedious process. If more than two constituents are present or more than two wavelengths are used, it gets even harder. A particularly efficient way of solving simultaneous equations is to use linear algebra, also known as matrix mathematics. This technique still requires many calculations, but the rules are straightforward and are perfectly suited for computers. In matrix terms, the previous equation can be formulated as

$$\begin{vmatrix} A_{\lambda 1} \\ A_{\lambda 2} \end{vmatrix} = \begin{vmatrix} K_{a,\lambda 1} & K_{b,\lambda 1} \\ K_{a,\lambda 2} & K_{b,\lambda 2} \end{vmatrix} \begin{vmatrix} C_a \\ C_b \end{vmatrix} + \begin{vmatrix} E_{\lambda 1} \\ E_{\lambda 2} \end{vmatrix},$$

or, more simply

$$A = KC + E.$$

In this case, A represents a (2×1) matrix of absorbances of the two selected wavelengths, K is a (2×2) matrix of the absorptivity constants, C is a (2×1) matrix of the concentrations of the two constituents, and E is the (2×1) matrix of absorbance error, or offset.

This model can be extended to performing calculations using many more wavelengths than just two. In fact, as long as the number of wavelengths used for the model is larger than the number of constituents in the mixtures, any number of wavelengths can be used. In fact, it is not unusual to use the entire spectrum when calibrating classical least squares models. In this case the matrices appear as follows:

$$\begin{vmatrix} A_{1,1} & \cdots & A_{n,1} \\ \vdots & & \vdots \\ A_{1,p} & \cdots & A_{n,p} \end{vmatrix} = \begin{vmatrix} K_{1,1} & \cdots & K_{1,m} \\ \vdots & & \vdots \\ K_{p,1} & \cdots & K_{p,m} \end{vmatrix} \begin{vmatrix} C_{1,1} & \cdots & C_{1,n} \\ \vdots & & \vdots \\ C_{m,1} & \cdots & C_{m,n} \end{vmatrix},$$

where A is a matrix of spectral absorbances, K is the matrix of absorptivity constants, and C is the matrix of constituent concentrations. (The E matrix is not shown for space reasons, but it has the same dimensionality as the A matrix.) The subscripts indicate the dimensionality of the matrix; n is the number of samples (spectra), p is the number of data points (wavelengths) used for calibration, and m is the number of constituents in the sample mixtures.

Using matrix algebra, it is trivial for a computer to solve these equations and produce the K matrix in the previous equation (the matrix of absorptivity coefficients). Just by the nature of matrix algebra, the solution gives the best fit least squares line(s) to the data. Once the equation is solved for the K matrix, it can be used to predict concentrations of unknown samples.

For those familiar with linear algebra, to solve for the K matrix requires computing the matrix equation:

$$K = AC^{-1},$$

where C^{-1} is the inverse of the constituent concentration matrix. Unfortunately, computing the inverse of a matrix requires that the matrix be square (having the same number of rows and columns). Unless the calibration set has exactly the same number of samples as constituents, this will not be true (remember, more samples are usually used to get the best representation of the true calibration equation).

This does not mean that the previous equation cannot be solved. An alternative to computing the true inverse of the C matrix is to compute its "pseudo-inverse" as follows:

$$K = AC'(CC')^{-1},$$

where C is the matrix transpose (pivot the matrix so that the rows become the columns) of the constituent concentrations matrix.

This method of quantitative analysis is known as K matrix, or classic least squares (CLS). It has the advantage of being able to use large regions of the spectrum, or even the entire spectrum, for calibration to gain an averaging effect for the predictive accuracy of the final model. One interesting side effect is that if the entire spectrum is used for calibration, the rows of the K matrix are actually spectra of the absorptivities for each of the constituents. These will actually look very similar to the pure constituent spectra.

However, this technique does have one major disadvantage: The equations must be calibrated for every constituent in the mixtures. Otherwise, the ignored constituents will interfere with the analysis and give incorrect results. This means that the complete composition of every calibration sample must be known, and that predicted unknowns must be mixtures of exactly the same constituents.

This limitation of the CLS model can be more easily understood by taking a closer look at the model equation:

$$A_{11} = K_{11}C_{11} + K_{12}C_{21} + E_{11}.$$

Notice that the absorbance at a particular wavelength is calculated from the sum of all the constituent concentrations multiplied by their absorptivity coefficients. If the concentration of any constituent in the sample is omitted, the predicted absorbance will be incorrect. This means that the CLS technique can be applied only to systems in which the concentration of every constituent in the sample is known. If the mixture is complex, or there is the possibility of contaminants in the unknown samples that were not present in the calibration mixtures, then the model will not be able to predict the constituent concentrations accurately.

In addition, there must be no significant interactions between the different constituents when they are mixed together. For example, if two liquids are mixed together, there may be some interaction that occurs to create a minute amount of a new compound. This third compound will have a spectrum of its own and, thus, a different set of absorptivity coefficients from the other two constituents. If the concentration of this reaction product (which is usually not known) is not included in the calibration, the calculated K matrix will not reflect the absorptivities of this constituent and the predictions of unknowns will be in error. Typically, the CLS calibration method is most useful when applied to samples that have minimal or no interconstituent interactions. Therefore, the primary application for CLS has been in the analysis of gas-phase samples.

CLS

Advantages	Disadvantages
• Based on Beer's law • Calculations are relatively fast • Can be used for moderately complex mixtures • Calibrations do not necessarily require wavelength selection; As long as the number of wavelengths exceeds the number of constituents, any number (up to the entire spectrum) can be used • Using a large number of wavelengths tends to give an averaging effect to the solution, making it less susceptible to noise in the spectra	• Requires knowing the complete composition (concentration of every constituent) of the calibration mixtures • Not useful for mixtures with constituents that interact • Very susceptible to baseline effects because equations assume the response at a wavelength is due entirely to the calibrated constituents

C. Inverse Least Squares Model

As discussed previously, if the concentrations of all the constituents in the mixtures are not known, there may be a significant error when solving for the absorptivity coefficients of the known constituents by the CLS approach. In real-world samples, it is very unusual to know the entire composition of a mixture sample. Sometimes, only the quantities of a few constituents in very complex mixtures of multiple constituents are of interest. Obviously, CLS calibration will fail for these samples because the full compositional chemistry is not known. One solution to this problem is to take advantage of algebra to rearrange Beer's law and express it as

$$C = \frac{A_\lambda}{\varepsilon_\lambda b},$$

or, by combining the absorptivity coefficient (ε_λ) and the path length (b) into a single constant as with CLS, express it as

$$C = PA_\lambda + E,$$

where, as before, C is the constituent concentration, A_λ is the absorbance at the wavelength λ, and E is a matrix of concentration (not absorbance!) errors. As with the CLS model, the E matrix can be thought of as the offset or "bias" of the model.

Due to the powers of mathematics, this seemingly trivial variation has tremendous implications for the experiment. This expression of Beer's law implies that the concentration is a function of the absorbances at a series of given wavelengths. This is entirely different from CLS, in which absorbance at a single wavelength is calculated as an additive function of the constituent concentrations. Consider the following two equations:

$$C_a = A_{\lambda 1} P_{a,\lambda 1} + A_{\lambda 2} P_{a,\lambda 2} + E_a$$

$$C_b = A_{\lambda 1} P_{b,\lambda 1} + A_{\lambda 2} P_{b,\lambda 2} + E_b.$$

Notice that even if the concentrations of all the other constituents in the mixture are not known, the matrix of coefficients (P) can still be calculated correctly. This model, known as inverse least squares (ILS), multiple linear regression (MLR), or P matrix, seems to be the best approach for almost all quantitative analyses because no knowledge of the sample composition is needed beyond the concentrations of the constituents of interest.

The selected wavelengths must be in a region where there is a contribution of that constituent to the overall spectrum. In addition, measurements of the absorbances at different wavelengths are needed for each constituent. In fact, in order to accurately calibrate the model, measurements of at least one different wavelength are needed for each additional independent variation (constituent) in the spectrum.

Again, for those interested in matrix algebra, the P matrix of coefficients can be solved by computing

$$P = CA^{-1},$$

but as with CLS, if the A matrix is not square, the pseudo-inverse must be used instead:

$$P = CA'(AA')^{-1}.$$

This model seems to provide the best of all worlds. It can accurately build models for complex mixtures when only some of the constituent concentrations are known. The only requirement is selecting wavelengths that correspond to the absorbances of the desired constituents.

Unfortunately, the ILS calibration approach does have some drawbacks. Due to the dimensionality of the matrix equations, the number of selected wavelengths cannot exceed the number of training samples. In theory, it should be possible to just measure many more training samples to allow for additional wavelengths, but this causes a new problem. The absorbances in a spectrum tend to all increase and decrease together as the concentrations of the constituents in the mixture change. This effect, known as collinearity, causes the mathematical solution to become less stable with respect to each constituent.

Another problem with adding more wavelengths to the model is an effect known as overfitting. Generally, starting from very few wavelengths and adding more to the model (provided they are chosen to reflect the constituents of interest) will improve the prediction accuracy. However, at some point, the predictions will start to get worse. When the number of wavelengths increases in the calibration equations, the likelihood that unknown samples will vary in exactly the same manner decreases. When too much information in the spectrum is used to calibrate, the model starts to include the spectral noise that is unique to the training set and the prediction accuracy for unknown samples suffers.

In ILS, the averaging effect gained by selecting many wavelengths in the CLS method is effectively lost. Therefore, wavelength selection is critically important to building an accurate ILS model. Ideally, there is a crossover point between selecting enough wavelengths to compute an accurate least squares line and selecting few enough so that the calibration is not overly affected by the collinearity of the spectral data.

Many of today's software packages that perform ILS (MLR) calibrations use sophisticated algorithms to find the "best" set of wavelengths to use for each individual constituent of interest. They attempt to search through the wavelengths and try different combinations to locate that crossover point. Luckily, the calculations involved in computing the P matrix are very fast. However, when you consider that a spectrum may have as many as 2000 wavelength data points, it is obvious to see that calculating ILS models for all possible combinations of wavelengths can be an excruciating task.

Inverse least squares in an example of a multivariate method. In this type of model, the dependent variable (concentration) is solved by calculating a solution from multiple independent variables (in this case, the responses at the selected wavelengths). It is not possible to work backwards from the concentration value to the independent spectral response values because an infinite number of possible solutions exist. However, the main advantage of a multivariate method is the ability to calibrate for a constituent of interest without having to account for any interferences in the spectra.

Because it is not necessary to know the composition of the training mixtures beyond the constituents of interest, the ILS method is better suited to more complex types of analyses not handled by the CLS approach. It has been used for samples ranging from natural products (such as wheat, wool, cotton, and gasoline) to manufactured products.

ILS

Advantages	Disadvantages
• Based on Beer's law • Calculations are relatively fast • Multivariate model allows calibration of very complex mixtures because only knowledge of constituents of interest is required	• Wavelength selection can be difficult and time-consuming; must avoid collinear wavelengths • Number of wavelengths used in the model limited by the number of calibration samples • Generally, a large number of samples are required for accurate calibration • Collecting calibration samples and measuring via a primary calibration can be difficult and tedious

IV. Eigenvector Quantitation Methods

There is a solution that gains the advantages of both the ILS and CLS techniques. In the previous example of a simple two-constituent mixture, there are ideally only two independent variations in the system. In other words, the spectrum of the mixture of A and B can be re-created by adding together fractions of the spectra of the pure compounds.

In real samples, there are usually many different variations that make up a spectrum: The constituents in the sample mixture, interconstituent interactions, instrument variations such as detector noise, changing environmental conditions that affect the baseline and absorbance, and differences in sample handling. However, even with all these complex changes occurring, there should be some finite number of independent variations occurring in the spectral data. Hopefully, the largest variations in the calibration set would be the changes in the spectrum due to the different concentrations of the constituents of the mixtures. If it were possible to calculate a set of "variation spectra" that represented the changes in the absorbances at all the wavelengths in the spectra, then this data could be used instead of the raw spectral data for building the calibration model. There should be fewer common variations than the number of calibration spectra (in most cases) and, thus, the number of calculations for the calibration equations will be reduced as well.

Presumably, the variation spectra could be used to reconstruct the spectrum of a sample by multiplying each one by a different constant scaling factor and adding the results together until the new spectrum closely matches the unknown spectrum. Obviously, each spectrum in the calibration set would have a different set of scaling constants for each variation because the concentrations of the constituents are all different. Therefore, the fraction of each "spectrum" that must be added to reconstruct the unknown data should be related to the concentration of the constituents.

The variation spectra are often called eigenvectors (a.k.a., spectral loadings, loading vectors, principal components, or factors) for the methods used to calculate them. The scaling constants used to reconstruct the spectra are generally known as scores.

Because the calculated eigenvectors came from the original calibration data, they must somehow relate to the concentrations of the constituents that make up the samples. The same loading vectors can be used to predict unknown samples; thus, the only difference between spectra of samples with different constituent concentrations is the fraction of each loading added (scores).

The calculated scores are unique to each separate principal component and training spectrum and can be used in place of absorbances in either of the classical model equations (CLS or ILS). Because the representation of the mixture spectrum is reduced from many wavelengths to a few scores, it seems best to use the ILS expression of Beer's law for calculating concentrations due to its ability to calculate concentrations among interfering species. Note, however, that the calculations maintain the CLS averaging effect by using a large number of wavelengths in the spectrum (up to the entire spectrum) for calculating the eigenvectors. Therefore, in effect, eigenvector models combine the best features of both the CLS and ILS methods together in the same calculation. This is the main reason why eigenvector models are generally better than classical models in both accuracy and robustness.

The trick in using these models is in how the eigenvectors are calculated. Note that these models base the concentration predictions on changes in the data and not on absolute absorbance measurements (which are used in all the classical models). In order to calculate the principal components analysis (PCA) model, the spectral data must change in some way. The best way to accomplish this is to vary the concentrations of the constituents of interest. As with the ILS model, there can be problems with colinearity. If the concentrations of two important constituents in the calibration samples are always present in the same ratio (for example, 2:1 of A to B, such as if dilutions were made from a single stock sample), the model will only detect one variation, not two! As far as the model is concerned, all the absorbance peaks of constituent A increase or decrease when constituent B also increases or decreases and vice versa. Thus, only one variation is detected: the changes in the spectrum of A + B. Therefore, it is very important when calibrating eigenvector models that the calibration data have concentrations of the individual constituents of interest present in evenly and randomly distributed ratios (Fig. 4).

For the following discussion of the different eigenvector methods, the set of synthetic spectral data shown in Fig. 5 is used for demonstrations.

A. Principal Component Regression

Principal component regression (PCR) uses this model of the spectral variation to calculate the calibration equations. One way to calculate all the possible variations in the spectra is to use a technique called PCA. As with the other quantitative techniques (ILS, CLS, etc.), PCA also requires a group of training spectra that represent the composition of the samples. These samples must contain the constituents of interest and span the range of expected variation (concentration) for the unknown samples.

Before PCA is applied to a training set, the data is commonly mean centered. This means that the mean spectrum (average spectrum) is calculated from all the calibration spectra and then subtracted from every calibration spectrum. Mean centering has the effect of enhancing the subtle

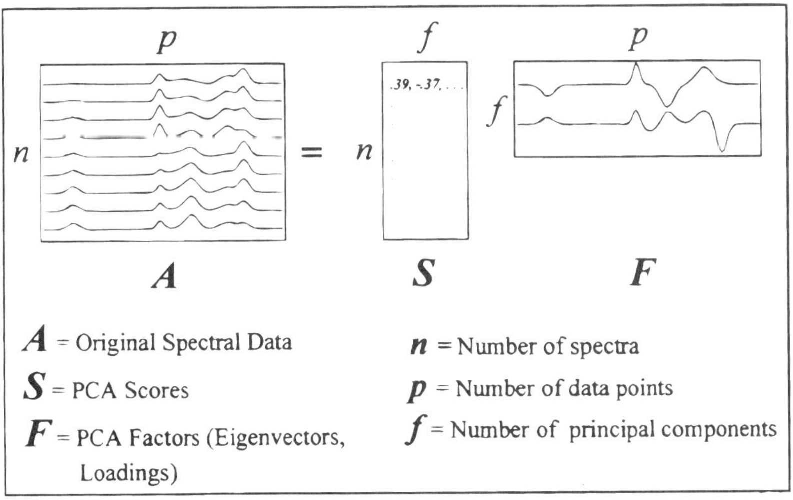

FIG. 4. PCA breaks apart the spectral data into the most common spectral variations (factors, eigenvectors, and loadings) and the corresponding scaling coefficients (scores).

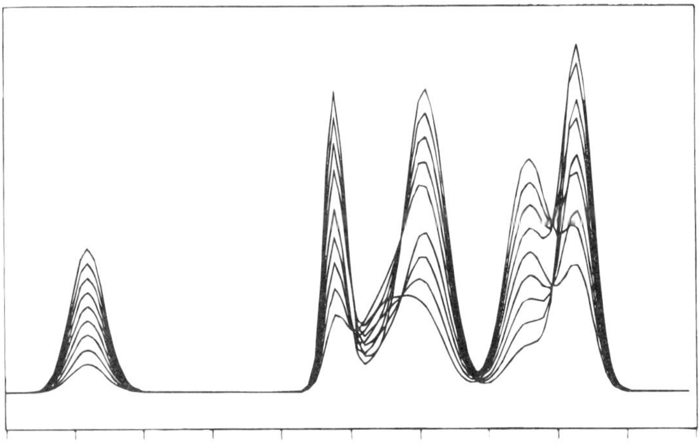

FIG. 5. A hypothetical training set of spectra with three independent constituents consisting of two Gaussian bands each.

differences between the spectra. Remember, eigenvector methods calculate the principal components based on changes in the absorbance data and not the absolute absorbance. Therefore, anything that improves the ability of the calculation to detect the differences between the calibration spectra will improve the model.

This actually makes a lot of sense when considered in the context of how PCA calculates the eigenvectors. Because the eigenvectors represent the changes in the spectral data that are common to all the calibration spectra, removing the mean simply removes the first most common variation before the data are even processed by the PCA algorithm.

PCA is effectively a process of elimination. By iteratively eliminating each independent variation from the calibration spectra in series, it is possible to create a set of eigenvectors (principal components) that represent the changes in the absorbances that are common to all. When the training data have been fully processed by the PCA algorithm, it is reduced to two main matrices: the eigenvectors (spectra) and the scores (the eigenvector weighting values for all the calibration spectra). The matrix expression of the model equation for the spectral data appears as follows:

$$A = SF + E_A,$$

where A is an $n \times p$ matrix of spectral absorbances, S is an $n \times f$ matrix of score values for all of the spectra, and F is an $f \times p$ matrix of eigenvectors. The E_A matrix is the errors in the model's ability to predict the calibration absorbances and has the same dimensionality as the A matrix. In the case of eigenvector analysis, the E_A matrix is often called the matrix of residual spectra. The dimensions of the matrices are representative of the data they hold; n is the number of samples (spectra), p is the number of data points (wavelengths) used for calibration, and f is the number of PCA eigenvectors. As will be shown later, this is actually a simplification of the true model equation (Fig. 6).

The model equation should look vaguely familiar, and in fact it is very similar to the CLS model for the spectral data. However, in this case, the spectral data are not constructed from the concentrations and absorptivity coefficient spectra but rather from the scores and eigenspectra. In fact, as with the CLS model, there is no limit to the number of wavelengths that can be used; therefore, all data up to the entire spectrum can be included in the model. However, this is only a model for the spectral data. The concentrations matrix C has not played a role in the calculations at all. Therefore, PCA alone cannot be used as a model for predicting constituent concentrations.

Remember that the eigenvectors represent the spectral variations that are common to all the calibration data. Therefore, using that information to calculate a regression equation will produce a model that is very useful for predicting concentration. The F matrix from PCA performs a similar task

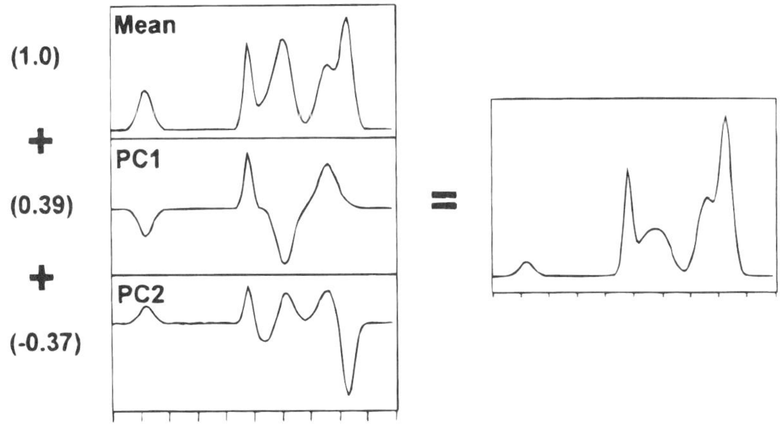

FIG. 6. By multiplying PC1 and PC2 (eigenvectors) by the set of representative scalar fractions (scores) and summing the results (along with the mean spectrum if the data were mean centered), the original calibration spectra can be re-created. The "spectral residual" is the difference between this reconstruction and the original.

to that of the K matrix in the CLS model; it stores the "constituent" spectral data. This does not mean that the rows of the F matrix are the spectra of the pure constituents: They are not. However, they cannot be used alone without the scores matrix S to represent the original data (as in the CLS, it needs the C matrix to perform the same function).

On the other hand, the scores in the S matrix are unique to each calibration spectrum, and just as a spectrum is represented by a collection of absorbances at a series of wavelengths, it can also be a series of scores for a given set of factors. Much like the classical models performed a regression of the concentration C matrix directly on the spectral absorbances in the A matrix, it is also possible to regress C against the scores S matrix.

In this case, the regression technique from the ILS model is obviously the best choice. This gives the model the best qualities of the ILS method, such as no *a priori* knowledge of the complete sample composition and some robustness in predictions with respect to contaminant constituents not present in the original calibration mixtures. The model equation is therefore

$$C = BS + E_C,$$

where C is the $m \times n$ matrix of constituent concentrations, B is an $m \times f$ matrix of the regression coefficients, and the S matrix is the scores from the PCA model. The dimensions of the matrices are n for the number of samples (spectra), m for the number of constituents used for calibration, and f for the number PCA eigenvectors. As with ILS, the B coefficients matrix can be solved by the regression:

$$B = CS'(SS')^{-1}.$$

Thus, the name for this type of model is principal components regression; it combines principal components analysis and inverse least squares regression to solve the calibration equation for the model. All that remains is to come up with a single unified equation that represents the PCR model. Therefore, rearranging the previous matrix model equation to represent the scores as a function of the spectral absorbances and the eigenvectors produces

$$S = AF'.$$

It is not necessary to use the inverse (or pseudo-inverse) of F to solve this equation. This is due to the fact that when PCA is used to solve the spectral model, the resulting F matrix of eigenvectors is a special type of matrix called an orthonormal matrix. This type of matrix has a very interesting quality: When the matrix is multiplied by its own transpose, the identity matrix is the result. Multiplying any matrix by the identity matrix is the same as multiplying a single number by 1. The result is always the number again. Therefore, to get the equation for the scores, both sides of the earlier PCA model equation for the spectral data were simply multiplied by the transpose of the F matrix.

Finally, by combining the concentration equation with the scores equation, the final PCR model equation emerges:

$$C = BAF' + E_C,$$

where C is the $m \times n$ matrix of constituent concentrations, B is an $m \times f$ matrix of the regression coefficients, A is an $n \times p$ matrix of spectral absorbances, and F is an $f \times p$ matrix of eigenvectors. The dimensions of the matrices are n for the number of samples (spectra), m for the number of constituents used for calibration, p for the number of data points (wavelengths) used for calibration, and f for the number of PCA eigenvectors.

The PCR calibration model is not completely free of problems, however. It is important to note that PCR is a two-step process; the PCA eigenvectors and scores are calculated and then the scores are regressed against the constituent concentrations using a regression method similar to ILS. Remember that the ILS method can build accurate calibrations, provided that the selected variables (in the earlier discussion, the variables were the responses at selected wavelengths) are physically related to the properties (constituent concentrations) they are regressed against. However, the PCA factors/scores are calculated independently of any knowledge of these concentrations. They merely represent the largest common variations among all the spectra in the training set. Presumably, these variations will be mostly related to changes in the constituent concentrations, but there is no guarantee this will be true.

In fact, many PCR models include more factors than are actually necessary because some of the eigenvectors are not related to any of the constituents of interest. Ideally, a PCR model should be built by performing a selection on the scores (much like a selection of wavelengths for an ILS model) to determine which factors should be used to build a model for each constituent. In practice, this is a difficult process, in terms of both developing the selection rules and making it simple to perform. Most commercial chemometric software packages do not support this type of PCR model.

In addition, like the ILS method, the predictive ability of the PCR model will suffer if the constituent concentrations are collinear. Again, this means that a relatively large number of the training set samples are required, and they must be tested by the primary calibration method to determine the "randomness" of the constituent concentrations.

PCR

Advantages	Disadvantages
• Does not require wavelength selection; any number can be used, usually the whole spectrum or large regions • Larger number of wavelengths gives averaging effect, making model less susceptible to spectral noise • PCA data compression allows using inverse regression to calculate model coefficients; can calibrate only for constituents of interest • Can be used for very complex mixtures because only knowledge of constituents of interest is required • Can sometimes be used to predict samples with constituents (contaminants) not present in the original calibration mixtures	• Calculations are slower than those of most classical methods • Optimization requires some knowledge of PCA; models are more complex to understand and interpret • No guarantee PCA vectors directly correspond to constituents of interest • Generally, a large number of samples are required for accurate calibration • Collecting calibration samples can be difficult; must avoid collinear constituent concentrations

1. *Calculating PCA and PCR*

This section is for those who are interested in knowing the mechanics of the PCA and PCR calculations (Refs. 8, 10, 11, 15). There are actually two different methods used to calculate the principal components of a set of data; the NIPALS algorithm and decomposition of covariance. The following descriptions of these algorithms assume that the matrices involved have the following dimensions: A is an $n \times p$ matrix of spectral absorbances, S is an $n \times f$ matrix of scores, F is an $f \times p$ matrix of eigenvectors, and λ is an $f \times f$ matrix of the eigenvalues. In this case, n is the number of samples (spectra), p is the number of data points (wavelengths), and f is the number PCA eigenvectors. When used, the subscripts on the matrices indicate the matrix row.

PCR involves a second step of performing a regression of the scores matrix S against the $n \times m$ matrix of constituent concentrations, C, to produce the matrix of calibration coefficients, B, which has the dimensions $f \times m + 1$. In these matrices, m is the number of constituents.

NOTE: The eigenvalues matrix λ was not discussed previously. The eigenvalues are the relative weights of how important each factor is for reconstructing the spectral data. The eigenvalues matrix is a diagonal square matrix. That is, it only has values on the diagonal and is zero everywhere else. The eigenvalues should always descend along the diagonal because the first eigenvector represents the largest variation, the second is the next largest, and so on.

2. *PCA—The NIPALS Algorithm*

This is the most commonly used method for calculating the principal components of a data set. It gives more numerically accurate results when compared with the decomposition of covariance method but is slower to calculate. The steps are as follows:

1. Set the eigenvector to the first spectrum: $F_i = A_1$
2. Calculate the eigenvalue: $\lambda_{i,i} = (\sum S_i^2)^{1/2}$
3. Normalize the eigenvector: $F_i = F_i/\lambda_{i,i}$
4. Compute the scores for the eigenvector: $S_i = AF_i'$
5. Check for convergence by comparing these scores to the scores from the previous pass for this eigenvector. If this is the first pass for the current eigenvector, or the scores are not the same, continue with step 6. If the scores are the same, skip to step 8.
6. Recompute the eigenvector: $F_i = A'S_i$
7. Go back to step 2.
8. If $i = f$, stop calculating, otherwise compute the residual matrix for the next eigenvector: $A = A - S_i F_i$
9. Increase eigenvector counter, $i = i + 1$, and go back to step 1.

3. *PCA—Decomposition of the Variance–Covariance Matrix*

This method is the faster of the two algorithms but tends to have numerical errors in the later eigenvectors when performed on computers using only single precision. This is due to the data compression step performed at the beginning to calculate the variance–covariance matrix Z (which has dimensions of $n \times n$). When using this algorithm, it is necessary to use double precision numbers, or to use it only to calculate the first few eigenvectors. The steps are as follows:

1. Compute the variance–covariance square matrix: $Z = AA'$
2. Initialize scores to an arbitrary number: $S_i = 0.1$
3. Compute new scores: $S_i = ZS_i'$
4. Calculate the eigenvalue: $\lambda_{i,i} = (\sum S_i^2)^{1/2}$
5. Normalize the scores: $S_i = S_i/\lambda_{i,i}$
6. Check for convergence by comparing these scores to the scores from the previous pass for this eigenvector. If this is the first pass for the current eigenvector, or the scores are not the same, go back to step 3. If the scores are the same, continue with step 7.

7. If $i = f$, go to step 9; otherwise, compute the residual variance–covariance matrix for the next eigenvector: $Z = Z - (S_i S_i')\lambda_{i,i}$
8. Increase eigenvector counter, $i = i + 1$, and go back to step 3.
9. Compute all the eigenvectors: $F = S'A$

4. *PCR Regression*

Once the PCA has been calculated from the spectral data, the concentration data can be regressed against the scores matrix using the inverse least squares method to generate the matrix of constituent calibration coefficients. A usual practice in performing PCR regression is to add an extra unit vector column to the scores matrix to allow for inclusion of an offset coefficient in the regression.

B. Partial Least Squares

Partial least squares (PLS) is another spectral decomposition technique that is closely related to PCR. However, in PLS, the decomposition is performed in a slightly different fashion. Instead of first decomposing the spectral matrix into a set of eigenvectors and scores and regressing them against the concentrations as a separate step, PLS actually uses the concentration information during the decomposition process (Fig. 7). This causes spectra containing higher constituent concentrations to be weighted more heavily than those with low concentrations. Thus, the eigenvectors and scores calculated using PLS are quite different from those of PCR. The main idea of PLS is to get as much concentration information as possible into the first few loading vectors.

In actuality, PLS is simply taking advantage of the correlation relationship that already exists between the spectral data and the constituent concentrations. Because the spectral data can be decomposed into its most common variations, so can the concentration data! In effect, this generates two sets of vectors and two sets of corresponding scores; one set for the spectral data and the other for the constituent concentrations. Presumably, the two sets of scores are related to each other through some type of regression, and a calibration model is constructed.

Now this is actually a bit of an oversimplification. Unlike PCR, PLS is a one-step process. In other words, there is no separate regression step. Instead, PLS performs the decomposition on both the spectral and concentration data simultaneously. As each new factor is calculated for the model, the scores are "swapped" before the contribution of the factor is removed from the raw data. The newly reduced data matrices are then used to calcu-

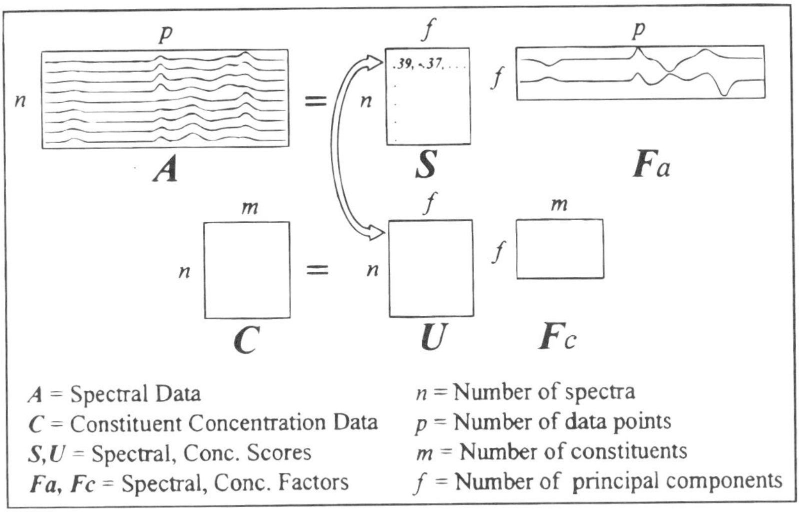

FIG. 7. PLS is similar to PCA/PCR. However, in PLS the constituent concentration data are included in the decomposition process. In fact, both the spectral and concentration data are decomposed simultaneously, and the scores (S and U) are "exchanged" as each new factor is added to the model.

late the next factor, and the process is repeated until the desired number of factors is calculated. Unfortunately, this makes the model equations for PLS significantly more complex than those of PCR. For those who are interested, the algorithms for calculating the PLS model eigenvectors and scores are shown in a later section.

As mentioned previously, one of the main advantages of PLS is that the resulting spectral vectors are directly related to the constituents of interest. This is entirely unlike PCR, in which the vectors merely represent the most common spectral variations in the data, completely ignoring their relation to the constituents of interest until the final regression step (Fig. 8).

There are actually two versions of the PLS algorithm: PLS-1 and PLS-2. The differences between these methods are subtle but have very important effects on the results. Like the PCR method, PLS-2 calibrates for all constituents simultaneously. In other words, the results of the spectral decomposition for both of these techniques give one set of scores and one set of eigenvectors for calibration. Therefore, the calculated vectors are not optimized for each individual constituent. This may sacrifice some accuracy in the predictions of the constituent concentrations, especially for complex sample mixtures. In PLS-1, a separate set of scores and loading vectors is calculated for each constituent of interest. In this case, the separate sets of

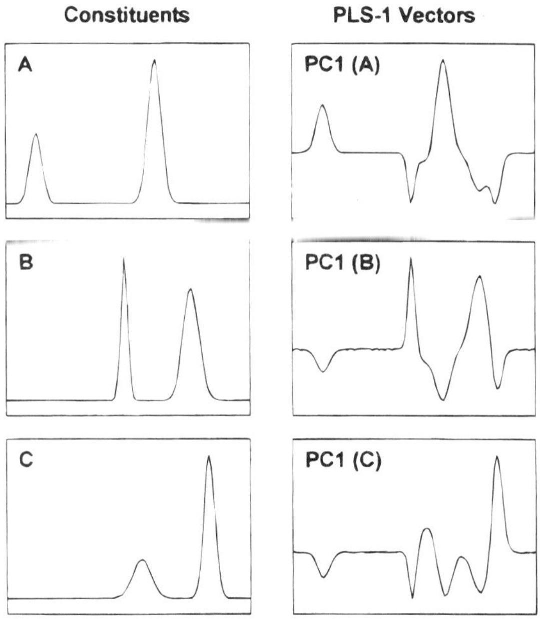

FIG. 8. The vectors generated by PLS (especially PLS-1) are more directly related to the constituents of interest than those from PCA. The left column shows the spectra of the "pure" constituents used to construct the data set in Fig. 5. The center column shows the first PLS-1 vector for each constituent calculated from the data set, whereas the right column shows the first two PCA vectors for the same data.

eigenvectors and scores are specifically tuned for each constituent and, therefore, should give more accurate predictions than PCR or PLS-2.

There is, however, a minor disadvantage in using the PLS-1 technique: the speed of calculation. Because a separate set of eigenvectors and scores must be generated for every constituent of interest, the calculations will take more time. For training sets with a large number of samples and constituents, the increased time of calculation can be significant.

PLS-1 may have the largest advantage when analyzing systems that have constituent concentrations that are widely varied. For example, a set of calibration spectra contains one constituent in the concentration range of 50–70% and a second constituent in the range of 0.1–0.5%. In this case, PLS-1 will almost certainly predict better than the other techniques. If the concentration ranges of the constituents are approximately the same, PLS-1 may have less of an advantage over PLS-2 and will definitely take longer to calculate.

Although all these techniques have been successfully applied for spectral quantitative analysis, the arguments in the literature generally show that PLS has superior predictive ability. In most cases, PLS methods will give better results than PCR, and PLS-1 will be more accurate than PLS-2. However, there are many documented cases in the literature in which certain calibrations have performed better by using PCR or PLS-2 instead of PLS-1. Unfortunately, there are no definite rules, and only good research practices can determine the best model for each individual system.

PLS

Advantages	Disadvantages
• Combines the full spectral coverage of CLS with partial composition regression of ILS • Single-step decomposition and regression; eigenvectors are directly related to constituents of interest rather than largest common spectral variations • Calibrations are generally more robust provided that calibration set accurately reflects range of variability expected in unknown samples • Can be used for very complex mixtures because only knowledge of constituents of interest is required • Can sometimes be used to predict samples with constituents (contaminants) not present in the original calibration mixtures	• Calculations are slower than most classical methods, especially PLS-1 • Models are more abstract, thus more difficult to understand and interpret • Generally, a large number of samples are required for accurate calibration • Collecting calibration samples can be difficult; must avoid collinear constituent concentrations

1. *Calculating PLS Eigenvectors and Scores*

This section is for those who are interested in knowing the mechanics of the PLS calculation. As mentioned previously, there are two variants of this algorithm known as PLS-1 and PLS-2. In fact, PLS-1 is a reduced subset of the full PLS-2. The algorithms have been combined here, with appropriate notes on where they differ. Note that a PLS-2 model of a training set with only one constituent is identical to a PLS-1 model for the same data.

The main difference between PLS and PCR is that the concentration information is included in the calculations during the spectral decomposition. This results in two sets of eigenvectors: a set of spectral "loadings" (Bx) that represent the common variations in the spectral data and a set of spectral "weights" (W) that represent the changes in the spectra that correspond to the regression constituents. Correspondingly, there are two sets of scores: one for the spectral data (S) and another for the concentration data (U).

The following description assumes that the matrices involved have the following dimensions: A is an $n \times p$ matrix of spectral absorbances, C is an $n \times m$ matrix of constituent concentrations, S is an $f \times n$ matrix of spectral scores, U is an $f \times n$ matrix of concentration weighted scores, Bx is an $f \times p$ matrix of spectral loading vectors, W is an $f \times p$ matrix of spectral weighting vectors, By is an $f \times m$ matrix of the constituent loading vectors, and V is a $1 \times f$ vector of the PLS model cross-products. In this case, n is the number of samples (spectra), p is the number of data points (wavelengths), m is the number of constituents, and f is the number PLS eigenvectors. When used, the subscripts on the matrices indicate a matrix row.

1. Set the weighting scores to a starting value: $U_i = C'_1$
   (For PLS-1 or single constituent models, use the desired constituent vector. For PLS-2, use the first constituent column vector.)
2. Calculate the spectral weighting vector: $W_i = U'_i A$
3. Normalize the weighting vector to unit length: $W_i = W_i/(W_i W'_i)$
4. Calculate the spectral scores: $S_i = AW_i$
   (For PLS-1 or single constituent model, set $By_i = 1$ and skip to step 9.)
5. Calculate the concentration loading vector: $By_i = S_i C$
6. Normalize the concentration loading vector to unit length: $By_i = By_i/(By_i By'_i)$
7. Calculate new weighting scores: $U_i = By_i C'$
8. Check for convergence by comparing new U_i scores to the previous pass for this vector. If this is the first pass for the current vector, or the scores are not the same, go back to step 2. If the scores are effectively the same, continue with step 9.
9. Calculate the PLS cross-product for this vector: $V_i = S_i U'_i/(S_i S'_i)$
10. Calculate the spectral loading vector: $Bx_i = S_i A$
11. Normalize the spectral loading vector by the spectral scores: $Bx_i = Bx_i/(S_i S'_i)$
12. Remove contribution of the vector from the spectral data: $A = A - S'_i Bx_i$
13. Remove contribution of the vector from the concentration data: $C = C - (S'_i By_i)V_i$
14. Increase vector counter, $i = i + 1$, and go back to step 1. Continue until all desired factors are calculated ($i = f$).

15. If performing PLS-1, reset A back to the original training set values and redo all steps using a different constituent in step 1. Note that this generates a completely different set of S, U, W, Bx, By, and V matrices for every constituent!

2. *Predicting Samples with a PLS Model*

The following are the calculational steps used to predict a spectrum against a PLS model. The variable descriptions are as previously discussed, except that Au is the $1 \times p$ vector of the spectral responses of the sample being predicted, and Cu is the $1 \times m$ vector of the predicted constituent concentrations. Initially, Cu is set to zero, and the vector counter i to 1.

1. Calculate the unknown spectral score for a weighting vector: $S_i = W'_i Au$
2. Calculate the concentration contribution for the vector: $Cu = Cu + (By'_i S_i V_i)$
3. Remove the spectral contribution of the vector: $Au = Au - (S_i Bx'_i)$
4. Increment the vector counter, $i = i + 1$, and go back to step 1. Continue until all desired factors are calculated ($i = f$).
5. If performing PLS1, reset the data in Au back to the original unknown spectrum values and repeat from step 1 with the next set of constituent vectors.

Note that the data remaining in the Au vector after all factors have been removed are the residual spectrum.

C. Determining the Number of Factors for the Model

One of the most difficult tasks in using PCR and PLS is determining the correct number of loading vectors (factors) to use to model the data. As more and more vectors are calculated, they are ordered by the degree of importance to the model (either by variance in PCA or concentration weighted variance in PLS). Eventually, the loading vectors will begin to model the system noise (which usually provides the smallest contribution to the data).

The earlier vectors in the model are most likely to be the ones related to the constituents of interest, whereas later vectors generally have less information that is useful for predicting concentration. In fact, if these vectors are included in the model, the predictions can actually be worse than if they were ignored altogether. Thus, decomposing spectra with these techniques and selecting the correct number of loading vectors is a very effective way of filtering out noise.

However, if too few vectors are used to construct the model, the prediction accuracy for unknown samples will suffer because not enough terms are being used to model all the spectral variations that compose the constituents of interest. Therefore, it is very important to define a model that contains enough vectors to properly model the components of interest without adding too much contribution from the noise.

Models that include noise vectors or more vectors than are actually necessary to predict the constituent concentrations are called overfit. Models that do not have enough factors in them are known as underfit.

Unfortunately, there is usually no clear indication of how many factors are required to move from "constituent" vectors into "noise" vectors and prevent both underfitting and overfitting. However, there are a variety of methods that can be used to aid in determining this value. One of the most effective is to calculate the prediction residual error sum of squares (PRESS) for every possible factor. This is calculated by building a calibration model with a number of factors, then predicting some samples of known concentration (usually the training set data itself) against the model. The sum of the squared difference between the predicted and known concentrations gives the PRESS value for that model:

$$\text{PRESS} = \sum_{i=1}^{n} \sum_{j=1}^{m} (C\text{p}_{i,j} - C_{i,j})^2,$$

where n is the number of samples in the training set, m is the number of constituents, Cp is the matrix of predicted sample concentrations from the model, and C is the matrix of known concentrations of the samples.

The smaller the PRESS value, the better the model is able to predict the concentrations of the calibrated constituents. By calculating the PRESS value for a model using all possible factors (i.e., first with one factor, then two, three, etc.) and plotting the results, a very clear trend should emerge.

However, as with everything in chemometrics, there are a variety of methods that can be used to optimize a model. The main issue is what data to use during the prediction step before calculating the PRESS.

1. *Self-Prediction*

This is the simplest method for testing a calibration model, but unfortunately it is not very useful. In this method, the models are built using all the spectra in the training set, then the same spectra are predicted back against these models. The problem with this approach is that the model vectors are calculated from these same spectra. Therefore, all the vectors calculated exist in all the training spectra. The PRESS plot will continue to fall as new factors are added to the model and will never rise. In effect, this gives the impression that all the vectors are constituent vectors, and there are no noise vectors to eliminate, which is never the case with real data (Fig. 9).

The only reason to use this method is that it is very fast. Because it requires only building the models once, predicting the samples can be done in one step. Sometimes, it is possible to select the number of factors as the place where the plot starts to "flatten out." However, this is an inexact measure and gives no indication of the true optimum number of factors for the model when predicting unknown samples.

2. *Cross-Validation*

Cross-validation is conceptually very simple to understand, but it is also the most calculationally intensive method of optimizing a model. In effect, cross-validation attempts to emulate predicting unknown samples by using the training set data. The procedure is as follows:

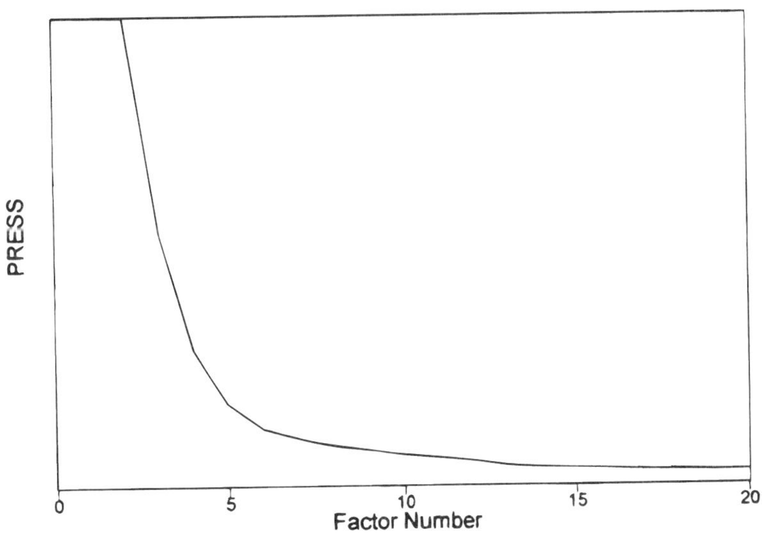

FIG. 9. A PRESS plot for a self-prediction validation of a training set of NIR diffuse reflectance spectra of 50 samples of wheat. Note that the PRESS value continues to decrease as new factors are added. There is no clear indication of the optimum number of factors for this model.

1. Select a sample (or a small group of samples if the training set is large enough) and remove the spectrum (spectra) and corresponding concentration data from the training set. Set the factor counter to $i = 1$.
2. Use the remaining training set samples to perform the decomposition and calibration calculations for factor i (loading vector).
3. Predict the concentrations of the removed sample(s) using the calibration equation from step 2 and calculate PRESS(i).
4. Increment the factor counter ($i = i + 1$) and repeat from step 2 until all desired factors ($i = f$) have been calculated and predicted.
5. Place the previously left out sample data back into the training set and select a different sample (or group). Return to step 1 and repeat the calculations. As each sample is left out, add the calculated squared residual error to all the previous PRESS values. Repeat until all samples have been left out and predicted at least once.

In Fig. 10, notice that from 0 to 7 factors the prediction error (PRESS) decreases as each new factor is added to the model. This indicates that the model is underfit and there are not enough factors to completely account for the constituents of interest.

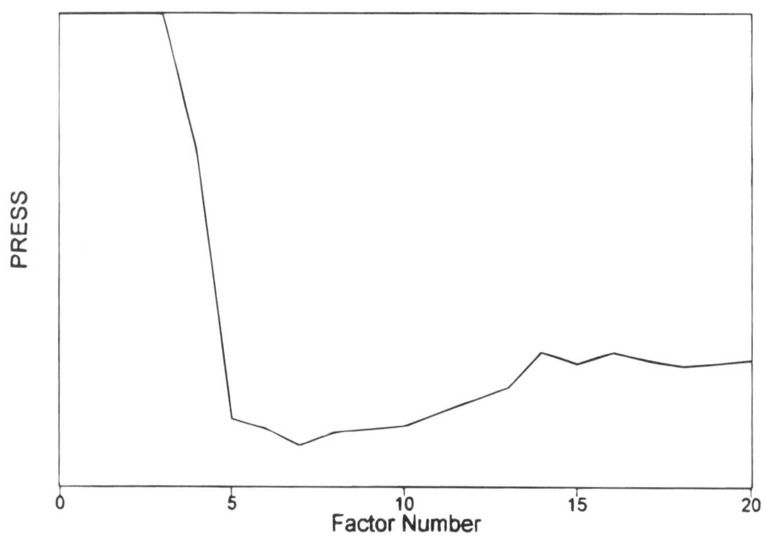

FIG. 10. A PRESS plot for a cross-validation prediction of a training set of NIR diffuse reflectance spectra of 50 samples of wheat. Note how the plot reaches a minimum (at approximately 7 factors), then starts to rise as more noise vectors are added to the model.

At some point the PRESS plot should reach a minimum and start to ascend again. At this point the model is beginning to add factors that contain uncorrelated noise that are not related to the constituents of interest. When these extra noise vectors are included in the model, it is overfit and its predictive ability is diminished.

There are two main advantages of cross-validation over all other methods. The first is in how it estimates the performance of the model. Because the predicted samples are not the same as the samples used to build the model, the calculated PRESS value is a very good indication of the error in the accuracy of the model when used to predict unknown samples in the future. The larger the training set and the smaller the groups of samples left out in each pass (optimally only one sample at a time, but this can be very time-consuming), the better this estimate will be. In effect, the model is validated with a large number of unknown samples (because each training sample is left out at least once) without having to measure an entirely new set of data (see Validation Set Prediction).

The second benefit of cross-validation is better outlier detection. Although this will be discussed in more depth in a later section, it will suffice to mention that cross-validation is the only validation method that can give complete outlier detection for the training set data. Because each sample is left out of the models during the cross-validation process, it is possible to calculate how well the spectrum matches the model by calculating the spectral reconstruction (see Fig. 6) and comparing it to the original training spectrum (via the spectral residual). If the predicted concentrations for a single sample are way off and the spectrum does not match the model very well but the rest of the data work very well, the sample is possibly an outlier. Identifying and removing outlier samples from the training set should always improve the predictive ability of the model.

It is very difficult to perform outlier detection on the training set data without performing a complete cross-validation. The results of the other validation methods mentioned here (self-prediction, leverage, and validation set) are generally not adequate because the predictions are based on a model built using every available sample. Any unique variations that are present in an outlier sample(s) are therefore built into the model. Thus, when the validation spectra (either the training set or a separate validation set) are predicted back against the model, it can appear to be working well. The accuracy of the predictions is actually worse than if those training samples were removed and the model was rebuilt.

Unfortunately, cross-validation is a very time-consuming process. It requires recalculating the models for every sample left out. However, there are a few somewhat acceptable shortcuts. If the number of samples in the training set is large enough, the number of samples rotated out in each pass can be more than one. This obviously does not give the best statistics for each sample, but it does speed the calculations and can be acceptable for determining the number of factors for the model.

In fact, in some cases, leaving out groups of samples at a time can be preferable to leaving out only one at a time. In training sets that contain replicate spectra of the same sample, the rotation should be performed on each standard sample and not on each spectrum. For example, if a training set of 50 spectra contains two spectra each of 25 known samples, then each pair of replicates should be left out together. This completely removes the contribution of that sample from the model before prediction. Otherwise, if a rotation value of 1 is used, there will always be a similar spectrum of the removed sample in the set and the sample will never be predicted as a true unknown.

Another trick is to perform a pseudo-validation set prediction (see Validation Set Prediction). If the training set is very large, setting the rotation to one-half the total number of samples effectively accomplishes the same goal. By building a model with half the training set data and predicting it with the other half, similar trends will appear in the PRESS plot. The added advantage is that all the collected training samples can ultimately be used to build the final calibration, making it more robust.

3. *Leverage Prediction*

This method is an attempt to compromise between a full cross-validation (which is very slow but gives the best estimate of the model's performance when it is applied to unknown samples) and a self-prediction (which is very fast but gives limited information about the predictive ability of the model).

In leverage validation, the models are built using all the training spectra, similar to self-prediction. However, when the samples are predicted, the scores are corrected for the individual sample leverage. The leverage is a measure of the importance of the sample to the overall model equations. Generally, samples at the high and low end of the constituent concentration range will have large leverages, whereas samples that lie closer to the mean concentrations will have low leverages. If a single sample has a leverage that is significantly larger than the rest of the training set, this can indicate that the spectrum is very different from the rest of the training set and does not represent the actual sample (this is called an outlier) (Fig. 11).

When this method is used, the models are calculated only once, which significantly increases the speed of the analysis. However, the leverage correction is applied to the scores assuming that they will then be used in a regression. Although this is fine for PCR models, remember that PLS

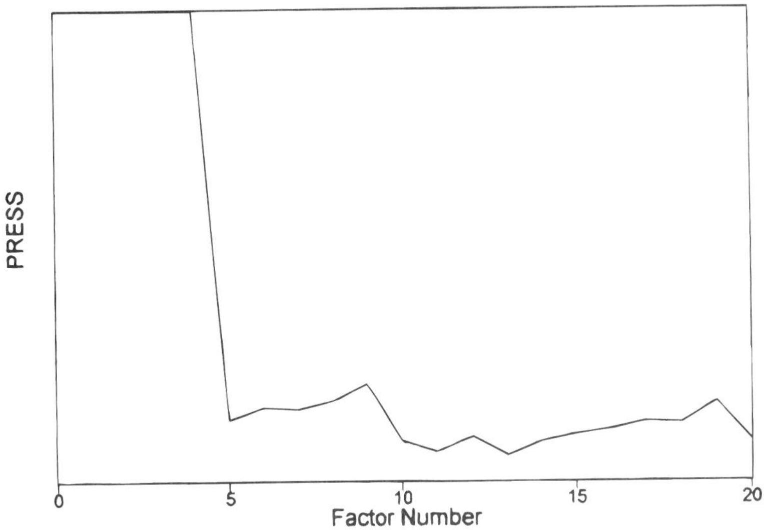

FIG. 11. A PRESS plot for a leverage validation of a PCR model built from a training set of NIR diffuse reflectance spectra of 50 samples of wheat. The minimum occurs at 13 factors; however, an argument could also be made for selecting 11. In either case, the results are higher than the 7 factors reported by cross-validation for the same data set.

models do not use a separate regression step. The outcome is that leverage validation works fairly well with PCR models but suffers the same problem as self-prediction when applied to PLS models (Fig. 12).

4. *Validation Set Prediction*

In this method, a new set of training spectra is measured under the same conditions as the training set and then calibrated by the primary method. These data are called a validation set. This new data set is then predicted against the calibration models built from the training set. This approach gives the best estimate of the model's performance because none of the samples in the validation set were used to build the model.

The downside to using a separate validation set is the time and cost involved in generating the data. When the validation set method is used, typically a very large set of training data (both spectra and concentrations) is measured. This is then split into two groups: a set of training data and a set of validation data. Depending on the available samples, the splits could be 80/20, 60/40, or even 50/50%, respectively, of the total number of

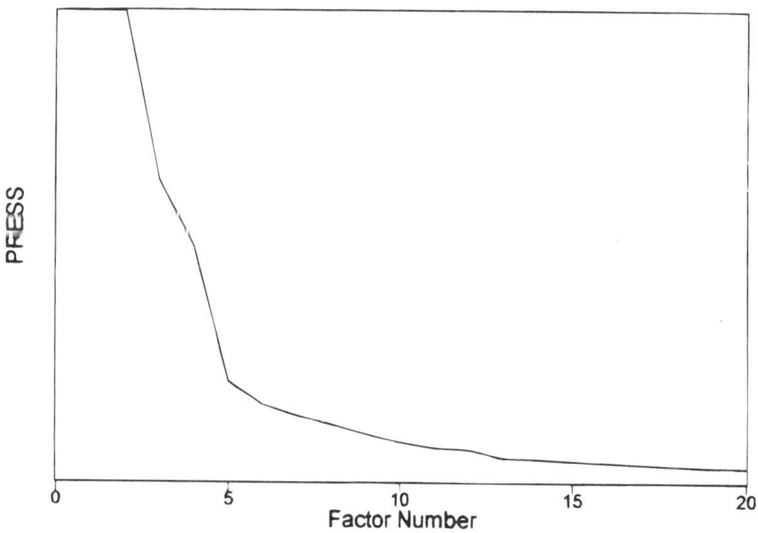

FIG. 12. A PRESS plot for a leverage validation of a PLS model built from a training set of NIR diffuse reflectance spectra of 50 samples of wheat. With this data set, no minimum is reached within the 20 factors calculated.

samples. After the model is built, the validation set is effectively "discarded." However, if the same data were used in a cross-validation, every sample collected could be used to both build the model and validate it.

The main advantage of validation set prediction is the ability to test the model's performance with a completely different data set than the calibration/training data. This is most important for determining the long-term stability of the model. If a model is constructed and used to predict samples, there is no guarantee that the spectrometer will continue to perform exactly the same way. There are many things that affect the spectrum of a sample that cannot be completly controlled: spectrometer/detector wear, sample handling, and environment (moisture and temperature). Collecting and predicting a validation set long after the model has been in use is one way of ensuring that the concentration predictions are still within the desired range of accuracy (Fig. 13).

5. *Selecting the Factors Based on PRESS*

To avoid building a model that is either overfit or underfit, the number of factors at which the PRESS plot reaches a minimum would be the obvious

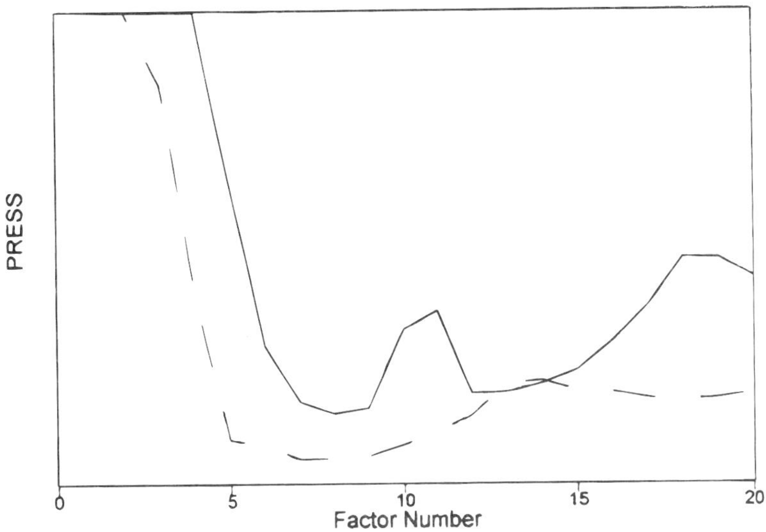

FIG. 13. A PRESS plot of a validation set prediction (solid line) of 50 samples and the cross-validation prediction (dashed line) from Fig. 10 of a training set of 50 samples of NIR diffuse reflectance spectra of wheat. The minimum in the validation set prediction is at 8 factors.

choice for the best model (except in the case of self-prediction). Although the minimum of the PRESS may be the best choice for predicting the particular set of samples, most likely it is not optimum for prediction of all unknown samples in the future.

Because there are a finite number of samples in the set used for prediction, in many cases the number of factors that gives a minimum PRESS value can still be overfit for predicting unknown samples. In other words, there is a statistical possibility that some of the noise vectors from the spectral decomposition may be present in more than one sample. These vectors can appear to improve the calibration by a small amount when, by random correlation, they are added to the model. However, if these exact same noise vectors are not present in future unknown samples (and most likely they will not be), the predicted concentrations will have significantly larger prediction errors than if those additional vectors were left out of the model.

A solution to this problem has been suggested in which the PRESS values for all previous factors are compared to the PRESS value at the minimum. The ratio between these values (also known as the F ratio) can be calculated and assigned a statistical significance based on the number of

TABLE 1

F Statistic Critical Values for the 99th Percentile ($\alpha = 0.01$)

	$\nu_1 = 1$	2	3	4	5	6	7	8	9	10	20	40	60	120	∞
$\nu_2 = 1$	4052	5000	5403	5625	5764	5859	5928	5981	6023	6056	6209	6287	6313	6339	6366
2	98.5	99.0	99.2	99.2	99.3	99.3	99.4	99.4	99.4	99.4	99.4	99.5	99.5	99.5	99.5
3	34.1	30.8	29.5	28.7	28.2	27.9	27.7	27.5	27.3	27.2	26.7	26.4	26.3	26.2	26.1
4	21.2	18.0	16.7	16.0	15.5	15.2	15.0	4.80	14.7	14.5	14.0	13.7	13.7	13.6	13.5
5	16.3	13.3	12.1	11.4	11.0	10.7	10.5	10.3	10.2	10.1	9.55	9.29	9.20	9.11	9.02
6	13.7	10.9	9.78	9.15	8.75	8.47	8.26	8.10	7.98	7.87	7.40	7.14	7.06	6.97	6.88
7	12.2	9.55	8.45	7.85	7.46	7.19	6.99	6.84	6.72	6.62	6.16	5.91	5.82	5.74	5.65
8	11.3	8.65	7.59	7.01	6.63	6.37	6.18	6.03	5.91	5.81	5.36	5.12	5.03	4.95	4.86
9	10.6	8.02	6.99	6.42	6.06	5.80	5.61	5.47	5.35	5.26	4.81	4.57	4.48	4.40	4.31
10	10.0	7.56	6.55	6.99	5.64	5.39	5.20	5.06	4.94	4.85	4.41	4.17	4.08	4.00	3.91
20	8.10	5.85	4.94	4.43	4.10	3.87	3.70	3.56	3.46	3.37	2.94	2.69	2.61	2.52	2.42
30	7.58	5.39	4.51	4.02	3.70	3.47	3.30	3.17	3.07	2.96	2.55	2.3	2.21	2.11	2.01
40	7.31	5.18	4.31	3.83	3.51	3.29	3.12	2.99	2.89	2.80	2.37	2.11	2.02	1.92	1.80
60	7.08	4.98	4.13	3.65	3.34	3.12	2.95	2.82	2.72	2.63	2.20	1.94	1.84	1.73	1.60
120	6.85	4.79	3.95	3.48	3.17	2.96	2.79	2.66	2.56	2.47	2.03	1.76	1.66	1.53	1.38
∞	6.63	4.61	3.87	3.32	3.02	2.80	2.64	2.51	2.41	2.32	1.88	1.59	1.47	1.32	1.00

samples used in the calibration set:

$$F \text{ ratio}_i = \frac{\text{PRESS}_i}{\text{PRESS}_{\text{min}}},$$

where i indicates the number of factors in the model.

This ratio is an indicator of the relative significance of each model to the model with the number of factors at the minimum of the PRESS. The number of factors at which the F ratio falls below a predefined significance level determines the optimum number of factors for a model used for predicting unknowns. This is easily determined by determining the point at which adding a new factor to the model causes the F test probability level to fall at or below 0.75 (Refs. 16, 17). This is applied by calculating the F ratio as described and looking the value up in a table of F statistic values similar to Table 1 (these can commonly be found in the back of a statistics book) for the $\alpha = 0.25$ significance level.

In order to use the F statistic tables properly, it is also necessary to know the degrees of freedom in both the numerator (v_1) and denominator (v_2) of the F ratio value. For F ratios based on PRESS values, the number of samples used to calibrate the model has been suggested as the proper value for both. Therefore, in the case of a cross-validation, the degrees of freedom would be the total number of sample in the training set minus the number left out in each group. For a validation set prediction, they would be the total number of samples in the training set.

Instead of looking up numbers in F statistic tables, another approach is to estimate the significance level directly from the F ratio value and the degrees of freedom. In Table 2 the F test column shows the probability levels of each factor estimated by a method used for the PRESS data in Fig. 10. It is easy to see that the minimum of the PRESS occurs at 7 factors suggesting that this is the optimum number of factors. However, when the F test is applied, the higher factor models fail the significance test after 5 factors. Therefore, using 5 factors to predict unknown samples is less likely to be overfit than using a model with 7 factors.

Applying the F test to PRESS values from a self-prediction generally does not work. This is due to the fact that the F test is primarily designed to find the statistically optimum number of factors for predicting samples that were not included when the model was built. In the self-prediction scheme, every sample is already included in the model, which gives no information on the performance of the model with true unknowns. This is merely one more reason why one of the other validation methods should be used to optimize the number of factors for the model.

For those who are interested in statistics, the F test probability estimates are based on the following equations. In this method, the F distribution is

TABLE 2

F RATIO VALUES FOR THE CROSS VALIDATION PRESS DATA IN FIG. 10

Factor	PRESS	*F* ratio	*F* test
1	107.7276	85.75637	1.0
2	8.863851	7.056051	1.0
3	5.214601	4.151073	0.999987
4	3.001495	2.389334	0.998189
5	1.462032	1.163848	0.699706
6	1.400733	1.115051	0.646483
7	1.256206	1	0.5
8	1.257433		
9	1.286249		
10	1.404102		
11	1.545576		
12	1.711533		
13	2.045521		
14	2.08551		
15	1.972338		
16	1.957134		
17	1.889865		
18	1.871311		
19	1.890717		
20	1.937525		

The F test probabilities were calculated using the following degrees of freedom: numerator, denominator = 49 (total No. samples − No. samples left out). Notice the sharp drop in the estimated *F* test significance at 5 factors

approximated by a chi-square (χ^2) distribution:

$$z = \left[\left(1 - \frac{2}{9v_2}\right)F^{1/3} - \left(1 - \frac{2}{9v_1}\right)\right]\left(\frac{2}{9v_2}F^{2/3} + \frac{1}{9v_1}\right)^{-1/2}$$

$$\text{Prob}\{F_{\text{actual}} > F\} = 1 - \tfrac{1}{2}(1 + c_1 z + c_2 z^2 + c_3 z^3 + c_4 z^4)$$

$$c_1 = .196854 \quad c_3 = .000344$$

$$c_2 = .115194 \quad c_4 = .019527$$

D. OUTLIER SAMPLE DETECTION

Equally important to choosing the optimum number of factors for the model is outlier detection. If one or more of the training samples are in error, it will cause errors in the calibration model and ultimately poor prediction results for unknowns. In short, the models attempt to account for all the variations in the training set data when they are calibrated. Outlier samples usually arise from some incorrect measurement, whether it is in the concentration data (i.e., errors in the primary calibration technique of transcription errors) or in the spectral data (i.e., spectrometer error, sample handling procedures, or environmental control such as temperature, humidity, etc.). Including outlier samples in the training set will introduce a bias to the final model. In effect, outlier samples will tend to "pull" the model in their direction, causing the predicted concentrations of valid samples to be less accurate (or even erroneous) than if the sample was completely eliminated from the training set.

1. *Concentration Residuals*

One powerful tool for outlier detection is the cross-validation procedure used to calculate the PRESS values described previously. When the optimum number of factors for the model has been determined, the predicted concentrations of each training sample from the sample rotation with the selected factor model can be used for outlier detection. The difference between the actual and predicted concentrations for a sample is known as the concentration residual:

$$R_c = C_{orig} - C_{pred}.$$

The model attempts to account for all the variations in the training data when the calibration calculations are performed. Therefore, the prediction error of most of the samples should be approximately the same. Samples that have significantly larger concentration residuals than the rest of the training set are known as concentration outliers.

This type of outlier generally arises when either the experimenter makes a mistake in creating the calibration mixtures or there was an error in the analysis of the sample from the primary calibration technique used to generate the calibration concentration values. Another possibility that frequently occurs is a transcription error. The analyst simply types in the wrong concentration value when building the computerized training set.

Looking at Fig. 14, it is clear to see that sample No. 31 is significantly different from the rest of the training set and most likely a concentration outlier. However, outliers in most data sets will not be as obvious as this. Although the human eye is excellent at discerning patterns in data, visual inspection is not always a valid basis for a decision of this type. What is really needed is a mathematical way to accurately determine the likelihood that a sample is really an outlier.

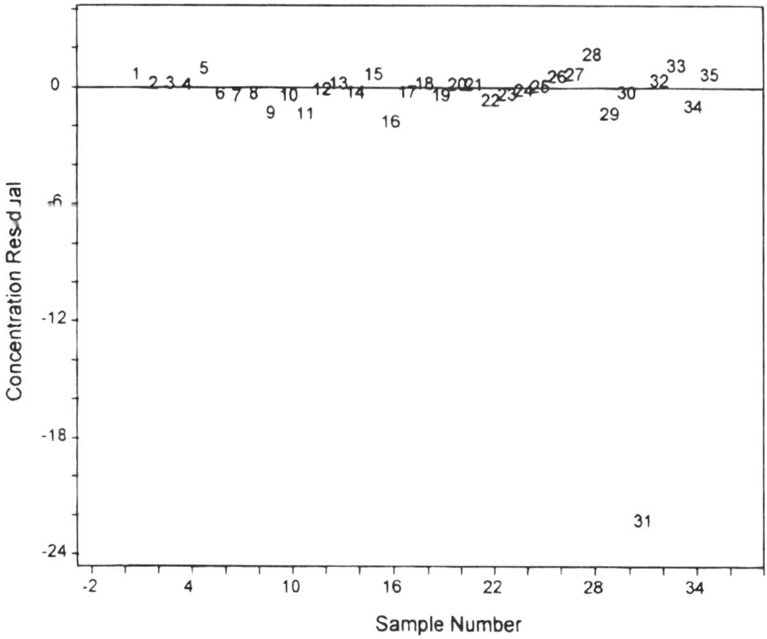

FIG. 14. A plot of concentration residual of the predicted hydroxyl number versus the sample number for a cross-validation of a PLS model built from 35 Fourier transform near-infrared spectra. Note that sample 31 has a significantly different residual than the remainder of the data set, indicating that it is probably an outlier.

The F test method for determining the optimum number of factors from a cross-validation PRESS analysis is also useful for determining the statistical significance of a sample's concentration residual with respect to the rest of the training set. In this case, the F ratio value is calculated by

$$F \text{ ratio}_i = \frac{(n-1)R_{c_i}}{(\sum_{j \neq i} R_{c_j})},$$

where i is the number of the sample being tested, n is the number of samples in the training set, and R_c is the concentration residual value of the sample predictions. In order to determine if the sample is an outlier, this F ratio can be looked up in an F statistic table or the same probability calculation used to estimate the significance of the PRESS values can be used. In this case, the degrees of freedom are one (1) for the numerator (v_1) and $(n-1)$ for the denominator (v_2). Generally, samples that exhibit probabilities of 0.99 ($\alpha = 0.01$) are considered outliers and should be removed from the training set before calculating the final calibration model.

2. *Spectral Residuals*

Another powerful tool in seeking out outlier samples is the spectral residual. This was discussed briefly in an earlier section. Similar to looking for concentration outliers, spectral outliers are detected by using a model for which the optimum number of factors has been determined by a cross-validation.

Remember that when each sample is predicted, a set of scores is found that best fits the model loading vectors to the sample spectrum. By using the calculated scores and the calibration loading vectors, a new model reconstructed spectrum can be calculated. This new spectrum is what the PLS or PCR model thinks the sample spectrum should look like. The spectral residual is the difference between this spectrum and the actual prediction spectrum and is calculated as

$$R_s = \sum_{k=1}^{p} (A_{\mathrm{orig}k} - A_{\mathrm{pred}_k})^2,$$

where p is the number of wavelengths (data points) in the spectrum, A_{orig} is the original spectrum absorbances, and A_{pred} is the model predicted spectrum absorbances.

As with concentration residuals, samples that have significantly higher spectral residuals than the rest of the training set may be outliers. Spectral outliers can be caused by many different factors, including inconsistent sample handling, changes in the performance of the instrument, or anything that contributes to a significant change in the spectrum of a given sample.

The spectral residual plot in Fig. 15 was chosen to illustrate an obvious outlier (sample No. 45). It is necessary to use the F test method to determine that the samples are indeed outliers in the same manner that is used for concentration outliers. For spectral residuals, the F ratio is calculated as

$$F\ \mathrm{ratio}_i = \frac{(n-1)R_{s_i}}{(\sum_{j \neq i} R_{s_j})},$$

where the subscript i indicates the number of the sample being tested, n is the number of samples in the training set, and R_s is the spectral residual values of the sample predictions.

Unfortunately, there is some debate over the actual number of degrees of freedom to use for spectral residuals. Some values have been suggested that seem to work well in practice: 1 for the numerator (v_1) and $n - f - 1$ for the denominator (v_2), where f is the number of factors in the model. Again, samples that exhibit probabilities of 0.99 ($\alpha = 0.01$) are considered outliers and should be removed from the training set before calculating the final calibration model.

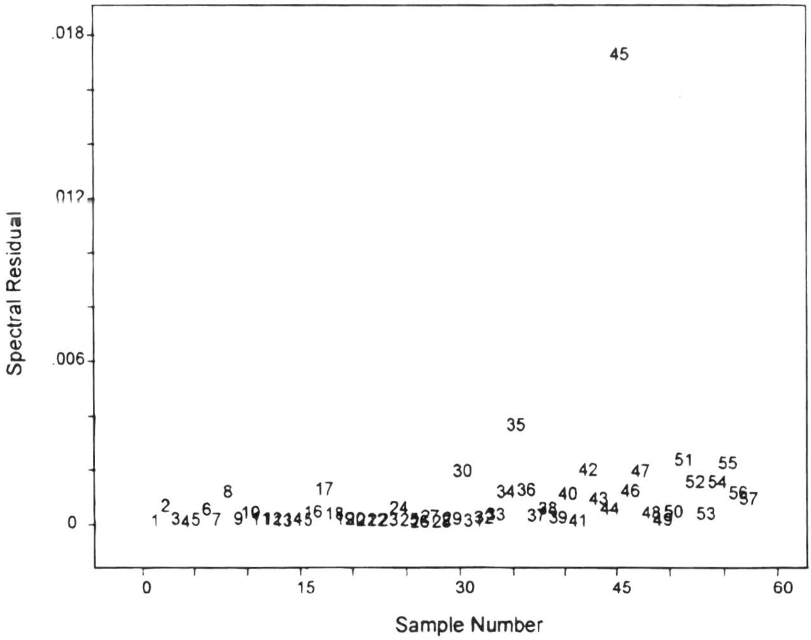

FIG. 15. A plot of the spectral residual versus sample number for a cross-validation of a PLS model of research octane number built from 57 NIR spectra of gasoline. Notice that sample No. 45 has a significantly different residual than the remainder of the set, indicating that it is a possible outlier.

3. *Cluster Analysis*

There are other methods of outlier detection that are more abstract but equally valuable. Cluster analysis is a method that is used to look for samples that have scores inconsistent with other samples in the training set. In this technique, the scores of one loading vector are plotted versus the scores of another vector for every sample in the training set. (Remember that the scores are the scalar values by which each loading is multiplied to reconstruct the original spectrum.) If all the samples in the training set are similar in composition and calibration value, the data points will tend to "cluster" about some mean value. If a sample point lies significantly outside this cluster, it indicates that the ratio of the two factor scores for this sample is inconsistent with the other spectra in the training set and it may be an outlier.

There is, however, one exception: Samples that lie at the ends of the calibration concentration range (i.e., the sample contains the highest or the lowest concentration of a constituent) can be expected to lie at the extreme limits of the cluster. An extreme sample will sometimes appear as an outlier, even though it may not be one at all.

Once again, it is desirable to have a more statistical measure of a sample's potential to be an outlier than simple visual inspection. For score clusters, it is possible to use a measure of the Mahalanobis distance (Figs. 16 and 17). This is calculated as the distance of the potential outlier sample point as measured from the mean of all the remaining points in the cluster. The distance is scaled for the range of variation in the cluster in all dimensions and then assigns a probability weight to the sample in terms of standard deviation. Any sample that lies outside of 3 standard deviations from the mean can be considered suspicious.

The Mahalanobis distance is also useful in qualitative analysis of spectral data for which the constituent concentrations are not known. This method, along with the mathematics, will be discussed in a later section.

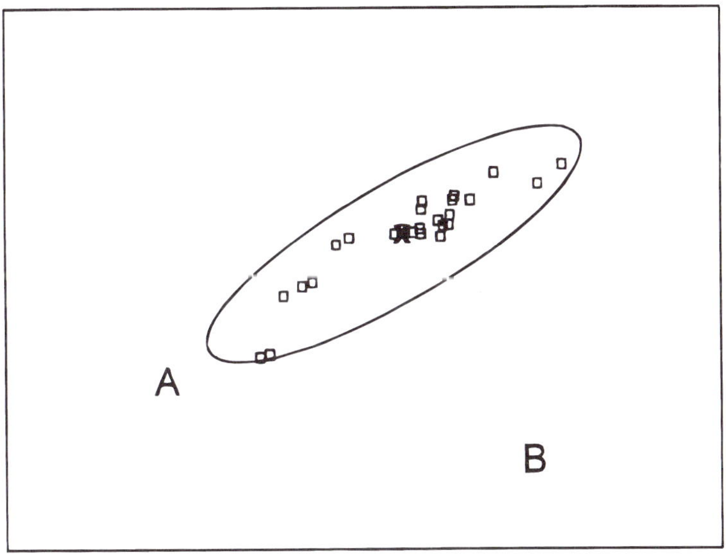

FIG. 16. The Mahalanobis distance of a point is measured from the mean point of the cluster (indicated by X). Unlike an absolute distance measurement, it takes into account the "shape" of the cluster. Although points A and B appear to be equidistant from the mean, in terms of Mahalanobis distance, A is much closer and therefore more likely to be a member of the cluster.

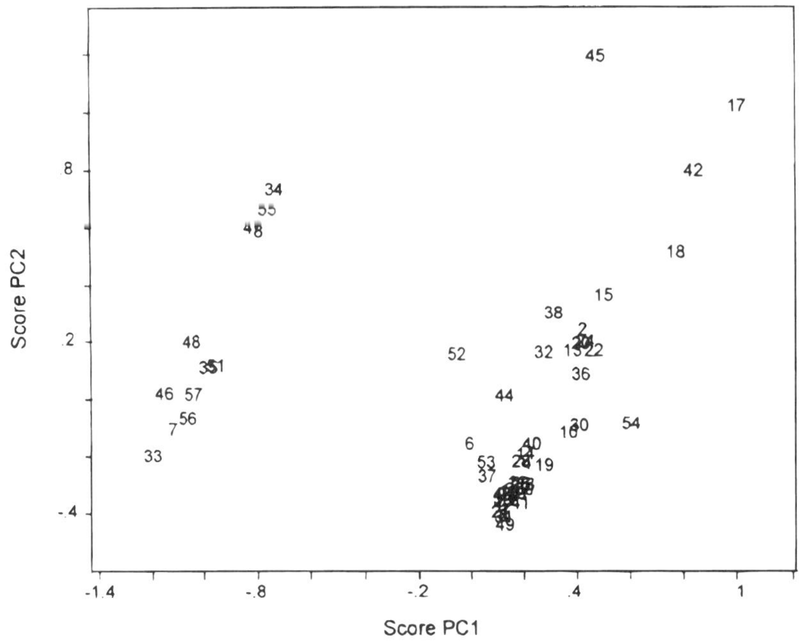

FIG. 17. Sometimes trends in the training data can be revealed by looking at score cluster plots. This plot shows the sample scores of the first two principal components of the training set data for the same model in Fig. 15. The data are clearly split into two separate clusters indicating two different types of samples. After careful examination of the sample concentration values, it was determined that the clusters do not represent the low and high concentrations of research octane number. It was later discovered that the spectra had been collected at different times by two different analysts.

4. *Leverage and Studentized t Test*

Another useful plot for identifying outliers is a plot of the Studentized concentration residuals versus the leverage value for each sample in the training set (Fig. 18). The leverage value (as discussed earlier) gives a measure of how important an individual training sample is to the overall model. The Studentized residuals give an indication of how well the sample's predicted concentration is in line with the leverage. If a sample has a very high leverage compared to the rest of the training set, it is not necessarily always an outlier. It could just be a sample at the high or low end of the concentration range. However, if a sample has both a high leverage and a Studentized residual that is very different from the rest of the data set, most likely it can be eliminated as an outlier.

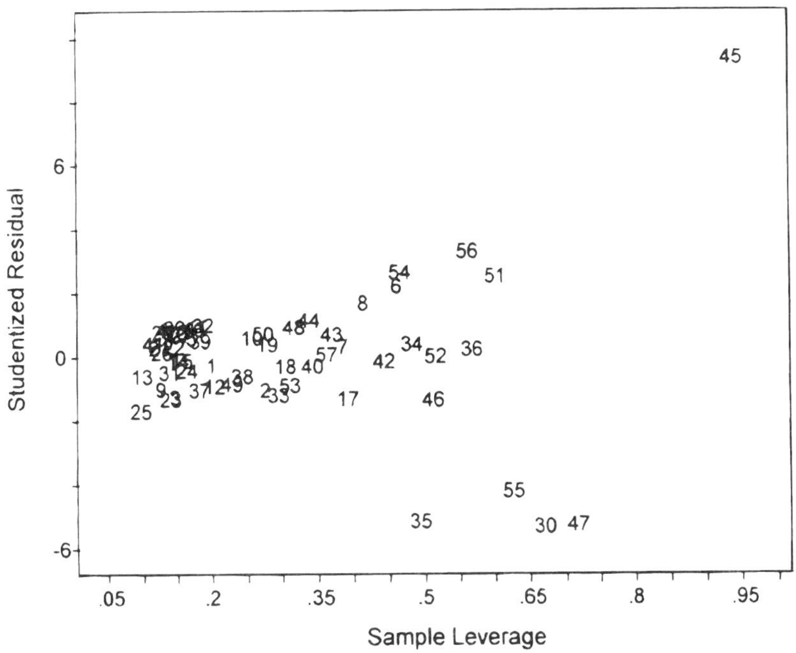

FIG. 18. A leverage vs studentized residual plot of same cross-validation prediction of the model in data from Fig. 15. Note that both the studentized concentration residual and the leverage of sample No. 45 are both significantly larger than the remainder of the training set. This is another confirmation that this sample is an outlier.

Sample leverages are calculated from the factor scores in PCA/PCR and PLS models. It is a relatively simple calculation:

$$H = S(S'S)^{-1}S'$$

$$\text{Leverage}_i = H_{i,i},$$

where S is the $n \times f$ matrix of sample scores, and H is an $n \times n$ square matrix (also known as the "Hat" matrix). As before, n is the number of samples in the training set, and f is the number of factors in the model. The subscript i is the sample number in the training set. Note that the individual sample leverages are the diagonal elements of the Hat matrix.

The Studentized residual is then calculated by

$$\text{St}_i = \frac{C_{\text{r}i}}{\left(\dfrac{\sum_{j=1}^{n} C_{\text{r}j}}{(n-f)}\right)^{1/2} (1 - \text{Leverage}_i)},$$

where C_r is the concentration residual of every sample in the training set.

E. Spectral Region Selection

As mentioned earlier, one of the best features of factor analysis-based models is the ability to use the entire spectrum for building a calibration. This gives the analyst advantages over other methods, mainly the averaging effect of using a large number of data points makes more robust calibrations and the ability to build models with little or no *a priori* knowledge of the spectra of the constituents of interest.

When setting up a chemometric model, it is very easy to simply select the entire range of the training spectra as the set of data to use for calibration. The PLS and PCR factor models will certainly be able to figure out the regions in the spectra that are most important for calibration based on the information in the training set. However, the performance of the calibrations can usually be improved by some intelligent masking of regions that are obviously useless. Remember, reducing the number of data points in the regions used also reduces the amount of computer memory and time necessary to perform all the calibration calculations.

Because there is no apparent penalty for using as many wavelengths as possible for calibrations, why not just use the entire spectrum? There are actually many good reasons. A very simple example is that regions of the spectrum in which the detector, the spectrometer source, or the optics are not effective. Including data from wavelengths below the detector cutoff is merely adding randomly distributed and uncorrelated absorbances to the factors. In many cases, analysts will have some idea of the important spectral bands for some constituents. This information, along with the chemical knowledge of the samples, should be considered when selecting spectral regions for inclusion in a calibration.

Another important consideration is the overall contribution of each wavelength to the spectra of the individual constituents. If there are regions of the spectrum where none of the constituents absorb, why bother including those wavelengths in the calibration? Including these wavelengths should have no benefit to the prediction because they contain no information about the constituents of interest. Generally, selecting only the highly correlated regions of the spectrum for calibration will improve the accuracy for predicting the constituents of interest. However, it comes at the expense of discovery of impurities in the samples (known as outlier samples). If the spectral bands of the impurities do not appear in the selected calibration regions, then there will be no indication that the predicted constituent values are potentially incorrect.

Finally, selecting specific regions can avoid the problem of detecting nonlinearity. If there are regions in the spectrum with very strong absorption bands (thus nonlinear with respect to Beer's law), it is usually best to choose the regions on either side, thus excluding that band. Although PCA and PLS factor analysis can correct for some nonlinearities, it cannot correct for regions of overabsorbance ("black sample").

There are mathematical methods that can aid the experimenter in selecting the best regions for calibration. One of these is to compute the correlation spectrum for each of the constituents of interest (Fig. 19). This is simply done by calculating the correlation of the absorbance at every wavelength in the training spectra to the concentrations of constituents. Regions that show high correlation are regions that should be selected, whereas regions that show low or no correlation should be ignored.

Correlation spectra can be viewed in one of two ways, depending on the type of region information desired. The first is probably familiar to almost all analysts: R^2, also called the coefficient of determination. This is classically used to determine the goodness-of-fit for a linear regression, where a value of 1 indicates a perfect fit. Similarly, when R^2 spectra are viewed, regions of the spectrum that are equal to or nearly equal to 1 indicate regions of correlation between the spectral absorbances and the constituent concentrations, whereas regions that are at or equal to zero are not correlated.

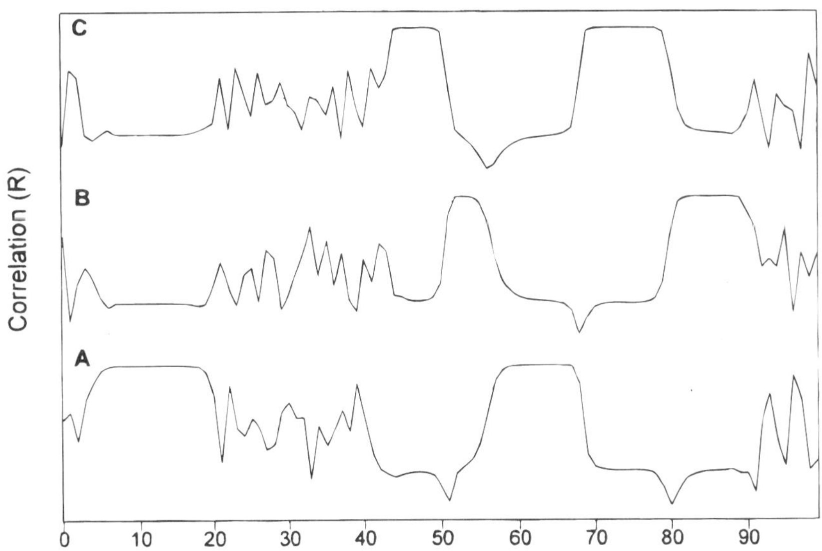

FIG. 19. The correlation spectra (R) of the constituents of the synthetic training set data in Fig. 5. Notice how the regions with the largest correlations correspond to the bands of the pure constituent spectra shown in Fig. 8.

The other method is to look at the spectrum of the linear correlation coefficient, also known as R (Fig. 20). This type of correlation plot indicates not only regions of the spectrum that are correlated to the constituents but also the type of correlation. The linear correlation coefficient varies between ± 1. Regions of the spectrum that give highly positive correlation values (near $+1$) are most likely due to the constituent. In other words, as the

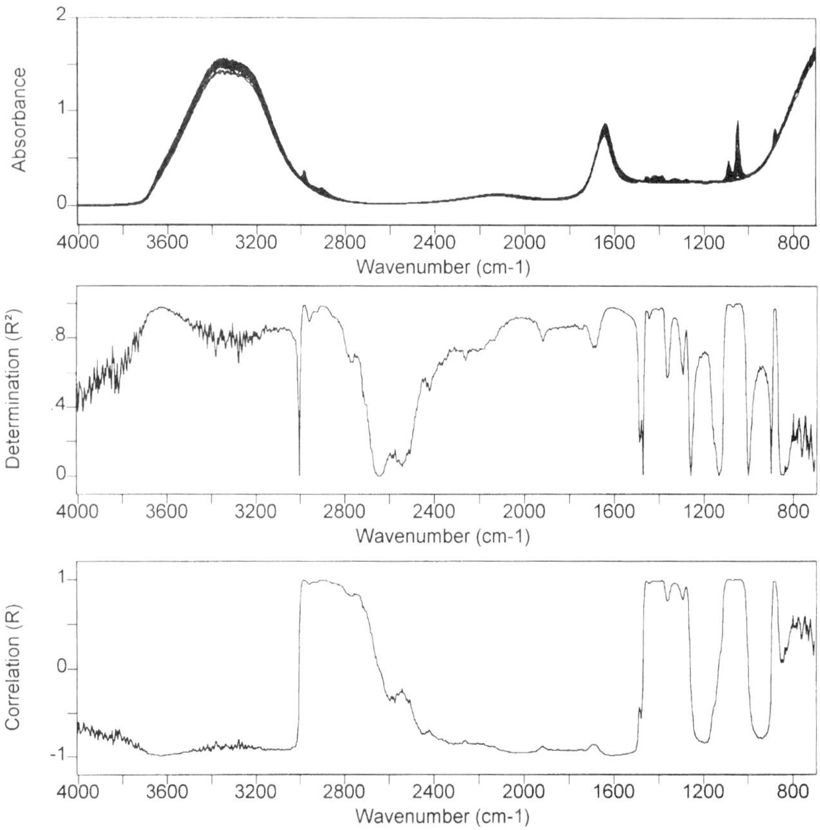

FIG. 20. Linear correlation (R) (*bottom*) and coefficient of determination (R^2) (*middle*) spectra of ethanol from a training set of 20 Fourier transform infrared, spectra (*top*) of ethanol/water mixtures. Notice the high-correlation bands at 1100–1020, 1450–1275, and 2980–2740 cm^{-1}. Note from the R spectrum that the remaining regions of the spectrum are negatively correlated. Because this is only a two-constituent mixture, these regions must be water, thus confirming that an increase in ethanol concentration gives a corresponding decrease in water.

concentration of the constituent increases, so do the absorbances at its characteristic wavelengths. However, there can also be regions where R indicates negative correlation (near -1). This does not mean the region is not useful for calibration. If anything, the opposite is true. A negative correlation indicates that an *increase* in constituent concentration caused a *decrease* in the spectral absorbance at those wavelengths or vice versa.

How can an increase in concentration cause the absorbances to decrease? It is actually quite simple. Remember that these methods are usually only applied to spectra of samples that are mixtures. Often, these mixtures are extraordinarily complex, as is the case with natural products such as wheat or gasoline. Generally, increasing the concentration of one constituent in a mixture "dilutes" the others. If the dilution occurs in a manner that is a ratio function of the increase in the constituent of interest, a negative correlation will appear. In most cases, these regions are as useful for calibration as the positively correlated regions.

Unfortunately, this is not completely true for all training sets. If the concentrations of the constituents of interest vary as a function of one another, or they vary as a function of other unknown constituents, the correlations will indicate regions that are not really useful. A perfect example of this is creating a training set by simply making dilutions of a single mixture. Even though the input data are actually from different concentrations of the constituents and the spectra all appear different, every wavelength in the spectrum will be highly correlated. Making dilutions of a mixture causes the concentrations of all constituents to increase/decrease together, not just the ones that are of interest for the calibration. Therefore, there is no unique information in the spectrum to allow the correlation to locate the constituents of interest. (Calibration set design is discussed in more detail in a later section.)

F. Data Preprocessing

One of the major problems in applying chemometric models to spectra is the fact that the acquired spectrum of a sample is dependent on many different, sometimes uncontrollable, factors. For example, samples of powdered solids are usually measured by diffuse reflectance. Light scattering off the particles causes every spectrum, even of remeasurements of the same sample, to look a little bit different due to the particale size distribution and alignment with the incident beam of light. Although the quantitative information related to the constituents is still contained within the spectral data, it may not be immediately apparent. Another example is if the path length of the samples cannot be controlled (such as measuring spectra of thin films). Remember that Beer's law describes a direct relationship between the absorbance and the path length.

Chemometric models can sometimes correct for these effects by adding extra loading vectors, but generally the models will perform better if they can be removed or at least minimized before running the data through the calculations. Because they are applied to the data before use in the model, they are often called preprocessing algorithms. There are a variety of methods that can be used to remove the nonconstituent-related aberrations in the data. Most algorithms are targeted at removing a specific interference [multiplicative scatter correction (MSC), for example, specifically attempts to remove the effects of light scattering]. Properly applying preprocessing requires understanding the interferences in the data and selecting the appropriate algorithm(s) to correct the effect(s).

1. *Correcting for Light Scattering in Reflectance Measurements*

As discussed in the previous section, samples that are measured using diffuse reflectance often exhibit significant differences in the spectra due to the nonhomogeneous distribution of the particles. In fact, multiple spectral measurements of different aliquots of the same sample can look completely different. In many cases, the scattering can be an overpowering contributor to the spectrum, sometimes accounting for most of the variance in the data (Fig. 21).

The degree of scattering is dependent on the wavelength of the light that is used; longer wavelengths are more adversely affected. Therefore, the scattering is not uniform throughout the spectrum. Typically, this appears as a baseline shift, tilt, and sometimes curvature, in which the degree of influence is more pronounced at the longer wavelength end of the spectrum.

2. *Multiplicative Scatter Correction*

This correction method assumes that the wavelength dependency of the light scattering is different from that of the constituent absorption (Ref. 45). Theoretically, by using data from many wavelengths in the spectrum it should be possible to separate the two.

This method attempts to remove the effects of scattering by linearizing each spectrum to some "ideal" spectrum of the sample. The key to this method is that all spectra must be corrected using the same ideal spectrum. Obviously, there is no way that the perfect spectrum can be collected that best represents all possible samples, so an estimate is used instead. There-

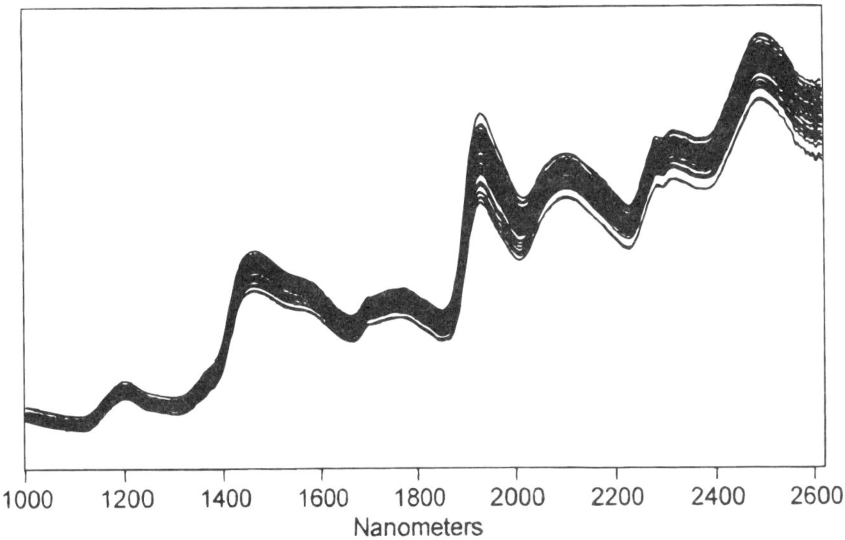

FIG. 21. A set of 50 log(1/R) near-infrared spectra of ground wheat samples measured using diffuse reflectance. The concentrations of the constituents of interest fall in a relatively narrow concentration range. However, note that the light scattering causes the spectra to appear quite different.

fore, MSC calculates the average spectrum from all the data in the training set and uses it as the ideal spectrum (Fig. 22).

Once the ideal average spectrum has been calculated, performing MSC is a relatively easy task. The spectral responses in each spectrum are used to calculate a linear regression against the corresponding points in the ideal spectrum. The slope and offset values from this regression are subtracted and ratioed respectively on the original training spectrum to give the MSC corrected spectrum:

Mean spectrum: $$\bar{A}_j = \sum_{i=1}^{n} A_{i,j}$$

Linear regression: $$A_i = m_i \bar{A} + b_i$$

MSC correction: $$A_{i(\mathrm{MSC})} = \frac{(A_i - b_i)}{m_i},$$

where A is the $n \times p$ matrix of training set spectral responses for all the wavelengths, $\bar{A}$ is a $1 \times p$ vector of the average responses of all the training set spectra at each wavelength, A_i is a $1 \times p$ vector of the responses for a

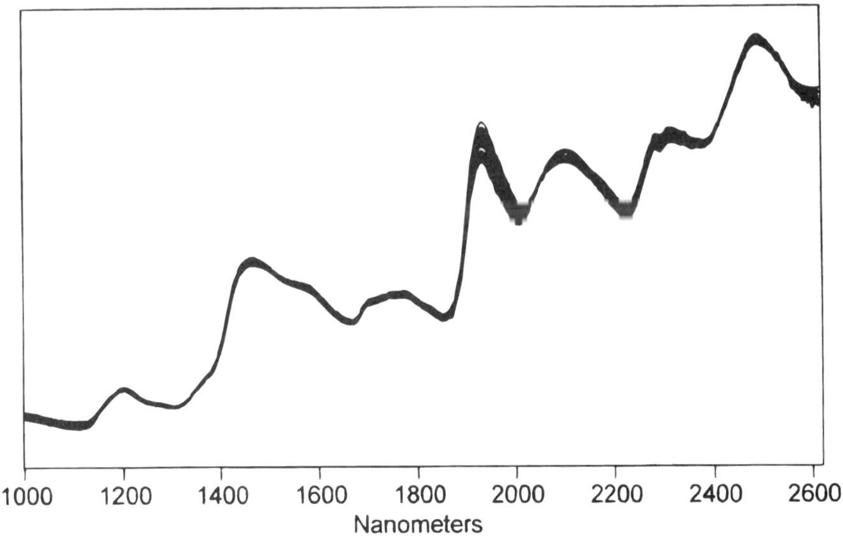

FIG. 22. The MSC corrected spectra of the same data in Fig. 21. Note that a significant portion of both the baseliine offset and the slope have been removed.

single spectrum in the training set, n is the number of training spectra, and p is the number of wavelengths in the spectra. The m_i and b_i values are the slope and offset coefficients of the linear regression of the mean spectrum vector $\bar{A}$ versus the A_j spectrum vector. One important fact to note is that if the spectral regions for the calibration model do not cover the entire spectrum, each individual region is MSC corrected separately. This is done to ensure that the correct slope is calculated because the endpoints of the regions may be disjointed in terms of response.

By adjusting the slope and offset of the sample spectra to the ideal average spectrum, the chemical information is preserved while the differences between the spectra are minimized. This is not to say that the MSC spectra represent the "true" spectra of the samples, but that the major source of random variance between them has been removed as much as possible.

One important fact to note; this is only applicable to spectra that have responses that are fairly linear in concentration. Therefore, any spectra collected in reflectance units should be converted to log(1/R) or Kubelka-Munk units before applying MSC. In addition, MSC works very well with spectra of samples that are chemically similar. If the appearance of the spectra in the training set are substantially different from one another due

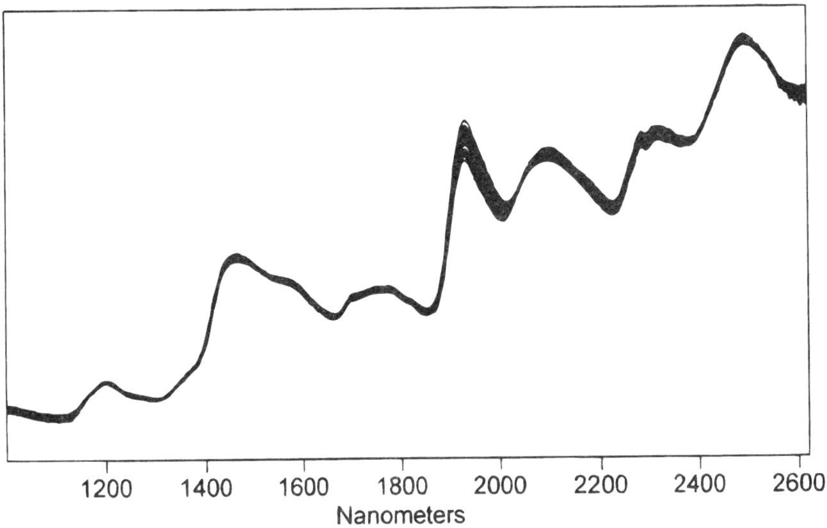

FIG. 23. The SNV corrected spectra of the same data in Fig. 21. Note that this is very similar to the results of MSC; however, no external reference ("ideal" spectrum) is required in order to calculate the correction.

to a wide range of variability in the sample composition, correcting to the mean spectrum will not give the desired results.

3. *Standard Normal Variate and Detrending*

Standard normal variate (SNV) correction and detrending (see Ref. 51) are other methods that attempt to remove the major effects of light scattering from the spectra. However, the calculational methods used are quite different.

This correction is actually two separate algorithms that are usually applied together. SNV is applied first to correct for the effects of the multiplicative interferences of scatter and particle size, similar to MSC. Detrending usually follows to attempt to remove the additional variations in baseline shift and curvilinearity typically present in diffuse reflectance spectra. Although, in some cases, applying SNV alone can be more useful for interpretation of the of the final model vectors. Detrending is not usually applied alone; it is either used after the SNV correction or not used at all.

SNV uses a different approach than MSC to correct for the scattering effects. Here, no external "ideal spectrum" is required (Fig. 23). Instead, the scattering is removed by normalizing each spectrum by the standard deviation of the responses across the entire spectral range:

$$\text{Mean response:} \qquad \bar{a}_i = \sum_{j-1}^{p} A_{i,j}$$

$$\text{SNV correction:} \qquad A_{i(\mathrm{SNV})} = \frac{(A_i - \bar{a}_i)}{\sqrt{\dfrac{\sum_{j=1}^{p}(A_{i,j} - \bar{a}_i)^2}{(p-1)}}},$$

where A is the $n \times p$ matrix of training set spectral responses for all the wavelengths, A_i is a $1 \times p$ vector of the responses for a single spectrum in the training set, $\bar{a}_i$ is the average of all the spectral responses in the vector, n is the number of training spectra, and p is the number of wavelengths in the spectra.

The next step is to apply detrending. As with SNV, each spectrum is treated independently of the others in the training set so that there is no external reference. It is a relatively simple calculation: A linear least squares regression is used to fit a quadratic polynomial to the responses in the spectrum. This curve is then subtracted from the spectrum to give the result. As mentioned earlier, the quadratic curvature component attempts to correct for the effects of particle size and sample packing (Fig. 24).

As with MSC, this method is applicable only to spectra that have responses that are fairly linear in concentration. Any spectra collected in reflectance units should first be converted to $\log(1/R)$. However, because each spectrum is corrected independently, this method may allow for greater variability in the training spectra composition than MSC.

4. *Correcting Sample Path-Length Differences*

Beer's law states that there is a direct and linear relationship between sample concentration, path-length, and the absorbance of light at a particular wavelength. Most chemometric models are built using samples that vary in concentration, but the path length is fixed (except for diffuse reflectance measurements, however, MSC or SNV are used to correct this type of data). Factor-based chemometric models are not limited by this requirement, although the performance of the models is certainly better when this is true.

Unfortunately, it is not always possible to collect spectra of either training samples or unknown samples with a constant path-length. For example, when measuring transmission spectra of thin films, it is very difficult to

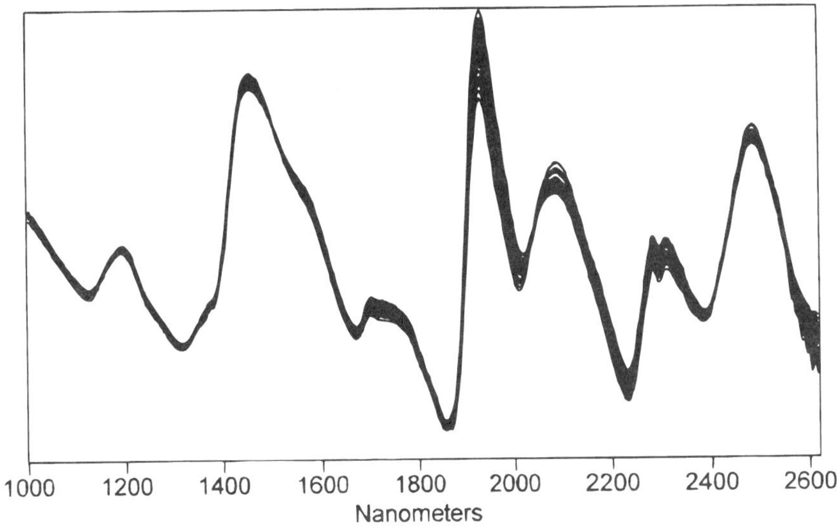

FIG. 24. Spectra of the data from Fig. 21 after SNV correction and detrending.

extrude polymers to a constant thickness every time. Obviously, if the path length varies in the sample, this will appear in the spectra as changes in response that are not correlated to concentration. Sometimes the factor-based models can correct for these effects if the range of different path-lengths is not too large. However, the model will most certainly work much better if the path-length effect can be removed altogether.

In addition to the methods listed here, MSC has also been applied to correct spectra of samples with indeterminate path-lengths, but not necessarily measured by diffuse reflectance (Ref. 52). Although it will not be as effective as either of the two specific path-length corrections discussed in the following sections (measured and thickness), it will usually give better results than no correction at all. Some success has also been shown in using MSC to correct the path-length effects in spectra measured using attenuated total reflectance.

5. *Measured Path-Length Calibration*

In some cases, it is possible to actually measure the exact path-lengths of the training samples. This information can be used during the model building step by including the path length as an extra constituent in the calibration. During the calibration calculations, all the concentration data are scaled to the entered path length for each sample. When an unknown sample is predicted, its path length is predicted at the same time. The concentrations are then unscaled by the predicted path length before they are reported (Ref. 52).

This method is useful in cases in which the path-lengths of the training samples are easy to acquire. This can be used to correct for samples in which the path length of the unknowns either cannot be (easily) measured or is expected to vary substantially over a period of time. The downside to this method is that accurately measuring the path lengths of every training sample can be a tedious process.

6. *Sample Thickness Correction*

This type of path-length correction is sometime called the "internal standard" method (Ref. 52). It is primarily used for samples that cannot be corrected by the measured pathlength method. One requirement for this method is that there must be an isolated band in every spectrum that arises from a constituent that does not vary in concentration in all samples, for both the training set and unknowns for prediction (Fig. 25).

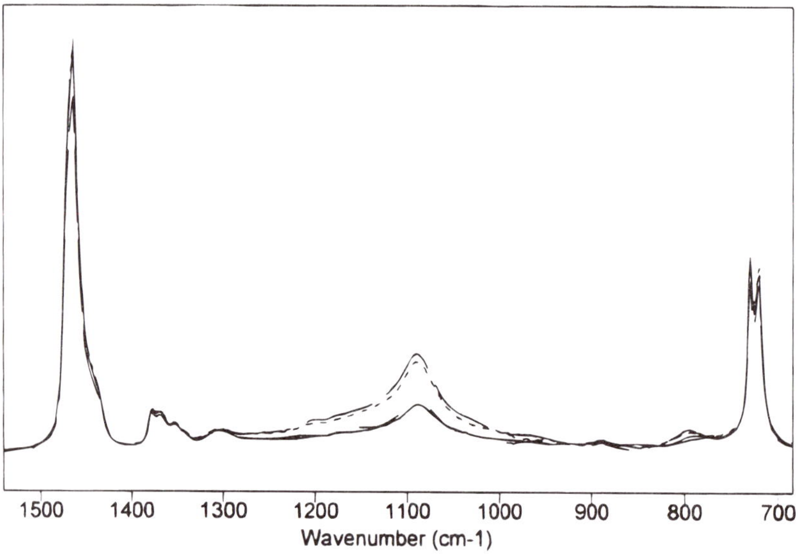

FIG. 25. Spectra of four polymer thin films. The constituent of interest is clearly the band in the middle: 2% (solid), 4% (dashed), 8% (dotted), and 10% (dash–dot). Note that the relative intensities of the constituent bands are not correct due to the varying thickness of the samples.

Because the chosen spectral band is assumed to be concentration invariant in all samples, an increase or decrease of the absorption of that band in the spectrum can be assumed to be entirely due to an increase or decrease in the sample path length. Therefore, by normalizing the entire spectrum to the intensity of the band, the path-length variation is effectively removed. The intensity of the band can be calculated as either the response at a single wavelength in the band (usually the peak maximum) or the integrated area.

One potential problem with this method is that it is extremely susceptible to baseline offset and slope effects in the spectrum. When calculating the path-length normalization constants, the spectra must be baseline corrected before creating the training set or a local baseline must be calculated "on-the-fly" under the thickness correction band for each spectrum. The latter approach is usually recommended because the former method requires separate manual baseline correction of each unknown spectrum before prediction against the calibration model.

Thickness correction is very useful for spectral measurements of samples for which the path-length cannot be guaranteed to be constant. However, it does require that the samples have a constant concentration constituent and that an isolated spectral band can be identified that is solely due to that constituent (Fig. 26).

Another use for thickness correction is to allow the calibration model to be path-length independent. Instead of normalizing to a small region (isolated band) in the spectrum, a much larger region or even the entire spectral range is used to calculate the integrated area. This allows correcting samples with nearly any path-length. The only requirement in using thickness correction in this manner is that the range of constituent concentrations must be relatively small. Large variations in the concentrations will cause the integrated areas of the spectra to vary mostly by concentration and not path-length differences. This will actually introduce nonlinearities in the spectra-constituent correlations and degrade the predictive ability of the model. However, if the concentration range is relatively small, this is a great way to build models that are insensitive to changes in path length of both the training spectra and the spectra of unknowns.

7. *Unit Area Normalization*

This method attempts to correct the spectra for indeterminate path length when there is no way of measuring it or isolating a band of a constant concentration constituent. In this approach, the spectra are normalized by calculating the area under the curve for the entire spectrum.

In effect, this method is the same as using the thickness correction on a large region of the spectrum as mentioned previously. However, here the

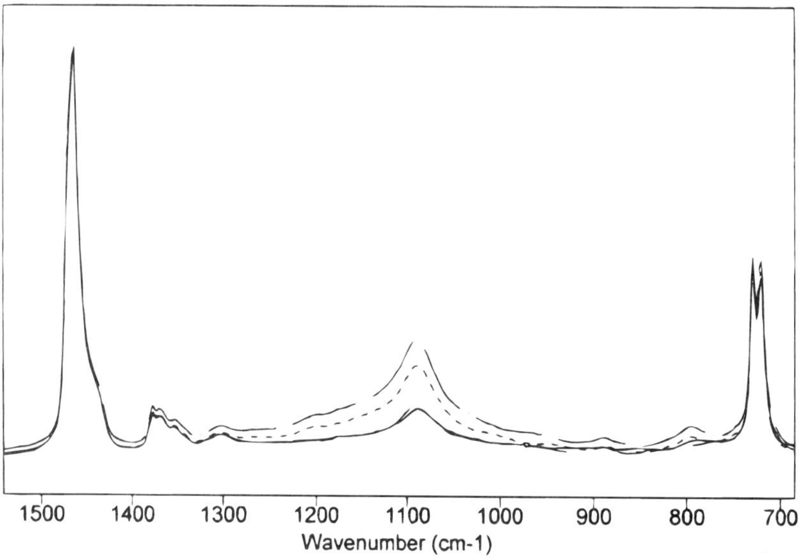

FIG. 26. Spectra from Fig. 25 after thickness correction. The integrated area of the band on the left (1525–1400 cm^{-1}) was used as the thickness correction factor. Notice that the relative intensities of the constituent bands now appear more in line with the known concentrations.

entire spectrum is always used rather than a large selected region. This method is very simple to implement but has some drawbacks. First, the concentration variations between all the training samples and unknown samples must not be too large, for the same reasons discussed for thickness correction. In addition, because this method uses the entire spectrum, the responses at all wavelengths in the spectrum must contain useful data. This means spectra that exhibit evidence of detector or optics cutoffs, or "black sample" (sample is too thick or too concentrated, causing complete absorbance of all light at some wavelengths), cannot be corrected. Finally, if the spectra do not have a constant baseline between all measurements, the integrated area will be calculated incorrectly. It is generally best to combine this method with some form of baseline correction.

8. *Correcting Baseline Effects*

As all spectroscopists know and have observed, spectrometers do not always collect data with an ideal baseline. Due to a variety of problems (detector drift, changing environmental conditions such as temperature, spectrometer purge, sampling accessories, etc.), the baseline of a given spectrum is not always where it should be. Beer's law assumes that the absorption of light at a given wavelength is due entirely to the absorptivity of the constituents in the sample; it does not account for "spectrometer error" or "sampling error." Therefore, in order to accurately calculate concentrations, it is necessary to remove the baseline effect introduced by the spectrometer.

As with most random variations in the spectral data, most chemometric models can compensate for these effects by adding extra factors or, if the variations are truly completely random, ignore them altogether. However, as with all preprocessing methods, a more robust model will usually result when the known interferences in the data are removed first.

There are a number of methods used by spectroscopists to remove baseline effects from the spectra they collect. The problem with most methods is that they require the spectroscopist to decide that the baseline is correct by visual inspection. In addition to being very subjective, most of these methods cannot easily be applied in the somewhat automated fashion required for a calibration model.

However, there are some methods that are reasonably automated to be used as part of a calibration model. The list of baseline correction methods presented in the following section is not exhaustive, and there are many other ways of autocorrecting the spectrum baseline as a chemometric preprocessing step.

9. *Linear Regression Baseline Fitting.* This is a very simple approach to baseline correction in that it requires no effort to set up. In this method, a least squares regression line is fit to the responses in each spectral region selected for calibration. This line is then subtracted from the response values in the region before using the data to perform the calibration model calculations (Ref. 52).

Unfortunately, this is not always the best approach, especially when the selected spectral regions are primarily large bands from the constituents of interest. It tends to work better when the entire spectrum is used or when the selected regions are very broad. In some cases, this method actually degrades the performance to the calibration models more than if no baseline correction was used at all. In general, this method should be used only in situations in which baseline aberrations are severe and a limited number of training sample spectra are available.

10. *Two-Point Linear Baseline.* Another approach is the tried-and-true method of selecting two baseline points in the spectrum, connecting them with a line, and subtracting them from the spectral responses. This is known as a two-point baseline correction.

The main problem with this method is selecting the two points. In the optimum case, the training sample spectra will always have at least two regions that are at different ends of the spectrum where there is no absorption. In the worst case, the entire spectrum may exhibit absorption, and defining a baseline point can be difficult if not almost impossible. Another problem is how to select baseline points from looking at a single spectrum and be sure that no band will suddenly appear at that wavelength in future spectra (Fig. 27).

Despite these limitations, there are some things that can be done to make this method of baseline correction more robust. For example, instead of selecting two single points for baseline correction, select a range of points in two parts of the spectrum. Then locate the wavelength point that has the minimum response in each range, and use these values as the two points. Another method is to calculate the average of the points in the selected baseline regions (Fig. 28). These methods will presumably solve the problem of peaks shifting near the selected points.

In some instances, it may only be possible to locate a single baseline point. In these cases, the method of selecting the point in the single baseline region (either average or minimum) is the same, but the baseline simply extends flat in both directions (Fig. 29).

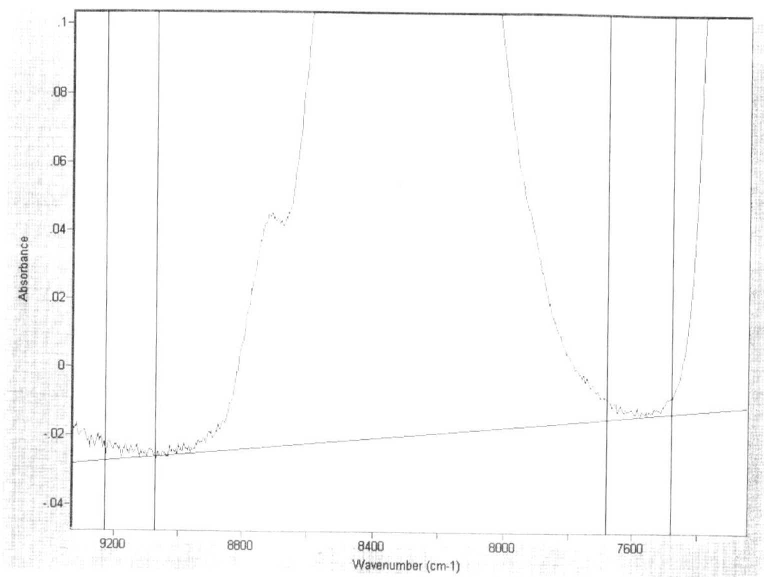

FIG. 27. Using wavelength ranges to select baseline points attempts to avoid incorrect baselines when the peaks may shift. This baseline was calculated using the minimum points in the two selected ranges.

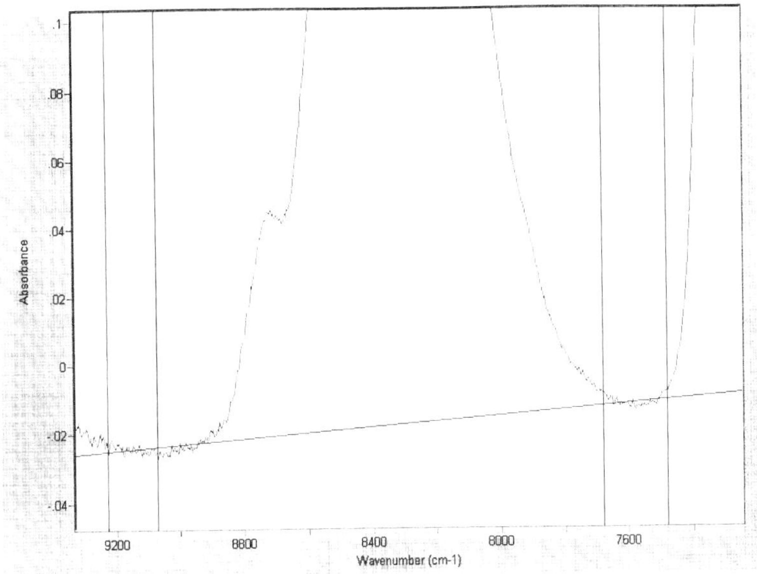

FIG. 28. The same data in Fig. 27 with the baseline calculated using the average of the points in each region.

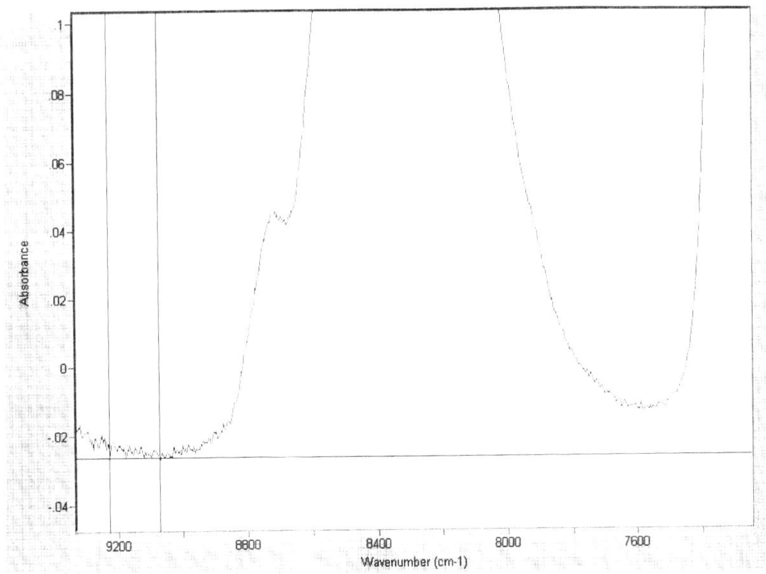

FIG. 29. When only one baseline region is available, the baseline can be extended from the single region.

First and Second Derivatives. One of the best methods for removing baseline effects is the use of derivative spectra. This method is one of the earliest methods used to attempt to correct for baseline effects in spectra solely for the purpose of creating robust calibration models (Refs. 46, 47). The first derivative of a spectrum is simply a measure of the slope of the spectral curve at every point. The slope of the curve is not affected by baseline offsets in the spectrum, and thus the first derivative is a very effective method for removing baseline offsets (Fig. 30). The second derivative is a measure of the change in the slope of the curve. In addition to ignoring the offset, it is not affected by any linear "tilt" that may exist in the data and is therefore a very effective method for removing both the baseline offset and the slope from a spectrum (Fig. 31).

There are many ways to calculate derivatives. One of the easiest is by using the method of simple differences. In this approach, the derivative at a given point is calculated by

$$A_i = A_{i+1} - A_i.$$

Unfortunately, this method is not always useful for calculating "real" derivatives. In fact, because it attempts to estimate the derivative from the between-point differences, in most cases it only succeeds in enhancing the

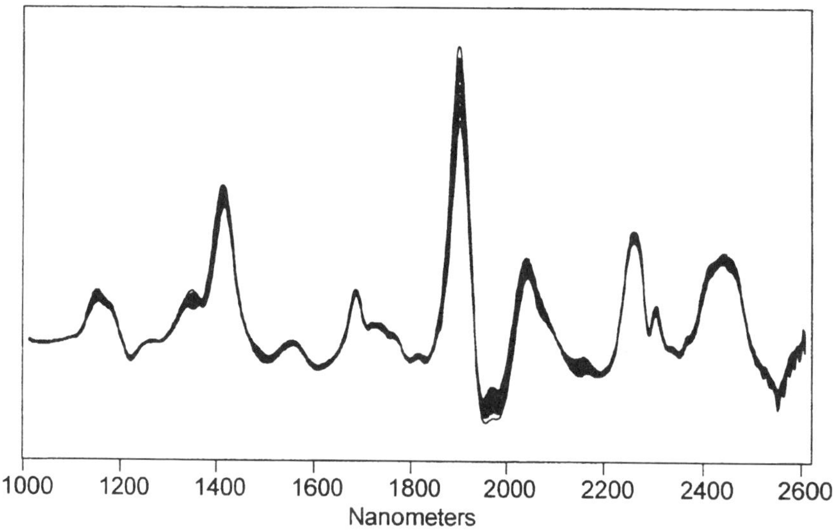

FIG. 30. First derivatives of the spectra in Fig. 21. Derivatives were calculated using the gap method with a gap value of 10 nm.

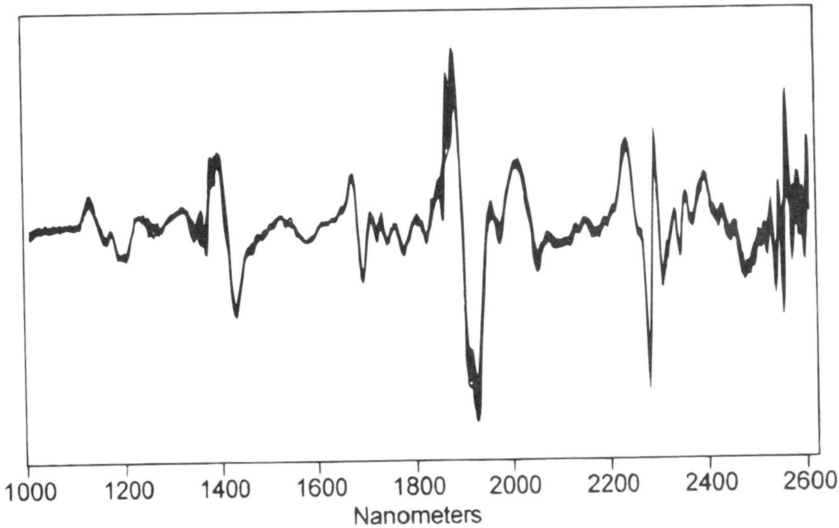

FIG. 31. Second derivatives of the spectra in Fig. 21. Derivatives were calculated using the gap method with a gap value of 10 nm.

noise in the spectrum.

There are better algorithms for calculating derivatives, including the gap method (Refs. 46, 47) and the Savitsky–Golay method (Refs. 48–50). Both these algorithms use information from a localized segment of the spectrum to calculate the derivative at a particular wavelength rather than the difference between adjacent data points. In most cases, this avoids the problem of noise enhancement from the simple difference method and may actually apply some smoothing to the data.

One problem in applying these methods is that they require an extra parameter: the size of the spectral segment to use for calculation of the derivative points. For the gap method, this is the size of the gap (usually measured in wavelength span but sometimes in terms of data points) between the difference points. The Savitsky–Golay method uses a convolution function, and thus the number of data points in the function must be specified. If the segment is too small, the result may be no better than using the simple difference method. If it is too large, the derivative will not represent the local behavior of the spectrum (especially for gap), and it will smooth out too much of the important information (especially for Savitsky–Golay). Although there have been many studies on the appropriate size of the spectral segment to use, a good general rule is to use a sufficient number of points to cover the full width at half height of the largest absorbing band in the spectrum.

The main disadvantage of using derivative preprocessing is that the resulting spectra are very difficult to interpret. As mentioned previously, the loading vectors for the calibration model represent the changes in the constituents of interest. In some cases (especially in the case of PLS-1 models), the vectors can be visually identified as representing a particular constituent. However, when derivative spectra are used, the loading vectors cannot be easily identified. In addition, the derivative makes visual interpretation of the residual spectrum more difficult, and thus locating the spectral absorbances of impurities in the samples cannot be done.

11. *Data Enhancement*

Due to the multivariate nature of factor-based chemometric models, the direct relationship between the spectral response and the constituent concentration (univariate) is not very important. These models do not examine the absolute relationship between these values, but instead they calculate the relative change in the spectra and attempt to correlate that to a corresponding change in the constituent concentrations. This is why the models tend to be so robust and why they can calibrate for the constituents of interest in the presence of many other interferences.

Due to this fact, there are some mathematical enhancements that can be applied to data that are to be used in a multivariate model that would render it useless for a univariate model. The purpose of these algorithms is to remove redundant information and enhance the important sample-to-sample differences that exist within the data.

Mean Centering. Mean centering is almost always applied when calculating any multivariate calibration model. This involves calculating the average spectrum of all the spectra in the training set and then subtracting the result from each spectrum. In addition, the mean concentration value for each constituent is calculated and subtracted from the concentrations of every sample.

$$\text{Mean spectrum:} \qquad \bar{A}_j = \sum_{i=1}^{n} A_{i,j}$$

$$\text{Mean centering:} \qquad A_{i(\text{MC})} = A_i - \bar{A},$$

where A is the $n \times p$ matrix of training set spectral responses for all the wavelengths, $\bar{A}$ is a $1 \times p$ vector of the average responses of all the training set spectra at each wavelength, A_j is a $1 \times p$ vector of the responses for a single spectrum in the training set, n is the number of training spectra, and p is the number of wavelengths in the spectra.

By removing the mean from the data, the differences between the samples are substantially enhanced in terms of both concentration and spectral response. This usually leads to calibration models that give more accurate predictions.

Variance Scaling. Variance scaling is used to emphasize small variations in the data by giving all values equal weighting. Variance scaling is calculated by dividing the response at each spectral data point by the standard deviation of the responses of all training spectra at that point. The concentration data are scaled likewise for each constituent. Note that variance scaling is only applicable after the data has already been mean centered.

Variance spectrum: $$A_{v_j} = \sqrt{\frac{\sum_{i=1}^{n} (A_{i,\, j(\mathrm{MC})})^2}{(n-1)}}$$

Variance scaling: $$A_{i(\mathrm{VS})} = A_{i(\mathrm{MC})} - A_{v},$$

where A is the $n \times p$ matrix of training set spectral responses for all the wavelengths, Av is a $1 \times p$ vector of the variance of the training set spectral responses at each wavelength, A_j is a $1 \times p$ vector of the responses for a single spectrum in the training set, n is the number of training spectra, and p is the number of wavelengths in the spectra.

This preprocessing algorithm is most useful when analyzing minor (low concentration) constituents that have spectral bands that overlap those of major (higher concentration) constituents. By giving all the information in the data equal weighting, the calibration errors in the model should be more consistent across all constituents.

G. Training Set Design

One of the most important things to remember when creating a good calibration model is the quality of the training set data. As with any quantitative calibration method, the predictive ability of the equations is only as good as the data used to calculate them in the first place. Control over such variables as collecting representative samples, an accurate primary calibration method, and appropriate sample measurements are critical for obtaining good results.

One of the most apparent drawbacks of multivariate calibration models is the comparatively large number of training set samples required. Some complex materials may require hundreds or even thousands of samples to be tested and spectra measured before a suitable set of training samples can be identified. Although the work to create a multivariate calibration model is more significant, they tend to be more robust than the simpler univariate methods and require less maintenance in the long run.

1. *Training Samples Should Be as Similar as Possible to Unknowns*

A common misconception in quantitative spectroscopic calibration is that the spectrum of a constituent looks pretty much the same when it is part of a mixture as it does when it is in the pure form. Unless the samples are very simple mixtures or they are being measured in gas phase, nothing could be further from the truth. Mixing constituents together causes all kinds of changes in the spectra that are not readily apparent. Factor-based models can compensate for these interconstituent interactions, however, only if the training set contains examples of it.

Most samples used for factor-based multivariate quantitative spectroscopic analysis are not simple mixtures. Otherwise, it would not be necessary to use these models and the calibration could be performed using a much simpler method. These models have the distinct advantage for complex samples because they can find the important information in the spectra and ignore the rest. In order to give the model the best chance to learn to recognize the information for the constituents of interest, it is important to train it using samples that emulate the unknowns as closely as possible.

Most experienced analysts using these methods collect actual samples from the plant, the field, or any other source of the material they expect to measure with the calibration model. These samples are then brought back to the lab and analyzed using other primary calibration methods (chromatography, wet chemical test, drying, etc.) to arrive at the constituent values. These data, along with the sample spectra, formulate the training set they will use for building the calibration model. Remember, one of the main advantages of factor-based multivariate methods is the ability to calibrate for individual constituents in samples with very complex compositions, provided the unknowns exhibit the same behavior as the training samples.

2. *Bracket the Expected Range of Constituent Values*

As in all quantitative methods, the constituent values for the training samples should span the expected range of all future unknown samples. Although the models can sometimes extrapolate outside the range of calibration, this is generally not a good idea for multivariate calibration, let alone any other calibration method. There is no method other than external validation that can determine how well a model will predict outside the original calibration range, and this is not a good measure of future performance.

Basically, this means that the constituent values in the training samples should be both larger and smaller than the expected values in unknown samples. By bracketing the range of concentration, the model will give the most accurate answer possible. Keep in mind that calibration points in the middle of the range are required as well.

In some situations, it can be difficult to get samples with both high and low concentrations of the constituents of interest. This is especially true of samples of natural products that tend to be very homogeneous from batch to batch and where the analyst has no control over the sample composition. As mentioned previously, it may require collecting hundreds or thousands of samples and testing them with the primary calibration method until a suitable set of samples can be identified for multivariate calibration.

3. *Using Enough Samples to Model the Data Variability*

In order to use multivariate calibration methods the training set must have at least as many samples as there are constituents of interest and usually many more. How many samples are required to build a good model? Unfortunately, there is no hard answer such as "use at least 10 samples for one constituent, 20 samples for two constituents, etc." The real answer is "use as many samples as it takes." A statistically significant number of samples is critical in both evaluating the analysis and obtaining a robust calibration model. The more data, the higher the confidence in the analysis and in the statistics.

Another reason to use a large number of samples for calibration is to allow more factors in the model. Due to the nature of the linear algebra used to solve the eigenvector decomposition of the spectra, the maximum number of factors that can be calculated for a given training set is limited by the smallest dimension of the data matrix. Therefore, a training set with only 10 samples can only calculate 10 factors; if one-sample-out cross-validation is used, then only 9 factors are possible For complex materials, this may not be enough to account for all the variability in the real samples that will be predicted as unknowns. As a side note, this also applies for the number of wavelengths selected for calibration. If a training set has 500 samples but the calibration regions have only 20 total spectral data points, then the maximum number of factors is limited to 20 as well.

As mentioned previously, most training sets will have many, many samples before an accurate, robust calibration can be built. Because these types of chemometric models examine the relative changes in the data, putting more samples in the training set allows the calculations to more easily identify which spectral information is important (real signal) and which is not (noise). Remember that the quality of the data is just as important as the quantity, if not more so. Simply piling a huge number of spectra together as a training set will not guarantee a better model than carefully measuring and qualifying a much smaller number for calibration.

4. *Avoid Constituent Collinearity*

What exactly is collinearity and why is it a problem in multivariate models? Collinearity is the effect observed when the relative amounts of two or more constituents are a constant throughout all the training samples. This causes so much trouble for multivariate models because of the way they correlate information. Remember that these models do not calibrate by creating a direct relationship between the constituent data and spectral response. Instead, they try to correlate the change in concentration to some corresponding changes in the spectra. When constituents are collinear, multivariate models cannot not differentiate them, and the calibrations for the constituents will be unstable.

For example, consider the case of creating artificial "standards" for calibration. A typical practice is to make one mixture with high concentrations of all constituents of interest and then make multiple dilutions of that one mixture to create the remaining samples. Although this approach will work fine for univariate methods and is used quite frequently for least squares regression models, it will completely fail for univariate models. The main problem is that there is no interconstituent variations in the data. When the concentration of one constituent increases, they all increase and vice versa (Fig. 32). Correspondingly, the spectra will have the same problem: The spectral responses of the constituents all increase and decrease in sympathy. To a multivariate model, this appears as one constituent regardless of how many were mixed together in the original high concentration standard. To an eigenvector-based model, only one factor will arise containing nearly all the variance in the data. Any sample predicted against this model that does not have exactly the same ratio of constituent concentrations as the mixed standards will be predicted completely wrong.

Although this example is a fairly obvious one, there are cases in which the data can be collinear seemingly without good reason. A simple visual aid to identifying this potential problem is to plot the sample concentrations of each constituent in the model against the others. If the points fall on a straight line, the concentrations are collinear. If the constituents were completely uncorrelated, they would form a nice symmetric square shape. However, in most cases, they will look more like a cluster of points (Fig. 33).

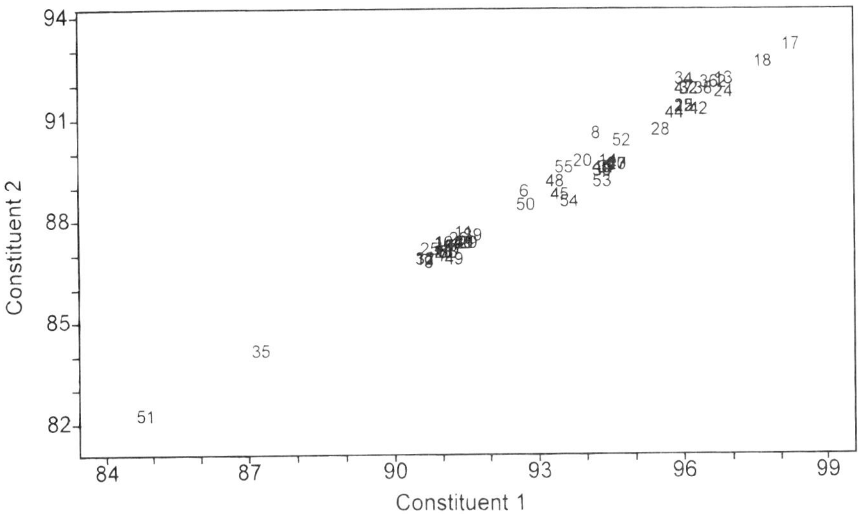

FIG. 32. This training set has constituent values that are very collinear. Notice the trend in the values for constituent 2; they increase as the corresponding values for constituent 1 increase. It will be very difficult for a multivariate model to distinguish these constituents.

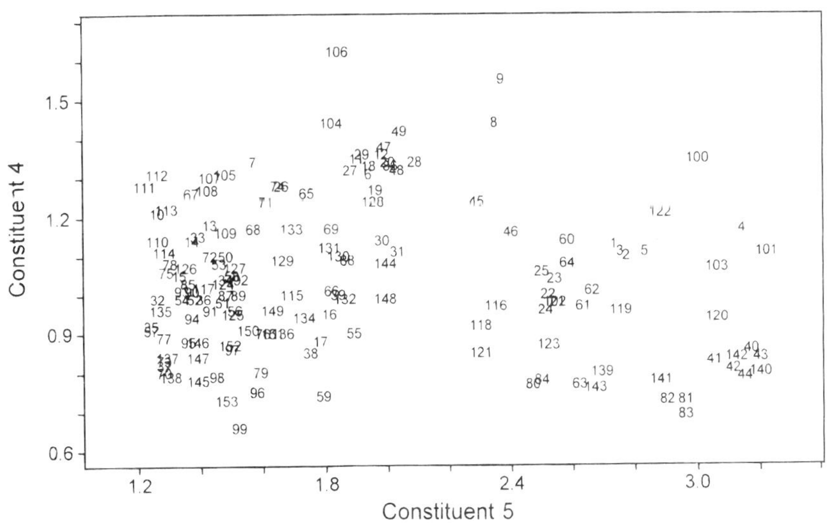

FIG. 33. Another training set with more evenly distributed constituent variations. There appears to be little to no collinearity in the values, which should lead to a better multivariate model.

SPECTROSCOPIC QUALITATIVE ANALYSIS

JAMES DUCKWORTH
Galactic Industries Corporation

There are many advantages of using spectroscopy as a detection technique for quality control of complex samples. It is fast, requires little or no sample preparation for most types of samples, and can be implemented at or near the source of the samples. However, often quantitative methods are being employed to simply gauge the suitability of the material being measured. In a significant number of cases, the only result that is desired is to know whether the sample falls within a defined range of allowed variability to determine if the material is of the desired quality. It is not always necessary to measure the quantities of the constituents in the sample to meet this goal.

The quantitative models discussed previously generally require a large number of training samples to build accurate calibrations. In turn, this requires a lot of initial work collecting all the samples and measuring the concentrations of the constituents by the primary method before the data can even be used for model building. The accuracy of the calibration is limited by the accuracy of the primary method used to get the concentration values. If the primary method is not very good, the multivariate model will not be very good either. If merely knowing that a sample is of a given quality is required and the quantity of the constituents is not needed, using a quantitative model adds a lot of extra work to simply determine if the sample is the same as the training set data.

ISBN 0-12-764070-3.

In addition, the quantities of the constituents are usually not the whole story when measuring product quality. Sometimes samples can be contaminated with other compounds and impurities. Generally, quantitative models will always predict reasonable values for the calibrated constituents, provided the spectra of the unknown samples are fairly similar to the training set. However, the reported concentrations alone will not indicate if the samples are contaminated.

In some cases, the primary quantitative information is simply not available to allow building a principal component regression (PCR) or partial least squares model. There may not be a primary calibration method available for the constituent of interest, or the samples may simply be too complex. However, the spectrum of a sample is unique to the composition of its constituents. Samples of the same or similar composition quality should have spectra that are very similar as well. Theoretically, it should be possible to tell the difference between a "good" sample and a "bad" one by comparing their spectra.

Unfortunately, the tolerances required for determining the differences between spectra in quality control applications cannot usually be met by simple methods such as visual inspection or spectral subtraction. In addition to requiring user interaction (and they are therefore subjective methods inappropriate for quality control), they cannot be easily used by anyone other than a trained spectroscopist. What is needed instead is an unambiguous mathematical method for spectral matching.

I. Discriminant Analysis

What has been described here is the basis for another area of chemometrics known as discriminant analysis. The purpose of these techniques is to classify samples into well-defined groups or categories based on a training set of similar samples without prior or with limited knowledge of the composition of the group samples. A good discriminant algorithm is one that can "learn" what the spectrum of a sample looks like by "training" it with spectra of the same material. For this reason, discriminant analysis is sometimes called pattern recognition.

The aim of discriminant analysis is to unambiguously determine the identity or quality of an unknown sample. There are two basic applications for spectroscopic discriminant analysis: sample purity/quality and sample identification/screening. In the capacity of sample quality checking, discriminant analysis methods can replace many quantitative methods currently used. In effect, the algorithm gives an indication of whether the spectrum of the "unknown" sample matches the spectra from samples taken

FIG. 1. Quality control/assurance application of spectroscopic discriminant analysis. The spectrum of the sample is compared against the model to determine if it matches the training data for the model. If the training set was constructed from spectra of samples that were of known quality, the model can accurately predict if the sample is of the same quality by matching the spectrum and giving a "yes" or "no" answer.

previously that were known to be of good quality. Some algorithms can even give statistical measurements of the quality of the match (Fig. 1).

When discriminant analysis is used in a product identification or product screening mode, the spectrum of the unknown is compared against multiple models. The algorithm will give an indication of the likelihood of the spectrum matching a model and the product can then be identified as a particular material. This mode of discriminant analysis is sometimes used for grading materials as well. For this application, each model is built from a set of samples that represent a particular grade/purity/quality of the material. When the unknown spectrum is predicted against the models, the material is classified as the closest match (or no match at all).

The analyst can control how the discrimination is calculated. Any samples in the training set become representative of the allowed form of the spectrum. For example, discriminant analysis could also be used to classify samples into chemical classes by making training sets of spectra of different compounds that share similar functional groups. As long as enough samples are used to represent the range of variability found in those types

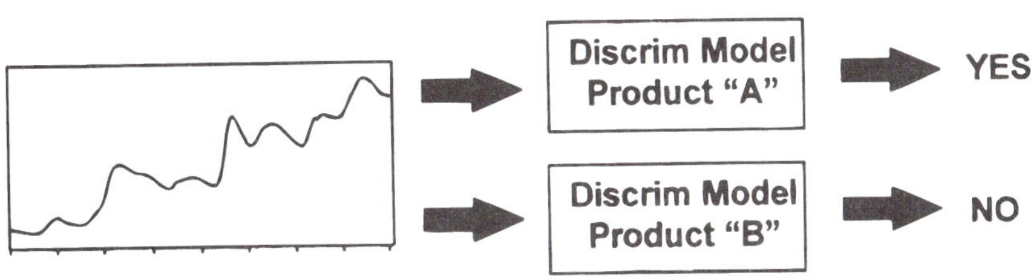

FIG. 2. Sample identification/screening application of spectroscopic discriminant analysis. The spectrum of the sample is compared to multiple models of different materials or different levels of quality of the same material. The models can predict the likelihood that the sample matches the training spectra they were constructed from, again giving a "yes" or "no" answer.

of compounds, unknowns could be chemically classified by comparing them to all the training sets and looking for a match (Fig. 2).

There are a vast number of useful analyses that can be solved by discriminant analysis. The main advantage these methods have is that they are generally easier to apply to spectroscopic analysis than quantitative methods because they do not need any primary calibration data to build the model. They give simple "pass" or "fail" answers as to how well the samples match by comparing them to training sets of the desired quality samples. They learn to recognize the spectra of materials based entirely on the spectral data itself without any other external information other than the analyst's logical grouping of the spectra into training sets.

II. Simple Spectrum Matching Methods

Many different methods have been developed for performing discriminant analysis on spectra. One class of algorithm that is already familiar to many spectroscopists is library spectral search. In these algorithms, the spectrum of an unknown sample is compared against many different spectra in a library of known compounds. By comparing the responses at all wavelengths in the unknown spectrum to the corresponding responses in a series of known (or "library") spectra, a list of the closest matches can be identified by ranking the known spectra by a calculated "hit quality index." Hit quality values close to zero indicate a very good match, whereas values near 1 indicate no match at all.

There are a variety of algorithms avaliable for this type of spectral matching. One of the most popular is the Euclidean distance method:

$$\text{Euclidean distance:} \qquad \text{HQI} = 1 - \left(\frac{\sqrt{\sum_{i=1}^{n} A_i L_i}}{\sqrt{\sum_{i=1}^{n} A_i^2} \sqrt{\sum_{i=1}^{n} L_i^2}} \right),$$

where A is the vector of the spectral responses of the unknown spectrum, L is the vector of the library or known spectrum, and n is the number of wavelengths in each vector. In order to get comparative results from the Euclidean distance method, the data must be preprocessed in a very particular way. Each spectrum must be normalized by subtracting the minimum response value from the vector and then scaling the vector by the maximum response. This will give a spectrum with responses that go from 0 to 1.

Another often used algorithm for simple spectral matching is the vector correlation method. It is similar to the Euclidean distance method but does not require that the spectra be normalized. In this method, each spectrum is centered around the mean response value to calculate the hit quality index.

Mean responses: $\bar{A} = \frac{(\sum_{i=1}^{n} A_i)}{n}, \quad \bar{L} = \frac{(\sum_{i=1}^{n} L_i)}{n}$

Correlation: $\text{HQI} = 1 - \left(\frac{[\sum_{i=1}^{n} (A_i - \bar{A})(L_i - \bar{L})]^2}{[\sum_{i=1}^{n} (A_i - \bar{A})^2][\sum_{i=1}^{n} (L_i - \bar{L})^2]} \right),$

where all variables are the same as those in the Euclidean distance calculation, with the addition of the mean values of the unknown spectral responses, $\bar{A}$, and the mean of the library spectral responses $\bar{L}$ (Fig. 3).

Many commercially available library search programs use these techniques to generate a list of the most likely matches of the unknown sample. However, there are many problems with this technique. First, search techniques simply identify samples as the materials from the closest matching spectrum in the library. If the library does not contain any spectra of the "true" compound, it will just report the best match it found regardless of whether it is really even the same class of material.

In addition, the spectral library search technique is sensitive only to the general spectral shapes and patterns and not to very subtle variations

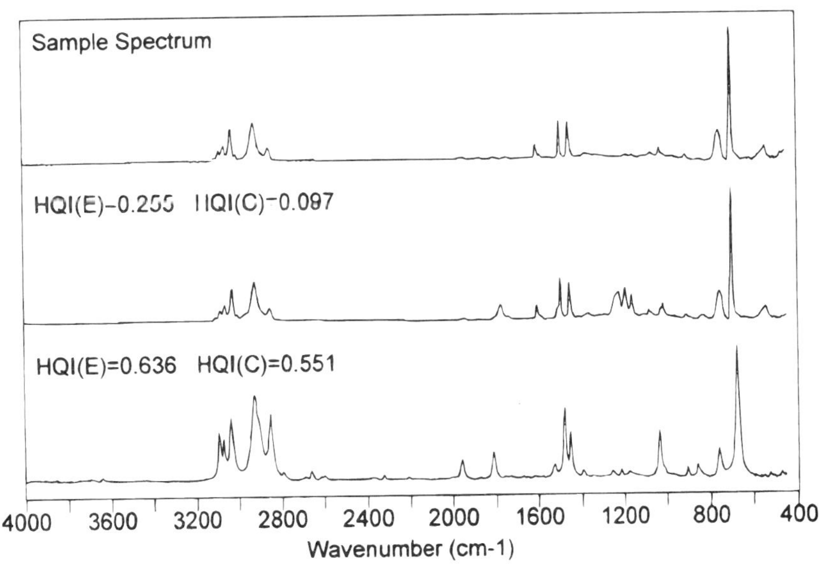

FIG. 3. Sample spectrum (top) was compared against two spectra using both Euclidean distance [HQI(E)] and correlation [HQI(C)] methods to determine a match.

within the sample. If the variations in between the spectra of a good sample and bad sample cannot easily be seen by visual inspection, chances are a spectral library search will not be able to do it either. The Euclidean distance search algorithm cannot be trained to recognize a range of variability in the data because the spectrum of the unknown is compared only to a single representative spectrum for each different class of material.

Another problem is that the spectra must have flat baselines in order for these methods to work properly. As seen earlier in the discussions of preprocessing methods for quantitative spectroscopy, there are methods for accomplishing this with little or no user interaction. However, these methods are very sensitive to baseline instabilities, and the correction applied must be very good to have any degree of success with these methods.

Finally, the hit quality index does not provide any absolute measure of the probability that the sample actually is the same as the library sample. The arbitrary scale of the hit quality values (0–1) does not give a very good statistical measure of the similarity of the spectra. In short, using only a single training spectrum to represent all possible samples in the future does not give the analyst any statistical assurance that the spectra are truly the same or different. It provides only a relative measure for all the library samples. For anyone who has tried simple library search techniques for spectrally similar samples, this result is all too obvious.

III. Other Methods

There have been many other methods put forth in the literature including K-nearest neighbor (Ref. 42), cluster analysis (Ref. 11), principal component analysis (PCA) factorial discriminant analysis (Refs. 43, 44), SIMCA (Ref. 36), and BEAST (Refs. 37–40).

Many of these methods use PCA as a spectral data compression technique. As discussed in Chapter 4, PCA produces a reduced representation of the training set spectra based on the variations between the samples (refer to Chapter 4, Eigenvector Quantitation Methods section), Remember that PCA decomposes a set of spectra into their most common variations (factors) and produces a small set of well-defined numbers (scores) for each sample that represent the amount of each variation present in the spectrum. Similar to using the scores for creating a quantitative calibration equation, they can also be used for discrimination because they provide an accurate description of the entire set of training spectra. However, many of the methods listed previously utilize only the first few significant factors for discrimination. In many cases, only the first two factors are used. Thus, only a limited portion of the total spectral information available for a class of material is actually used; the rest is simply discarded.

Despite all the methods available, the remainder of this chapter will focus on a method that uses a technique called the Mahalanobis distance to measure the spectral similarity.

IV. The Mahalanobis Distance Method

The actual mathematics of the Mahalanobis distance calculation has been known for some time. In fact, this method has been applied successfully for spectral discrimination in a number of cases (Refs. 33–36). One of the main reasons the Mahalanobis distance method was chosen is that it is very sensitive to intervariable changes in the calibration data. In addition, the distance is measured in terms of standard deviations from the mean of the training samples. Not only does the calculation give a very sensitive discrimination but also the reported matching values give a statistical measure of how well the spectrum of the unknown sample matches (or does not match) the original training spectra.

The mathematical basis behind the Mahalanobis distance measurement is really quite simple, but it is much easier to understand when it is explained graphically. Consider a set of spectra of different samples of the same material as shown in Fig. 4.

When measuring samples of the same material, it is expected that the spectra will be very similar to one another. However, no two spectra will be exactly the same. Each spectrum would be slightly different due to spectrometer drift, differences in sample handling, changing environmental conditions such as humidity, as well as batch to batch variations in the sample material.

However, because the spectra are all of the same material, the relative intensities at all the wavelengths should remain approximately the same. The peaks will tend to rise and fall together throughout the entire spectrum. To demonstrate this, measure a series of spectra of different samples of the same material to form a training set. Then, select any two wavelengths in the spectrum (preferably at or near the tops of two major bands) and plot the responses at the first wavelength versus the responses at the second. What should arise is a plot similar to that shown in Fig. 5.

Notice that the points tend to form an elliptical cluster indicating the subtle differences between the spectra (in terms of baseline shift, path length, and concentration). The mean position of the cluster is unique to the particular material of interest because the intensities at these two wavelengths

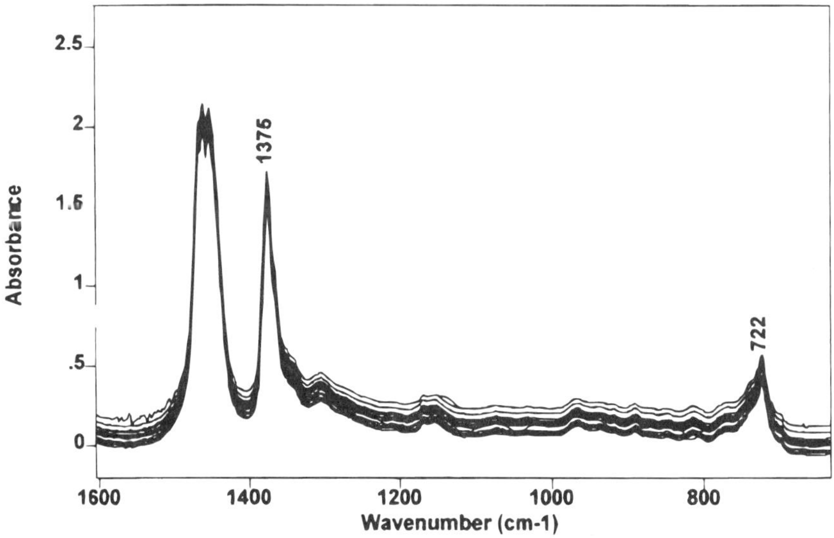

FIG. 4. Multiple spectra of the same compound collected on different dates using the same Fourier transform infrared spectrometer. Notice the subtle variations in the spectral bands and the large variations in baseline.

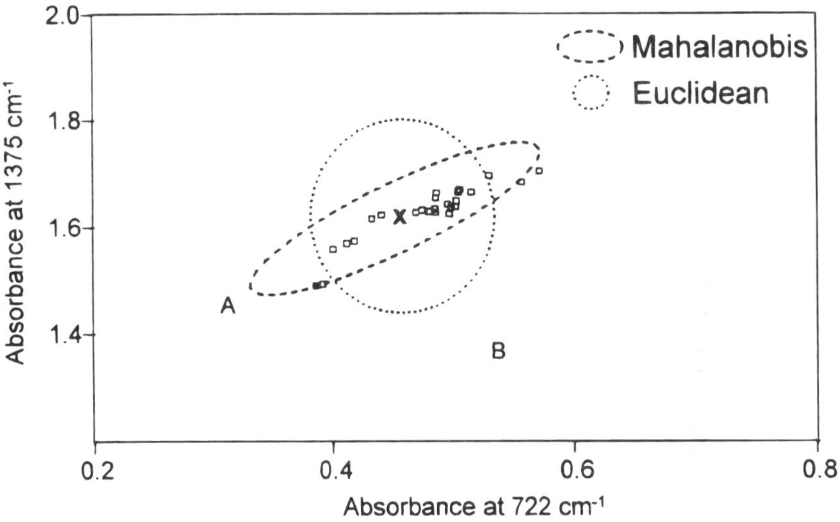

FIG. 5. Absorbance of two selected wavelengths plotted against each other. The values are from multiple spectra of the compound shown in Fig. 4. The mean (X) point of the group should be unique for the material. The elliptical cluster shape formed by the group points is typical of spectra from the same material, with most samples clustered near the mean.

would be different for a different material. This theory can be tested by taking a series of spectra of a different compound and plotting the intensities at the same wavelengths together with the previous data. An additional group of points would form on the graph but would be centered at a different position. Each of these groups is unique to the particular material that created it. This gives a very simple method of determining the similarity of an unknown spectrum to one of the groups. Just measure the spectrum of the unknown, plot the intensities of the selected wavelengths, and see if the point falls "near" the mean point of one of the groups. If the unknown point is close enough, the sample can be classified as being the same material. If the point is far away, it does not match, and the sample may be a different material or may not have the same purity as the training set data.

The approach seems relatively straightforward, but how is the concept of being near a group actually defined? As mentioned earlier, visual inspection is not a viable method for real-world discriminant analysis applications. What is needed is a mathematical equation to measure "nearness" of the unknown point to the mean point of the group(s).

One such measurement technique has already been mentioned: the Euclidean distance. As with using the responses at all wavelengths in a spectrum to perform a simple single spectrum match, the same formula can be applied to calculating the distance of the unknown point to the group mean point. In this case, the vectors would have only two data points each: the selected wavelengths. This would be a fine method except for two simple facts. First, as mentioned previously, the Euclidean distance does not give any statistical measurement of how well the unknown matches the training set. In addition, the Euclidean distance measures only a relative distance from the mean point in the group. It does not take into account the distribution of the points in the group.

For reference, an example Euclidean boundary has been superimposed on the group points in Fig. 5. In addition, two hypothetical unknown sample points A and B have been added. Notice that although the training set group points tend to form an elliptical shape, the Euclidean distance describes a circular boundary around the mean point. By the Euclidean distance method, sample B is just as likely to be classified as belonging to the group as sample A. However, sample A clearly lies along the elongated axis of the group points, indicating that the selected wavelengths in the spectrum are behaving much more like the training group than those same wavelengths in the spectrum of sample B. Clearly, the Euclidean distance method does not take into account the variability of the values in all dimensions and is therefore not an optimum disciminant analysis algorithm for this case.

The Mahalanobis distance, however, does take the sample variability into account. Instead of treating all values equally when calculating the distance from the mean point, it weights the differences by the range of variability in the direction of the sample point. Refer to the Mahalanobis boundary that has been superimposed on Fig. 5 and this concept becomes much clearer. The Mahalanobis distance constructs a space that weighs the variation in the sample along the axis of elongation less than that in the shorter axis of the group ellipse. In terms of Mahalanobis measurements, sample A will have a substantially smaller distance to the mean than sample B because it lies along the axis of the group that has the largest variability. Therefore, sample A is far more likely to be classified as the same material as the group. Mahalanobis distances look at not only variations (variance) between the responses at the same wavelengths but also interwavelength variations (covariance). The Mahalanobis group defines a multidimensional space whose boundaries determine the range of variation that are acceptable for unknown samples to be classified as members.

Another advantage of using the Mahalanobis measurement for discrimination is that the distances are calculated in units of standard deviation from the group mean. Therefore, the calculated circumscribing ellipse formed around the cluster actually defines the one sigma (σ) or one standard deviation boundary of that group. This allows the analyst to assign a statistical probability to that measurement. In theory, samples that have a Mahalanobis distance of 3σ or greater have a probability of 0.01 or less and can be classified as nonmembers of the group. Samples that have distances less than 3 are then classified as members. In practice, others have found a Mahalanobis distance of 10–15 works better as a maximum variance for classification (Ref. 33). However, the determination of the cutoff value depends on the application and the type of samples.

Just like many multivariate quantitative methods, the Mahalanobis distance can solve for multiple dimensions simultaneously. The Mahalanobis group can therefore be extended to more than two dimensions by simply selecting more wavelengths. This is generally a good idea because this will attempt to compensate for variations in other regions of the spectrum.

Calculation of the Mahalanobis equation is relatively straightforward. It starts by calculating the mean point of the selected wavelengths:

$$\bar{A}_j = \frac{(\sum_{i=1}^{n} A_{i,j})}{n},$$

where A is an $n \times p$ matrix of the spectral responses for the training set data, $\bar{A}$ is a $1 \times p$ vector of the average responses, n is the number of samples, and p is the number of wavelengths selected. The j subscript indicates the wavelength index. The spectral responses of each individual training sample are then mean centered:

$$A_{j(\mathrm{MC})} = A_j - \bar{A},$$

and the centered data are then used to calculate the final Mahalanobis matrix:

$$M = \left(\frac{A'_{(\mathrm{MC})} A_{(\mathrm{MC})}}{(n-1)} \right),$$

where M is the $p \times p$ Mahalanobis matrix.

To predict the Mahalaonbis distance of an unknown sample, the responses at the same wavelengths are placed in a $1 \times p$ vector A_{unk} and then used in the following equation:

$$D^2 = (A_{\mathrm{unk}} - \bar{A})M^{-1}(A_{\mathrm{unk}} - \bar{A})',$$

where D^2 is the square of the Mahalanobis distance (in terms of standard deviations) of the spectrum A_{unk} from the mean of the training set.

Unfortunately, this method is not perfect. In fact, there are a number of drawbacks. As with the inverse least squares (ILS) quantitative analysis method, this approach to discriminant analysis relies on selecting a subset of wavelengths to represent the entire spectrum. Again, if any impurities or aberrations appear in the spectra of the unknowns that do not appear at the selected wavelengths, the discriminant analysis will determine that the sample matches the group, when in fact it does not!

A simple solution would appear to be simply to select more wavelengths. For that matter, why not select every data point in the spectrum? The reason is that the Mahalanobis model tends to become overfit very quickly as more wavelengths are added. This is logical only when the method of calculating Mahalanobis matrix is considered. Because all the inter-wavelength variations are considered just as important as matching the corresponding wavelengths, the likelihood of an unknown sample having the same relative intensity values at all selected wavelengths decreases substantially (Ref. 34). In the worst case, using too many wavelengths can cause good samples to be misclassified as not in the group. In practice, using more than approximately 10–15 wavelengths can lead to misclassification of known samples. In other words, samples that should be classified as members of that group are rejected as nonmembers. There have been procedures put forth for optimum wavelength selection (Ref. 33) based on all the groups used for comparison. However, this can be a time-consuming and computationally intensive process.

In order to ensure that all impurities and other anomalies in the unknowns are detected, the discrimination method needs to be able to use the entire spectrum, or at least large regions, instead of selected wavelengths. However, spectra are usually collected with many data points and certainly more than the 10–15 variable limit of the Mahalanobis distance method. So how is it possible to combine these apparently opposing necessities into one method for spectral discrimination? As with quantitative analysis methods, the answer lies in reducing the spectral data into its component variations with principal component analysis.

V. Coupling Mahalanobis Distance with PCA

As mentioned in Chapter 4, PCA is a very effective data-reduction technique for spectroscopic data. To review, PCA decomposes the training set spectra into mathematical spectra (called loading vectors, factors, principal components, etc.) that represent the most common variations to all the data. A set of scaling coefficients (scores) for each factor can be calculated for every spectrum in the training set. When the scores are multiplied by the loading vectors, and the results summed, the original spectra are reconstructed. By knowing the set of loading vectors, the scores will represent the spectra as accurately as the original responses at all the wavelengths.

Replacing the selected wavelengths in the ILS quantitative method with PCA scores created a new method called PCR. It avoided the problem of overfitting by selecting too many wavelengths and allowed for full spectral coverage, thus circumventing two of the major drawbacks of the ILS method.

Similarly, replacing the selected wavelengths in the Mahalanobis distance equation with PCA scores gives this discriminant analysis method the same advantages. Again, using PCA as a data-reduction technique allows full spectral coverage for all samples and alleviates the need for wavelength optimization. Because PCA also reduces data into a smaller set of representative numbers (known as scores), the problem of overdiscrimination can be avoided while still using entire spectra or spectral ranges. The Mahalanobis matrix equation then becomes

$$M = \left(\frac{S'S}{(n-1)}\right),$$

where M is an $f \times f$ Mahalanobis matrix, S is the $n \times f$ matrix of training sample PCA scores, n is the number of samples, and f is the number of PCA factors. The prediction equation becomes

$$D^2 = (S_{\text{unk}})M^{-1}(S_{\text{unk}})',$$

where D^2 is the square of the Mahalanobis distance (in terms of standard deviations) of the spectrum represented by the scores S_{unk} of the unknown sample spectrum. For a detailed description of the PCA algorithm and a more in-depth discussion of how it works, please refer to the Eigenvector Quantitation Methods section of Chapter 4.

A. Improving the Sensitivity

There is one problem with basing the discrimination purely on the PCA factor scores. It is that any "impurities" or extra aberrations that are in the unknown spectra, but were not present in the training samples, will not appear in the scores calculations. As long as the rest of the variations in the spectrum are consistent with the training set, the model will predict the sample as a match. Therefore, the information in the unknown spectrum that is not compensated by the PCA factors must be considered in order to make an accurate assessment of the match.

As mentioned previously, one incidental advantage of PCA is that a spectrum can be mathematically "reconstructed" by multiplying the spectrum scores by the set of primary factors and summing the results. The reconstructed spectrum can be subtracted from the original spectrum to determine how well the PCA model is performing for the sample. The result of this subtraction is known as the spectral residual. If the spectral residual is a relatively flat line (or just noise) near zero, then the model is able to account for all the variations in the spectrum. However, if the spectral residual has additional peaks or significantly more noise than expected, then the model is not completely predicting the information in the sample.

By calculating the sum of the squares of the spectral residuals across all the wavelengths, an additional representative value can be generated for each spectrum. The spectral residual is effectively a measure of the amount of each spectrum left over in the secondary or "noise" vectors. This value is the basis of another type of discrimination method known as SIMCA (Refs. 13, 36). This is similar to performing an F test on the spectral residual to determine outliers in a training set (see Outlier Sample Detection in Chapter 4). In fact, one group combined the PCA–Mahalanobis distance method with SIMCA to provide a biparametric method of discriminant analysis (Ref. 41). In this method, both the Mahalanobis distance and the SIMCA test on the spectral residual had to pass in order for a sample to be classified as a match.

However, another approach is to combine the PCA scores and spectral residuals for each spectrum and use them all for the Mahalanobis group matrix calculations. These values are effectively included in the Mahalanobis group calculations by adding an extra "score" for each spectrum that contains the sum squared spectral residuals for each spectrum. Therefore, the relationship between the PCA scores and the spectral residual is considered when a sample is predicted against the Mahalanobis matrix.

Including the sum squared spectral residual as an additional discriminating factor for the Mahalanobis group is an important extra step that improves the sensitivity of the unknown sample classifications. It not only sets the maximum allowed variation in the factors but also limits the range of variation in the residual for a sample to be classified as a member of the group. This is particularly important in quality control applications in which it is important not only to verify the identity of a material but also to determine if it contains substantial impurities different from those in the training data.

For an example, refer to Figs. 6–8. The sample spectra in these figures were predicted against Mahalanobis discrimination models built using the three methods discussed so far: selected wavelengths, PCA scores, and PCA scores with spectral residuals. Notice that the wavelength method accurately predicts the sample that is supposed to match ("in spec") and the sample that is not a match ("out spec"). The PCA scores-only method performs similarly. However, both methods misclassify the "contaminated" sample as a match. The PCA scores with spectral residuals method accu-

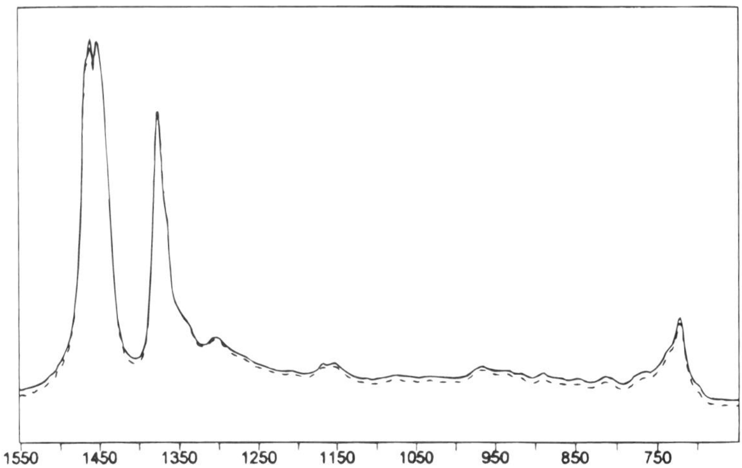

FIG. 6. Spectra of samples for discrimination against the training set in Fig. 4. The solid line is a spectrum of the same material "in spec"), whereas the dotted line is a spectrum of a very similar but not the same material ("out spec"). The spectra have not been baseline corrected and actually are nearly the same after the baseline is removed.

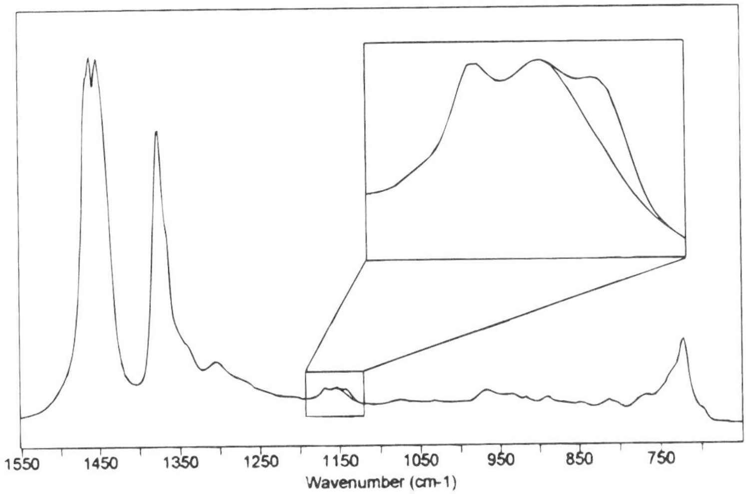

FIG. 7. Spectra of a sample of the same material as that of the training set shown in Fig. 4. One sample is the same in spec spectrum from Fig. 6, whereas the other is contaminated with a very small band at approximately 1140 cm^{-1} ("contaminated").

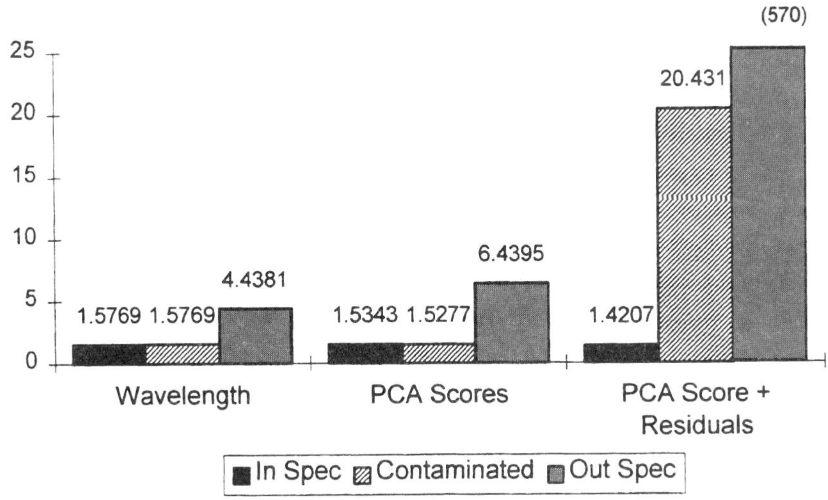

FIG. 8. Predicted Mahalanobis distances of the spectra shown in Fig. 6 and Fig. 7 by the three methods discussed in the text. The in spec values are from the good sample, the contaminated values are from the sample with the extra band, and the out spec values are for the sample of the similar but not same material.

rately predicts all samples. In addition, the sensitivity to the out spec sample is drastically increased. This indicated that the material is spectrally very similar to the group samples; however, there is some extra information left in the spectral residual that helps discrimination.

Another method used to improve the sensitivity of the analysis is to normalize the predicted distances by the root mean squared group size (RMSG) (Ref. 33). Depending on the number of training samples and the variations in the sampling technique, the points that form the Mahalanobis ellipse may be "scattered" more for one group than another. This causes each group to have a different "size" in the multidimensional Mahalanobis space. In order for the Mahalanobis distance from one group to be compared to the distance from another group, this size difference must be removed. This is easy to understand by remembering the fact that Mahalanobis distances are measured in terms of standard deviations. If the standard deviations for the two groups are different, no distance cutoff value can be determined that will effectively work for both groups. The RMSG is simply determined by calculating the root mean square of the predicted distances for every training sample against the whole training set. Normalizing the predicted sample distances by the RMSG causes all groups to have the same weight in the predictions.

Finally, one of the best ways to improve the sensitivity of the discrimination is to create separate training sets for each group. Some methods in the literature pool all samples together for calculating the PCA and then try to calculate a Mahalanobis matrix for the subsets of scores that belong to individual classes of materials (Refs. 33, 43, 44). This has the advantage of performing all calculations in the same score space and requires only one PCA to calculate all the factors. However, it has some serious disadvantages as well. Certainly, creating a new training set for every material or group requires collecting a lot more spectra. On the other hand, using separate training sets creates PCA factors that are unique to each group used for classification. Because the factors are going to be slightly different for each group (even for very spectrally similar materials), the likelihood of group overlap and, thus, misclassification is reduced. In addition, because each group is calculated separately, inclusion of a new classification group in the analysis does not require reoptimizing the entire model for all groups; a new model is simply built for the new group.

1. *Calculating Mahalanobis Distance by PCA with Residuals*

The following information is provided for those who are interested in the complete calculations of the (Mahalanobis distance by PCA with residuals method). This discussion assumes that the training data have already been reduced to the component PCA factors and scores. For a complete description of the algorithms available for PCA, refer to the Calculating PCA and PCR section of Chapter 4.

The basic model for PCA of the spectral data matrix is

$$A = SF + E,$$

where A is the $n \times p$ matrix of training spectra, S is an $n \times f$ matrix of the scores, F is an $f \times p$ matrix of the PCA factors, and E is an $n \times p$ matrix of the residual error not modeled by PCA. The dimensions are n for the number of spectra, p for the number of wavelengths, and f for the number of PCA factors.

The spectral residual is therefore

$$E = A - SF.$$

The sum squared residual for each row of the spectral residual matrix is then calculated as

$$R_i = \sum_{j=1}^{p} E_{i,j}^2,$$

where R_i is the sum squared residual for the spectrum number i. This vector of residuals is then mean centered and then appended as an extra column to the PCA scores matrix:

$$\text{Mean center residuals:} \qquad R_{ci} = R_i - \left(\frac{\sum_{j=1}^{n} R_j}{n}\right)$$

$$\text{Append to scores:} \qquad S_r = \begin{vmatrix} S_{1,1} & \cdots & S_{1,f} & R_{c1} \\ \vdots & \ddots & \vdots & \vdots \\ S_{n,1} & \cdots & S_{n,f} & R_{cn} \end{vmatrix},$$

where S_r is the new $n \times f + 1$ residual augmented scores matrix. The calculation of the Mahalanobis matrix is then done on this matrix rather than just the PCA scores matrix:

$$M = \left(\frac{S_r' S_r}{(n-1)}\right).$$

The root mean squared group size in then calculated so that predicted samples can be normalized. This first requires predicting all training samples against the Mahalanobis matrix:

$$D_i^2 = (S_{r_i}) M^{-1} (S_{r_i})',$$

where D_i is the predicted distance of training sample number i. The predicted distances are then used to calculate the RMSG normalization factor:

$$\mathrm{RMSG}^2 = \frac{\sum_{i=1}^{n} D_i^2}{(n-1)}.$$

Whenever an unknown sample is predicted, effectively the same set of steps are used; however, the mean of the training group residuals is subtracted from the residual of the unknown, and the predicted distance is divided by RMSG before reporting the value.

B. Determining the Number of Factors for the Model

One of the biggest problems in using PCA spectral decomposition for discriminant analysis is identifying the correct number of factors to use for the models. In the case of quantitative analysis methods, there is always a set of secondary benchmarks to compare the quality of the model: the primary calibration data. By performing a prediction residual error sum of squares (PRESS) analysis, it is very easy to determine the number of factors by calculating the prediction error of the constituent values at every factor. The smaller the error, the better the model.

However, with discriminant analysis methods, the only information available is the set of training spectra. There is no external set of data to compare the model's predictive capabilities. The basic problem therefore becomes how to include enough factors in the model that will give the appropriate discrimination without including extra noise factors that are unique to the training set data and are not likely to be found in the spectra of unknown samples. The set of vectors that represent the true variations in the data are often called the primary factors, whereas the remaining noise vectors are known as the set of secondary factors.

1. *PCA Eigenvalue Methods*

Fortunately, there are ways of determining the number of factors by looking at the igenvalues of the PCA factors. One fact that has been left out in all the discussions of PCA is that the data are not broken down into just two sets on values (scores and factors) but rather into three. The third set of values is the eigenvalues. Due to the way the PCA decomposition is calculated, the scores and factors generally span span a data range of ± 1. If the scores and factors were the only representation of the data, all the principal component spectra would have the same relative intensities in the samples. Clearly, this is not the case; some components will vary larger than others.

Therefore, the eigenvalues are actually a measure of the importance of each factor to reconstructing the real spectra.

Typically, the eigenvalues of the first few factors are much larger than those of the remaining factors. The trick is to determine the relative importance of each factor in the model by comparing the PCA eigenvalues. There are a number of methods in the literature. This discussion will focus on some simple views of the eigenvalue data.

Simply looking at a plot of the eigenvalues might lead to a model that has too few factors. Empirically, from examination of the eigenvalue plot in Fig. 9, it appears that a model with four factors would probably work fine because the values seem to be very small from this factor on up. However, in actuality, this would build a model that is significantly underfit for this data set.

Instead, methods that use the eigenvalues to calculate some comparative statistics are needed. One such method is to calculate the total percentage variance for each factor. When the maximum number of factors is calculated (which is equal to the total number of samples in the training set), all the variance in the data is accounted for, or 100%. Because the PCA factors

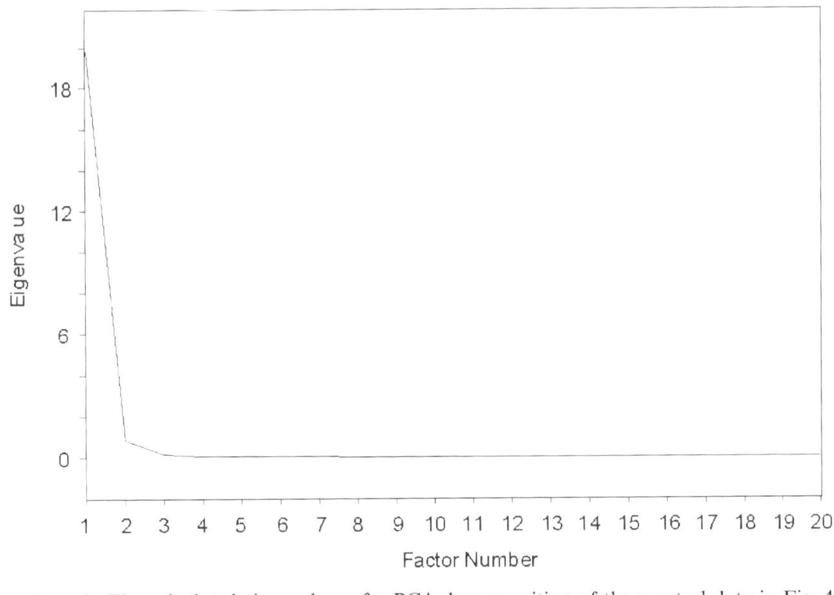

FIG. 9. The calculated eigenvalues of a PCA decomposition of the spectral data in Fig. 4. Notice that the first eigenvalue is substantially larger than the rest, and that the values fall off rapidly at the higher numbered factors.

represent the variations in the data, and the eigenvalues are the relative weights of each of the factors, the eigenvalues can also be thought of as the amount of variance in the data that is represented by that factor. By summing the eigenvalues, an estimate can be made of how much variance is accounted for by the PCA factors:

$$\mathrm{TPV}_i = \frac{\sum_{j=1}^{i} \lambda_j)}{(\sum_{j=1}^{f} \lambda_j)},$$

where TPV_i is the total percentage variance accounted for by a model with i PCA factors, λ (is the $1 \times f$ vector of eigenvalues, and f is the total number of factors for the data set.

When looking at a plot of the total percentage variance versus the number of factors, a model that accounts for at least 99.9% of the total variance will generally give good predictive ability for similar samples (Fig. 10). However, this is not a hard and fast rule. Depending on the nature of the variations in the data, some models will work with less total variance and some with more.

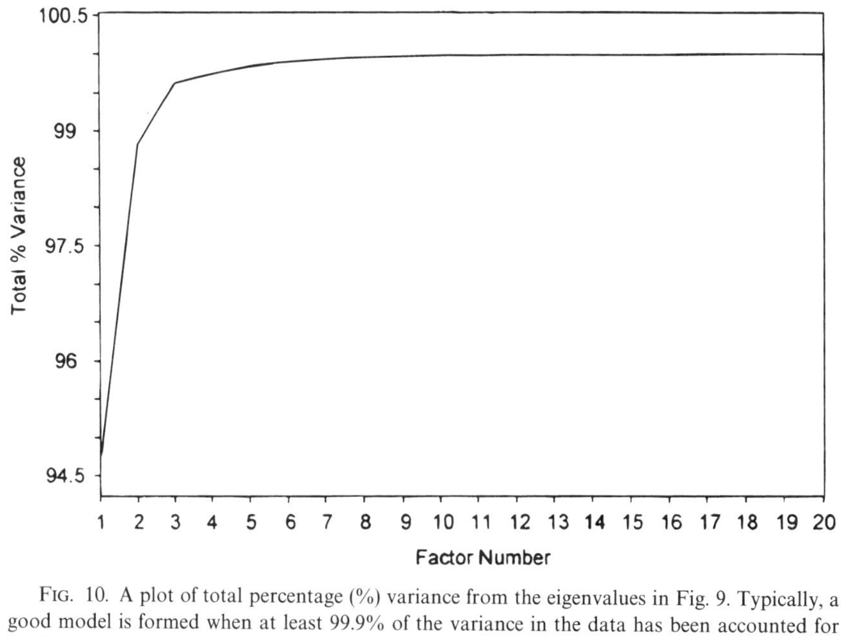

FIG. 10. A plot of total percentage (%) variance from the eigenvalues in Fig. 9. Typically, a good model is formed when at least 99.9% of the variance in the data has been accounted for in the factors. For this data set, this occurs at 7 factors, although it is difficult to see in this plot.

Another method is through the use of "indicator functions." Similar to a PRESS for quantitative analyses, these functions will usually give a minimum at the optimum number of primary factors or they will show a "leveling" once the optimum has been reached. One of the more useful functions is called Malinowski's indicator (Refs. 8–10) (Fig. 11). Much like the total percentage variance, it is calculated from the PCA eigenvalues:

$$\mathrm{RE}_i = \frac{\sqrt{\sum_{j=i}^{f} \lambda_j}}{(d_1 - 1)(d_2 - i)}$$

$$\mathrm{MI}_i = \frac{\mathrm{RE}_{i-1}}{(d_2 - i)^2},$$

where RE_i is the real error function at factor number i, MI_i is the value of Malinowski's indicator function at factor number i, λ is the $1 \times f$ vector of eigenvalues, and f is the total number of factors for the data set. The d_1 and d_2 values are the dimensions of the original spectral data matrix used for the PCA. The d_1 value is the smaller of n, the number of samples, and p, the number of spectral data points selected, and d_2 is the larger.

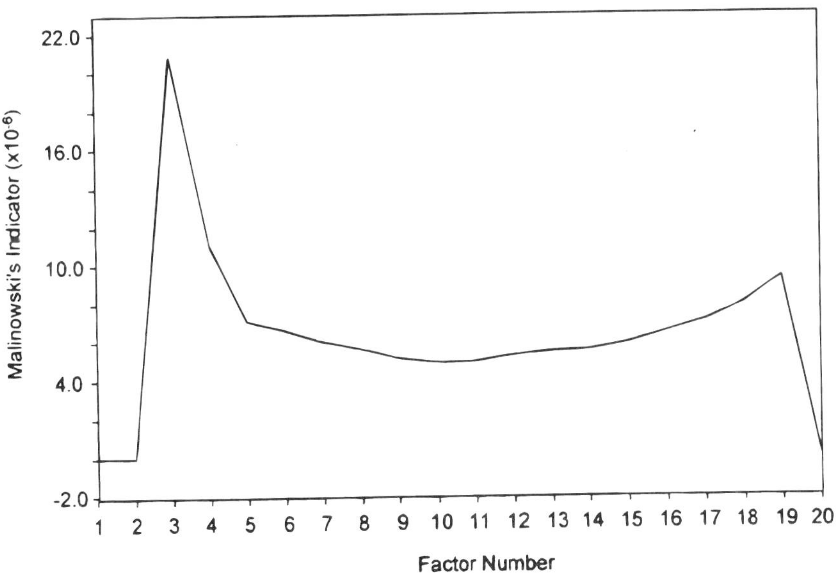

FIG. 11. Plot of Malinowski's indicator function for the eigenvalues in Fig. 9. Notice that the minimum occurs at 10 factors. This is 3 more factors than indicated by the total percentage variance plot in Fig. 10.

Although indicator functions such as Malinowski's are useful in helping determine the optimum number of factors, they generally tend to create models that are overfit when used for discriminant analysis.

Another method [which was also proposed by Malinowski (Ref. 9)] is to calculate the statistical significance of the factors by performing an F test on the eigenvalues. This method is very similar to the f test on the F ratio values from a PRESS analysis for optimizing a quantitative model. The F test is not actually applied directly to the eigenvalues but rather to what Malinowski called the reduced eigenvalue (REV):

$$\mathrm{REV}_i = \frac{\lambda_i}{(n-i-1)(p-i-1)},$$

where REV_i is the reduced eigenvalue at factor number i, λ is the $1 \times f$ vector of eigenvalues, f is the total number of factors, n is the number of samples in the training set, and p is the number of spectral data points selected. Effectively, the data set eigenvalues are normalized by the degrees of freedom in the data set to arrive at REV.

To arrive at the optimum number of factors, the F ratio and F test calculations are then performed on REV (Table 1). (See Selecting the Factors Based on PRESS section in Chapter 4 for details on this calculation.) However, in this case, the significance level is set at $\alpha = 0.01$. Therefore, all factors with a probability greater than or equal to 0.99 are maintained as primary factors, and the remaining factors are assumed to be noise or the set of secondary factors.

2. *Cross-Validation Distance Prediction*

Another approach is to perform a cross-validation much like that for quantitative models. However, rather than predicting the constituent values as each sample is rotated out (and there are none to predict in discriminant analysis anyway), the Mahalanobis distance of each sample is predicted at every factor. The cross-validation procedure is basically the same; remove a sample or set of samples, construct a Mahalanobis matrix for one factor, two factors, etc., and then predict the sample(s) left out against it. The samples are then returned to the training set, and a new set is removed. The process is continued until every sample has been rotated out once.

Remember that the Mahalanobis distance is normally distributed and measures the distances in terms of standard deviations from the group mean point. Therefore, good samples should be at least 3 Mahalanobis distances away or less to be classified as a member of the group. Assuming that all the samples in the training set are good (no outliers), then all samples should give a predicted distance of 3 or less when rotated out. It

TABLE 1
F RATIO VALUES ON THE CALCULATED REV FOR THE EIGENVALUES IN FIG. 9[a]

Factor	REV	*F* ratio	*F* test
1	0.001623	49988.69	1.0
2	7.26E-05	2234.558	1.0
3	1.49E-05	459.1997	1.0
4	2.29E-06	70.4749	1.0
5	2.01E-06	61.80122	1.0
6	1.14E-06	35.28593	0.999997
7	8.88E-07	27.34555	0.999988
8	5.26E-07	16.1824	0.999798
9	2.96E-07	9.119959	0.997072
10	1.86E-07	5.729409	0.984148
11	1.80E-07	5.547444	0.982481
12	1.83E-07	5.638769	0.98334
13	1.15E-07	3.549242	0.943741
14	8.34E-08	2.567879	0.895077
15	7.77E-08	2.393017	0.882256
16	5.76E-08	1.772096	0.820263
17	4.35E-08	1.33844	0.754295
18	3.25E-08	1.0	0.681533
19	3.41E-08		
20	3.47E-08		

[a] The *F* test probabilities were calculated using the following degrees of freedom: Numerator = 1, Denominator = (No. samples − 1) × (No. data points − 1). This would indicate that nine factors are appropriate for this model.

should therefore be possible to determine the number of factors by calculating the average predicted distance for all the samples at each factor and locating the point at which the value goes above 3 (Fig. 12).

C. OUTLIER SAMPLE DETECTION

As with quantitative models, outlier samples in the training set can have an unwanted influence on the discrimination ability of the model. Many of the same techniques (spectral residual plots and cluster plots) used in quantitative models can also be used to check for outliers in discriminant models. However, keep in mind the purpose of the discriminant analysis experiment: to build a model that can accurately match a spectrum to the training group but allow enough variation in the model to compensate for the natural variations seen in real samples.

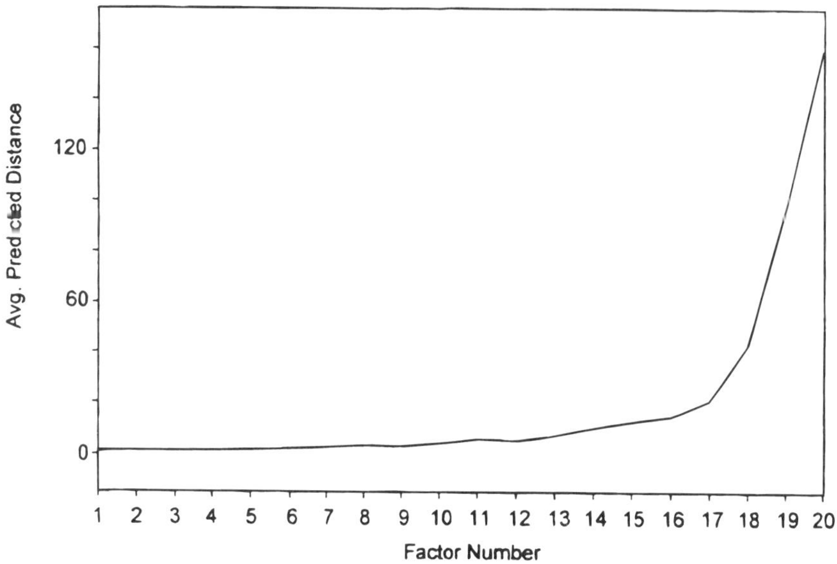

FIG. 12. Average predicted Mahalanobis distance for a cross-validation of the spectra in Fig. 4. The plot reaches a value of 3.3 distances at 8 factors. However, the value at 9 factors is actually 3.0. An argument could be made for either number being correct.

Any data that are included in the training set in effect become part of that allowed variation. If there are a few spectra of samples that are substantially different from the rest (which will tend to cluster tightly around the group mean), then they may appear as outliers by many of the statistical tests. However, if these samples are known to be good, then they should not be removed from the training set just because they fail the statistical tests.

There is one additional method to use in determining outliers in discriminant analysis models: to look at a plot of the predicted Mahalanobis distances (either from a cross-validation or self-prediction) to see if any samples stand out (Fig. 13).

D. Spectral Region Selection

One of the most difficult tasks is deciding what spectral region(s) to include in the analysis. Generally, there are no proven guidelines for choosing the best regions and, for the most part, entire spectra work well. Remember that discriminant analysis is effectively a spectrum-matching technique. Any information that is included in the training set spectra will be considered an "allowed" variation for unknowns.

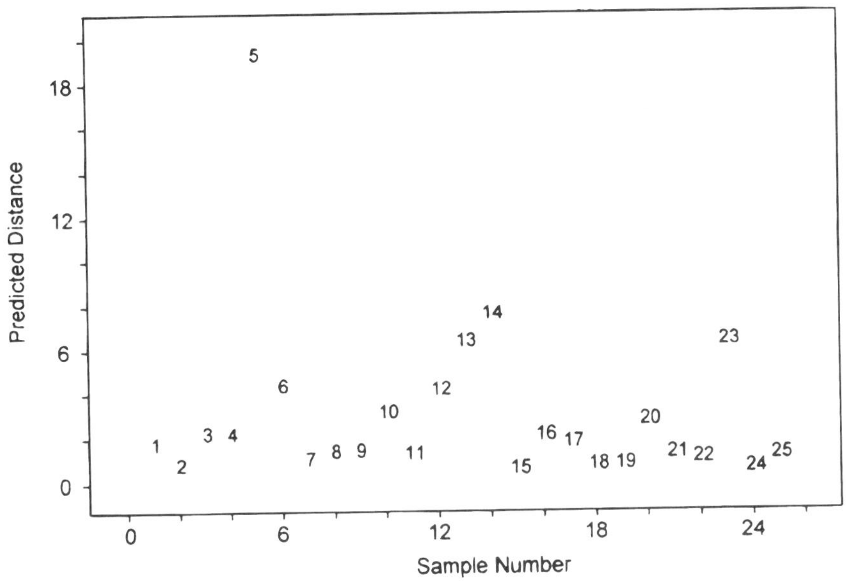

FIG. 13. Predicted Mahalanobis distances from a cross-validation of the training spectra in Fig. 4. The model was created with nine factors. Notice that sample No. 5 appears to be substantially different from the rest. However, upon further examination of the data, the spectrum of this sample had the largest baseline shift. Most likely the sample is fine and was left in the training set for building the final model.

There are, however, some general rules on regions to exclude. Any region that will not be consistent in future spectra should be removed from the data before performing the model calculations. Examples of some typical regions to exclude are regions below the detector/optical cutoff point, very strongly absorbing regions where nonlinearity in the spectrometer response might occur (i.e., >2 A), regions of strong interference (i.e., water vapor in the mid-infrared), or even regions where the signal-to-noise ratio is poor.

There are some techniques that can aid in determining regions where groups of spectra are most "different" from each other. In other words, locating the portions of the spectrum where the sample is varying the most from run to run. One method is to calculate a standard deviation spectrum. This is effectively accomplished by subtracting the average spectrum from every spectrum in a training set and then calculating the standard deviation at every wavelength. Regions that have large positive peaks are regions where the spectra are varying and will contribute the most to the PCA factors for the group (Fig. 14).

Remember that the purpose of using the Mahalanobis distance discrimination method is to calculate a model matrix that gives an allowed range of

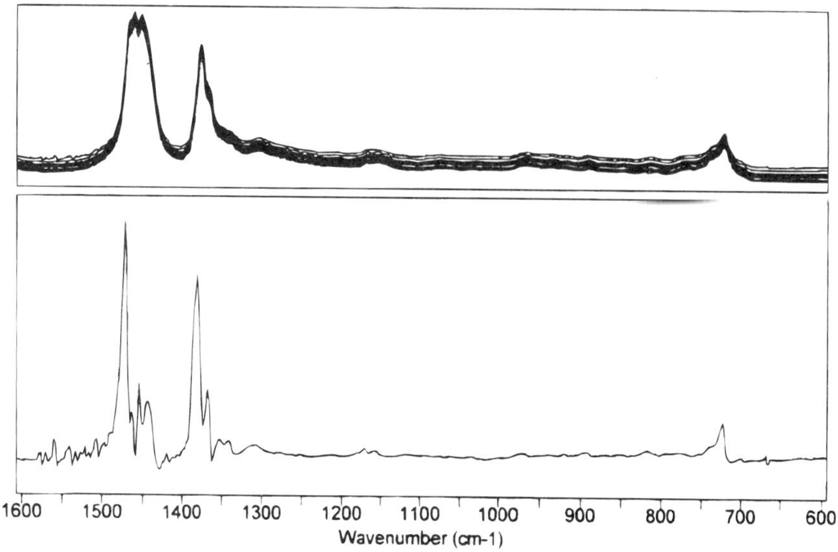

FIG. 14. Training set spectra (top) from Fig. 4 with the standard deviation spectrum (bottom). The large bands in the standard deviation indicate that most of the variation in the spectra is taking place in the three major bands at 1470, 1380, and 725 cm^{-1}.

variation in the data. Choosing regions of little or no variation can result in a Mahalanobis group that is too restrictive and will not find any matching data outside the training set. However, the standard deviation spectrum is merely an indicator of changes occurring between the spectra. As with selected wavelength models, a PCA–Mahalanobis model that includes only small portions of the spectrum is likely to miss important features (such as impurities) that can lead to misclassification.

Section 2

REFLECTANCE MEASUREMENTS OF SOLIDS

REFLECTANCE SPECTROSCOPY: AN OVERVIEW OF CLASSIFICATION AND TECHNIQUES

ART SPRINGSTEEN
Labsphere, Inc.

Light incident on a substance can interact with the material in six ways (Fig. 1). It can be absorbed, transmitted either in a regular fashion or scattered, or reflected diffusely, specularly (also known as regular reflection), or retroreflected back to the source. This chapter will cover the techniques of measurement of reflectance and, to a small extent, diffuse transmittance where it is associated with diffuse reflectance in a technique called transflectance. Retroreflection, although a natural phenomenon—a cat's eye shining in the beam of a headlamp is an example of retroreflection in nature—is not commonly measured using spectrophotometers and those specialized instruments to do such measurements are beyond the scope of this chapter.

Reflectance spectroscopy concerns the measurement of four distinct types of materials and their interaction with light. Specular materials reflect the predominant amount of radiation at an angle equal and opposite to the incident radiation. Diffusely reflective materials scatter light over a wide range of angles, with the perfectly diffuse (or Lambertian) scatterer exhibiting a cosine response to the incident radiation. Gonioapparent

ISBN 0-12-764070-3/$25.00.

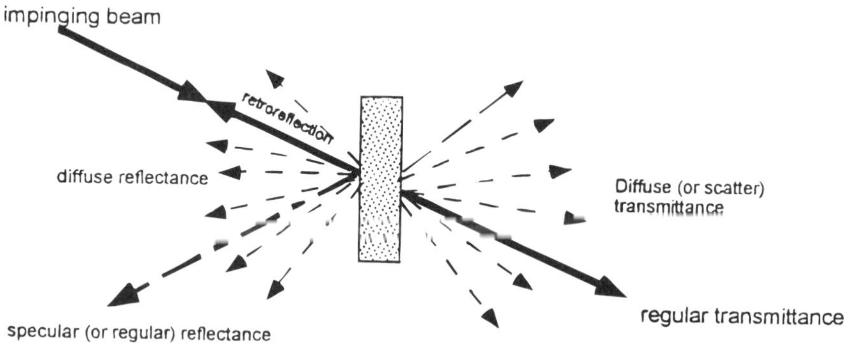

FIG. 1. Light incident on a substance.

materials have both a specular and a diffuse component (as do most materials) but these substances have particular scattering components only at certain angles of incidence or collection due to particulate inclusions within the materials. Finally, a discussion of transflectance is included to cover those materials that both diffusely reflect and diffusely transmit incident radiation.

Specular reflectance is almost certainly the first type of reflectance to be studied by the ancients. The ability to make a highly reflective surface to observe one's appearance or the mythological "burning glasses" used by the Greeks to destroy the Spartan fleet depended on a highly polished surface of metal to reflect light in a nonscattering manner. Early mirrors were generally polished surfaces of brass or bronze, perhaps coated with a layer of silver to increase their efficiency at reflecting visible light. In recent times, the ability to produce highly reflective specular surfaces for use in photocopiers, lasers, and other modern optical instrumentation has made the accurate measurement of specular reflectance of tantamount importance to optical engineers.

Specular reflectance measurements fall into two categories. Relative reflectance requires the use of a reference mirror to produce accurate measurements. Absolute specular measurements do not require a reference mirror but instead use various optical techniques to bring the reflected beam to a detector. Both of these measurements have their advantages and pitfalls that must be considered in choosing the measurement technique.

I. Relative Specular Measurements

The most common type of device for measurement of specular reflectance is a near-normal incidence, near-normal collection accessory for a spectrophotometer. This accessory is generally two mirrors set at angles that allow illumination at an angle as close to 0° as physically possible with collection at the same equal and opposite angle (Fig. 2). Typically, reference mirrors are calibrated in this same configuration, because the uncertainties due to polarization, near-angle scatter, etc. are minimized. Typical angles used by National Laboratories include 6, 8, and 10°. These minor differences in angle are essentially meaningless in the measurement of specular materials.

Specular surfaces can also be measured by using at integrating sphere at 8° incidence or collection geometry. The measurement technique is identical to the near-normal technique described previously. A reference mirror is used to establish the instrument baseline, and then the sample mirror replaces the reference and a scan is obtained. The reflectance of the sample is the product of the measured value for sample times the actual reflectance for the reference. This technique is quite convenient to use and is as accurate as the multiple mirror method. Care must be taken, however, to ensure that the incidence or collection angle is not 0°, because a specular-excluded measurement of a mirror will certainly give the incorrect answer for reflectance!

These devices are used by establishing a baseline using a reference mirror of known reflectance, replacing that mirror with the sample mirror, and comparing the two mirrors. This technique is applicable to both first- and second-surface mirrors, but the reference mirror must be the same type as the sample to be measured or inaccurate measurements will be obtained.

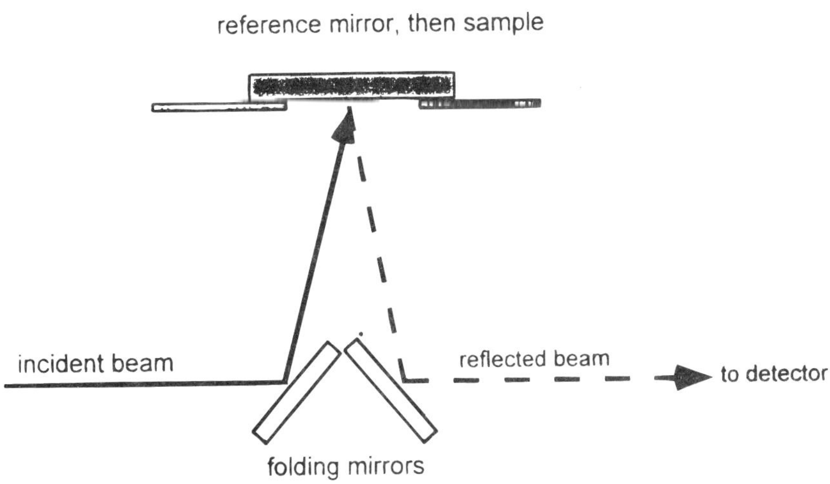

FIG. 2. Near-normal specular reflectance accessory.

Obviously, the accuracy of the technique depends on the reference mirror, which in the case of first-surface mirrors tends to degrade in reflectance over time.

Typically, aside from the near-normal angles, fixed-angle relative specular accessories are available at 30, 45, and 60° angles. These accessories are usually instrument independent and are available from a wide variety of sources. Accessories for use in ultraviolet (UV), visible (VIS), and near-infrared spectroscopy (NIR) are generally equipped with optical components that are first-surface aluminum mirrors. For infrared application, gold mirrors are preferred. When measurements are made at angles further from near-normal, errors due to polarization tend to appear. Most spectrophotometers produce incident radiation that is polarized to some extent both from the sources and from dispersing elements such as diffraction gratings. Although polarization is not a factor with the measurement of uncoated first-surface mirrors, any nonmetallic surface, such as a magnesium fluoride, antireflectance, or thin-film coating, a second-surface mirror, or nonmetallic sample will often show large changes in reflectance with changes in the polarization state of the beam. It is therefore often necessary to provide some form of depolarization of the incident beam or to quantitate the polarization of the instrument to compensate for the resulting change in reflectance.

A more specialized type of relative specular accessory is used for determining the thickness of coatings or of impurities on specular surfaces. The so-called grazing angle accessory projects the incident beam onto the surface of the sample at angles of 80° or greater. This increases the path length through any coating or impurity that happens to be on the surface of the reflecting surface and allows better quantitation of the surface due to the increased path length. Although this technique is most commonly used in the infrared, applications in the VIS–NIR are becoming more common with the availability of grazing angular specular devices designed for instruments that measure in this region of the spectrum.

Typical relative specular reflectance accessories are shown in Fig. 3.

It is often desirable to measure a number of angles on one instrument. For this application, variable-angle relative accessories can be used. Typically, the angles can either be fixed (7.5, 30, 45, and 60°), with the incident and exiting beam being collected by fiber inputs (Fig. 4) and outputs, or be continuously variable by use of a series of moving mirrors. In either case, care must be taken to ensure that the reference mirror has been accurately calibrated at the same angle at which the sample is being measured and that polarization is considered in the measurement. It is the author's experience that many of the relative variable-angle accessories on the market are poorly designed, for both the control of the angle of incidence on the

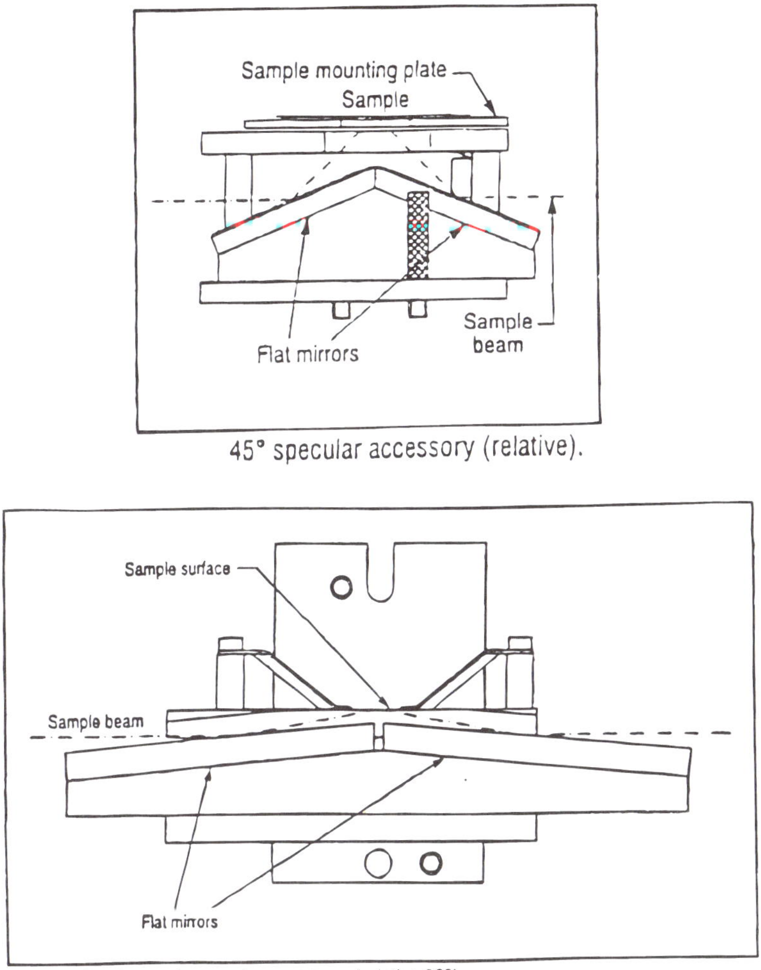

45° specular accessory (relative).

Grazing angle specular accessory (relative 80°).

FIG. 3. 45/80° specular accessories.

sample and the efficiency of collecting the reflected beam. Also, accurate measurement of standard mirrors at angles off near-normal are difficult to obtain, and thus users typically assume that the reflectance of the reference at near-normal incidence is identical to that at higher angles. For an uncoated mirror, this may be a good approximation, but for coated surfaces that are typically used, large errors can be introduced by making this assumption.

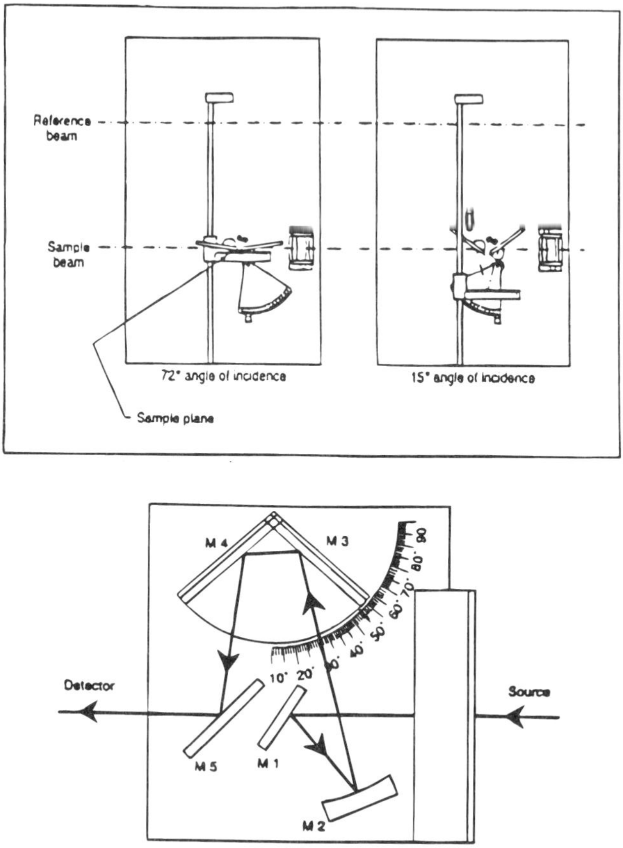

FIG. 4. Optical diagram: variable angular specular devices.

II. Absolute Specular Reflectance

Absolute specular reflectance measurements are defined as the measurement of a material in which the only change from the measurement of the baseline to the measurement of the sample is that the sample changed in the optical chain. All angles of incidence and the total path length of the measurement are maintained throughout the measurement. This concept means that no artifact standard is necessary for the measurement, thus canceling a major source of uncertainty in the measurement of specular materials.

Traditionally, absolute specular measurements have been made at near-normal incidence, usually at approximately 8°, using either of two geometries. The so-called V-W accessory (Fig. 5), in which the sample beam is

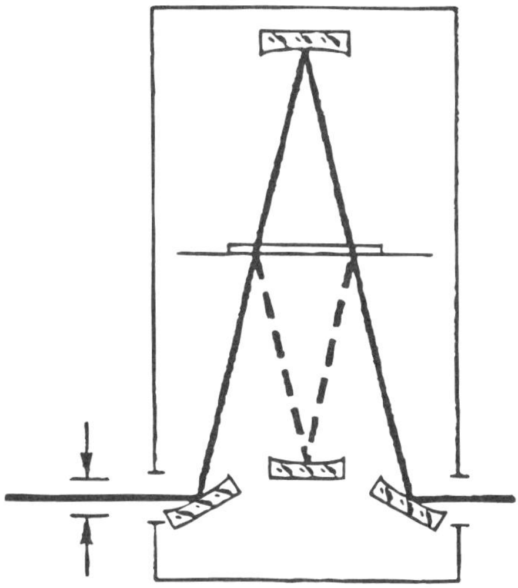

FIG. 5. Optical diagram: V–W absolute specular accessory.

incident on the sample twice, gives the square of the actual reflectance, which must be considered in the calculation of actual reflectance. Typically, a measurement is made in the "V" position, the sample to be measured is inserted, and the sample stage is rotated to the "W" position with the sample in place. Although V-W accessories are optically easy to design and are the "traditional" absolute specular accessory, the double bounce off the sample leads to a number of nontrivial problems. First, the assumption is made that the sample is totally uniform over the surface and that both spots have the same reflectance. The second problem, and by far the most important, is that for extremely low reflectance materials, such as anti-reflectance coatings or dark specular materials, the square of the reflectance may be a very small number and thereby began to be in a competition with the noise level of the instrument. For example, a coating with a reflectance of 0.2% at a certain wavelength (not atypical in an antireflectance coating) would measure at 0.000,004 on a V-W accessory. This is well below the noise levels of most spectrophotometers, so it would be difficult to determine the reflectance from the noise. Still, for typical high-reflectance samples, the V-W accessory is the most typical and accurate measurement of absolute reflectance.

The second geometry to measure absolute specular reflectance is the V-N geometry (Fig. 6). In this measurement geometry, the only difference is a single bounce off the sample at a particular angle. Although obviously an

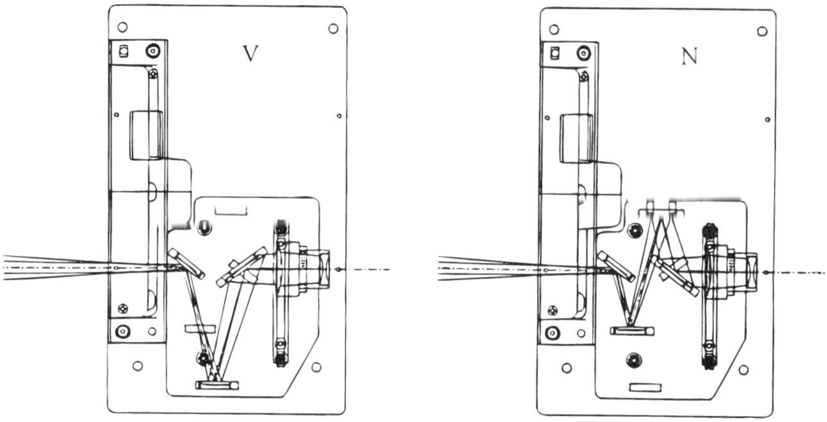

FIG. 6. Optical diagram: V–N near-normal specular accessories.

improvement over the V-W design, the devices are difficult to design and manufacture compared to the V-W geometry. The advantage is that only a single bounce off the sample occurs, eliminating errors due to low signal in low-reflectance materials. Another advantage is that, recently, V-N devices for other than near-normal angles have been designed. Figure 7 shows optical layouts for 45 and 60° V-N accessories.

The ability to measure specular samples at a variety of angles without the use of a reference has long been a goal of material scientists. During the past 5 years, accessories for spectrophotometers to measure absolute variable-angle specular reflectance have become available. Available in either double- or single-bounce configurations, the concepts are the same. A small integrating sphere is used as a collection device to establish the baseline for the instrument. A sample is then inserted and moved to the measurement angle of interest while the sphere, which is mechanically tied to the sample holder, automatically rotates to the desired collection angle (Fig. 8). The reference beam of the spectrophotometers (because these devices are predominately used on double-beam UV–Vis–NIR instruments) is also ported into the sphere by means of moving mirrors or a fiber optic reference channel. The resulting measurement is the absolute specular reflectance at the angle of choice. These accessories can usually measure from approximately 10° up to approximately 70° incidence angle.

Although the term absolute implies highly accuracy, in reality many absolute measurements are no more or are less accurate than a relative measurement. Absolute accessories are highly sensitive to errors due to mis-

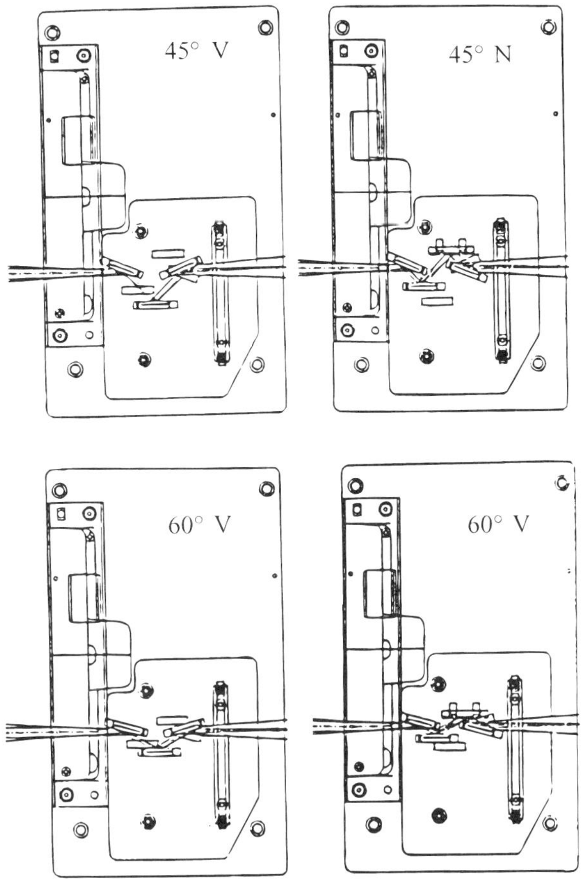

FIG. 7. Optical diagrams: 45 and 60° V–N absolute specular accessories.

alignment of the accessories. If the beams for the baseline measurement and the sample measurement are not identically and exactly aligned on the detector(s), errors will occur in the measurement. Even using an integrating sphere to alleviate the problem of beam/detector alignment, as is done in the variable-angle absolute accessories, is not necessarily a sure cure, because hot spots on the sphere wall will cause different view factors to come into play and give associated errors of measurement. Furthermore, even the slightest scatter from the sample mirror due to dust or imperfections can cause sample signal to be lost, leading to incorrect reflectance results.

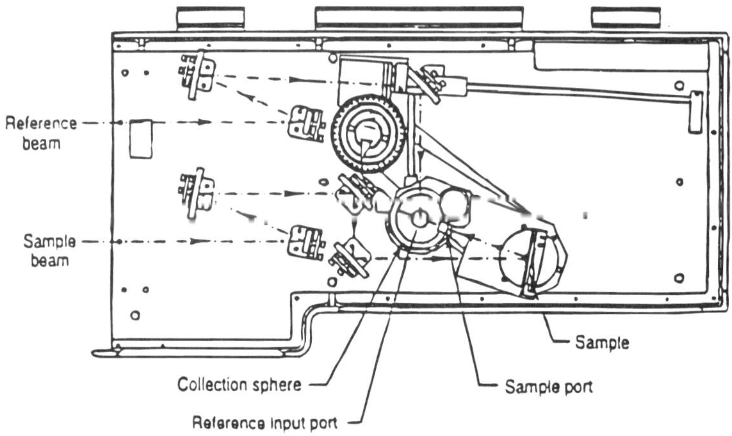

FIG. 8. Optical Layout of VAASR accessory: Perkin–Elmer RSA-PE-19s.

III. Sample Preparation and Handling

First-surface mirrors are extremely sensitive to scratching and contamination, so samples should be handled with extreme care. Cotton or latex gloves should be used at all times in handling such sensitive materials. Because first-surface mirrors are almost impossible to clean once dirtied, it is imperative to avoid fingerprints or other contamination. *Although such mirrors can be cleaned by spraying with a detergent such as Fantastik, followed by rinsing with distilled water, these mirrors cannot be considered standard-grade mirrors.*

The measurement of curved mirrors was once thought to be almost impossible. The measurements are still extremely difficult and all measurements should be regarded with some suspicion. The author knows of no method to measure such mirrors by relative techniques. Measurement using flat mirrors as references in an integrating sphere is possible, but the changes in view factor in the sphere lead to systematic errors. An absolute technique using a sphere as a collector and using accessories as the general-purpose optical bench (to be discussed later) is a possibility but, again, measurements should be regarded as informational and not necessarily correct.

IV. Diffuse and Total Reflectance

Our perception of the world—every object we see—is the result of the reflectance of light from that object reaching an extremely sensitive, if somewhat wavelength-limited, photodetector: the eye. For that reason alone, the measurement of reflectance in the laboratory is important to those in business, the arts, science, and industry. It enables us to quantify what we see, which is an important concept because our human detector systems vary widely from individual to individual and from condition to condition.

Diffuse reflectance spectroscopy is a versatile tool that allows us to accurately measure the flux per wavelength of light reflected in a scattered manner from a sample. Whether the sample be almost purely diffuse, as in a lightly packed powder, predominately specular, as in a burnished metal surface, or something in between, such as a glossy paint sample, diffuse reflectance spectroscopy tells much about the physical and chemical characteristics that are not available by other analytical means. In addition, diffuse reflectance spectroscopy in the "visible" region of the spectrum—that area between approximately 360 and 760 nm where our eyes are sensitive—allows quantitation of color measurement for biological, pharmaceutical, commercial, and artistic applications.

Typically, diffuse reflectance is measured by two methods.The most common is to use an integrating sphere that is mated with a spectrophotometer, allowing the measurement of both transmitted and reflected scatter. The second is by use of the so-called biconical geometry, in which a mirror focuses a beam of light at a small point on a sample and the scattered reflected light is caught by a parabolic or similar curved mirror that directs the beam to a detector. Biconical devices do not collect scatter at all angles, so anisotropic materials are not measured accurately. For this reason, biconical devices are generally thought of as qualitative in nature, whereas an integrating sphere is both qualitative and quantitative.

The advantage of a biconical accessory is that it allows one to get a spectrum on a far smaller sample than is required with an integrating sphere.

V. Integrating Sphere Measurements of Diffuse Reflectance

An integrating sphere is simply a hollow ball that has been coated (or is fabricated from a material) with a very highly reflective and diffuse surface on the inner diameter. Traditionally, these coatings were magnesium oxide or barium sulfate for the UV–VIS–NIR, although now almost all integrating sphere for use in reflectance spectroscopy are produced from Spectralon, a very stable, very diffuse high-reflectance material developed by Labsphere, Inc. This material is not affected by most conditions that affected previous coatings and has far better long-term stability over a wider wavelength range. In the near-IR to mid-infrared, integrating spheres are generally fabricated from aluminum with a diffuse gold coating (Labsphere InfraGold or InfraGold-LF) or occasionally from packed sulfur powder.

The wavelength range of an integrating sphere reflectance device is determined by three major factors: the reflectivity of the coating or material from which it has been fabricated, the detectors used in the sphere, and the light source/monochromator/interferometer/beam splitter range of the host instrument. This chapter will not deal with instrument design, so sources and monochromators/interferometers will not be discussed. The following table shows typical ranges of detectors and sphere materials.

Sphere material	Range	Comments
	Typical integrating sphere materials	
Magnesium oxide	≈360–1500 nm	Degrades in UV over time; now seldom used
Barium sulfate (packed)	≈250–2000 nm	Difficult to prepare; fragile; seldom used
Barium sulfate coating	≈300–1600 nm	Inexpensive; subject to changes in reflectance with humidity
PTFE powder (packed)	≈200–2500 nm	Highest reflectance in UV–VIS–NIR; difficult to prepare; fragile
Spectralon	≈200–2500	Durable, very slightly lower in reflectance than (machined fluorocarbon) packed PTFE
Diffuse gold (InfraGold)	≈1–20 μm	High reflectance; not very lambertian beyond 10 μm
Arc sprayed gold	≈2–>50 μm	Very Lambertian; not as high reflectance as InfraGold
Packed sulfur	≈2–20 μm	Fragile; high reflectance but seldom used; degrades over time
	Typical integrating sphere detectors	
Photomultiplier tube	<190–850 nm	Extended range may go out to 900 nm; sphere coating limits on low end
Silicon photodiodes	200–1100 nm	Not as sensitive as PMT over UV–VIS
Lead sulfide (cooled or uncooled)	850–3000 nm	range better S/N range if thermoelectrically cooled
Lead selenide	1–5 μm	Not commonly used; can go to 6 μm if cooled
Indium gallium arsenide	800–1700 nm	Used in some NIR systems
Mercury cadmium telluride	1.2–1.7 μm	Liquid nitrogen cooled; can get longer range with modifications
DTGS	1–20 μm	Not as sensitive as MCT but better range; not suitable for large spheres

As with any instrument, an integrating sphere system can perform no better than the most limiting component, so consideration in designing systems must take into account both detector range and sphere materials. It must also be considered that a sphere is by its very nature an attenuating device, and the larger the sphere, the more the signal is attenuated. For example, a small integrating sphere of 60 mm diameter, even if made of highly reflective Spectralon ($R \approx 99\%$ in the 350- to 1200-nm range), will attenuate signal by about 0.7 A; with a 150-mm sphere (a typical sphere for high-end UV–VIS–NIR instruments and many color instruments), the attenuaion will be in the order of 2 A. This can lead to severe signal-to-noise problems in very low reflectance or transmittance samples. Fortunately, almost all materials that reflect by scattering radiation do so at levels above 1% reflectance, so this attenuation problem does not manifest itself for these measurements.

VI. Integrating Sphere Geometries

A. Directional Illumination–Diffuse (Hemispherical) Collection

The most common geometry for integrating sphere measurements is directional illumination with hemispherical collection. A beam of usually monochromatic light from a spectrophotometer is focused onto the sample, which lies on a line tangent to the inside radius of the integrating sphere. The reflected radiation is collected by the sphere, which offers a 180° field of collection. The detector(s) lies at another point on the inside tangent of the sphere and collects the diffusely reflected radiation. Usually baffles made from the same material as that of the sphere coating are placed within the sphere to prevent "first strike" radiation from hitting the detector—the purpose of an integrating sphere is to integrate the signal—which may lead to erroneously high readings. The design of integrating sphere systems for spectrophotometers is somewhat complex and is beyond the scope of this chapter.

Total hemispherical reflectance is measured by having the sample beam incident on the sample at near-normal incidence, generally 8 or 10°. The incidence angle should be such that the specular component of reflectance is not lost out the entrance port to the sphere, nor should it be so high that the angle affects sample reflectance. A baseline is run against a sample of known reflectance, generally a diffuse reflectance standard that approximates the perfect reflecting diffuser, then the sample is run in the same configuration. The ratio of the measured sample to the reference material is the total hemispherical reflectance factor of the sample. For more accurate measurements, a "sphere zero" factor should also be included in the calculation. The sphere zero takes into account any incident radiation that does not initially strike the sample (the halo effect). To obtain the sphere zero, an efficient light trap is placed in the sample position after a baseline is run and a spectrum is run. On a well-designed instrument with well-designed optics, this number should be very close to zero (in the order of 0.1–0.2%R), which makes the sphere zero correction superfluous.

The following calculates the hemispherical reflectance, R_f.

$$R_f = \frac{(R_{\text{measured}}) - (\text{sphere zero})}{(R_{\text{reference}}) - (\text{sphere zero})}.$$

This measurement technique is applicable for both single- and double-beam configurations. Single-beam spheres will produce an error due to multiple reflectance off the sample after the first strike, called the single-beam substitution error. Substitution error, sometimes referred to as the single-beam sample absorption error, is a systematic, predictable, and nonrandom error inherent in single-beam integrating spheres measuring reflectance or transmittance. The error is caused by the difference between the throughput of the sphere when the reference makes up a portion of the wall and the throughput when the sample is substituted for the reference. For reflectance measurements the throughput is usually lower when the sample is present because a reference material of high (nearly 100%) reflectance is usually used. For transmittance measurements the throughput is usually higher when the sample is present because an open port (zero reflectance when viewed from inside the sphere) is usually used as a reference.

When the sample and reference are of similar reflectance the substitution error is negligible; at worst, it may reach up to 10%. In quality control applications in which a threshold value is used, the substitution error can be incorporated into this threshold. For chemical analysis in which the locations of absorption peaks, not necessarily their exact absorbance values, are important the error can also be ignored. Where correction of the substitution error is necessary, a number of correction techniques are available.

Instruments that chop between the sample beam and the reference beam have both beams present in the sphere at essentially the same time. In this case, because the sample beam and reference beam each "see" the same sphere, there is no substitution error.

When room permits, single-beam spheres can be designed with an additional "dummy" port. The background correction is performed with the sample to be measured in the dummy port and the physical reference in the sample port. A scanis then run with the positions of the two switched. In this comparison method the average reflectance (hence the throughput) of the sphere remains unchanged from the reference scan to the sample scan. This technique requires that a separate reference scan be run for each sample unless the samples are all very similar, in which case the residual substitution error caused by sample-to-sample differences can be ignored.

If a reference is used that is very close in reflectance to that of the sample, the substitution error is negligible. For samples with very low reflectance or transmittance, the substitution error will also be very small. For a sample with zero reflectance or transmittance, there is no substitution error at all. For low reflectance and transmittance samples, the substitution error is so small that it probably falls within the random noise of the instrument.

For single-beam spheres without dummy ports in applications in which substitution error is a concern, the spheres can be calibrated with a set of standards such as one of Labsphere's Reflectance Standards Sets, which has been measured on a sphere without substitution error. With these standards, a table of measured versus actual readings can be generated and used to correct for substitution error (Fig. 9).

In a double-beam sphere, the substitution error is eliminated by having both sample and reference on the sphere concurrently and by alternating the measurements between sample and reference, as shown in Fig. 10.

Double-beam integrating spheres can be designed in a number of ways to efficiently collect both diffusely reflected and transmitted radiation. A proper sphere design minimizes port size and port fraction (the amount of nonreflecting surface), adequately screens the detector(s) from the view of first-strike radiation, and should be as symmetrical as possible around the sample and reference beams (Hanssen and Snail). Two typical designs are shown in Figs. 11 and 12. Figure 11 shows a traditional design in which the sample and the reference beam are introduced on parallel paths and hit the sample and reference ports at an angle that is usually between 8 and 10°. The light trap port, with removable light trap, placed in the center between the two transmittance or input ports, serves to collect the specular component from both sample and reference. Although this design is ideal for

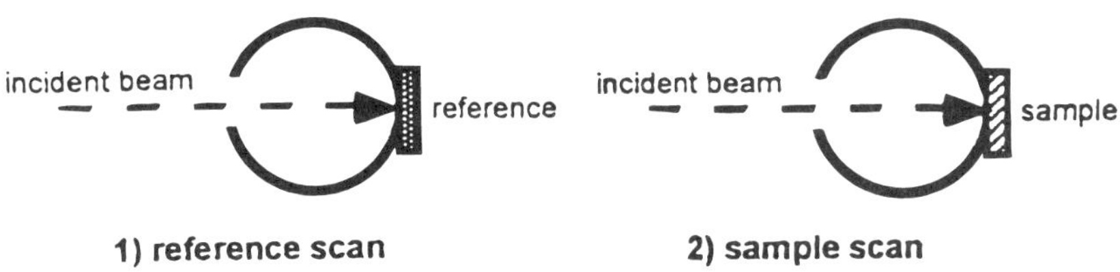

FIG. 9. Single-beam sphere measurement technique.

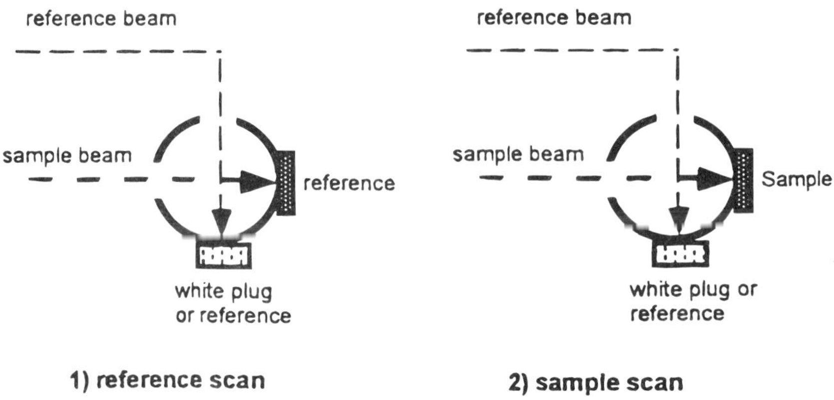

FIG. 10. Double-beam sphere measurement technique.

measurement of diffuse reflectance, the close spacing of the ports and the inability to utilize a center mount make the sphere a bit less versatile than the crossing design shown in Fig. 12.

Although this design does have its limitations, the great advantage is that there is total symmetry between the reference and sample beams, so the viewing function of the detectors (which are placed in the center of the circle of this top view in Fig. 12) is equal. It has been argued that this design allows for the highest accuracy of measurement of diffuse, specular, and mixed reflectance samples. This design is currently used by the National Institute of Standards and Technology (NIST) and a number of military research labs as the standard geometry of measurement for reflectance.

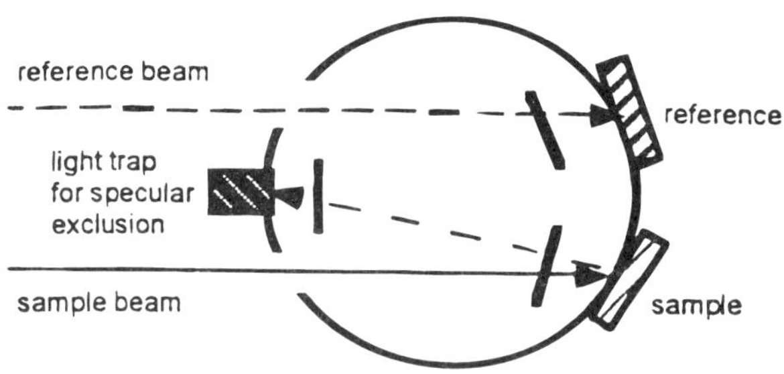

FIG. 11. Double-beam integrating sphere with noncrossing beams (top view). Solid lines are baffles.

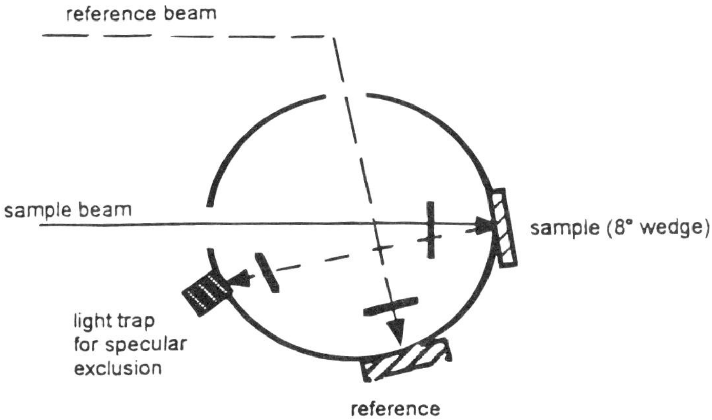

FIG. 12. Double-beam sphere with crossing beams. Solid lines are baffles.

The most common design currently used involves having the reference and sample beams cross paths, with the sample beam passing over the center of the sphere and the reference beam a bit off the center axis. Again, specular subtraction is obtained by using a removable light trap at 16° off the incident beam, with the sample reflectance port usually wedged at an 8° angle. (An option is to have two sample holders—one at 8° for measurement of total reflectance and another at 0° so that the specular component is reflected out the input port. It is the author's experience that the latter option—two sample holders—is not as efficient in accurately performing specular excluded measurements.) The major advantage is that a center mount can now be used by installing it through the top of the sphere. Center mounts are useful for measuring a number of physical properties, including reflectance at non-near-normal angles of diffuse materials and transflectance (also known as interactance), which is a direct measurement of total absorbance.

Another design for double-beam integrating sphere systems is the so-called "downlooking" sphere (Fig. 13). In such a device the sample beam enters not from a port along the equator of the sphere but rather from the top, with the sample sitting at the bottom and the detector(s) on the equator of the sphere. The reference beam, if there is one, enters at the equator as in Fig. 12. This design is the most common design for integrating spheres used in the mid-IR, in which the sphere accessory is mated to a Fourier transform infrared spectrometer, although UV–VIS–NIR versions are also available. Typically, this configuration is used for measurement of mineralogical samples, very large samples as are often presented in

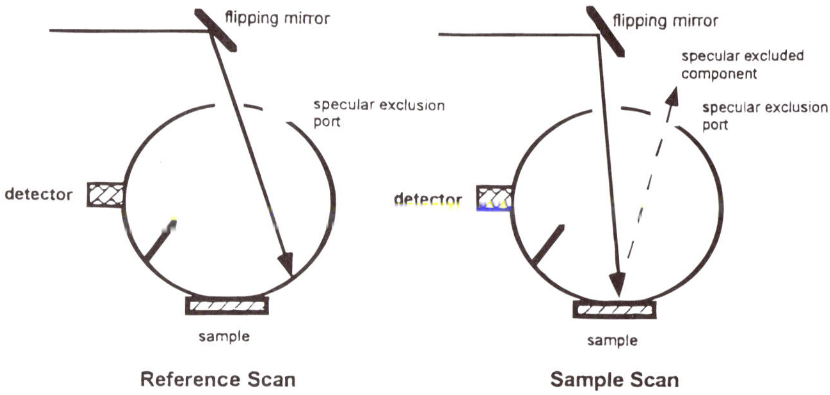

FIG. 13. Downlooking sphere in reference and sample measurement mode (side view).

aerospace and remote-sensing applications, and if the sphere is routinely used to measure powdered or semiliquid samples in the reflectance mode. Although downlooking spheres can be made to accommodate center mounts, they are not ideal for applications requiring the routine measurement of center-mounted samples.

This geometry (that of the third Taylor method) is said to be absolute because the only difference in sphere throughput between the reference and sample scans is the initial reflectance off the sample in the sample scan.

A final geometry exists for directional illumination, diffuse collection integrating sphere measurements. The 30/T geometry (30° incidence, total hemispherical collection) is specified by the American Society for Testing and Materials (ASTM) for the measurement of acoustical ceiling tile reflectance. The author knows of no other applications that specify this integrating sphere geometry.

VII. Nonintegrating Sphere Methods

Another method of measurement of diffuse reflectance does not require an integrating sphere. In this method, the sample beam is focused onto the sample by means of ellipsoidal or spherical mirrors and collected by another ellipsoidal mirror, at either 180 or 90° from the incident beam. This method has been used for many years in the infrared, since proposed by Fuller and Griffiths (1978, 1980). The advantage is very high collection efficiency and the ability to measure very small samples. Such geometry is known by a number of names, including biconical, Praying Mantis (a trade-

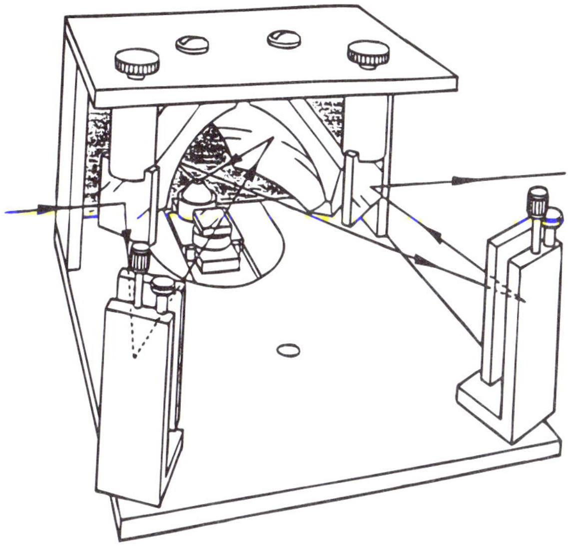

FIG. 14. Praying Mantis diffuse reflectance accessory. (Reprinted courtesy of Harrick Scientific Inc.)

mark of Herrick Scientific), or The Collector diffuse reflectance accessories (a trademark of Spectratech, Inc.).

For the 180° geometry, a problem exists with the inclusion of the specular component of reflectance, which can be a major source of error in the infrared. The problem can be alleviated by the inclusion of a small baffle that blocks the specular beam but allows the diffuse component to pass through unhindered (Figs. 14 and 15).

Recently, this measurement technique has been applied to UV–VIS–NIR spectrophotometers. Using two ellipsoidal mirrors at 90° to one another,

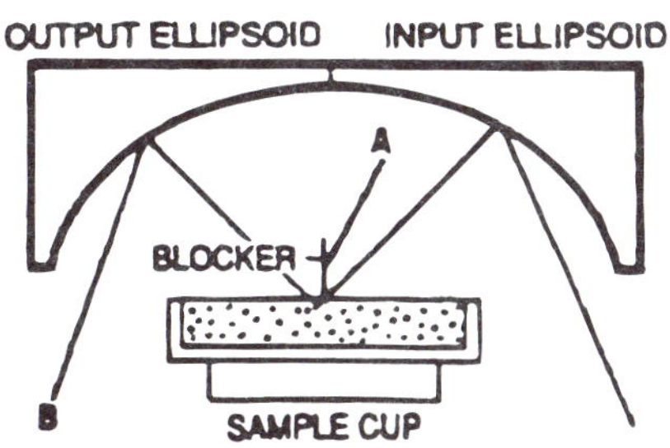

FIG. 15. Diffuse reflectance accessory with specular blocking baffle. (Reprinted courtesy of Harrick Scientific Inc.)

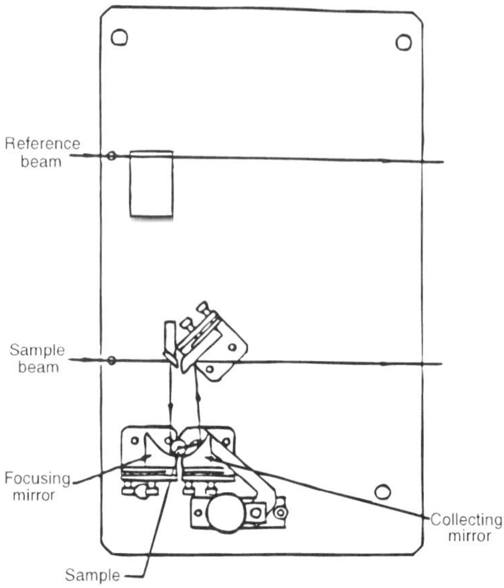

FIG. 16. 90° biconical geometry.

the specular component is also eliminated. This device allows measurement of very small samples (>1 mm) to be measured with ease. Such diffuse reflectance devices are invaluable in the measurement of small samples such as individual fibers, small stains, small amounts of crystalline substances and other applications such as are required in forensic laboratories (Fig. 16). Although these measurements are not as quantitative as those made using integrating sphere geometry, they are an irreplaceable asset in the qualitative analysis of solids, pastes, and even liquid samples.

VIII. Diffuse/Directional Measurement of Reflectance

The reverse geometry of directional diffuse measurement of reflectance illuminates the sample with diffuse illumination using either an integrating sphere or a hemiellipse and images the sample onto the detector or into the spectrometer. This geometry offers some advantages over directional diffuse geometry in that the illumination is polychromatic and that the signal reflected from the sample then passes through the monochromator or spectrometer. In UV–VIS applications, this allows any fluorescent component to be analyzed and separated from the purely reflected component.

Another major advantage is speed. Because a diode array processes all the spectral data at one time and there is no scanning, measurements can be made in as little as 1 sec. If the source is sufficiently strong, one can also get a very good signal-to-noise ratio with a relative short integration time, especially in the color region. For this reason, many of the commercial color analyzing spectrophotometers incorporate this design.

The disadvantage to this method is that usually a narrower wavelength range can be analyzed because the source limits the range of measurement—diffuse illumination can use only one source in the instrument, whereas a scanning spectrophotometer may use two sources to cover a wider wavelength range. Additionally, in the case of UV–VIS instruments, the detectors are most commonly silicon photodiode arrays, which also limit range. Recent advances in detector technology have provided lead sulfide or InGaAs arrays for measurement in the NIR, but these devices are still quite new and have not met with a great deal of success due to the lack of linearity of such arrays.

In a typical UV–VIS diffuse/directional instrument, the source, generally a tungsten halogen lamp, is placed inside an integrating sphere. The diffuse illuminated sample is imaged onto a slit and then to a polychromator and diode array. By having two sets of sample holders, one with a wedge and one at 0°, both specular-included and specular-excluded measurements can be made. The 0° sample holder ensures that the sample is not illuminated at 0° incidence and is thus specular excluded. A typical accessory for diffuse/directional measurement of reflectance is shown in Fig. 17.

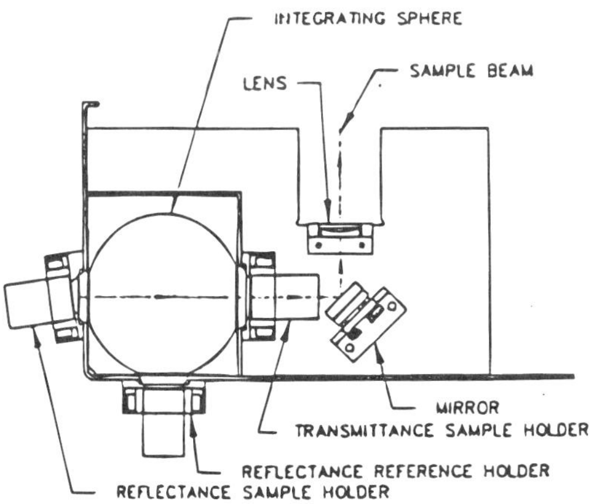

FIG. 17. Diffuse/directional integrating sphere accessory.

The wavelength range of such an instrument can be extended into the ultraviolet—a tungsten/halogen lamp is generally only effective to approximately 360 nm—by use of a xenon arc lamp to replace the tungsten lamp. The difficulty with such a replacement is that arc lamps are somewhat unstable and gaining a good stable baseline can become a difficult problem unless the source can be monitored by a second channel (Fig. 18).

A second problem that can be overcome is single-beam substitution error, as discussed previously. By use of a so-called "dummy port," the overall throughput of the sphere can be equalized between reference and sample scans by placing the sample in the dummy port while measuring the reference, then changing the position of sample and reference for the actual sample measurement scan. This dummy port is also illustrated in Fig. 17.

A recent development for diffuse/directional measurement of reflectance in the mid-infrared is the hemiellisoidal directional accessories and instruments independently developed by L. Hanssen at NIST and by Surface Optics Corporation. In this geometry, the source is placed at one foci of a highly specular hemiellipse and the sample at the other foci. A reverse geometry also exists when the sample is illuminated by a directional beam with the detector lying at the other foci of the ellipse (Figs. 19 and 20). The sample is imaged into a spectrometer as with the sphere method. The advantage is said to be a great increase in efficiency of collection. The disadvantage is the cost required to produce such instruments. The hemiellipse

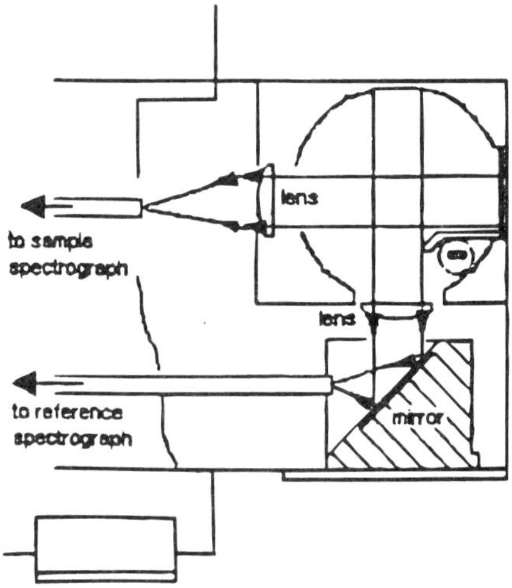

FIG. 18. Diffuse/directional reflectance instrument with source monitor.

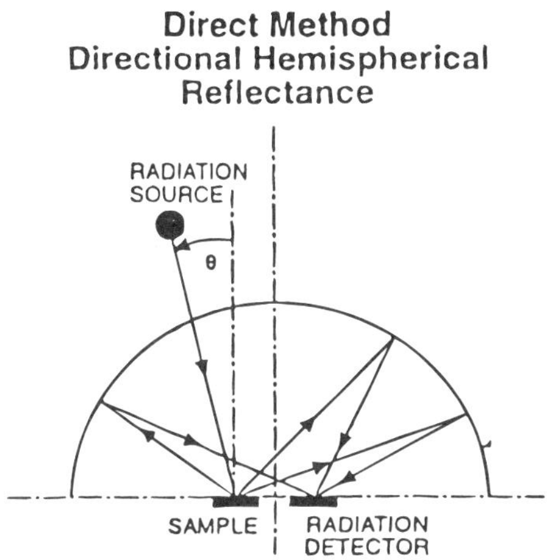

FIG. 19. Diffuse/directional hemiellipsoidal geometry for reflectance measurements. Radiation beam illuminates sample at angle θ; radiation is scattered into hemisphere collected and detected. (Reprinted courtesy of Surface Optics Corp. © 1997.)

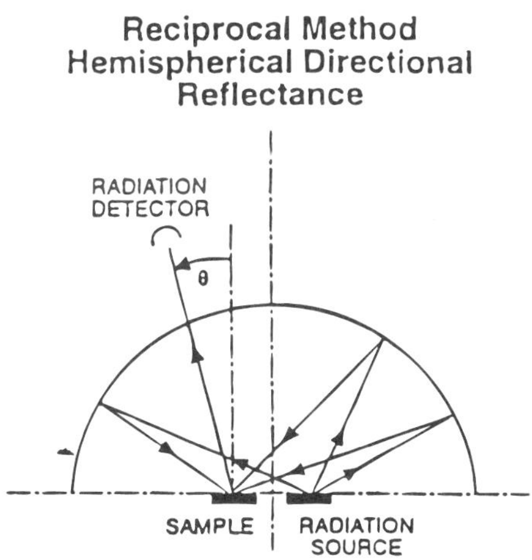

FIG. 20. Directional/diffuse hemiellipsoidal geometry for reflectance measurements. Radiation from the source uniformly illuminates sample over hemisphere; scattered radiation beam detected at angle θ. (Reprinted courtesy of Surface Optics Corp. © 1997.)

must be diamond turned with great precision and have a highly reflective specular coating, generally gold. Such devices are also quite effective in measuring emmitance properties of materials in the infrared. This methodology is used by a national laboratory for the measurement of diffuse reflectance.

IX. Directional/Directional Diffuse Reflectance Measurements

Another method of measuring a component of diffuse reflectance is by using a directional/directional geometry. In this type of measurement, the sample is illuminated by a usually collimated light source, either mono- or polychromatic, and the light is collected by a detector or series of detectors that image the sample at a nonspecular angle. The most familiar geometry for this type of measurements is 0/45 or 45/0, where the first number specifies the angle of incidence of the illuminating beam and the second number specifies the mean collection angle. Because a photon is indiscriminant as to its direction, these two geomteries are equivalent under a principle called Helmholtz reciprocity. Specifications for illumination and collection angle tolerances can be found in International Commission on Illumination (CIE) Document 15.2 (Fig. 21).

Directional/directional geometry offers a number of advantages over sphere-based diffuse reflectance devices. The primary advantage is that there is no sphere to become dirty or contaminated, which is why directional/directional geometry is often used in on-line applications in

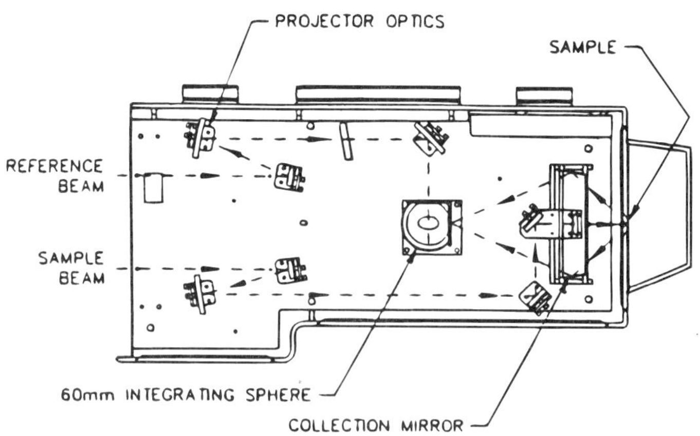

FIG. 21. Reference-grade 0/45 geometry reflectance accessory.

which dirty or high-humidity conditions may affect sphere performance. Another advantages is that the geometry is automatically specular excluded; thus, the gloss component of a material does not have to be measured independently. This is particularly important in color applications, in which a high degree of gloss can lead to problems in color matching. Directional/directional geometry is also advantageous because devices can be made, with some clever optical design, quite inexpensively and quite small in size.

The primary disadvantage is that the reflectance is highly dependent on the texture of the sample. Textiles, paper, and rough materials will show changes in the apparent reflectance just by rotation of the sample, which will not be observed using an integrating sphere. It is often said that directional/directional geometry is more a measurement of appearance than total reflectance, but the technique is still very useful if the types of samples measured are carefully controlled (Fig. 22).

The illumination for directional/directional measurements may be provided by a light pipe, focusing or collimating mirrors, or by fiber optics. The detector assembly usually employs some type of imaging device (lenses or mirrors) to keep the field of view of the detectors within the specified angular requirements.

Fiber probes can also be used for directional/directional measurements. The illumination and collecting fibers may be concentric, mixed fiber bundles, half-moons, or other configurations. These allow the reflectance probes to be very small and efficient in the collection of scattered light. Many of the new NIR probes for quality control inspection of powders *in situ* employ this geometry. The primary disadvantage is that it is impossible

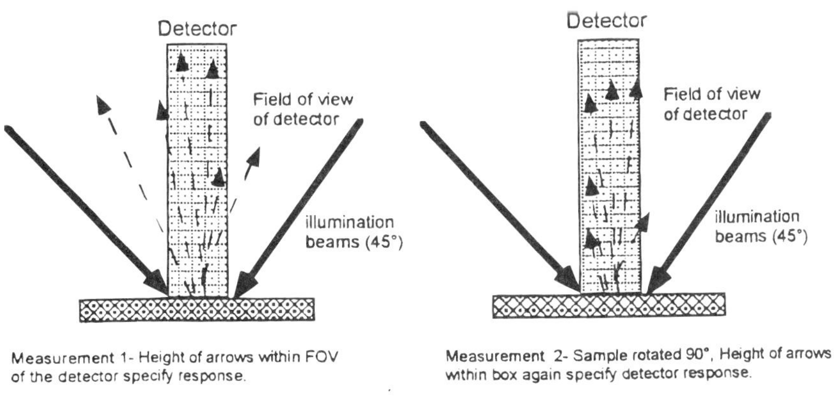

FIG. 22. Geometrical dependence of reflectance in directional/directional measurements.

to eliminate the specular component; however, for qualitative analysis this usually presents no problem.

X. Goniospectrophotometric Measurement of Reflectance

A goniospectrophotometer is an instrument that measures flux, either reflected or transmitted, as a function of illumination and collection angle per wavelength of light. This section will deal with the measurement of only the reflected component of light. Goniospectrophotometers generally measure only in a single plane, whereas instruments that measure bidirectional reflectance distribution function—a more complete version of goniophotometry—measure in multiple planes of incidence and collection. The measurement of reflectance by bidirectional reflectance distribution function (BRDF) is covered in Chapter 9.

Goniospectrophotometric measurements are essential in the measurement of gonioapparant materials. In such materials, subsurface structures, caused by inclusions of coated mica particles (pearlescents), metal flakes (generally flat aluminum particles), or coated graphite particles, are incorporated with an underlay of pigment in a clear layer of a polymeric resin. The resulting material not only changes overall reflectance with angle but also the spectral properties—most noticeably color—also change with incident and viewing angle. In addition, newer anticounterfeiting measures for security of credit cards (holograms) and bank notes (gonioapparent inks and holograms) also require a multiangle analysis system to properly measure reflected color.

The matching of color in such gonioapparent coatings is a commercially important measurement. There is some difficulty in applying coatings so that colors match; slight changes in physical application of the coating can cause the gonioapparent pigment to not properly align, which may not be apparent if one considers only total reflectance but is quite visibly obvious to the consumer. Consequently, a number of measurement geometries have been developed to properly evaluate reflectance (and hence color) over multiple angles.

A number of solutions to this problem have been suggested. Although there is currently no ASTM or CIE standard for multiangle geometry, instrument manufacturers and paint producers have worked together to develop instruments that serve the purpose of color matching. All have the linchpin of 0/45 geometry as a basis, so the incident angle is always either 0° or, more commonly, 45°, with collection at 45° or 0°. The rest of the angles are measured off the specular angle. A typical 45° incident geometry,

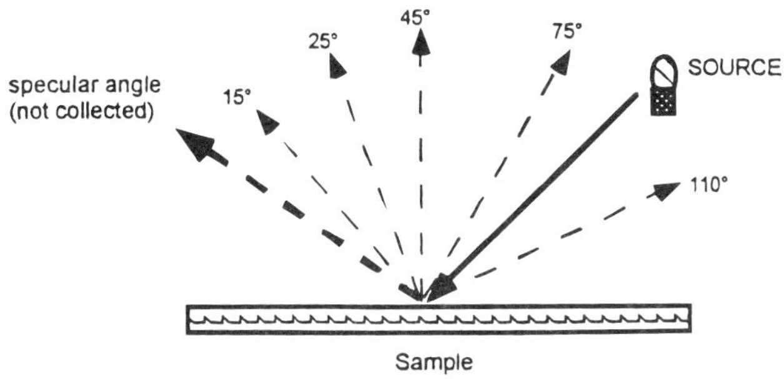

FIG. 23. X-Rite multiangle spectrophotometer geometry. Solid arrow indicates illumination angle (45°); dashed arrows indicate collection angles from the specular angle.

that is used by X-Rite, is shown in Fig. 23. This is a bit unusual in that the device collects at five angles; typically, three collection angles (25, 45, and 110°) are used.

Another option is to illuminate at a number of angles, generally through means of fiber optic sources, and collect at other angles as shown in Fig. 24. This principle has been used by Zeiss in their multiangle instrument.

A recent development by Labsphere allows for continuous variation of both the illumination and the collection angles over a wide range of wavelengths. Although the geometries described previously are all limited to the color range (360–780 nm), the Labsphere instrument (Fig. 25) is joined with a UV–VIS–NIR spectrophotometer and theoretically has a range of from 250 to 2500 nm, although the initial issues were only designed for use in the 300- to 850-nm range. By rotating the sample, a fair approximation of BRDF of scattering samples can be obtained.

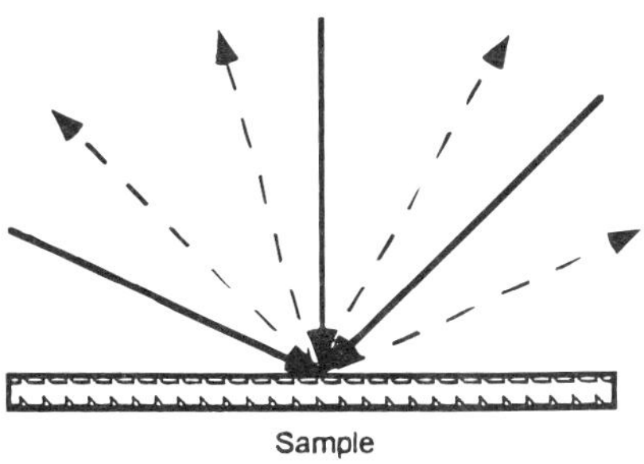

FIG. 24. X-Rite multiangle spectrophotometer geometry. Solid arrows indicate illumination angles; dashed arrows indicate collection angles.

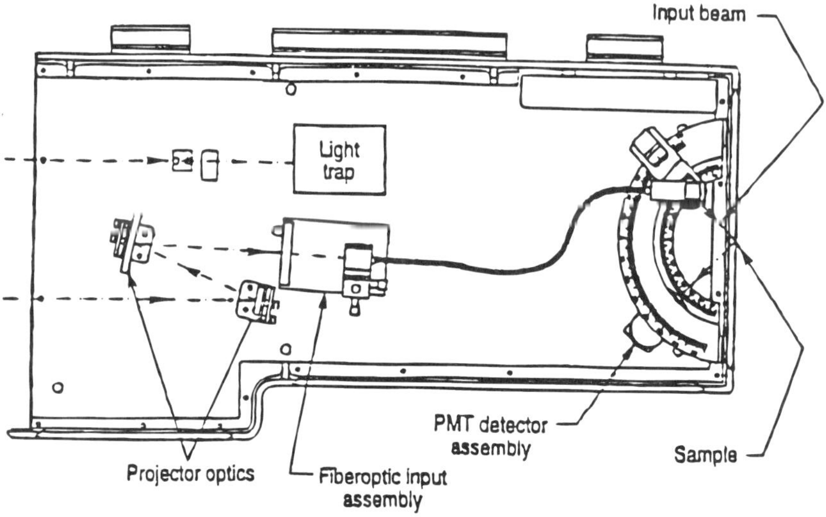

FIG. 25. Variable angle incidence and collection reflectance accessory.

XI. Reflectance Measurement Techniques

One of the first rules for measurement of reflectance is that the scattering properties of the reference should match those of the sample to be measured. Specular materials should be measured versus standard mirrors. Additionally with specular materials, a first-surface specular reference should be used as a reference for measuring another first-surface specular material and a second-surface reference should be used for measurement of other second-surface specular materials.

For scattering materials, the choice of references is a bit more complex. ASTM and the CIE specify the "perfect reflecting diffuser" as the reference for measurement of diffuse reflectance. Although the perfect reflecting diffuser is only a theoretical concept, there are a number of materials that closely approximate this theoretical material. The subject of reference materials for reflectance is covered in Chapter 8.

A problem arises when the material to be measured combines both substantial specular components but is primarily diffuse. Typical samples of this type are glazed tiles and textured metal surfaces. References for this type of measurement do exist and should be used when such measurements need be made, especially if the device used is an integrating sphere.

When using powder cells, consideration must be made for the specula component of the window surface (usually quartz or glass). This problem can be overcome by using a reference in an identical cell. Pressed powder samples can also present similar problems because the pressing usually gives a slightly specular surface to the material to be measured. This can be overcome by measuring the sample in the specular-excluded mode or by measuring both specular-excluded and total reflectance mode and subtracting the two measurements to determine the actual specular component.

Another problem arises in integrating sphere measurements when the sample cannot be placed flush against the port of the sphere. If the sample is specular, the reference should be recessed by an identical distance as the sample to be measured. For materials that are primarily diffuse in character, we have determined the measured reflectance for a lambertian material decreases by approximately 3% (absolute) per millimeter of distance of the sample from the sphere up to approximately 4 mm. Beyond that, it is difficult to predict. Measurement of the reference at a similar distance helps but is still very inexact.

Solid samples can generally be measured in the comparison method if the user can place the sample flush to the port of the integrating sphere. If the sample is translucent, it should be backed with a light trap or a material of low reflectance (a black-painted plate or black-felted material) to prevent transmitted light from entering back through the sample and giving a false high-reflectance reading.

Powdered samples may be pressed into pellets and measured in the normal fashion. Problems may arise because reflectance is a function of the packing density and because a specular surface on the pressed pellet may give inaccurate readings. A more efficient and accurate method of measurement is to use a powder cell. Powder sample holders (generally a Teflon, black Delrin, or aluminum block with quartz or glass windows) give accurate, reproducible measurements. Reference sample can be made to fit sample holder. This type of holder is also effective for measuring pastes and thick slurries. Another method is to use the biconical reflectance devices described previously, although these measurements tend not to be as quantitative as those using powder cells and appropriate references. Loosely packed powders may be measured in a cuvette but the reference should be measured in identical cuvettes.

Fabric samples can be stretched across a frame (a small embroidery hoop is excellent for this application) to give a flat, nonwrinkled surface. Backing the sample with light trap ensures that external light does not affect measurements. Yarn or pile samples can be bunched, bound, and then cut flat across the pile to give a flat surface to present to the sphere port.

Small solid samples should be masked (with identical mask for reference) so that the masked sample fills the sample port. The measurement of small samples is always slightly problematic. Ideally, the sample should not be overfilled, so a small spot kit (a set of lenses or focusing mirrors to reduce beam size) is used in most integrating sphere accessories. Care should be taken in making masks for the sample. These are usually thin metal or cardboard pieces that have circular or (more rarely) rectangular holes matched to the size of the sample. The masks should be painted with a very low reflectance black paint—black anodize is not sufficient because it is highly reflective outside of a narrow range in the UV–VIS. The reason for the black masks is that all sample that does not hit the sample should be absorbed, not averaged in with the reflectance of the sample. Additionally, if metal masks are used, they should be chamfered or knife edged to avoid a tunnel-like effect that can result in artificially low readings for reflectance. Additonal care should be taken when measuring highly diffuse and somewhat translucent substances (bulk rather than surface scatters) that the illuminating beam not come to close to the edge of the sample. If this occurs, incorrect (low) reflectance readings will be obtained because of the so-called translucent blurring effect (Fig. 26). Biconical or similar reflectance accessories are ideal for the measurement of small samples because they generally use a very small area of illumination.

Highly scattering liquids can be measured in reflectance using cuvettes or powder sample holders as discussed previously. Again, there is the problem with the transmitted component getting back into the sphere, causing artificially high reflectance readings. This problem can be alleviated by using cuvettes that are blackened on two sides and backing with a light trap or nonreflecting material or by surrounding the cuvette on three sides with black felt or a similar low-reflectance material.

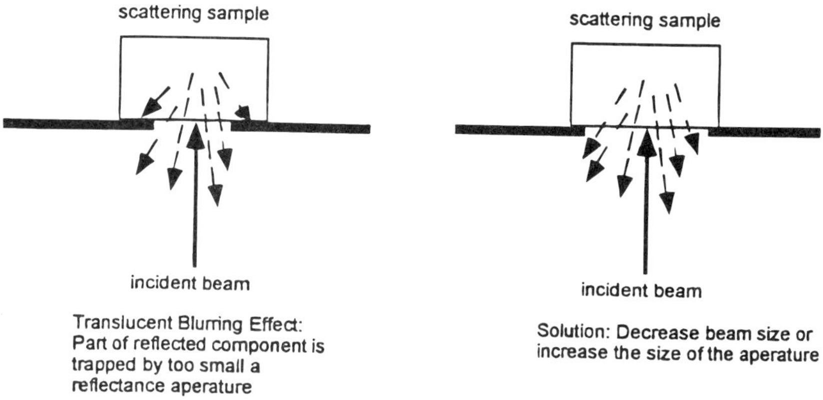

FIG. 26. Translucent blurring effect.

References

Retroflection Overview

Johnson, N. L. Colorimetry of retroreflectors. *Die Farbe* **37**, 137 (1990).

Specular measurements

Anthon, E. W., Villenueva, L., Van Horn, C., and Seddon, R. I. *Automated Precision Reflectometer of First Surface Mirrors I: Optical Head.* SPIE 1990 International Symposium on Optical and Optoelectronic Engineering, San Diego, CA (1990).

Ball, D. W. *Spectroscopy* **9**(6), 22(1994).

Ball, D. W. *Spectroscopy* **10**(1), 22(1995).

Clarke, F. J. J. Metrology and standards at NPL for the infrared region, in *Advances in Standards and Methodology in Spectrophotometry*. Elsevier, New York (1987).

Grum F., and Wightman, T. Absolute spectral reflectance of Mirrors. *Color Res. Appl.*, **8**(1) (1983).

Zwinkels J. C., and Dodd, C. X. *Appl. Optics* **28**(12), 2381 (1989).

Diffuse Measurements

Springsteen patents

Arecchi, A., and Carr, K. *A Guide to Integrating Sphere Theory and Applications*, Labsphere Technical Guide (1997).

Budde, W., and Dodd, C. X. Absolute measurements in the D/0° geometry. *Die Farbe* **19**, 94 (1970).

Clarke, F. J. J., and Compton, J. A. Correction methods for integrating sphere measurements of hemispherical reflectance. *Color Res. Appl.* **11**(4) (1986).

Freeman, G. H. C. Reference instruments for spectrophotometry at NPL,in *Spectrophotometry, Luminescence and Colour; Science and Compliance*. Elsevier, New York (1995).

Gindele, K., Köhl, M., and Mast, M. Spectral reflectance measurements using an integrating sphere in the infrared. *Appl. Optics* **24**(12) (1985).

Shetter, D., Oppenheim, U. P., and Devis, A. D. Absolute measurement of diffuse reflectance in the α/d configuration. *Appl. Optics* **30**(22) (1991).

Sheffer, D., Oppenheim, U. P., and Devis, A. D. Absolute Reflectometer for the mid infrared region. *Appl. Optics* **29**(1) (1990).

Storm, S., and Springsteen, A. W. *Quantitation of Single Beam Substitution Error*. Labsphere Technical Note (1996); *Quantitation of Single Beam Substitution Error*, Springsteen, Storm, Ricker, and Dwyer, Presented at Int'l Diffuse Reflectance Conference, Chambersburg, PA, August 1996.

Willey, R. R. *Results of a round-Robin Measurement of Spectral Emittance in the Mid-Infrared*, SPIE Paper No. 307–20, SPIE, The Hague (1987).

Hemiellipsoidal Instrument

Hanssen, L. M. New instrument development at National Institute of Standards and Technology for spectral diffuse reflectance and transmittance measurements,in *Spectrophotometry, Luminescence and Colour; Science and Compliance*, Elsevier, New York (1995).

Biconical Diffuse Reflectance

Complete Guide to FT-IR Accessories and Supplies, Spectratech Catalog (1993)

Griffiths, P. R., and deHaseth, J. A. *Fourier Transform Infrared Spectrometry*. Wiley-Interscience, New York (1986).

Fuller, M. P., and Griffiths, P. R. *Anal. Chem.*, **50**, 1906 (1978).

Fuller, M. P., and Griffiths, P. R. *Appl. Spectrosc.*, **34**, 533 (1980).

Directional–Directional

Clarke F. J. J., and Parry, D. J. Helmholtz reciprocity, its validity and application to reflectometry. *Lighting Res. Appl.*, **17**, 1 (1985).

Measurement Techniques/Sample Preparation

Brunsting, A., Hernaiz, R. S., and Dossman, A. J. Small area measurements of diffuse reflectance from 400–700 nm., *Appl. Optics* **24**(14) (1986).

Verrill, J. F. Advances in spectrophotometric transfer standards at the National Physical Laboratory, in *Spectrophotometry, Luminescence and Colour; Science and Compliance*, Elsevier, New York (1995).

General

ASTM Standards on Color and Appearance Measurement, 5th. Edition. American Society for Testing Materials (1996).

Budde, W. Calibration of reflectance standards. *J. Res. Nat. Bureau Standards*. **80**(4) (1976).

Burgess, C. and Mielenz, K. D. (Eds.), *Advances in Standards and Methodology in Spectroscopy*. Elsevier, New York (1987).

Coblentz, W. W. The diffuse reflecting power of various substances. *Bull. Natl, Bureau Standards*. **9**, 283 (1913).

Hunter, R. S., and Harold, R. W. *The Measurement of Appearance*, 2nd. Ed, Wiley-Interscience, New York (1987).

Kortüm, G. *Reflectance Spectroscopy*. Springer-Verlag, Berlin (1969).

Springsteen, A. W. *A Guide to Reflectance Spectroscopy*. Labsphere Technical Guide (1992).

Weidner, V. R., and Hsia, J. J. *Spectral Reflectance*, NBS Publication No. 250–8. (July 1987).

SPECTROSCOPY OF SOLIDS

JERRY WORKMAN, JR.
Kimberly-Clark Corporation,
Analytical Science and Technology

I. Introduction

The spectroscopy of solids is defined as the qualitative or quantitative measurement of the interaction of electromagnetic radiation (emr) with atoms or molecules in the solid state. The emr interacts as scattering, absorption, reflectance, or emission with solid matter. A variety of spectrometer configurations are used to optimize the measurements of electromagnetic radiation as it interacts with solid matter. This chapter provides an overview of

ISBN 0-12-764070-3/$25.00.

the instrumentation and measurement phenomena relative to the spectroscopy of solids. More exhaustive treatments of the subject material can be obtained from Bracey (1960), Clark (1960), Knowles and Burgess (1984), or Somorjai (1969). Tiny advances in this science occur daily, but the classical topics included within this chapter remain relatively constant.

II. The Physics of Light Interaction with Solid Materials

Light interacts with solid materials as scattering, absorption, transmission (transmittance), reflectance (both regular and diffuse), and diffraction. The purpose of spectroscopy is to quantify or qualify these interactions by the use of a variety of photon-producing and photon-detection devices. The physics of these interaction phenomena and devices will be presented in this chapter.

Solids, as black body radiators, emit light that can be characterized by its radiated power, spectral profile, and photon flux. These concepts are described in this chapter. The spectroscopy of solids is a vast field encompassing many volumes and thus the scope of this chapter is constrained to the basic mathematical concepts and relationships related to the spectroscopy of solids. For additional discussion on the interaction of light with solids see Blaker (1970), Ditchburn (1965), Fogiel (1981), Goodman (1985), Jenkins and White (1957), Johnson and Pedersen (1986), or Mach (1926).

A. Scattering

In scattering, such as Rayleigh scattered light, the intensity of the scattered energy (I_{RS}) is inversely proportional to the fourth power of the wavelength of the incident light energy (λ), as given by

$$I_{RS} \propto \frac{1}{\lambda^4}.$$

The intensity ratio of scattered light incident onto a particle surface is given by

$$\frac{\bar{I}_S}{I_0} = \frac{1}{r^2}\left(\frac{2\pi}{\lambda}\right)^4 \alpha^2 \sin\theta,$$

where $\bar{I}_S$ is the scattered radiation intensity flux per unit time (seconds), I_0 is the incident beam radiation intensity, r is the average particle radius, λ is the wavelength of the incident radiation, α is the proportionality factor (if a surface is isotropic, the α is by definition identical for all angles of reflectance); and θ is the reflected light angle versus normal incidence. Note:

$$\alpha = \frac{\text{The electric moment of the specimen}}{\text{The instantaneous field strength of the specimen}}.$$

For dilute gases, α is defined by the following (Kortum, 1969, pp. 72–81):

$$\alpha = \frac{n^2 - 1}{4\pi N},$$

where n is the refractive index of the gas particles and N is the particle density per cm^3. For additional discussion on scattering see Crosignani *et al.* (1975), Kortum (1969), Kubelka and Munk (1931), Schuster (1905), or Wendlandt and Hecht (1966).

B. Absorption and Transmittance

Light passing through an absorbing material is attenuated as a function of the absorption coefficient (a) in cm^{-1} of the material and the thickness (t) of the material in cm following the relationship:

$$I = I_0 e^{-at}.$$

The fraction (in T units) of a collimated beam of light of wavelength (λ) transmitted through a material of index of refraction (n) and a thickness (t) is given by

$$I_T = \frac{I_0}{1 + \dfrac{4r^2}{(1 \quad r^2)} \sin^2\left(\dfrac{2\pi nt}{\lambda}\right)},$$

where $r^2 = R$ (the reflectance) given for normal incidence as

$$R = \frac{(n - 1)^2}{(n + 1)^2}.$$

For additional references on absorption and transmission spectroscopy, see Braun (1987) or Clark (1960).

C. Reflectance

The reflectance of light at normal incidence passing through a material of refractive index (n_1) into a second material of refractive index (n_2) is given by the following equation:

$$R = \frac{(n_2 - n_1)^2}{(n_2 + n_1)^2}.$$

For films or coatings of refractive index (n_f) deposited on glass with a refractive index of (n_g) as is the case with antireflective coated optics, the reflection (R) from the coated surface is calculated using a simplified formula:

$$R = \frac{(n_g - n_f^2)^2}{(n_g + n_f^2)^2}.$$

D. Optical Density

The optical density (OD) of a material is defined by the mathematical relationship:

$$\mathrm{OD} = \log_{10}\left(\frac{I}{I_0}\right) = \frac{1}{2.303}\ln\left(\frac{I_0}{I}\right),$$

where I_0 is the intensity of the incident light beam, and I is the intensity of the transmitted beam. The OD of the material in which the light is passing through is given by the previous expression.

E. Diffraction

Diffraction can be loosely defined as the bending of light around objects. The defining principle for diffraction is Huygen's principle, which informs us that every point of a wave front can be considered a secondary point from which waves propagate in multiple directions. A single narrow slit can demonstrate the alternating light and dark patterns shown by diffraction.

The secondary waves propagated during diffraction have an intensity proportional to

$$1 + \cos\frac{\theta}{2},$$

where θ is the propagation angle of the secondary wave in relationship to the original (or primary) wave. This relationship defines the intensity for waves propagated in the direction of the original wave as 1 and 0 intensity for the wave propagated at 90° to the original wave. The central bright band in a diffraction pattern is approximately 20 (or more) times brighter than the next (or adjacent) bright band in a typical diffraction pattern resulting from a narrow slit. The following references discuss the phenomenon of diffraction in detail: Ball (1971), Cohen (1966), and Guild (1960).

F. Emission

The power (p) radiated or emitted by a black body radiator is given by the relationship:

$$p = kT^4,$$

where k is the Stafan–Boltzmann law proportionality constant = $5.670 \times 10^{-5}\ \mathrm{W/m^2}$, and $T = {}^\circ\mathrm{K} = ({}^\circ\mathrm{C} + 273)$.

The spectral profile of this emitted radiation is given by Planck's hypothesis as shown in the following equations:

$$E_\mathrm{p} = h\nu,$$

where E_p is the photon energy (in joules) at a given frequency (ν, in units of sec^{-1}), and h is Planck's constant = 6.6256×10^{-3} J · sec. (equivalent to 6.6256×10^{-27} erg · sec). Because $\nu = c/\lambda$, it follows that

$$E_\mathrm{p} = \frac{hc}{\lambda}.$$

Thus, using this equation the energy of each photon at a particular wavelength is determined.

Furthermore, because 1 W = J/sec, the number of photons (N_p) per second in an emission problem is calculated as

$$N_\mathrm{p} = \frac{1}{E_\mathrm{p}},$$

where N_p is the number of photons per second (photon flux).

Thus, emission light sources are generated by heating metal filaments or metal surfaces to obtain the appropriate emission spectral characteristics in power, spectral profile, and photon flux. See Fogiel (1981), Johnson and Pedersen (1986), and Ware (1971) for additional discussion on the emission phenomenon.

III. Devices Used for Spectroscopy of Solids

A. Photometers

Photometers are spectrometers characterized as instruments that use interference filters to select the wavelength incident to the sample specimen. See Bracey (1960) and Braun (1987) for additional discussions related to photometer design.

B. Dispersive Spectrometers

Dispersive spectrometers (or spectrophotometers) rely on the use of a monochromator(s) to disperse specific frequencies of light from a broadly emitting light source incident to the sample specimen. For additional details on dispersive instruments, see Ball (1971), Braun (1987), or Cohen (1966).

C. Interferometers

Interferometers utilize the interference phenomenon to produce an interferogram at a specific field of view. The interferogram is subjected to Fourier transformation mathematics to create a measurement spectrum of a sample specimen. The basic components of a classical Michelson interferometer are demonstrated by Chandler (1951) and Steel (1983).

D. Luminometers

A luminometer is used to measure the luminescence from a sample by directing light of a specific frequency onto the sample specimen and then collecting the luminescence for a specific frequency onto a detector. The major components of such a device are discussed by Curie (1963), Fogiel (1981), or Kallmann and Spruch (1962).

E. The Integrating Sphere

The purpose of an integrating sphere detector system is to provide a collection device for reflected, divergent, and scattered light from a sample. Whenever it is desirable to capture the total reflected light from a sample, the integrating sphere must be used.

The integrating sphere concept dates to approximately 1892, when W. Sumpner wrote a paper describing such a devise [*Proc. Phys. Soc. London* **12**(10)]. The integrating sphere consists of a hollow sphere or hemisphere coated with a highly Lambertian (diffusely reflecting) surface. An important characteristic of an ideal sphere is that the intensity of the reflected energy at any part of the sphere surface is proportional to the total energy entering the sphere. This relationship is ideally independent of the specific point of entry to the sphere and is independent of the incident angle of the energy into the sphere. The main considerations in the use of an integrating sphere include sphere throughput, sphere efficiency, average reflectance at the sphere wall, and sphere error.

The throughput (τ) of an integrating sphere, sometimes referred to as another measure of sphere efficiency, is given by the following expression

(Note that τ should be maximized to provide the maximum throughput):

$$\tau = \frac{A_e R_w}{A_S}\left(\frac{1}{\left[1 - R_w\left(\frac{1 - A_{T-e}}{A_S}\right)\right]}\right),$$

where A_e is the area of the sphere exit port (to the detector) (given by πr^2), R is the reflectivity of the sphere wall (e.g., 1.00 = 100%), A_S is the total area of the sphere (e.g., $4\pi r^2$), and A_{T-e} is the total area of all sphere ports minus the area of the exit port(s).

Sphere efficiency (E_S) is given by the relationship

$$E_S = \frac{A_{ep}}{(1 - \bar{R})A_S},$$

where A_{ep} is the area of the exit port (as πr^2), A_S is the total sphere area (as $4\pi r^2$), and $\bar{R}$ is the average reflectance of the sphere wall including the entrance and exit ports and the sample. The $\bar{R}$ is determined using the relationship:

$$\bar{R} = \frac{R_S A_\sigma + R_w(4\pi r^2 - \sum_{i=1}^{n} A_n)}{4\pi r^2},$$

where R_S is the reflectivity of the sample (e.g., 1.00 = 100%), A_σ is the spherical area of the sample port (as πr^2), R_w is the reflectivity of the sphere wall coating (1.00 = 100%), and r is the radius of the sphere. The expression $\sum_{j=1}^{n} A_n$ the sum of the spherical area of all ports on the sphere, where n is the number of ports, and πr^2 is the spherical area of each port, where r is the radius of the port.

For any integrating sphere, the sphere error (E_S) due the variance in sphere efficiencies between the sample and the standard when both are measured at the identical port is given by

$$E_S = \frac{\varepsilon_s}{\varepsilon_r} - 1 = -\frac{(R_r - R_s)\left(\frac{A_\sigma}{A_S}\right)}{1 - R_w\left(\frac{d}{A_S}\right) - R_\sigma\left(\frac{A_\sigma}{A_S}\right)},$$

where ε_s is the sphere efficiency when measuring a sample, ε_r is the sphere efficiency when measuring the reference reflectance standard material, R_r is the reflectance of the reference standard material (e.g., 1.00 = 100%), R_σ is the reflectance of the sample, A_S is the total sphere area, R_w is the reflectance of the sphere wall coating, and $d = A_S - A_{ent} - A_{exit} - A_\sigma$, where A_{ent}, A_{exit}, and A_σ are the areas of the entrance, exit, and sample ports as πr^2.

In addition to the previous expressions, the total power received at the detector (P_d) as a function of input power (P_i) is given using the following expression:

$$P_d = \frac{E_S A_d}{\pi\left(\frac{d_{exit}^2}{4} + \delta^2\right)},$$

where E_S is the sphere efficiency (as calculated previously), A_d is the detector window area, d_{exit} is the exit port diameter, and δ is the distance from the exit port entrance to the detector window; see Kortum (1969) or Wendlandt and Hecht (1966).

IV. Applications for Spectroscopy of Solids

For details on the application of spectroscopy to solids see Kortum (1969), Somorjai (1969), Wendlandt and Hecht (1966), or Woodruff and Delchar (1986).

A. Total Transmittance/Absorption Spectrometry

The total transmittance (or transmission) of light through solids or liquids is given by the following Beer–Lambert law relationship:

$$I_t = I_0 \times 10^{-ecd},$$

where I_t is the intensity of the transmitted light at a specific wavelength, I_0 is the incident beam intensity at that wavelength, e is the molar extinction coefficient at a wavelength for absorbing molecules within the material, c is the concentration of the absorbing molecules, and d is the thickness of the material. Note: For the previous nomenclature, ecd is the total absorption. This law is most often seen in the form:

$$\frac{1}{T} = 10^{A},$$

and by taking log 10 of each side:

$$\log \frac{1}{T} = A.$$

Absorbance is related to the concentration of the absorbing molecules as:

$$A = ecd.$$

Transmission spectroscopy is most useful with clear materials, such as thin films and crystalline plates. It is not particularly useful for opaque or turbid materials or for highly scattering materials.

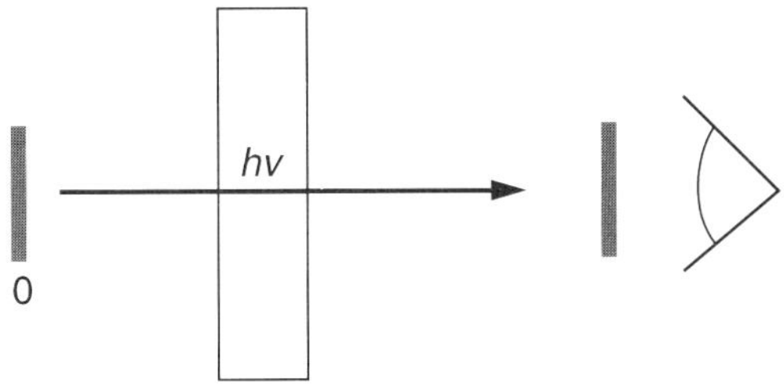

FIG. 1. Direct transmittance measurement geometry for solid samples.

The measurement should not be used when intermolecular interactions vary with concentration of absorbing molecules or solvent or when refractive index changes drastically with concentration. Turbidity will drastically reduce the value of transmission measurements, as will significant color changes in the material. This measurement geometry is referred to as direct transmission (transmittance) and is illustrated in Fig. 1.

B. DIFFUSE TRANSMITTANCE

The diffuse transmittance is defined as the total transmitted light passing through (and interacting with) a noninfinite thickness of a diffusely reflecting medium, with this medium being composed of multiple diffusely reflective surfaces. This measurement geometry is represented in Fig. 2.

C. SPECULAR (REGULAR) REFLECTANCE

Specular, external (or regular) reflectance (R_r) is described by the use of the Fresnel equation as

$$R_r = \frac{(n_2 \quad n_1)^2}{(n_2 + n_1)^2},$$

where n_2 and n_1 are the refractive indices of two materials, with n_2 being the material for which regular reflectance is measured and n_1 being the

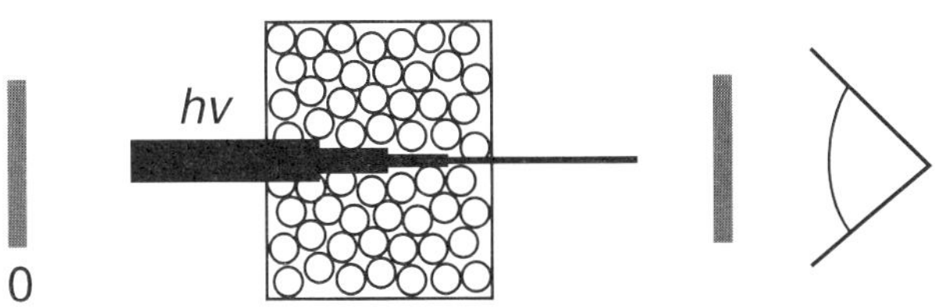

FIG. 2. Diffuse transmittance measurement geometry for solid samples.

surrounding medium. For a material in air, this relationship can be expressed as

$$R_r = \frac{(n_m - 1.0003)^2 + n_m^2 \varepsilon^2}{(n_m + 1.0003)^2 + n_m^2 \varepsilon^2},$$

where n_m is the refractive index of the material, and ε is absorption (or extinction) coefficient of the material.

Applications for external (or regular) reflectance include surface measurements for metals or semiconductors. The technique is used to measure the dielectric function of solids, for characterization of thin films, and to relate the reflectivity of a material surface to its electronic and/or surface structure. Thin film thickness (t, in units of wavelength) is measured using this technique combined with the mathematical relationship:

$$t = \frac{f}{2\sqrt{(n_m^2 - \sin^2 \alpha)} \times d},$$

where f is the number of interference fringes between initial and final fringe interval, d is the wavelength difference for the fringe count interval, and α is the angle of incidence.

The refractive index of a thin film (n_t) is calculated using

$$n_t = \sqrt{\left[\frac{(\sin^2 \alpha_1 d_1^2 - \sin^2 \alpha_2 d_2^2)}{(d_1^2 - d_2^2)}\right]},$$

where α_1 is the first angle of incidence, d_1 is the spacing between the fringes of the interference pattern created using incident light at angle α_1, α_2 is the

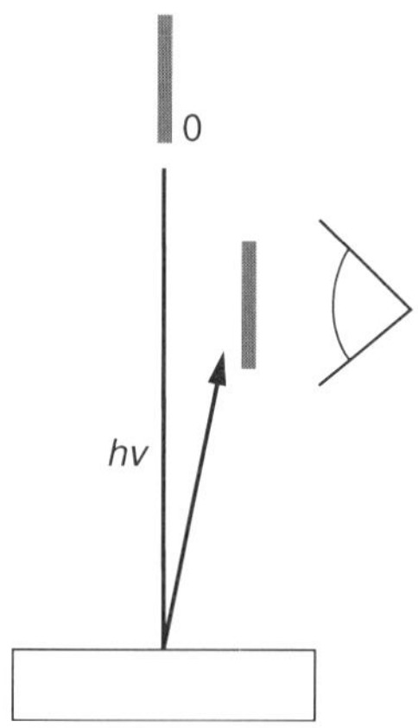

FIG. 3. Specular reflectance measurement geometry for solid samples.

second angle of incidence used to create a second interference pattern, and d_2 is the spacing between fringes of the second interference pattern created by light at angle of incidence α_2. Specular (or regular) reflectance is illustrated in Fig. 3.

D. Diffuse Reflectance

Reflectance is termed *diffuse* where the angular distribution of reflected energy is independent of the incident angle. J. H. Lambert was the first to describe this phenomenon (Lambert, *Photometrica Augsburg*, 1760) for a perfectly remitting (reflecting) surface. The technique of diffuse reflectance is one of the most widely used and will be described in detail in this chapter. The geometry for diffuse reflectance measurements is illustrated in Figs. 4–6.

The Lambert cosine law is given as

$$\frac{dI_r/df}{d\omega} = \frac{CS_0}{\pi} \cos \alpha \cos \vartheta = B \cos \vartheta.$$

The radiation flux (in W/cm^2 per unit surface area) is calculated using

$$\frac{dI_e}{df} = S_0 \cos \alpha,$$

where α is the angle of incidence of a parallel beam of light, and S_0 is the irradiation intensity in W/cm^2 for normal incidence ($\alpha = 0$).

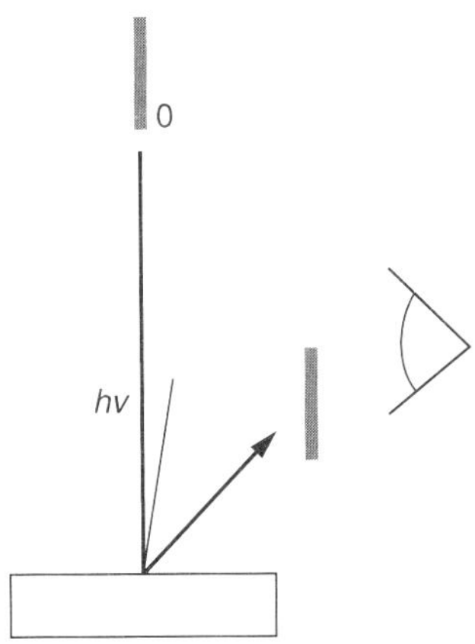

Fig. 4. Diffuse reflectance measurement geometry for solid samples.

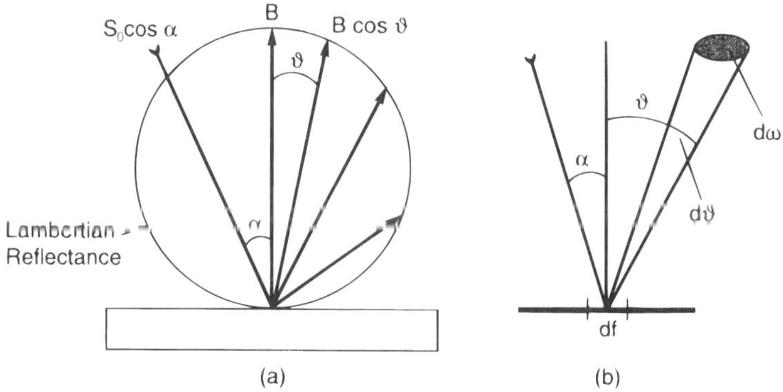

FIG. 5. The angular distribution of reflected light on a matte surface following the Lambert cosine law; the terms used in the Lambert law are illustrated.

The remitted (reflected) radiation flux per unit surface area of an irradiated disc depends on the cosine of the angle of observation for the radiating surface and is proportional to the solid angle (Fig. 5), thus the relationship

$$\frac{dI_r}{df} = B \cos \vartheta \, d\omega,$$

where ϑ is the angle of observation for the radiating surface from 0° to 90°, $d\omega$ is the area of observation (in cm^2) for the irradiating surface at some

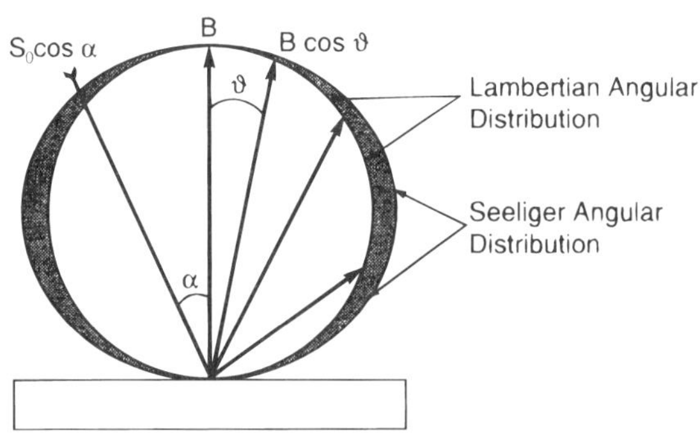

FIG. 6. The angular reflected energy distribution for the Lambert cosine law versus the Seeliger diffuse reflectance law.

distance from the surface, df is the unit area illuminated (in cm^2) by the incident energy, and B is the proportionality factor given by

$$B = \frac{dI_r/df}{\cos \vartheta \, d\omega},$$

and is referred to as the radiation density (surface brightness) factor in units of $W/\omega/cm^2$.

When an integrating sphere is used for a Lambertian (diffuse reflecting surface), the following equation applies:

$$d\omega = \sin \varphi \, d \, \vartheta \, d\phi,$$

where $d\omega$ is the solid angle for the sphere, φ is the angular range over which the energy is integrated from 0 to $\pi/2$, ϕ is the azimuth (in radians $= 360/2\pi = 57.29°$) from 0 to 2π for the hemisphere of the remitted (reflected) radiation.

The total remitted radiation strength integrated is by

$$\int_0^{\pi/2} \int_0^{2\pi} B \cos \vartheta \sin \vartheta \, d \, \vartheta d \, \phi = \pi B,$$

where π represents the angular range over which the energy is integrated (in radians; $\pi r = \frac{1}{2}$ circumference of a circle), and 2π represents the azimuth between the angles of illumination and collection (in radians; $2\pi r =$ circumference).

Seeliger (in 1888) derived a diffuse reflectance law based on the assumption that the radiation striking the surface of a powder will penetrate into the interior of the powder sample and thus does not represent a perfect remitting (Lambertian) surface (Fig. 6).

The emitted radiation in this case is the sum of all the individual elements dV of the penetrated powder volume. The solution for determining the total emitted radiation is found using

$$\frac{dI_r/df}{d\omega} = K \frac{\cos \alpha \cos \vartheta}{\cos \alpha + \cos \vartheta},$$

where the constant K includes both S_0 (the irradiation intensity for normal incidence) and the absorption constant k.

Lambert states that the remitted radiation is independent of the observation angle ϑ. In Seeliger's relationship, if the angle of incidence is $\alpha = 0°$ (for normal or perpendicular incidence) and $\vartheta = 0°$ or $\vartheta = 90°$ (observation angles), we see that for $\vartheta = 0°$ the radiation density is $K/2$, and for $\vartheta = 90°$ the radiation density is K. Therefore, in the Seeliger relationship the radiation density of remitted light increases with increasing observation angle to a first approximation.

Comparing the angular distribution of Lambert radiation to that of Seeliger, P. Bouguer (in *Traie' d'optique surla gradation de la lumie're*, Paris, 1760) carried out the first investigations to explain in theoretical terms the diffuse reflectance occurring at macroscopic surfaces. His assumption was that diffuse reflectance occurs as the regular reflectance of crystal faces whose surface planes are statistically distributed at all angles.

The following discussion applies for diffuse reflectance at absorbing materials and the specific interactions of sample absorption on reflectance. Two "processes" occur when incident energy interacts at a reflecting surface that also exhibits selective absorption. (i) Selective absorption occurs in the interior of the sample, and (ii) preferential reflection occurs at the sample surface following the Fresnel laws at the Bouguer elementary mirror surfaces.

The combination of these processes yields the spectral composition of the remitted (reflected) radiation. Deviations from the Lambert law increase with increasing ϑ and α for both absorbing and nonabsorbing samples; deviations in which strong absorption occurs are greater than those in which little or no absorption occurs; therefore, $R_{\text{regular}} = f(\varepsilon)$, where ε is the absorptivity of the sample at the wavelength of interest.

The larger the absorption index the larger the regular reflectance. The absorption is affected by particle size and is reduced with a reduction in particle size.

Arthur Schuster (1905) derived a reflectance function for isotropic scattering as

$$\frac{(1 - R_\infty)^2}{2R_\infty} = \frac{k}{s},$$

where R_∞ is the diffuse reflectance, k is the absorption coefficient, and s is the scattering coefficient. This equation is derived from the diffuse reflectance approximation for a sample layer of thickness τ approaching infinity:

$$R_\infty = \frac{J_{(\tau=0)}}{I_0} = \frac{1 - (k/(k + 2s))^{1/2}}{1 + (k/(k + 2s))^{1/2}},$$

where J is the radiation flux in the negative direction (resulting from scattering light), and k is the absorption coefficient. Von Paul Kubelka and Franz Munk (1931) developed the most general and useful reflectance function equations. Kubelka and Munk derived a description of reflectance (R_∞) similar to Schuster's:

$$\frac{(1 - 2R_\infty + R_\infty^2)}{2R_\infty} = \frac{(1 - R_\infty)^2}{2R_\infty} = \frac{k}{s}.$$

Reflectance, then, is a function of k/s. This equation can be reduced to a form useful for quantitative work as described by Olinger and Griffiths (1992).

$$\frac{(1 - R_\infty)^2}{2R_\infty} = \frac{k}{s} = \frac{(\ln 10)ac}{s},$$

where a is absorptivity of analyte, c is concentration of absorbing analyte, and s is the scattering coefficient.

Gerald Birth (1971) has added to the knowledge of reflectance measurements through extensive work. He reduced the Kubelka–Munk theory to several simple equations relating the absorption and scattering coefficients of a sample system to its diffuse reflectance, transmission, and scatter/pathlength terms. The reduction of the Kubelka–Munk theory is as follows: If $S = 2s$ (the scattering coefficient), and $K = 2k$ (the absorption coefficient), and we express

$$a = 1 + \frac{K}{S}$$

and

$$S_d = S \times d,$$

the simplified equations as presented by Birth are

$$R_0 \cong \frac{S_d}{aS_d + 1}$$

$$T \cong \frac{1}{S_d + 1}$$

and

$$S_d = \frac{R_0}{1 - aR_0},$$

where R_0 is the diffuse reflectance of the sample with a black (absorbing) background, T is the diffuse transmittance of the sample, d is the sample thickness, k is the absorption coefficient of the sample, and S is the scattering coefficient of the sample.

Given the previous equations for R_0, T, and S_d, one can derive the following cases given the absorptive and scattering properties of a sample.

Case I: Because $a \to 1$ and $S_d \to 0$, the resultant equation reduces to

$$R_0 \cong \frac{S_d}{aS_d + 1} \cong \frac{S_d}{S_d + 1},$$

$$T \cong \frac{1}{aS_d + 1} \cong \frac{1}{S_d + 1},$$

and

$$S_d \cong \frac{R_0}{1 - aR_0} \cong \frac{R_0}{1 - R_0}.$$

Because $T = 1 - R_0$ and $R_0 = 1 - T$, then

$$S_d \cong \frac{R_0}{1 - R_0} \cong \frac{1 - T}{T}.$$

Case II: When $a \to 0$ and $S_d \to 1$, then

$$R_0 \cong \frac{S_d}{aS_d + 1} \cong \frac{S_d}{0 \cdot S_d + 1} \cong \frac{S_d}{1},$$

$$T \cong \frac{1}{aS_d + 1} \cong \frac{1}{0 \cdot S_d + 1} \cong 1,$$

and

$$S_d \cong \frac{R_0}{1 - aR_0} \cong \frac{R_0}{1 - 0} \cong \frac{1 - T}{1}.$$

Case III: When $a \to 0$ and $S_d \to 0$, then

$$S_d \cong \frac{R_0}{1 - aR_0} \cong \frac{R_0}{1 - 0} \cong \frac{1 - T}{1}.$$

Case IV: When $a \to 1$ and $S_d \to 1$, then

$$S_d \cong \frac{R_0}{1 - aR_0} \cong \frac{R_0}{1 - 1 \cdot R_0} \cong \frac{1 - T}{T}.$$

Cases I and IV and II and III are alike pairs; therefore, when $a \to 1$ and $S_d \to 0$ (or 1),

$$S_d \cong \frac{1 - T}{T}.$$

When $a \to 0$ and $S_d \to 0$ (or 1),

$$S_d \cong \frac{1 - T}{1} \cong 1 - T.$$

Rewriting the equations to include the absorption coefficient term (k) the relationships become as follows:

1. For low absorptive samples:

$$\frac{k}{s} \cong \left(\frac{\frac{k}{1-T}}{1}\right) \cong \frac{k}{1-T}.$$

2. For highly absorptive samples:

$$\frac{k}{s} \cong \frac{1}{\frac{1-T}{T}} \cong \frac{kT}{1-T}.$$

In practical cases, Kubelka–Munk, $1/T$, absorbance ratios, derivatives, and other transforms have been applied to data with scattering properties. Continued research on the interaction of radiation reflected from different types of solid surfaces would be extremely useful. A summary of the measurement characteristics and their effects on reflected energy from solid surfaces is delineated in Table 1.

E. Reflectance Measurements of Solid Surfaces Using Polarized Light

The smaller the incident angle of light onto the surface of a solid material, the less is the degree of interaction (absorption) with the surface. The electronic vector (polarization) of the incident light determines the amount of interaction with a solid surface. When the electronic field vector (i.e., dipole moment) of the surface molecules in a solid is perpendicular to the field direction vector of the incident light, little interaction with the surface takes place. When the electronic field vector of the surface molecules is parallel to the field direction vector of the incident energy, an increased energy absorption interaction occurs. This orientation is termed parallel-polarized light.

When light energy is polarized in a parallel direction to the surface of a sample but perpendicular to the solid surface electronic field (i.e., dipole moment of the molecule) light will not interact with the surface and the resultant spectrum is featureless. Light that is perpendicularly-polarized exhibits a field electronic vector perpendicular to the solid sample surface. Perpendicularly-polarized light interacts with the solid sample surface, with the interaction increasing as the angle of incidence. The greatest absorption spectrum would thus occur for a solid sample surface by taking an external

TABLE 1
SUMMARY OF SOLID SAMPLE CHARACTERISTICS AND THEIR EFFECTS ON REFLECTANCE MEASUREMENTS OF SOLIDS (SEE FIGS. 5 AND 6)

Measurement characteristic	Effect(s) on reflected energy
Particle size of sample	Smaller particle size yields greater radiation density increasing with the observation angle as ($B/\cos\alpha$), and more intense glossy peaks. Decrease particle size does not yield an increase in the diffuse reflectance in most cases. The relative remittance (reflectance) increases with reduced particle size.
Packing density of sample	The greater the packing density the greater the surface gloss.
Sample preparation method	Affects remitted energy
Wavelength of incident light	Scattering efficiency is inversely proportional to wavelength; refractive index of sample is proportional to an increase in wavelength.
Refractive index of sample	Fresnel (regular) reflection increases as the refractive index of the material increases. Note: the refractive index of a sample media increases with the wavelength of incident light.
Crystalline form of sample	Surface orientation distribution determines energy distribution for reflected energy.
Surface character of sample	Smooth surfaces give more uniform reflectance and produce greater glossy reflection.
Where $\alpha = \vartheta$	Glossy points are observed, becoming more intense with increasing α.
Angle of observation	A reduction in observed energy occurs with an increase in the angle of observation with respect to the incident angle.
Absorption characteristics of sample	Preferential reflection at the sample surface due to selective absorption; the larger the absorption, the larger the regular reflectance from a surface. Absorption decreases with particle size.
Pressure on sample	High pressure combined with small particle size yields greater surface glossiness. Glossy peaks where $\alpha = \vartheta$ increase in breadth with pressure due to crystal orientation

reflectance spectrum using perpendicularly-polarized light at a large incidence angle.

Polarized light for UV–VIS–NIR measurements is often produced using a Glan–Taylor prism. The total reflectance for s- (R_s) and p-polarized (R_p)

light changes as a function of the angle of incidence. For a discussion of infrared dichroic measurements using polarized infrared energy see Appendix E.

F. Illumination

Illumination is defined as the energy of light (ε) striking a surface of specific unit area per unit time. This definition is shown using the expression:

$$\varepsilon = \frac{I_S \cos \alpha}{d^2},$$

where ε is the illumination (or light energy) in lumens (L)/mm^2, I_S is the source intensity in candlepower, α is the angle between the source light rays and a unit vector normal to the illuminated surface, and d is the distance (in mm) from the source to the illuminated surface.

Note: For unit conversions, 1 L = 7.958×10^{-2} spherical candlepower units, and 1 L/ft.^{2} = 1.0 foot-candles.

G. Illuminance and Contrast

Contrast is defined as the ratio of the difference between the maximum illuminance (e.g., in units of lux) and the minimum illuminance of a surface exhibiting two or more distinct levels of brightness (as the numerator) and the sum of the maximum and minimum illuminance. Contrast is often specified for interference-diffraction patterns given the alternating light and dark rings created by interference at a narrow slit. Contrast is given as

$$\text{Contrast} = \frac{I_{\max} - I_{\min}}{I_{\max} + I_{\min}}.$$

H. Luminance

Luminance is defined as the *luminous flux per unit area per solid angle*. If light is radiated equally from a light source in all directions, it is radiated at a solid angle (Ω_S) equal to 4π steradians (sr). Thus, the total luminance (L) of such a source, in units of W mm^{-2} sr^{-1}, is given by

$$L = \frac{P}{A\Omega_S},$$

where P is the radiated power from the source (generally in units of W), A is the area of the radiant filament (e.g., in mm^2), and Ω_S is the solid angle of radiance. For a filament source radiating in all directions this value is 4π sr.

I. LUMINESCENCE

The luminescence phenomenon is the emission of radiation resulting from an electron in an excited energy state falling to a lower energy state. The frequency of the emitted radiation depends on the difference in energy levels between the excited states and lower energy state. For luminescence to occur, one or more electrons from an atom or molecule in the ground state are excited to a higher energy state. The electron in the unstable excited state will return to the ground state and in the process emit energy as a photon. Luminescence is measured quantitatively on a perpendicular axis to the direction of the energy source. The intensity (I_Λ) of a luminescent material is measured using the relationship:

$$I_\Lambda = \beta I_0 c,$$

where β is a proportionality constant, I_0 is the intensity of the light source radiation, and c is the concentration of the luminescent atoms or molecules in a sample specimen.

J. FILM THICKNESS

A simplified formula for the spectrometric determination of film thickness under the measurement conditions in which light from the spectrometer passes from ambient air onto a single film surface and in which the underlying substrate reflects some of that light back to the spectrometer detection system is given below. In this case, the measurement spectrum demonstrates an interference pattern consisting of a continuous wave function whose period is dependent on the frequency of the energy striking the film surface. For these conditions the thickness of the film (τ) in centimeters is given by

$$\tau = \frac{N}{2(\nu_2 - \nu_1)} \times \frac{1}{\text{refractive index of the film}},$$

where N is the total number of complete waves (on the interference pattern). Note: One complete wave is measured from minima to minima or from maxima to maxima. The symbol ν_1 represents the lowest frequency over which the number of complete waves is measured, and ν_2 represents the highest frequency over which the number of complete waves is measured. For example, if three complete waves (maxima to maxima) are measured from 2000 to 1800 cm^{-1} on a single thin film having a refractive index of 2.0 and where an interference pattern superimposed on a spectrum, the film thickness is $3/2(2000 - 1800)$ cm $\times \frac{1}{2} = 0.00375$ cm (or 37.5 μm).

Figure 7 is typical of normal film thickness problems. Three samples of the same film material are measured using near-infrared spectroscopy. The

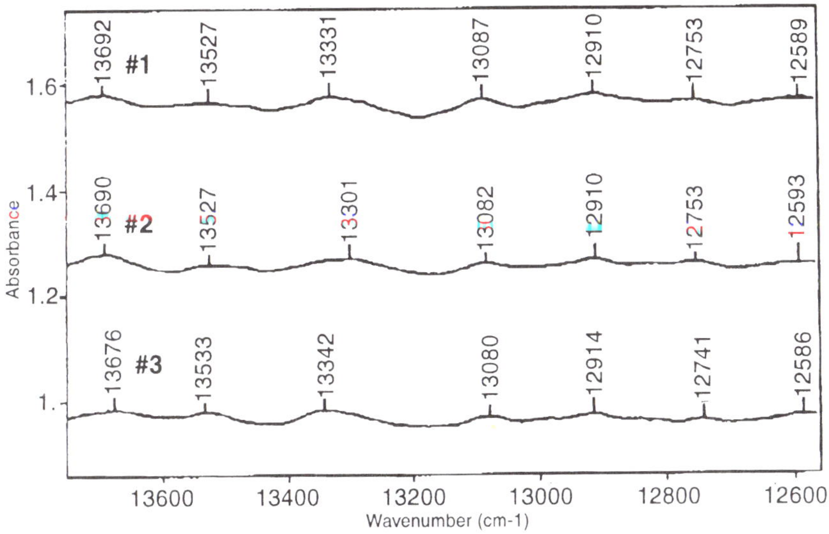

FIG. 7. Normal film thickness problem.

three examples are shown following baseline correction and after labeling the peak maxima for an ideal section of the spectrum in which the most useful interference pattern is demonstrated. The thickness of sample 1 is $6/2(1103) \times \frac{1}{2} = 0.00136$ cm $= 13.6$ μm; sample 2 is $6/2(1097) \times \frac{1}{2} = 0.001365$ cm $= 13.7$ μm; and sample 3 is $6/2(1090) \times \frac{1}{2} = 0.001375$ cm $= 13.8$ μm.

V. Conclusion

The measurement of solids using spectroscopy involves a myriad of physical phenomena that have been summarized within a few short pages. The reader is referred to the chapter on optical spectrometers for a more detailed description of the instrumentation used to make measurements on solid matter.

References

Ball, C. J. (1971). *An Introduction to the Theory of Diffraction*, pp. 134. Pergamon, Oxford, UK.

Birth, G. (1971). *Doctoral dissertation*, pp. 72–76. Purdue University, West Lafayette, IN.

Blaker, J. W. (1970). *Optics II—Physical and Quantum Optics*, pp. 91. Barnes & Noble, New York.

Bracey, R. J. (1960). *The Techniques of Optical Instrument Design*, pp. 316. The English Univ. Press, London.

Braun, R. D. (1987). *Introduction to Instrumental Analysis*, pp. 1004. McGraw-Hill, New York.

Chandler, C. (1951). *Modern Interferometers*, pp. 502. Hilger & Watts, Glasgow.

Clark, G. L. (Ed.) (1960). *The Encyclopedia of Spectroscopy*, pp. 787. Reinhold, New York.

Cohen, J. B. (1966). *Diffraction Methods in Material Science*, pp. 357. Macmillan, New York.

Crosignani, B., DiPorto, P., and Bertolotti, M. (1975). *Statistical Properties of Scattered Light*, pp. 226. Academic Press, New York.

Curie, D. (1963). *Luminescence in Crystals*, pp. 332. Wiley, New York.

Ditchburn, R. W. (1965). *Light*, 2nd ed., pp. 833. Blackie, London.

Fogiel, M. (Ed.) (1981). *The Optics Problem Solver*, pp. 817. Research & Education Association, New York.

Goodman, J. W. (1985). *Statistical Optics*, pp. 550. Wiley, New York.

Guild, J. (1960). *Diffraction Gratings as Measurement Scales*, pp. 211. Oxford Univ. Press, London.

Jamieson, J. A., McFee, R. H., Plass, G. N., Grube, R. H., and Richards, R. G. (1963). *Infrared Physics and Engineering*, pp. 673. McGraw-Hill, New York.

Jenkins, F. A., and White, H. E. (1957). *Fundamentals of Optics*, pp. 637. McGraw-Hill, New York.

Johnson, C. S., Jr., and Pedersen, L. G. (1986). *Problems and Solutions in Quantum Chemistry and Physics*, pp. 429. Dover, New York.

Kallmann, H. P., and Spruch, G. M. (Eds.) (1962). *Luminescence of Organic and Inorganic Materials*, pp. 664. Wiley, New York.

Knowles, A., and Burgess, C. (Eds.) (1984). *Practical Absorption Spectrometry*, pp. 234. Chapman & Hall, New York.

Kortum, G. (1969). *Reflectance Spectroscopy*, pp. 366. Springer-Verlag, New York.

Kubelka, V. P., and Munk, F. (1931). *Z. Tech. Physik.* **12**, 593.

Mach, E. (1926). *The Principals of Physical Optics*, pp. 324. Dover, New York.

Olinger, J., and Griffiths, P. (1992). In *Handbook of Near-Infrared Analysis* D. Burns and E. Ciurczak, (Eds.). Dekker, New York.

Schuster, A. (1905). *Astrophys. J.*, **21**, 1.

Smith, R. J. (1980). *Electronics: Circuits and Devices*, 2nd ed., pp. 494. Wiley, New York.

Somorjai, G. A. (Ed.) (1969). *The Structure and Chemistry of Solid Surfaces*, Wiley, New York.

Steel, W. H. (1983). *Interferometry*, 2nd ed., pp. 308. Cambridge Univ. Press, Cambridge, UK.

Strobel, H. A., and Heineman, W. R. (1989). *Chemical Instrumentation: A Systematic Approach*, 3rd ed., pp. 1210. Wiley, New York.

Ware, W. R. (1971). Transient luminescence measurements. In *Vol. 1. Creation and Detection of the Excited State* A. A. Lamola, (Ed.), pp. 213–300. Dekker, New York.

Wendlandt, W. W., and Hecht, H. G. (1966). *Reflectance Spectroscopy*, pp. 298. Wiley, New York.

Wist, A. O., and Meiksin, Z. H. (1986). *Electronic Design of Microprocessor Based Instruments and Control Systems*, pp. 287. Prentice Hall, New York.

Woodruff, D. P., and Delchar, T. A. (1986). *Modern Techniques of Surface Science*, pp. 453. Cambridge Univ. Press, Cambridge, UK.

STANDARDS FOR REFLECTANCE MEASUREMENTS

ART SPRINGSTEEN
Labsphere, Inc.

Reflectance is how we see the world, whether with our own eyes or through modern instrumentation that allows us to see beyond the rather limited range of wavelengths that make up human vision. The ability to see color, to visualize the radiation—no matter what the wavelength—from a building, an airplane, or a planet depends on the measurement of how that radiation is reflected. Until the 20th century, all measurement of reflectance was done by the eye, but the age of instrumentation has allowed us to "see" over a much wider range of the spectrum. These measurements do not come without some inherent problems because there are few convenient methods of measuring the reflectance of objects in an absolute sense (absolute being defined as a direct, nonreferenced measurement). Thus, one of the overriding factors in producing accurate measurement of reflected radiation is the need for stable, easily used standards.

Standards can be roughly grouped into two major groupings, specular, or mirror-like, and diffuse, or scattering. Specular standards are usually highly polished surfaces, with the reflecting layer a metal, typically silver, aluminum, gold, or beryllium. In many cases, this metal surface is over-

ISBN 0-12-764070-3/$25.00.

coated with a relatively inert, transparent, nonporous layer of material, such as magnesium fluoride or aluminum oxide, that serves to protect the fragile metal surface.

Diffuse or scattering samples can have many appearances or compositions. Materials such as ceramic tiles, packed powdered metal salts or polymers, painted chips, simple molded plastics, or more esoteric materials such as plasma-sprayed metals or sintered porous polymers, are all successfully used as standards for diffuse reflectance.

The overriding concern for any reflectance standard is its stability; perhaps it seems obvious that the foremost thought for any standard is that it give the same values over a period of years. In some cases, this is simply not possible—almost any of the commercial colored reflectance standards that are in the color gamut from yellow to red will likely be thermochromic to some extent. These materials must be used at the same temperature at which they were calibrated or their absorption spectra will shift; even radiation from the measurement instrument is enough to cause measurable changes in the reflectance spectrum. In other cases, materials are used on a disposable basis. A typical laboratory will use either packed barium sulfate or polytetrafluoroethylene (PTFE) powder as a primary standard for diffuse reflectance measurements. These well-characterized materials are quite chemically stable, but packed powders develop a static charge during their pressing and will attract dust, hair, or lint from the air, thus changing their reflectance properties over time. Such materials are also highly absorbent of airborne fumes such as hydrocarbons, which affect their reflectance in the ultraviolet region of the spectra. These disadvantages can be overcome by frequently producing new standards of fresh, uncontaminated material.

For those of us who are involved in reflectance measurements, there are a number of "rules of thumb" that are traditionally followed. Perhaps the most important (and not often always followed!) is to match the scattering properties of the standard with those of the material being measured. This is particularly important for reflectance measurements using hemispherical geometry. Thus, the measurement of specular surfaces should be made against a standard mirror. In fact, it is even recommended that the type of specular reference should be similar to samples—first surface vs first surface and second surface vs second surface. Similarly, whenever possible, diffuse or scattering samples should be measured against a highly diffuse reference. Both the International Commission on Illumination (CIE) and American Society for Testing Materials (ASTM) recommend the perfect reflecting diffuser, a hypothetical reflecting material that does not absorb light of any wavelength and reflected radiation is in a perfect cosine response to incident radiation. Such a material does not exist, but a number of standard materials that approach this ideal will be discussed.

I. Specular Standards

An ideal specular standard is one that perfectly reflects all incident radiation back at an angle and opposite to the incident angle with no scatter. Such a material is a theoretical concept because all materials show absorbencies over certain wavelengths. Although no such perfect specular materials exist, there are many materials that approach this criteria over a specific wavelength range. Typically, such materials are highly polished metallic surfaces, usually overcoated with an evaporated film to protect the metallic surface from scratching or oxidation.

Two types of specular references are used as standard materials. The first type, called first-surface mirrors, reflect incident radiation off the primary contact surface. These are typically vapor deposited or electroplated metals—gold, silver, rhodium, or aluminium on metallic or nonmetallic (glass, quartz, or some other material that is thermally and mechanically stable) substrates. An important consideration is the flatness of the surface to be coated. The material must have minimal roughness and the surface must be as flat as possible to avoid measurement error.

A. Ultraviolet, Visible, and Near-Infrared Specular Standards

Commercial standards of this type are readily available. Primary standards are available from national laboratories such as the National Institute of Standards and Technology (NIST) (SRM 2003 and SRM 2011) and the National Physical Laboratory (NPL, United Kingdom) as well as secondary grade standards from Labsphere (SPRS series) among others. These materials are first-surface aluminum mirrors, generally overcoated with a protective layer of evaporated magnesium fluoride, and aged to give a stable specular surface. For use in the near-infrared (NIR) range of the spectrum, first-surface gold mirrors are also frequently used. Typically, any of these standards should be recalibrated on a frequent schedule because they are not amenable to cleaning and can be easily damaged during the measurement process. Typical reflectance data for these mirrors are usually reported at fairly wide intervals from 250 to 2500 nm due to the relative flat spectral character, although data are usually reported over a much more narrow range where the aluminum absorption occurs. Data at very narrow wavelength intervals are available through the National Research Institute of Canada and other sources. Calibration of primary standard mirrors is usually performed in an absolute geometry using so-called V-N or V-W devices or other optically equivalent methods.

Typical reflectance curves for the NIST standard mirrors are shown in (Fig. 1). First-surface silver mirrors are becoming increasingly rare because they are typically prone to oxidation. Rhodium mirrors, although highly

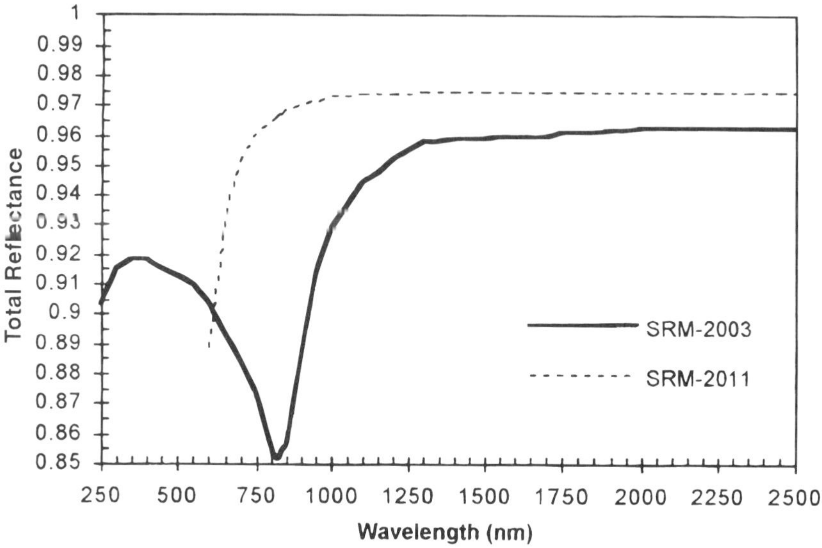

FIG. 1. Reflectance of NIST Standard Mirrors 2003 × (first-surface aluminum) and 2011 (first-surface gold).

stable, do not exhibit the high reflectance desired in a standard mirror.

There is also available a low-reflectance specular standard. A highly polished black glass substrate, calibrated from 250 to 2500 nm (0.25–2.5 μm) is available from NIST under the designation SRM-2026. This material is similar to the so-called Black Carrera glass formerly manufactured by Pittsburgh Plate Glass. A similar reflectance material can be obtained by using the first surface of an optically polished quartz flat coated on the rear surface with an index matching fluid and a suitable absorptive material such as carbon black.

B. Mid-Infrared Specular Standards

Two commercial sources of mid-IR specular standards are available. NIST offers SRM-2011, a first-surface gold on glass mirror, and NPL offers a similar mirror. At this time, the NIST mirror is only available calibrated in the 600–2500 nm (0.6–2.5 μm) range, whereas the NPL mirror can be obtained with a calibration extended to 20,000 nm (20 μm).

C. Second-Surface Standards

Second-surface standards are produced by depositing a thin but opaque film of aluminum or silver on the back of glass or quartz and then coating with a layer of inert, impervious material. Such standards are similar to those of mirrors manufactured commercially for home and automotive use. These standards are highly stable and should be used when performing measurements on other second-surface specular reflectors. These materials are most commonly used in the visible (VIS) and occasionally NIR regions of the spectrum. A commercially available standard may be obtained from NIST under the designation SRM-2023x.

II. Diffuse Reflectance Standards

A. High-Reflectance UV–VIS–NIR Standards

Both the CIE and ASTM have agreed that the ideal standard for measurement of diffuse reflectance is the perfect reflecting diffuser. This is a conceptual ideal material that exhibits no absorbances over its range of use that reflects all radiation in a cosine response to the incident angle of the illuminating beam. Although no such material exists, many attempts have been made over the years to find a material that is close to that concept.

W. Erb of the Physikalisch–Techniche Bundesanstalt has characterized an ideal standard as having the following properties: It is (1) transportable or, if not transportable, then easily reproduced with high reliability; (2) stable with respect to time, radiation, temperature, atmosphere, etc.; (3) homogeneous, with a smooth surface; (4) diffusely reflecting; (5) spectrally nonselective; (6) nontransparent (not translucent); (7) nonfluorescent; and (8) easy to handle.

Early materials used to realize this concept included magnesium carbonate, chalk, or magnesium oxide, either packed as a powder or collected as a smoke from burning high-purity magnesium ribbon. Both magnesium carbonate and chalk have relatively low reflectance for an ideal white material. Magnesium oxide was a great improvement but suffered from not being particularly stable. Burning magnesium produced a very white grade of magnesium oxide, but the material was not stable due to the concurrent formation of small amounts of magnesium nitride in the burning process. This magnesium nitride slowly decomposed in the presence of atmospheric moisture to form oxides of nitrogen that remained in the packed powder. These oxides lowered the reflectance of the material, particularly in the UV and blue region of the spectrum. An improvement on magnesium oxide came in the form of very high-purity barium sulfate, a material that is still widely used in reflectance measurements.

Barium sulfate is not without its problems. First, an extremely high-purity barium sulfate must be produced that is free from the typical iron impurities found in commercial grades. Second, the reflectance of the material is dependent on the thickness, particle size of the starting material, and the density of the packed powder. Typically, this powder is packed to a density of 2.0 g/cm^3. In addition, the pellets packed from even the finest barium sulfate are extremely brittle and prone to absorbing impurities from the atmosphere or gathering dust and lint due to the static charge induced by pressing the material. Commercial grades of barium sulfate that are suitable for use as standards include Zeiss's Barium Sulfat für Weisstandard, Eastman Kodak's White Reflectance Standard Grade, and Wako Pure and Fine Chemical Grade (Japan). It has been observed that the Eastman Kodak product does exhibit a small degree of blue luminescence when irradiated at approx 360 nm. The author has found that there are also commercially available pharmaceutical grades of barium sulfate that equal the materials described previously for reflectance and do not show any luminescent properties (Fig. 2).

Although barium sulfate is a suitable reflectance standard for many uses, the material exhibits some absorbance bands in the NIR, along with absorbances in the UV if any impurities are present. In 1976, Franc Grum of Eastman Kodak and Max Saltzman of UCLA reported to the CIE a new white standard for reflectance. The material was PTFE (also known by the

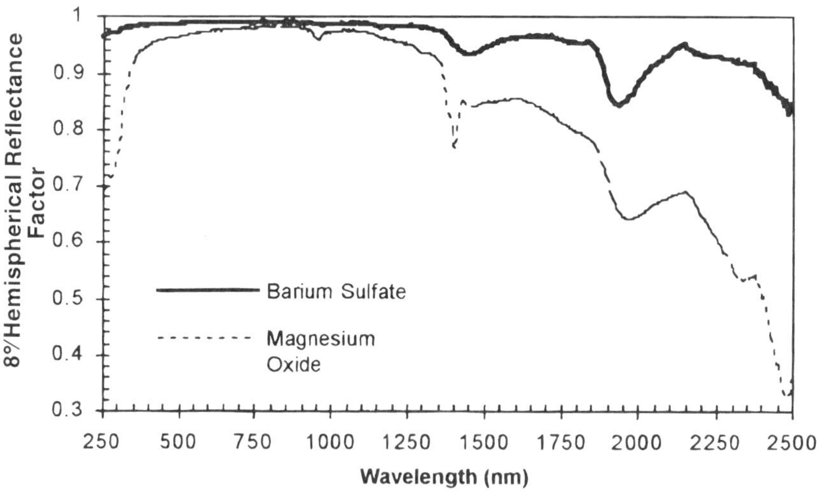

FIG. 2. 8° Hemispherical reflectance factor for packed barium sulfate and magnesium oxide powders.

Dupont trade name of Teflon, or in Grum and Saltzman's original work, Allied Halon G-80) powder, packed to a density of 1.0 g/cm^3. Grum and Saltzman's work was rapidly followed by papers by Jack Hsia and Vic Weidner (1981) of the National Bureau of Standards (now NIST) that thoroughly characterized packed PTFE. The initial Weidner and Hsia paper reported absolute reflectance data for packed PTFE powder from 200 to 2500 nm (0.2–2.5 μm) and established this material as the standard of choice for diffuse reflectance measurements in the UV–VIS–NIR region of the spectrum. A follow-up study involving a laboratory intercomparison of results showed that even as much as a 20% error in compression of the powder gave excellent results in consistency of reflectance (Fig. 3).

Packed PTFE exhibits extremely high reflectance and is close to a perfect Lambertian scatterer over a very wide reflectance range (190–2500 nm/0.19–2.5 μm). The material is nonhydroscopic and is very easily prepared from commercially available sources. The powdered PTFE is also reasonably inexpensive—currently about $8.00/pound in bulk.

Packed PTFE is not without its problems as a standard. The packed material is quite soft and is easily damaged by placing it with too much pressure against the reflectance port of the measurement instrument. In addition, the standards readily attract dust due to the formation of static charges on the powder induced during the packing process. Although not

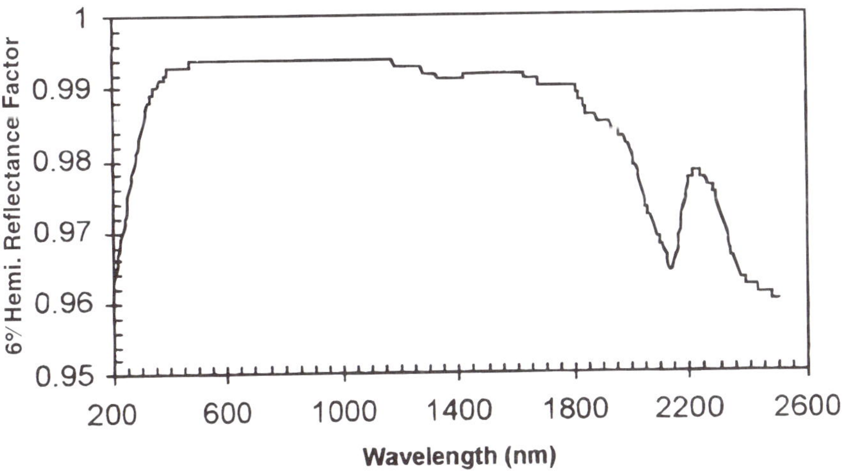

FIG. 3. 6° Hemispherical reflectance for packed PTFE powder (Halon G-80) from Weidner and Hsia (1981).

typically performed in many laboratories, it is good practice to make new standards on a daily basis.

Good packed PTFE standards must also be packed at a consistent, uniform density with a press and holders made exclusively for that purpose. Such presses have been described in the ASTM standard method for production of packed standards and are available from a number of commercial sources. The source of the PTFE is also crucial—not all PTFE powders are equivalent in particle size and purity. The ASTM again specifies certain materials, but testing in the author's laboratory has found that materials other than those specified by ASTM are also acceptable substitutes for grades of PTFE that are recommended but no longer commercially available.

A final problem with packed PTFE standards is their translucency. Although Grum and Saltzman (1976) state that a 2-mm-thick pellet of packed PTFE is opaque, it has been shown that such material is in fact quite translucent and will exhibit a phenomenon called the "translucent blurring effect." Although this effect is not important if the illumination spot on the standard is small and centered compared to the size of the measuring port, it can lead to significant errors if the illumination spot size is large or comes too close to the edge of the measurement port (Fig. 4).

Another route to stable high-reflectance standards is the use of ceramic tiles or white glass materials. Such materials are usually used only in the visible region of the spectrum due to absorbances shown by the pigments in the UV and NIR. Typically, the tiles are a ceramic baked onto a metallic substrate, with the white pigment usually being titanium dioxide. Such standard materials are provided by Erie Ceramics and the OMH. These standards are used heavily in the color measurement instrumentation in which durability is the key factor in their selection. A similar standard, under the appellation of SRM-2019, has been provided by NIST for use in the

Aperture small compared to illumination spot; some of scattered radiation blocked by aperture

Aperture large compared to illumination spot; no scattered radiation blocked by aperture

FIG. 4. Small aperture (left) exhibits translucent blurring effect, whereas large aperture compared to illumination area (right) eliminates error caused by this effect.

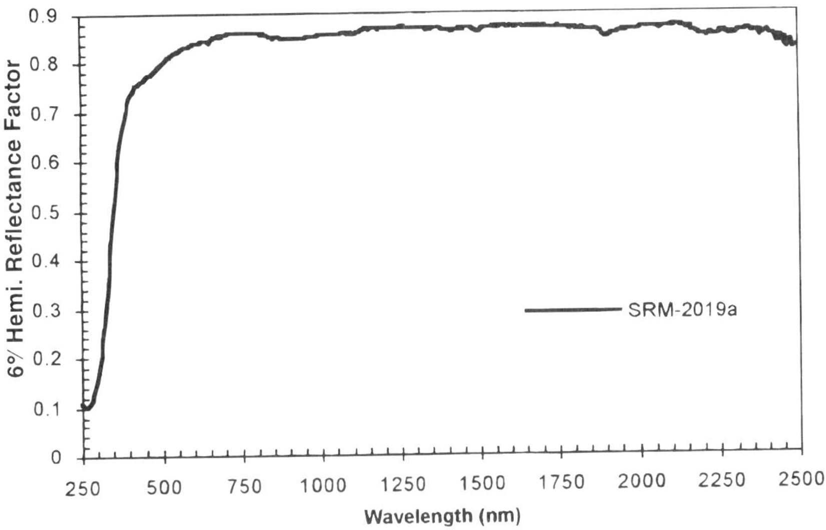

FIG. 5. 6° Hemispherical reflectance of NIST standard tile 2019x.

VIS–NIR region of the spectrum (Fig. 5). White glass materials, commonly known as "opal glass"—the most familiar of these are the so-called Russian Opal (also known as MS-20 milk glass) and Japanese opal glass—are also frequently used by manufacturers of color measurement instrumentation (Fig. 6).

Tile and glass standards exhibit some properties that render them less than ideal as diffuse reflectance standards. Many have a limited useful wavelength range, and most have a fairly significant specular component due to the smooth finish that gives them their durability. This specular component, unless compensated for in instrument design, can lead to significant errors in measurement of reflectance of nonglossy materials.

A few of the opal glasses may be obtained with a matte or diffuse surface. Although this alleviates some of the problems associated with the specular component, the materials are easily contaminated and are very difficult to clean once soiled.

A material that meets most of the criteria for an ideal diffuse high-reflectance standard is a sintered PTFE material with the trade name of Spectralon. Early work by V. Weidner and J. Hsia showed that sintered pucks of Halon-G-80 exhibited high-reflectance properties. In 1986, Springsteen of Labsphere, Inc. produced the first commercial grades of sintered PTFE for use as standards and remote sensing targets. The material is now perhaps the most commonly used reflectance standard for reflectance instrumentation worldwide.

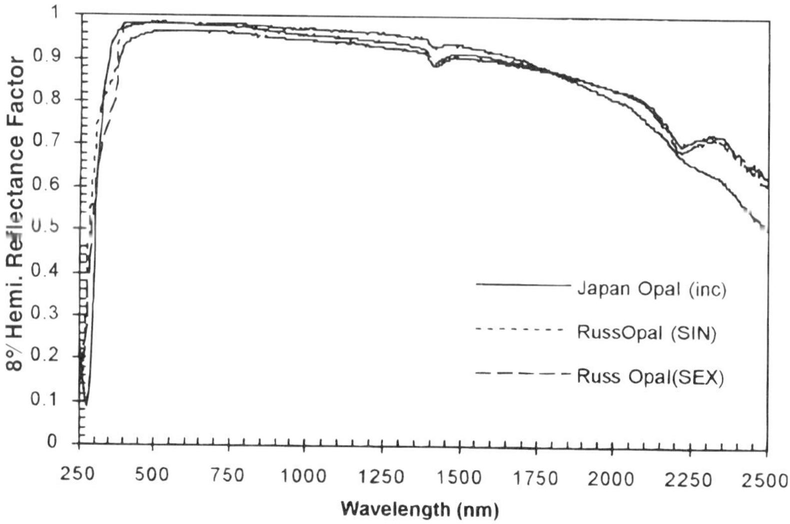

FIG. 6. 8° Hemispherical reflectance factor for Russian opal (specular included and excluded) and Japanese opal glass.

Spectralon exhibits almost perfectly Lambertian reflectance over most of its usable range of 200–2500 nm. The total hemispherical reflectance is generally greater than 95% from 250 to 2500 nm and in the vicinity of 99% from 350 to 1500 nm. The material also shows no strong absorbance bands anywhere in its range of use. Spectralon is also stable to >300°C and may be heated and cooled with no changes in reflectance properties.

Perhaps the most significant feature of Spectralon is its ability to be shaped and refinished without the loss of its reflectance properties. The material can be made as large targets or blocks and machined to shape. Refinishing the material can be accomplished, with no loss of the reflectance properties, by washing or mild sanding.

The characteristics that give Spectralon its unique optical properties are also those that give it weaknesses. The material is highly porous to hydrocarbons (it is totally hydrophobic when clean) and thus will absorb these materials with significant change in reflectance where the contaminants absorb. The absorption is reversible by heating the material, especially under mild vacuum. Spectralon is also somewhat translucent and exhibits the translucent blurring effect. This effect can be eliminated by doping the Spectralon with a small percentage of high-purity barium sulfate at a small cost in the upper end of reflectance (Fig. 7).

A recent development for applications requiring very large (in some cases as large as 3 m^2) reflectance standards for remote applications is that of

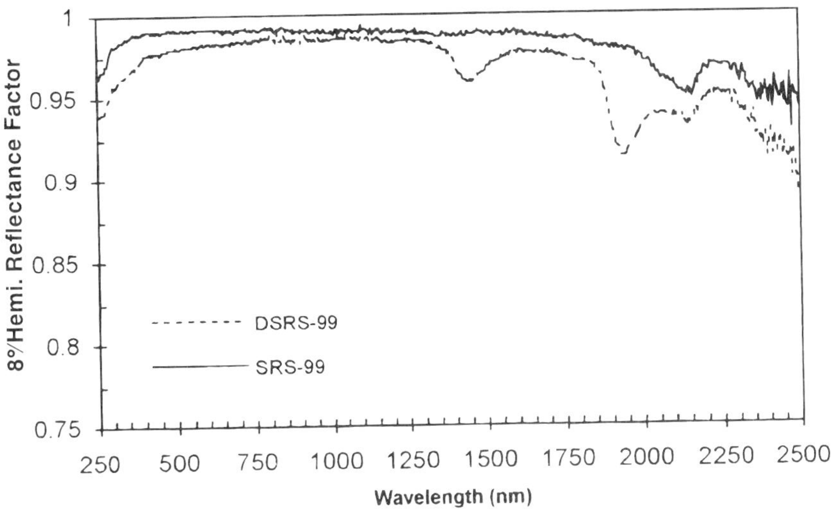

FIG. 7. Spectralon SRS-99 and barium sulfate-doped spectralon.

materials called SORIC screens. These large, woven fabric panels have been used for calibrants for telescopic and machine vision systems. Although their texture limit their applications, SORIC screens have uses as standards where no other materials are feasible (Fig. 8).

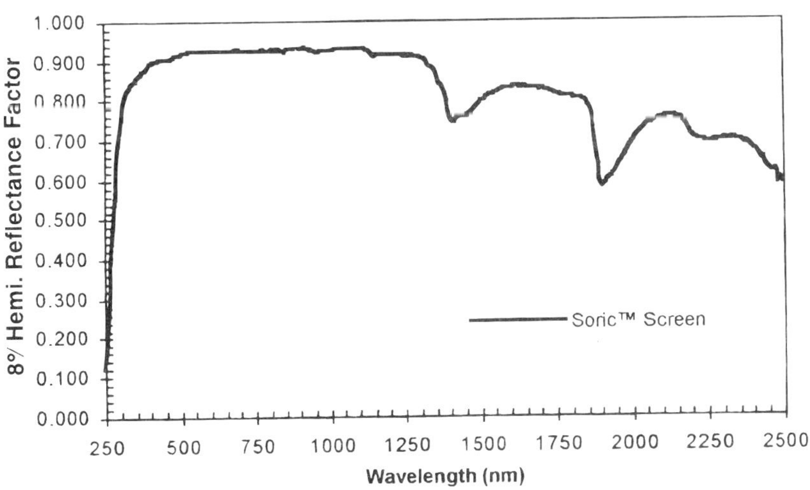

FIG. 8. 8° Hemispherical reflectance of SORIC screen standard.

III. Mid-Infrared Diffuse High-Reflectance Standards

Diffuse reflectance standards for use in the mid-infrared are a fairly recent development. Materials used in the UV–VIS are not acceptable due to absorbances shown by most polymers and minerals in the mid-IR region of the spectrum. It is also difficult to produce Lambertian standards in the mid-infrared. Highly Lambertian standards used in the UV–VIS–NIR are typically bulk reflectors; that is, they get their Lambertian properties from a combination of surface and internal scattering of the incident light. With one exception, this is not possible in the mid-IR, so scattering properties must come from making a properly rough surface structure. To produce a surface with extremely high roughness, high reflectance, and chemical stability is a problem that has only recently been solved.

Work by Clarke at NPL and Wiley (1987) has established the need for high-reflectance standards for integrating sphere and hemiellipsoidal measurements in the 2.5 to 20 μm region of the spectrum. Work by Nash (1986) at Jet Propulsion Laboratory, following early work by Kronstein and Kraushaar, suggested that packed, powdered sulfur was a potentially useful standard for mid-IR reflectance. A slightly more stable, if more difficult to prepare, material suggested as a standard was vapor-deposited gold on silicon carbide paper. Neither of these materials is ideal as a standard due to the fragility of the surfaces.

The first robust reflectance standard for mid-IR applications was produced by Frank J. J. Clarke at the National Physical Laboratory. The material is a flame-sprayed aluminum of sufficiently large particle size to produce a highly Lambertian surface. The drawbacks of this material are that the reflectance is in the range of 80%R, a bit low for a high-reflectance material, and that aluminum does form a surface oxide that shows absorbances in the spectral range of interest.

Materials that more closely fit the definition of an ideal standard are gold-plated surfaces on coarse metallic substrates. The two such commercially available materials were developed by Springsteen and coworkers at Labsphere. Infragold is a multilayer metallic on a mechanically roughened metallic substrate. This material exhibits a spectrally flat reflectance between 94 and 96% through the range 1–25 μm. The material is Lambertian at lower wavelengths but becomes increasingly specular as the wavelength increases.

To produce a more Lambertian surface, the Labsphere group produced a material that is an amalgamation of the previously discussed two technologies. The resultant material, called Infragold-LF, is a multilayer metallic coating, with the final layer being pure electroplated gold, over a substrate of arc or plasma-sprayed base metal. Infragold-LF exhibits excellent Lam-

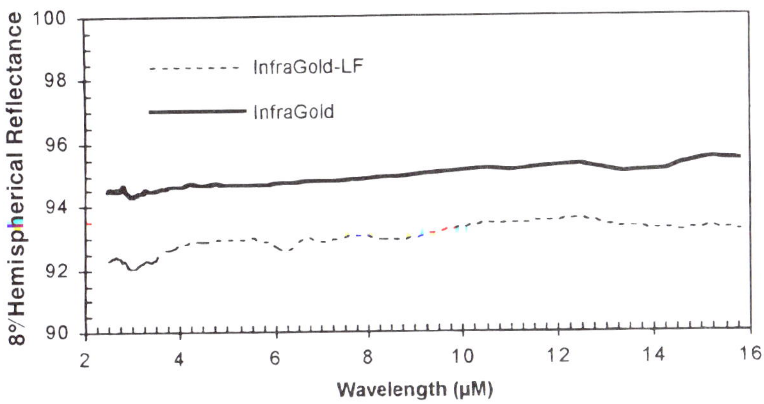

FIG. 9. 8° Hemispherical reflectance of InfraGold and InfraGold-LF.

bertian properties well into the mid- and even far-IR, with a spectrally flat reflectance slightly lower than that of Infragold. Because both Infragold and Infragold-LF have a reflective surface of pure gold, they are not prone to oxidation and are immune to all but the most corrosive chemicals (Fig. 9).

IV. Gray Scale Standards for UV–VIS–NIR

Gray scale standards of reflectance are the equivalent of neutral-density filters in transmittance spectroscopy. These materials are used in determining the linearity of detector systems for reflectance spectrophotometers. Although these materials are common in the field of photography and densitometry in the form of gray scale cards produced by Kodak and other film manufacturers, the equivalent in diffuse reflectance standards has been, until recently, difficult to produce or obtain.

The ideal gray scale material is nonglossy, uniform over its surface, and spectrally flat. That is, the reflectance does not change appreciably over the standard's range of use. Although photographic or paper-based gray scale meets the first two criteria but not the latter, ceramic tiles and plastic plaques have been used in the visible region of the spectrum tiles, due to the metal oxides used to produce the gray scale, generally have much higher reflectance in the near-IR than in the visible. Plastic plaques are usually pigmented with a mixture of carbon black and titanium dioxide. The titanium dioxide renders them unfit standards below 400 nm due to strong absorbance by the pigment. Consequently, specially formulated materials are necessary to produce standards that have a wide wavelength range of spectral flatness. Work by Lindberg showed that admixtures of finely powdered barium sulfate and carbon black produced standards that were quite uniform, spectrally flat, and nonglossy. The difficulty was in the preparation and lot-to-lot reproducibility.

A potential solution to the diffuse reflectance gray scale problem was suggested by Weidner *et al.* (1986) of the National Bureau of Standards. The NBS scientists blended increasing dilutions of carbon black in PTFE powder and then pressed and sintered it to form pellets of durable, refinishable solids with reflectance varying on the concentration of carbon black. Unfortunately, the Weidner *et al.* made no attempt to quantitate or codify the dilutions to produce commercial standards. Working concurrently, Springsteen was able to produce similar materials but with a degree of predictability not achieved by the group at NBS, and a patent was awarded for the work. The materials, known under the trade name of Spectralon gray scale, can be produced in reflectance values ranging from approximately 1–95%R in sizes normal for use in standard reflectance spectrophotometers to large standards for use in remote sensing applications. These materials have gained wide acceptance in the calibration of near-IR spectrophotometers used in the pharmaceutical and food industry. This use has recently been published as an ASTM standard (Fig. 10).

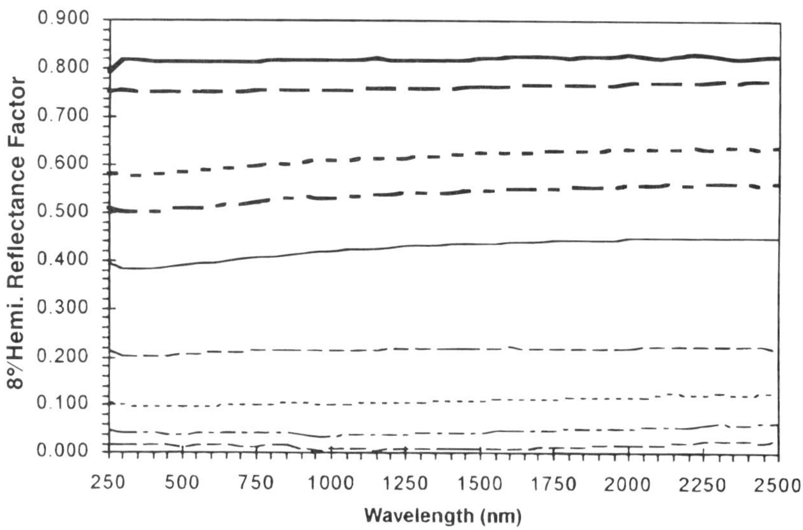

FIG. 10. 8° Hemispherical reflectance factor for a nine-step Spectralon gray scale. Nominal reflectances from bottom: 2, 5, 10, 20, 40, 50, 60, 75, and 80%.

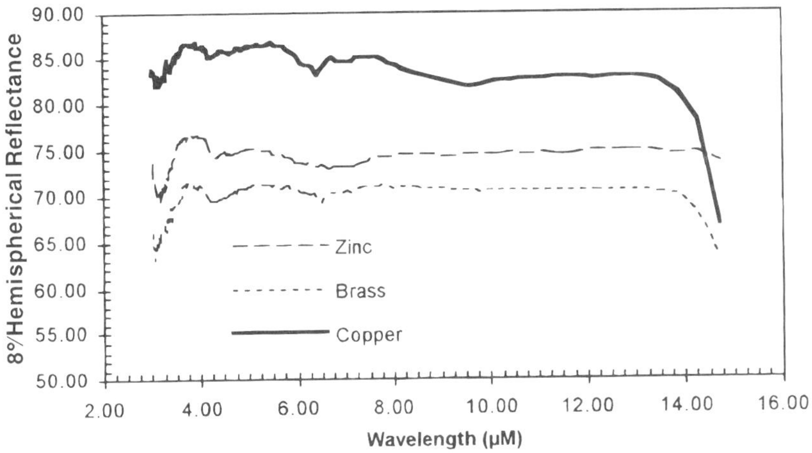

FIG. 11. 8° Hemispherical reflectance for mid-infrared gray scale (arc copper, zinc, and brass).

V. Mid-Infrared Gray Scale Standards

The equivalent of gray scale for mid-IR spectrophotometers has only recently become a topic of interest. With the increasing use of integrating spheres for use in measurement of geological specimens, a method for determination of the linearity of response of these systems has become important. Recent work by Salisbury (personal communication) at Johns Hopkins University Center for Earth and Terrestrial Science and A. Springsteen of Labsphere has shown that arc or plasma-sprayed metals, such as zinc, bronze, or brass, may serve as stable, highly diffuse standards for calibration of both integrating sphere and field instruments used for remote sensing applications (Fig. 11).

VI. Color Standards

The topic of color standards is a subject broad enough for its own monograph. Traditional color standards date back centuries for comparison of paint pigments, the color and clarity of oil and beer (Lovibond comparators), precious gems, and a variety of other applications in which color was a determining factor in the value of objects. Such visual comparison standards are still widely used in industry. Paper-based standards from Pantone, Munsell, and Colorcurve, along with paint chips and molded plastic materials from the same manufacturers, form the basis of color standardization in the textile, automotive, and plastics industries but are outside of the scope of this chapter.

More robust standards than paper are generally used in the standardization of spectrophotometers and spectrocolorimeters. These include ceramic tiles, such as those from the British Ceramic Research Association (BCRA), that are available in both calibrated and uncalibrated form from the NPL and other associated laboratories. Similar tiles are available from Erie Ceramics but must be sent to an outside agency for calibration. These tile sets provide both highly saturated and pastel colors, along with a number of nonneutral grays. The BCRA sets also provide a metameric pair to allow the user another method of checking instrument accuracy.

A similar set of standards, but on a smaller size scale and in plastic rather than ceramic, are available from F. T. Simon, Inc. The pigment-doped acrylic plates are available calibrated at either 0°/45° or hemispherical geometry.

Both ceramic tiles and plastic standards are glossy, which may be an undesirable feature in standardization of sphere-based instruments. The specular component of these standards, unless completely removed, can lead to errors in measurement of diffuse materials. Nonglossy standards are available but have other disadvantages. The BCRA tiles have recently been prepared in a matte (nonglossy) format but are not easily cleaned if soiled. Lapshere Spectralon color standards are nonglossy and easily cleaned but are not as spatially uniform, especially if measuring small areas of view as were the standards mentioned previously.

An undesirable feature of color standards, especially in the yellow/orange/red area of the spectrum, is the tendency to exhibit spectral shifts with a change in temperature. This property, known as thermochromism, has been studied by Verrill *et al.* (1995) at the NPL and a search for nonthermochromic standards has been undertaken. Recent attempts by Labsphere, with a series of pastel colors related to the original Labsphere color standards plus painted materials from Germany, indicate that nonthermochromic standards are possible if a high degree of color saturation is not needed.

VII. Wavelength Calibration Standards

Calibrations of reflectance systems for wavelength accuracy and linearity are usually performed in a transmittance mode using filters containing various rare earth oxides, most commonly holmium oxide or didymium oxide (in reality, a mixture of neodymium and praseodymium oxides). Another widely used technique, although not particularly convenient for most spectrophotometer systems, is to use the emission lines of either mercury or deuterium arc lamps to calibrate the entire spectrophotometer system.

During the past 5 years, standards that may be used to calibrate reflectometer systems indirectly in reflectance mode have been developed. All these standards incorporate stable rare earth oxides, either as powders or as the oxides protected by a matrix of ceramic or polymer. The initial standard of this type was produced by NIST under the designation SRM-1920. The standard was a mixture of holmium, erbium, and dysprosium oxide in a quartz-covered powder cell and was meant for usage in the near-IR only. J. Verrill at NPL has produced a holmium oxide-doped glazed ceramic tile for use as a standard. This material suffers from some peak broadening and shifting, likely due to the high temperature of process of the ceramic.

A similar set of materials has been produced by Springsteen and coworkers at Labsphere. Known as Spectralon WCS standards, the patented materials are each of the three rare earth oxides used in NIST SRM-1920 sintered into a fluorocarbon matrix. As with the Verrill ceramic standard, the positions of some of the peaks are slightly shifted from that of the unprocessed powders. Although the cause of these peak shifts is unknown, it is likely due to some change in the lattice of the oxides when heated to the temperatures needed to process the ceramics or sintered fluorocarbons.

A recent standard that gives excellent agreement with the powdered oxide peak values in a stable, diffuse matrix has been developed by M. Trygstad and K. Norris of NIRSystems (personal communication), working with Ricker and Springsteen of Labsphere. The material combines the three rare earth oxides in high concentration with a fluorocarbon resin. The admixture is then subjected to extremely high pressure but at low temperature. Peak values of the rare earth oxides coincide very well with the NIST standard (Fig. 12).

Another method used in determining the wavelength accuracy of NIR spectrophotometers is to use the overtone peaks of polystyrene in the transmittance mode, much as is done in mid-infrared spectroscopy. It has been suggested that polystyrene can also be used in the reflectance mode to provide a similar calibration standard.

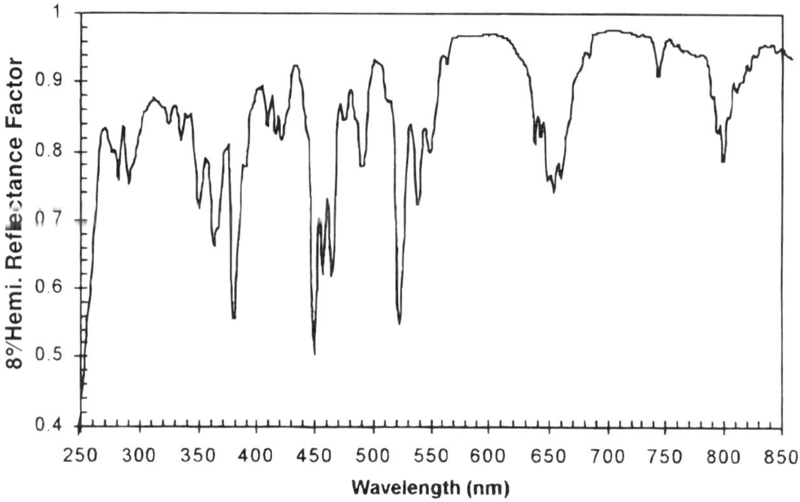

(a)

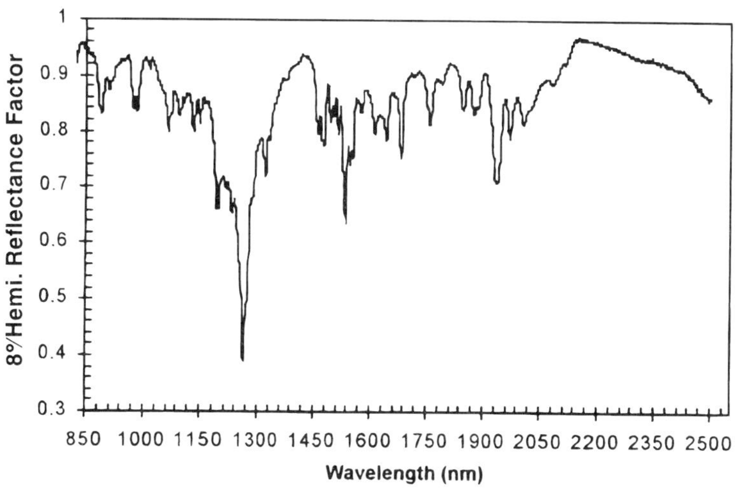

(b)

FIG. 12. (a) 8° Hemispherical reflectance factor, mixed rare earth oxides in pseudo-spectralon, UV–VIS region. (b) 8° Hemispherical reflectance factor, mixed rare earth oxides in pseudo-spectralon, NIR region.

References

General References

Carter, E. C., and Billmeyer, F. W. (1989, revision D. C. Rich) *ISCC Technical Report 89-1* "Guide to Material Standards and Their Use in Color Measurement," Inter-Society Color Council. *Probably the best source for use and theory of reflectance standards for color measurement. It was updated in the Fall of 1996 by D. C. Rich and A. Springsteen.*

C. Burgess and K. D. Mielenz (Eds.) (1987). *Advances in Standards and Methodology in Spectrophotometry.* Elsevier, New York. *The state of the art in references in 1987. Still cogent today because book describes much of the testing that went into many of the current standard reference materials distributed by National Laboratories.*

C. Burgess and D. G. Jones (Eds.) (1995). *Spectrophotometry, Luminescence and Colour, Science and Compliance.* Elsevier, New York. *An update and further work of the Burgess and Mielenz book.*

ASTM Committee E-12 (1994). *ASTM Standards on Color Measurement and Appearance, 4th Edition.* ASTM. *A compendium on measurement techniques and use of artifact standards. An invaluable source for the use of transfer standards in spectrophotometry.*

Springsteen, A. W., Leland, J. E., and Ricker, T. M. (1994). *Guide to Reflectance Materials and Coatings*, Labsphere technical guide. *In-house (but commercially available) technical guide that discusses physical and optical properties of Labsphere-provided reflectance standards, which have been widely accepted and distributed through national laboratories.*

N. M. Trahey (Ed.) (1995). *N.I.S.T. Special Publication 260, Standard Reference Material Catalog 1995–96.* U.S. Dept. of Commerce. *Source of NIST available standard reference materials.*

Optical Radiation Measurement Services (1995). National Physical Laboratory Publication 5K/NE/1/95. *Description of artifact standards and measurements offered by NPL*, United Kingdom.

Weidner, W., and Hsia, J. J. (1987). *N.I.S.T. Publication 250-8, Spectral Reflectance.* U.S. Dept. of Commerce. *Information on absolute measurement of standards and preparation methods of diffuse standards.*

Erb, W., and Budde, W. (1979, Fall). Properties of standard materials for reflectance. *Color Res. Appl.* **4**(3). *Historically important with reflectance data on materials that are outdated but still occassionally used as standards of reflectance.*

Erb, W. (1975). Requirements for reflection standards and the measurement of their reflection values. *Appl. Optics* **14**, 493. *Discusses the optimal properties of materials and their use as reflectance standards, along with the state of the art in 1975.*

Budde, W. (1976). Calibration of standard materials for reflectance. *J. Res. NBS* **80A**(4). *Description of measurement of absolute reflectance and the state of the art in 1976 (it has not changed much).*

Verrill, J. F. (1994). *Measurement Standards for On-Line Metrology.* Abstract of a paper presented at On-Line Optical Measurement Conference. *Overview of artifact standards for transmittance and reflectance available from National Physical Laboratory*, United Kingdom.

Specular Standards

Clarke, F. J. J. (1992). *Infrared Reflectance Standards from N.P.L.* NPL Publication No. DES h 056. *Description of gold mirror standards and diffuse mid-IR standards available from NPL.*

Clarke, F. J. J. (1992). *Absolute Regular Reflectance Standards for the Thermal Infrared Region*, N.P.L. Publication No. DES h 076. *Description of measurement techniques for mid-IR specular samples.*

Zwinkels, J. C., Noël, M., and Dodd, C. X. (1994). Procedures and standards for accurate spectrophotometric measurements of specular reflectance. *Appl. Optics* **33**, 7933. *Probably the best overview of measurement techniques, error analysis of measurement, and stability of specular standards.*

National Institute of Standards and Technology. *Certificates of Reflectance for Calibrated Specular SRM 2003 and SRM 2011. Technical specifications for perhaps the most commonly used and referenced specular standards.*

Diffuse White Standards

Clarke, F. J. J., Garforth, F. A., and Parry, D. J. (1983). Goniometric and polarization properties of white reflection standard materials. *Lighting Res. Technol.* **15**, 133. *Description of barium sulfate, magnesium oxide, and white standard tiles and the means of measurement. Perhaps the most thorough treatment of effects of polarization on reflectance.*

Grum, F., and Luckey, G. W. (1968). Optical sphere paint and a working standard for reflectance. *Appl. Optics* **7**, 2289. *Initial work, later superceded by Grum and Wightman (1977), on barium sulfate as standard.*

Grum, F., and Wightman, T. E. (1977). Absolute reflectance of Eastman white reflectance standard. *Appl. Optics* **16**, 2775. *Corrects inaccuracies of measurement in 1968 article; shows round-robin results with three national laboratories (NRC, NBS, and PTB).*

Grum, F., and Saltzman, M. (1976). *A New White Standard of Reflectance.* Proceedings of the 18th Session C.I.E., C.I.E. Publication No. 18, p. 91. *First description of the use of packed PTFE powder as a transfer and working standard for UV–VIS–NIR diffuse reflectance.*

Weider, V. R., and Hsia, J. J. (1981). Reflection properties of pressed polytetrafluoroethylene powder. *J. Opt. Soc. Am.* **71**, 856; and Weider, V. R., Hsia, J. J., and Adams, B. (1985). Laboratory intercomparison study of pressed poly-tetrafluoroethylene powder reflectance standards. *Appl. Optics* **24**, 2225. *Along with Grum and Saltzman, the definitive works on PTFE as diffuse reflectance standard.*

Trytten, G., and Flowers, W. (1966). Optical characteristics of a proposed reflectance standard. *Appl. Optics* **5**, 1895; and Watson, R. D. (1991). Spectral and bidirectional reflectance of pressed vs unpressed fiberfrax. *Appl. Optics* **10**, 1685. *Field standard for VIS–NIR remote sensing.*

Hsia, J. J. (1976). *Optical Radiation—The Translucent Blurring Effect—Method of Evaluation and Estimation*, N.I.S.T. Technical Note No. 594-12. U.S. Government Printing Office, Washington, DC. *The analysis of the translucent blurring effect/*edge-*loss for ceramics and glasses.*

Diffuse Infrared Standards

Clarke, F. J. J., and Larkin, J. A. (1985). Emissivity determined from hemispherical reflectance and transmittance throughout the thermal infrared spectrum. *High Temperatures–High Pressures* **17**, 89. *Description of new method of mid-IR diffuse measurements leading to mid-IR diffuse standards.*

Clarke, F. J. J., and Larkin, J. A. (1988). Improved techniques for the N.P.L. hemispherical reflectometer. *Proc. S.P.I.E.* **917**, 7. *Overview of measurement techniques for absolute measurement of mid-IR diffuse reflectance.*

Wiley, R. R. (1987). Results of a round-robin measurement of spectral emittance in the mid-IR. *Proc. S.P.I.E.* **10**, 807. *First comprehensive study of Infragold and other diffuse mid-IR standards.*

Nash, D. B. (1986). Mid-infrared reflectance spectra (2.3–22μ) of sulfur, gold, KBr, MgO, and halon. *Appl. Optics* **25**, 2427. *Discusses and characterizes a wide variety of possible standards for mid-IR reflectance.*

Stuhlinger, T. Bidirectional reflectance distribution function of gold-plated sandpaper. *Optical Eng.* **22**, SR–142; and Lea, T. K., and Schotland, R. M. (1988). Infrared retroreflecting lidar calibration targets. *Appl. Optics* **27**, 208. *Discuss gold-plated sandpaper, once widely used as standard for mid-IR diffuse reflectance.*

Gray Scale Diffuse Reflectance

Weidner, V. R. (1986). Gray scale of diffuse reflectance for the 250–2500 nm wavelength region. *Appl. Optics* **25**, 1265. *Discussion of potential diffuse reflectance standards from carbon black and Halon powder.*

Snail, K. (Naval Research Laboratory), Salisbury, J. (Johns Hopkins Dept. of Earth and Terrestial Science), and W. Lynn (Systems Research Laboratory) personal communication with A. Springsteen. *Labsphere has developed novel mid-IR standards with evaluation provided by the people and laboratories mentioned.*

Color Standards

Malkin, F., and Birtles, J. F. (1967). *Proc. Symp. Colour Measurements Ind. London* 260. *Description of the evolution of the B.C.R.A. tiles for color measurement.*

Clarke, F. J. J., and Malkin, F. (1981) Development of a new series of ceramic colour standards. *J. Soc. Dyers Colourists* **97**, 503. *Description of current BCRA Series II color tiles supplied by NPL.*

Compton, J. A. (1984). The thermochromic properties of the ceramic colour standards. *Color Res. Appl.* **9**, 15. *Evaluation of thermochromism of BCRA series I tiles. Theory and practice also applicable to all other existing color standards to some degree.*

Fillinger, L., Lukacs, G., and Endor, G. (1978). Thermochromism of color standards. *Hun. Sci. Instr.* **44**, 25; Fairchild, M. D., and Grum, F. (1985). The thermochromic properties of ceramic reference tiles. *Appl.* Optics **24**, 3432; and Verrill, J. F., Knee, P. C., and O'Halloran, J. (1995). *A study of the uniformity and thermochromism of surface colour standards.* Abstract of paper presented at C.I.E. International Meeting, Berlin. *Recent studies of surface uniformity and thermochromism of currently used color standards (translucent blurring effect).*

Springsteen, A., U.S. Patent #5,462,705 (1995).

Wavelength Calibration Standards

Weidner, V. R., Barnes, P. Y., and Eckerle, K. L. (1986). A wavelength standard for the near infrared based on the reflectance of rare-earth oxides. *J. Res. Natl. Bureau Standards* **91**, 243. *Origin of NIST standard reference material SRM 1920 (wave cals).*

Trygstad, M. (Perstorp Analytical) and Norris, K. (Perstorp Analytical), personal communication with Springsteen, A., and Ricker, T. M. (Labsphere). *Labsphere and Perstorp have been researching new wavelength calibration standards to be used in NIR systems and other NIR instruments.*

REFLECTANCE MEASUREMENTS OF DIFFUSING SURFACES USING CONIC MIRROR REFLECTOMETERS

KEITH A. SNAIL
Optical Sciences Division,
Naval Research Laboratory

LEONARD M. HANSSEN
Optical Technology Division,
National Institute of Standards & Technology

I. Introduction

Accurate measurements of the spectral reflectance of diffusing surfaces are important to numerous scientific and engineering disciplines, including the fields of solar energy materials, color science, analytical chemistry, spacecraft thermal engineering, military low observable materials, and remote

ISBN 0-12-764070-3.

sensing. Such reflectance measurements are typically made by comparing the surface to be tested to a standard of known reflectance. The techniques used to calibrate these standards are referred to as absolute methods. In 1969, the International Commission of Illumination (CIE) (1) recommended that reflectance standards based on smoked magnesium oxide should be replaced with a perfect reflecting diffuser. Because no perfect reflecting diffusers exist, it then became necessary to calibrate reflectance standards for diffuse surfaces with an absolute reflectometer.

For diffusely scattering surfaces, the four most common approaches to performing reflectance measurements are based on the use of a goniometer, the Kubelka–Munk technique, an integrating sphere and a conical collecting mirror. Hohlraums (2, 3) were investigated in the 1950s for measuring reflectance. These devices were not studied further due to significant self-emission from the sample and the difficulty of modulating the hohlraum's radiation source. Goniometric measurements involve tilting and rotating the sample or detector about the sample center in order to measure the bidirectional reflectance distribution function (BRDF) at selected points on the hemisphere above the sample. The time required to perform a goniometric measurement has decreased significantly in the past 20 years due to advances in computer-controlled instrumentation. Nevertheless, goniometric measurements are still time consuming due to the large number of data points required, which also results in a large total uncertainty. Hence, the goniometric technique is still rarely used. The Kubelka–Munk theory (4, 5) is widely used in the near-infrared (NIR) and visible to analyze the reflectance of powdered samples. Several techniques for determining the absolute reflectance of solid opaque diffusing samples using the Kubelka–Munk theory have been proposed and tested (6, 7). These approaches have not been widely used due to the requirement for multiple measurements and their high uncertainty level (6, 7). Methods based on the integrating sphere are by far the most common in use today. An extensive literature has been built up over the years on integrating spheres for reflectance measurements. Two reviews of absolute methods for reflectance measurements including extensive descriptions of sphere based methods were published in the 1970s by Budde (6) and by Ooba *et al.* (7).

This review is limited to reflectometers based on specular conic mirrors and it includes several unanticipated results. To our knowledge, this chapter represents the first comprehensive review of such devices. Previous reviews of reflectance instrumentation contain only a limited discussion of conic mirror devices (6, 7). It has been prompted by new developments that expand the scope of conic mirrors used for reflectometry. During the past decade, inexpensive PC- and workstation-based raytracing programs have become available that allow one to easily model the performance of both conic mirror and integrating sphere reflectometers. In this review we present raytracing results for the three primary conic mirror reflectometers as well as several recently derived magnification formulae. A thorough review of the literature on conic mirror reflectometers is also presented, followed by a systematic review and analysis of the major sources of measurement error.

All the reflectometer designs and equations discussed herein can be applied after minor modifications to transmittance measurements of diffusely scattering materials. In doing so, one must account for any increase in the spatial extent of the beam radiation that occurs when the beam traverses a transmitting material that exhibits volume or first-surface scattering. Similarly, when measuring the reflectance of opaque samples with significant volume scattering the beam spot size must be sized appropriately.

II. Terminology

The terminology used throughout this chapter will follow that of the CIE (8), with several minor modifications. The key definitions are noted below:

1. Luminous and radiant quantities will be used interchangeably. Luminous quantities refer to measurements performed with a detector having a spectral response equivalent to the CIE standard photometric observer (1), whereas radiant quantities refer to spectrally resolved measurements. The measurement geometries used for luminous and radiant measurements are identical.
2. The following are the most commonly used measurement geometries for calibrating reflectance standards and for performing reflectance measurements relative to standards.

1. Normal/45°	0/45
2. 45°/normal	45/0
3. Directional/hemispherical	$\theta/2\pi$
4. Hemispherical/directional	$2\pi/\theta$
5. Hemispherical/hemispherical	$2\pi/2\pi$

In this notation, the first term refers to the geometry of illumination, whereas the second term refers to the geometry of light collection. 2π refers to a hemispherical solid angle, whereas 0, 45, and θ refer to small solid angles centered around the specified angle. Geometries 1 and 2 are used primarily for colorimetric measurements and will not be discussed here. The hemispherical/hemispherical reflectance is not easily measured directly; however, it can still be an important quantity. For instance, the hemispherical/hemispherical

TABLE 1
SYMBOL DEFINITIONS

Symbol	Definition
A_o	Sample optics or limiting aperture projected area
A_D	Detector area
a	Hemiellipsoid semimajor axis length
b	Hemiellipsoid semiminor axis length
c	Sample center position in hemiellipsoid and hemisphere
δ	Hemisphere dimensionless parameter ($\equiv r_s/R$)
ε	Hemiellipsoid eccentricity ($\equiv c/a$)
ε'	Hemisphere eccentricity ($\equiv c'/R$)
ε''	Dual-paraboloid eccentricity ($\equiv r_s/2f_s$)
f_D	Focal length of detector or source paraboloid
f_s	Focal length of sample paraboloid
$f/\#$	F number of focusing optics; the ratio fits focal length to its diameter
F	Multiple reflectance enhancement factor
F_o	Sample optics loss factor
F_m	Radiative transfer factor for detector to conic mirror
F_s	Radiative transfer factor for sample overfilling
λ	Wavelength
L_o	Distance between sample ($\theta/2\pi$) or source ($2\pi/\theta$) and limiting aperture of the sample optics
L_p	Separation distance of dual paraboloids
$L(\theta)$	Radiance
M_x	Maximum linear magnification along x axis
M_y	Maximum linear magnification along y axis
$M_{y,\,HE}$	Maximum linear magnification of hemiellipsoid along y axis
$M_{y,\,DP}$	Maximum linear magnification of dual paraboloid
$\langle n\rangle$	Average number of reflections on compound parabolic concentrator (CPC) mirror for $\theta < \theta_c$
$\langle n'\rangle$	Average number of reflections on CPC mirror for $\theta > \theta_c$
P_{sc}	Total source emitted power
P_s	Sample power ($2\pi/\theta$) passing through viewing port
P_r	Source power ($2\pi/\theta$) measured through sample aperture with same optics used to measure P_s, view port plugged
ρ	Directional hemispherical ($\theta/2\pi$) reflectance; composed of diffuse (ρ_d), and specular (ρ_r) components
ρ'	Hemispherical/hemispherical ($2\pi/2\pi$) reflectance
ρ_D	Detector directional hemispherical reflectance
ρ'_D	Detector hemispherical/hemispherical reflectance
ρ_d	Diffuse reflectance ($\equiv \rho - \rho_r$)
ρ_r	Specular or regular reflectance ($\equiv \rho - \rho_d$)
$\rho_{r,\,m}$	Conic mirror specular reflectance
$\rho_{r,\,cpc}$	CPC mirror specular reflectance
ρ_s	Sample directional hemispherical reflectance
ρ'_s	Sample hemispherical/hemispherical reflectance
ρ_{sc}	Source directional hemispherical reflectance
ρ'_{sc}	Source hemispherical/hemispherical reflectance
r_s	Beam spot radius ($\theta/2\pi$) or viewed radius ($2\pi/\theta$) on sample
R	Hemisphere radius
σ	Root mean square surface roughness
θ	Polar angle
θ_c	CPC critical angle, beyond which the throughput is zero for meridional rays
T_s	Throughput (ratio of output flux to input flux) of conic mirror reflectometer for sample measurement
T_r	Throughput of conic mirror reflectometer for reference measurement
TIS	Total integrated scatter ($\equiv \rho_d/\rho_r$)

reflectance explicitly appears in the throughput equation of integrating sphere and conic mirror reflectometers. Geometries 3 and 4 are the most frequently used geometries. For $\theta/2\pi$ reflectometers having a fixed illumination angle, a near-normal ($2^\circ \leq \theta \leq 10^\circ$) angle, rather than normal incidence, is frequently chosen in order to include specularly reflected light, which would otherwise be lost through the beam port. Similarly, a nonzero θ collection angle would be selected for $2\pi/\theta$ instruments.

3. The *directional/hemispherical reflectance* is the fraction of flux incident within a small solid angle about the direction θ that is reflected into the hemisphere above the surface and will be denoted by the symbol ρ. By the Helmholtz reciprocity principle (9, 10), a measurement in the $2\pi/\theta$ geometry will yield the same reflectance value. In practice, actual measurement geometries only approximate the ideal geometry due to the presence of ports and other factors. The directional/hemispherical reflectance is composed of a diffuse and a specular or regular component.
4. The term *diffuse reflectance* (ρ_d) will be used to refer to a directional/hemispherical reflectance measurement in which the specular or regular component (ρ_r) is not included.
5. A *perfect reflecting diffuser* is defined as an idealized object with a reflected radiance independent of the angle of viewing (i.e., a Lambertian reflector) and a reflectance equal to 1.
6. The term *reflectance factor* is defined as the ratio of the radiant flux reflected from a sample to that reflected in the same geometry by a perfectly reflecting diffuser that is identically irradiated. This term is often used instead of reflectance, along with an indication of the deviations of the light collection cone from 2π sr.

The symbols used in this review and their definitions are listed in Table 1.

III. Historical Overview

The use of hemispherical mirrors for measuring directional/hemispherical reflectance predates the use of integrating spheres for the same purpose by a few years. Paschen (11) placed lamp and platinum black-coated detectors at the center of a hemispherical mirror in order to enhance the detector's absorptance and used this arrangement to determine the constants in Wien's radiation law. Royds (12) positioned a sample and detector at the conjugate foci of a hemispherical mirror in order to measure the reflectance of blackened thermal detectors. In the acknowledgments of a 1911 paper

Royds (13) gives Paschen credit for the idea of measuring reflectance with a hemispherical mirror; however, the exact date of this insight is not given. Royds (13) and Coblentz (14) developed and characterized reflectometers that used small-diameter (5–10 cm) silvered hemispherical mirrors and blackened thermopile detectors. Royds' work was published several years before that of Coblentz; however, reflectometers employing a hemispherical mirror are frequently referred to as Coblentz hemispheres. This is due to Coblentz's treatment of several measurement errors that Royds neglected and his measurement of the reflectance of a variety of diffusing materials.

Shortly after Coblentz's work appeared, a number of scientists in the United States developed visible reflectometers based on the integrating sphere. For the next 40 years, the measurement of directional/hemispherical reflectance in the visible and near-IR was performed almost exclusively with integrating spheres. In 1931, Beckett (15) analyzed the multiple reflections that can occur between the sample and detector in a Coblentz hemisphere and developed a technique for correcting for this error. Shortly after World War II, Sanderson (16) developed a hemispherical mirror reflectometer to characterize several new low-emittance paints, and during the 1950s and 1960s a steady succession of papers (17–24) appeared on hemispherical mirror devices, most of which included some infrared capability. The use of Coblentz hemispheres to perform infrared measurements was motivated by the low throughput of integrating sphere coatings available at the time. In 1964, Brandenberg (25) quantified the reduction in aberrations of a hemiellipsoidal mirror compared to a hemispherical mirror, and in the following year Blevin and Brown (26) reported on the first hemiellipsoidal reflectometer. Subsequently, many groups built and tested hemiellipsoidal instruments (27–31). A $\theta/2\pi$ conic mirror reflectometer based on two on-axis, hemiparaboloidal mirrors was built and tested (32, 33) in the 1950s. A dual-ellipsoidal (34) mirror design also appears possible but has not been tested to our knowledge. A $2\pi/\theta$ reflectometer utilizing two paraboloidal mirrors (one off-axis) with unequal focal lengths has been built and tested (35).

As discussed in detail in the following section, the key challenge in designing a $\theta/2\pi$ conic mirror reflectometer is to develop a black, large-area, spatially uniform detector with an angle-independent response. A low reflectance is required to attenuate multiple reflections between the detector and sample, and a spatially uniform response is needed because of the aberrations of the conic mirror. Similarly, a $2\pi/\theta$ conic mirror reflectometer requires a black (i.e., perfect emitter), large-area, spatially uniform Lambertian source. In the past, these considerations have prompted many modern users of hemispherical (36) and hemiellipsoidal (37) mirrors to select the $2\pi/\theta$ geometry. However, recent improvements in detector designs have made the $\theta/2\pi$ geometry viable.

The widespread use of conic mirror reflectometers in the visible has been limited by the small but significant scatter (see Section V,H) from mirror surfaces. Conic mirror reflectometers remained the most commonly used method for measuring directional/hemispherical reflectance in the mid- to far-infrared ($\lambda > 2$ or 3 μm) until the early 1980s. Since the advent of more durable, high-reflectance diffuse gold coatings (38, 39) for integrating spheres in the late 1970s, reflectance measurements in the infrared ($\lambda = 2$–14 μm) are increasingly being performed with integrating spheres. The newest diffuse gold integrating sphere coatings (40–42) exhibit near-lambertian bidirectional reflectance distribution functions (BRDFs) and reflectances above 95% at 10 μm.

Direct comparisons between conic mirror and integrating sphere reflectometers are highly dependent on the characteristics of the individual reflectometers. The measurement error associated with each reflectometer type is frequently dominated by different phenomena. The throughput of a conic mirror is typically a factor of 100 times that of an IR integrating sphere. Nevertheless, the overall signal-to-noise ratio (SNR) of an integrating sphere reflectometer can be higher depending on the detector characteristics. For example, in the infrared, cryogenically cooled, high-responsivity, photoconductive detectors can be used with integrating spheres. This is because the spatial nonuniformity typically associated with these detectors will not affect the measurement accuracy significantly. On the other hand, as described in Section V,B, it could lead to substantial measurement errors in conic mirror devices.

A fundamental difference between the conic mirror and integrating sphere devices is in how absolute reflectance values are determined. Absolute measurements in conic mirror reflectometers are direct and simple. Sources of error, as discussed in Section V, need to be accounted for. However, absolute reflectance values from measurements made with integrating sphere reflectometers are based on integrating sphere theory. Also, integrating sphere theory is based on a variety of assumptions including that of a perfect Lambertian inner wall coating. The effects of deviations from the assumptions of the theory are difficult to quantify (43). This dependence of absolute results on the sphere theory restricts the use of most integrating sphere reflectometers in the infrared primarily to relative measurements.

Currently, the large body of theoretical examinations of measurement errors associated with integrating spheres is unmatched by a comparable body of knowledge for conic mirror devices. The intent of the next section is to heuristically review the major sources of measurement error in collecting mirror reflectometers and provide a foundation for further improvements to the performance of conic mirror instruments.

IV. Measurement Geometries

Conical or collecting mirror reflectometers utilize hemispherical, hemiellipsoidal, or dual-paraboloidal mirrors for collecting diffuse radiation. Most of the conic mirror reflectometers built before 1970 were designed to operate in a relative mode, comparing light reflected off of a sample to that reflected off of a standard reference. Absolute reflectance measurements require three steps: (i) a direct measurement of the beam power at the sample (the reference measurement), including attenuation by the conic mirror; (ii) a direct measurement of the reflected power (the sample measurement); and (iii) corrections for any significant errors. Significant error sources are described in Section V.

We begin by defining the measurement geometry of a hemispheroidal mirror reflectometer shown in Fig. 1. The mirror is centered on the Z axis with its open side lying in the X–Y plane. The semimajor axis length is denoted by a, the semiminor axis length by b, and the symmetric positioning of the sample and detector (or source) at $Y = c$ and $Y = -c$, respectively. For a hemiellipsoidal mirror, $a \neq b$, and the sample location, c, is at one focus. The eccentricity ε of the hemiellipsoidal mirror is given by c/a. For a hemispherical mirror, $a = b = R$, where R is the radius of the hemisphere. For comparison purposes, a quantity $\varepsilon' = c/R$, analogous to the eccentricity, is defined for the hemisphere.

For the $\theta/2\pi$ measurement mode, a detector is located at $Y = -c$, whereas for the $2\pi/\theta$ measurement mode a source is located at $Y = -c$. The optics (not shown in Fig. 1) used to illuminate $(\theta/2\pi)$ or view $(2\pi/\theta)$ the

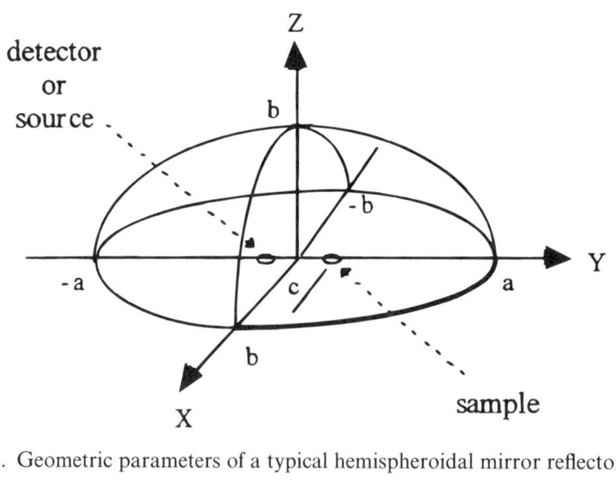

FIG. 1. Geometric parameters of a typical hemispheroidal mirror reflectometer.

sample can be either suspended inside the conic mirror (internal optics) or outside a port in the conic mirror (external optics). The sample is assumed to be centered at $Y = c$. However, the spatial extent of the aberrations of an image formed at $Y = -c$ from a finite size source at $Y = c$ is not symmetric about $Y = -c$. In order to optimally underfill a detector (or source) its center should be appropriately shifted to $Y < -c$ (25).

The measurement geometry of a dual-paraboloidal mirror reflectometer is shown in Fig. 2 in cross section. The orientation of the top (sample) paraboloid is similar to that of the hemispheroidal mirror with its open side lying in the X–Y plane. Because of circular symmetry about the Z axis, only the Y–Z cross section is shown. The axes of the two paraboloids are collinear, with the sample centered on the axes. The sample and detector (or source) are located at $Z = 0$ and $Z = -L_p$, respectively. The radius of the sample paraboloidal mirror is given by $2f_s$, where f_s is its focal length. For comparison purposes, a quantity $\varepsilon'' = r_s/2f_s$, analogous to the eccentricity, is defined for the sample paraboloid, where r_s is the sample radius. The lower paraboloid has a focal length f_D and radius $2f_D$. The separation between the focal points of the two paraboloids is denoted by L_p.

For the $(\theta/2\pi)$ measurement mode, a detector is located at $Z = -L_p$, whereas for the $(2\pi/\theta)$ measurement mode, a source is located at $-L_p$.

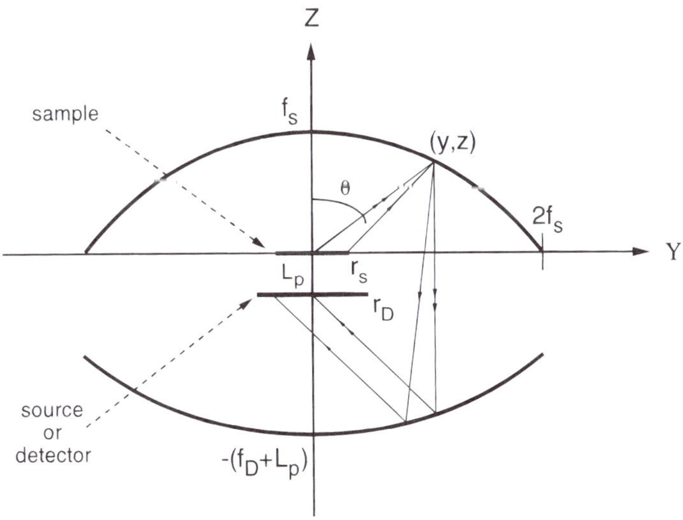

FIG. 2. Dual-paraboloidal mirror reflectometer geometric parameters. The upper paraboloid has a focal length f_s and radius $2f_s$. The lower paraboloid has a focal length f_D and radius $2f_D$.

Again, the optics used to illuminate ($\theta/2\pi$) or view ($2\pi/\theta$) the sample are not shown in Fig. 2. The lower paraboloid may, in general, be selected to have a different focal length than the sample paraboloid. This will affect both the angular range and the spatial extent of illumination on (or collection from) the detector (or source).

V. Sources of Error

Conical mirror reflectometers may exhibit systematic errors due to a variety of sources. As shown in detail below, the size of the resulting errors in measurement can be substantial. Hence, a thorough list of the sources of error are first provided, and then in Sections V,A–V,H a detailed analysis of each error source is given as well as a summary of the approaches taken to minimizing the size of the errors.

Conical mirror reflectometers may exhibit systematic errors due to the following phenomena:

1. Magnification effects that result in detector overfilling ($\theta/2\pi$) or nonuniform sample illumination ($2\pi/\theta$)
2. Detector ($\theta/2\pi$) or source ($2\pi/\theta$) spatial nonuniformity
3. Detector ($\theta/2\pi$) or source ($2\pi/\theta$) angular response
4. Beam optics losses from the optics used to view ($2\pi/\theta$) or illuminate ($\theta/2\pi$) the sample
5. Interreflections between the sample and detector ($\theta/2\pi$) or source ($2\pi/\theta$)
6. Sample and detector ($\theta/2\pi$) or source ($2\pi/\theta$) position misalignment
7. Shadowing losses from sample or optic supports
8. Mirror imperfections such as surface roughness, slope errors, and spatially nonuniform reflectance

If a conic mirror reflectometer is not designed to minimize systematic errors, the magnitude of the associated corrections and uncertainties can be so large as to render it unsuitable for calibrating reflectance standards.

A. Magnification Effects

The image size on the detector ($\theta/2\pi$), and hence the magnification of the mirror, depends on the BRDF of the sample. For Lambertian samples, the magnification will be a maximum. Knowledge of the maximum linear magnification of a hemispheroidal or dual-paraboloidal mirror is required to size a reflectometer's detector ($\theta/2\pi$) so as to collect all the reflected radiation or to size the radiation source ($2\pi/\theta$) so as to uniformly (44) illuminate the sample. For cases in which an analytical solution is not available, the maximum magnification of a conic mirror reflectometer can be determined by raytracing. In many cases the tracing of two extremal rays from the edges of the beam spot on the sample will suffice to determine the maximum linear magnification. One would like to use a conic mirror with the smallest magnification possible so as to minimize the source size ($2\pi/\theta$) or the detector size ($\theta/2\pi$). For most reflectometers, the detector noise will scale (45) as $\sqrt{A_D}$, where A_D is the detector area.

1. *Hemispherical Mirror*

The maximum linear magnification of a hemispherical mirror along the axis containing the detector/source and sample is approximately given (25) by

$$M_{y,\,\mathrm{HS}} \approx \frac{1}{(1 - 2\varepsilon')^2}. \tag{9.1}$$

However, because of spherical aberration, Eq. (8.1) significantly underestimates the true magnification for $\varepsilon' > 2r_s/R$, where r_s is equal to the viewed radius ($2\pi/\theta$) or the beam spot radius ($\theta/2\pi$) on the sample. The exact expression, derived by Snail (46), which takes spherical aberration into account, is

$$M_{y,\,\mathrm{HS}} = \frac{\dfrac{2\varepsilon'^2}{\delta} - (2\delta - 1)}{(2\delta - 1)^2 - 4\varepsilon'^2}, \tag{9.2}$$

where δ is equal to r_s/R. If the sample is fully illuminated, then $\varepsilon' = \delta$ corresponds to the sample and detector just touching at $y = 0$. Under this condition, Eq. (9.2) reduces to $1/(1\text{-}4\varepsilon')$. For $2\pi/\theta$ instruments, the ratio of the source radius to the viewed sample radius should be equal to or greater than the maximum magnification in order to ensure uniform illumination over the viewed area on the sample (44). In this case, the sample will be overilluminated.

The magnification along the x axis, M_x, in Fig. 1 is equal to the square root of M_y for both the hemisphere and the hemiellipsoid (25). Hence, when using a monochromator with a spheroidal mirror, one can minimize the detector ($\theta/2\pi$) or source ($2\pi/\theta$) area by aligning the image of the monochromator's slit on the sample so that it is perpendicular to the spheroidal mirror axis containing the sample and detector/source.

2. *Hemiellipsoidal Mirror*

Brandenberg (25) has shown that the maximum linear magnification along the semimajor axis of a hemiellipsoidal mirror is approximately equal to

$$M_{y,\,\mathrm{HE}} \cong \left(\frac{1+\varepsilon}{1-\varepsilon}\right)^2. \tag{9.3}$$

Raytracing analyses (16, 17) indicate that for a circular source, Eq. (9.3) sets a lower bound for the linear magnification. Equation (9.3) can be written as a power series in $\varepsilon(M_{y,\,\mathrm{HE}} = 1 + 4\varepsilon + 8\varepsilon^2 + 12\varepsilon^3 + \cdots)$. The maximum magnification of a hemisphere (Eq. (9.3) can also be expanded in a power series with $\delta = \varepsilon'$. The resultant series $(M_{y,\,\mathrm{HS}} = 1 + 4\varepsilon' + 16\varepsilon'^2 + 64\varepsilon'^3 + \cdots)$ differs from the hemiellipsoid result in the second order and higher terms. For $\delta = \varepsilon' = \varepsilon = 0.05$, the magnification of the hemisphere is only 2.3% higher than that of the hemiellipsoid.

3. *Dual-Paraboloidal Mirror*

The maximum magnification of a dual-paraboloid system has been evaluated by Hanssen (46) for the case of equal focal-length parabolas and coincident foci. If we define a quantity $\varepsilon'' = r_s/2f_s$, then the maximum linear magnification of a dual-paraboloidal mirror with $f_s = f_D$ and $L_p = 0$ is given exactly by

$$M_{y,\,\mathrm{DP}} = \left(\frac{1}{1-2\varepsilon''}\right). \tag{9.4}$$

Because of the circular symmetry of the dual-paraboloidal geometry, the magnification of a circular beam spot centered on a diffuse sample can be characterized by a single value (i.e., $M_{x,\,\mathrm{DP}} = M_{y,\,\mathrm{DP}}$). The series expansion of Eq. (9.4) $(M_{x,\,\mathrm{DP}} = M_{y,\,\mathrm{DP}} = 1 + 2\varepsilon'' + 4\varepsilon''^2 + 8\varepsilon''^3 + \cdots)$ has a first order term coefficient of only half that of the other conical mirror systems.

Two other properties of the dual-paraboloidal design are worth mentioning. First, if f_D is greater than f_s, the maximum incidence angle on the detector is reduced and consequently the magnification increases in order to conserve étendue. Second, the larger range of incidence angle (0–45°) on the two paraboloids compared to those on the hemisphere and hemiellipsoid may introduce a polarization-related error, which is not discussed in this review.

4. *Comparison*

A comparison of Eqs. (9.2), (9.3), and (9.4) reveals the maximum magnification of the dual-paraboloidal mirror to be reduced by about the square root of the values obtained for the hemispherical and hemiellipsoidal mirror. A comparison calculated from Eqs. (9.2)–(9.4) of the maximum magnification of the three primary types of conic mirror reflectometers is shown in Fig. 3 (46). The magnification of the hemiellipsoid and dual-paraboloid mirrors are significantly less than the hemisphere for $\varepsilon = \varepsilon' = \varepsilon'' > 0.1$.

In practice, other considerations may preclude $\varepsilon = \varepsilon' = \varepsilon''$ for comparison purposes. For instance, if the largest physical dimensions of the conic mirrors are set equal ($R = a = 2f_s$) and if the detector size is fixed, the beam spot size ($\theta/2\pi$) on the sample required to fully illuminate the detector will be different for each conic mirror type. Hence, one may have $\varepsilon'' < \varepsilon = \varepsilon'$. The SNR may be a more useful figure of merit, for comparison purposes, than the magnification. In comparing the SNR of the three designs, one must consider the changes in signal level associated with the beam spot size variations as well as the lower throughput of the dual-paraboloidal design due to self-shadowing by the sample and the second reflection on the detector paraboloid.

B. Detector/Source Spatial Uniformity

A systematic measurement error can be introduced if the response of the detector ($\theta/2\pi$) or the radiation source output ($2\pi/\theta$) is not spatially

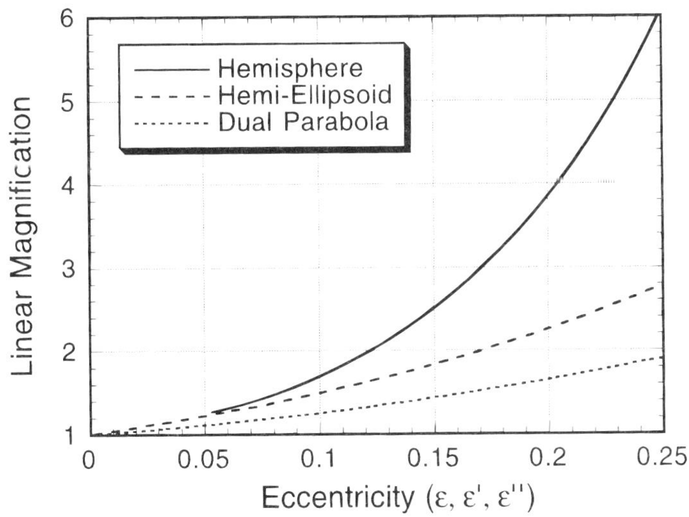

FIG. 3. Maximum linear magnification from Ref. (46) for a hemispherical, hemiellipsoidal, and dual-paraboloidal mirror for the geometries discussed in the text. For the hemisphere, the ratio of the sample to hemisphere radius was set at 0.05.

uniform. This is because individual samples will reflect the incident light with different BRDFs. As described in Section V,A, the resulting illumination of the detector will be distributed differently both spatially and in angular distribution. The change in the illumination pattern on the spatially nonuniform detector results in measurement error. The equivalent is true for the $2\pi/\theta$ case. Dunn *et al.* (27) reported that directional/hemispherical reflectance measurements performed with a $\theta/2\pi$ hemiellipsoidal mirror and a thermopile or Golay cell detector were in error by 40–50% due to variations in the areal sensitivity and angular response of the detector. A scan of a 1.6-mm-diameter beam over the detector surface showed variations in response of up to 50%.

For $\theta/2\pi$ mode reflectometers, two approaches have been used to minimize spatial uniformity errors. The first approach involves using a small integrating sphere in conjunction with the reflectometer's detector. Dunn (48) developed a small sulfur-coated integrating sphere for use with an IR hemiellipsoidal mirror. Today, roughened gold coatings are available (40) with near-Lambertian properties and reflectances above 95% from 2 to 14 μm. Two disadvantages of using an integrating sphere are the relatively low throughput of diffuse gold spheres (1–5%) and the interreflection error caused by reradiation from the beam port (see Section V,E).

An alternate approach involves the use of spatially uniform detectors. Blevin and Brown (26) developed a gold-film bolometer with a broadband response that was uniform to within 2% over an area of 9 × 4 mm for a 0.25-mm beam spot. The detector used a NaCl window and had a minimum detectable power of 4 nW for a bandwidth of 1 Hz. Hanssen and Snail (49) measured the spatial uniformity of a windowless, 14 × 14-mm gold–black coated pyroelectric ($LiTaO_3$) detector at 10.6 μm. For a 0.25-mm beam spot, the response was uniform to within $\pm 1.7\%$ over 13 mm, excluding one narrow scratched region where a 10–15% deviation was observed. The minimum detectable power was 1 μW (standard uncertainty) for a bandwidth of 1 Hz (50).

We are not aware of any published data on the uniformity of sources used in $2\pi/\theta$ mode conic mirror reflectometers. Estimating the size of measurement errors due to spatial nonuniformities may require a raytracing analysis.

C. Detector/Source Angular Response

1. *Directional Hemispherical Reflectometers*

For $\theta/2\pi$ mode reflectometers, the angular response of detectors (51) associated with Fresnel effects, obscuration from mounts, and/or cold shields can cause the reflectometer's throughput to be nonisotropic. For example, the angular response of a gold–black coated pyroelectric ($LiTaO_3$) detector is shown in Fig. 4 for unpolarized radiation at 10.6 μm (49). Minimizing the angular response errors associated with this detector would require a means of restricting the incidence angle on the detector to less than 30–40° or a means of ensuring that the detector was illuminated in a Lambertian fashion independent of the sample BRDF.

For $\theta/2\pi$ mode reflectometers, three approaches to minimizing angular response errors have been proposed. Blevin and Brown (26) and Dunn *et al.* (27) reported on reflectometers that used a hemiellipsoidal mirror with the sample and detector normals parallel to the major axis and the ellipsoidal mirror truncated along a plane at $y = c$. This gives a maximum incidence angle at the center of the detector of $\tan^{-1}[b\sqrt{(1-\varepsilon^2)}/2c]$. For the ellipsoid parameters used, maximum detector incidence angles of 19 and 24° were achieved. Both groups used ellipsoidal mirrors with an eccentricity of approximately 0.7, giving a transverse magnification of approx. 6. Similarly, a dual-paraboloidal mirror with unequal focal lengths has a maximum incidence angle at the center of the detector of $\tan^{-1}(2f_s/f_D)$. Raytracing analysis indicates that a dual-paraboloid ($f_s = 15$ cm, $f_D = 67$ cm) with a maximum incidence angle of 24° on the detector and a 1 cm diameter beam spot on the sample would have a magnification of approx. 4.5. Assuming that the

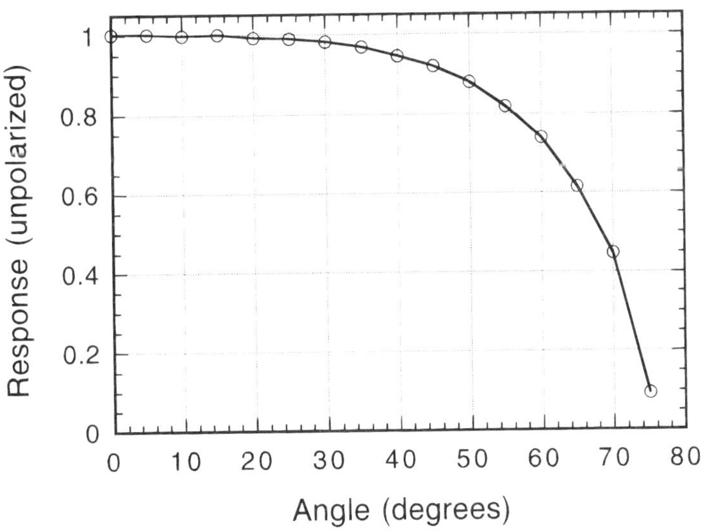

FIG. 4. Response of a gold–black coated pyroelectric ($LiTaO_3$) detector at 10.6 μm versus incidence angle for unpolarized radiation.

detector noise scales with the detector radius, this dual-paraboloid mirror reflectometer would have a signal-to-noise ratio 33% better than that of the hemiellipsoid reflectometers described previously.

A second approach is to place an integrating sphere at the detector or source focus of any of the conic mirror arrangements. The disadvantages of this approach are the low throughput of integrating spheres (e.g., 1–5%), the interreflection error that reradiation from the beam port introduces (see Section V,E), and the variations in angular response one encounters in integrating spheres designed for direct reflectance measurements.

A third approach (47) is to place an inverted compound parabolic concentrator (CPC) in front of a detector with a response that is spatially uniform. The CPC restricts the incidence angle on the detector to a range such that the variation in the detector's Fresnel response is small (e.g., 1–3%). Raytraces of parallel light incident at four different angles on an inverted CPC with $\theta_C = 30^\circ$ are shown in Fig. 5.

The predicted angular response of a hemiellipsoid with a photomultiplier (PMT) with and without a CPC ($\theta_C = 30^\circ$) is shown in Fig. 6 based on the

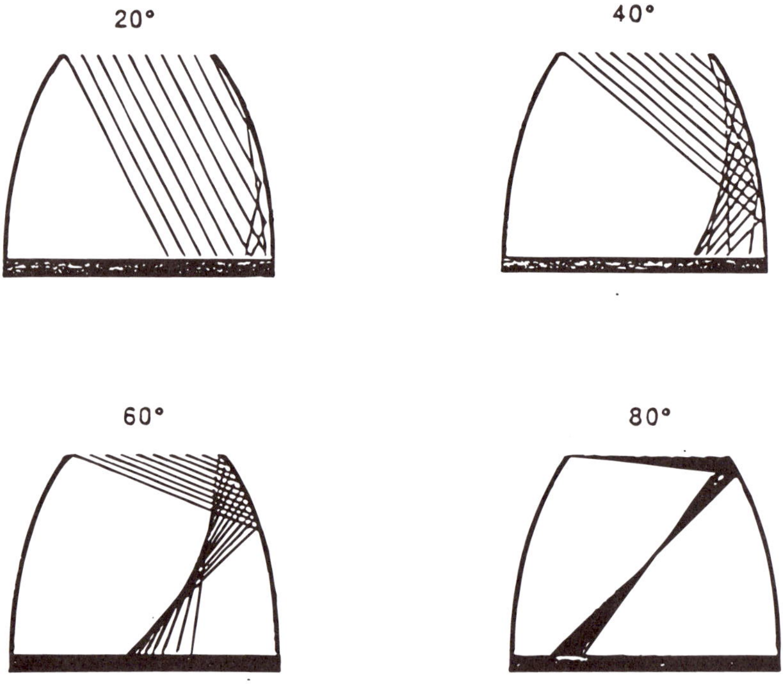

FIG. 5. Raytrace of an inverted compound parabolic concentrator (CPC), with $\theta_c = 30^\circ$, for meridional ray incidence angles of 20, 40, 60, and 80°. Note the limited incidence angle on the detector surface for 80° incidence on the CPC.

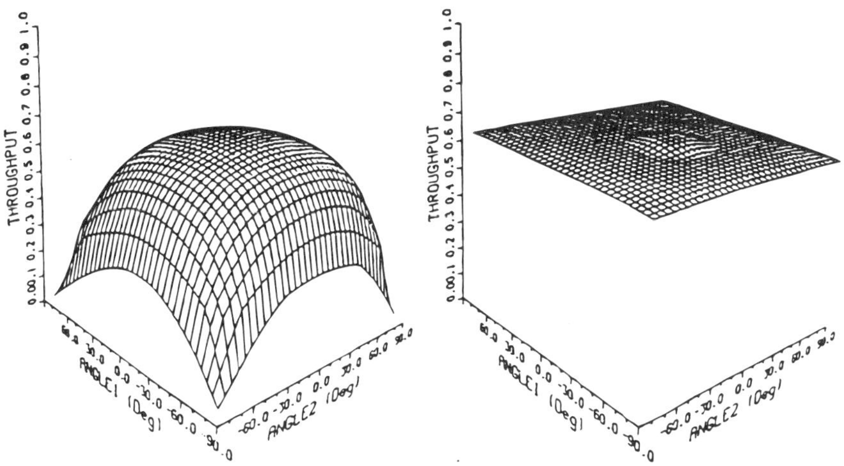

FIG. 6. Angular response of a hemiellipsoid reflectometer utilizing a bare photomultiplier tube (PMT) (left) or an inverted compound parabolic concentrator (CPC) with a PMT (right).

PMT angular response data of Edwards *et al.* (51). All mirror reflectances were set equal to 0.97, the eccentricity $\varepsilon = 0.1$, and $r_s/a = 0.1$. The slight rounding of the CPC/PMT response at large angles is due to skew rays that undergo multiple reflections. The dimple in the CPC/PMT response at $\{\text{angle 1, angle 2}\} = \{0°,0°\}$ is due to rays passing directly through the CPC without reflection. The variation in throughput caused by these effects can be reduced by using a high-reflectance coating on the CPC.

2. *Hemispherical Directional Reflectometers*

For $2\pi/\theta$ mode reflectometers, a non-Lambertian radiation source may introduce a measurement error. Wood *et al.* (31) described a typical cavity blackbody source that exhibits a monotonic drop in emitted radiance (52) with angle. The percentage decrease in radiance from the normal incidence value is 3 to 4% at 45° and 6 to 7% at 70°. Beyond 80°, the falloff is very rapid due to direct viewing of the cavity wall near the exit aperture of the source. If a Lambertian sample is illuminated with radiance having an angular dependence $L(\theta)$, where $L(0) = 1$, then the ratio of the measured reflectance to the true reflectance will be given by

$$2 \cdot \int_0^{\pi/2} L(\theta) \cdot \sin\theta \cdot \cos\theta \cdot d\theta. \tag{8.5}$$

If a Lambertian sample was illuminated with the source radiance profile given in Ref. (31), the sample reflectance would be measured to be 0.92 of the actual value.

In 1986, the reflectance of a roughened gold coating used with IR integrating spheres was measured from 0.3 to 2.0 μm with an integrating sphere and from 2 to 20 μm with a $2\pi/\theta$ mode hemiellipsoidal reflectometer (53). At the crossover wavelength a drop in the measured reflectance of 5–8% occurred in going from the integrating sphere results to the hemiellipsoidal reflectometer results. One would expect that the reflectance of this coating would increase monotonically with wavelength, however. A non-Lambertian $2\pi/\theta$ source is the most likely explanation for these results. Angular response errors need to be minimized when designing a reflectometer because corrections for this error are dependent on the BRDF of the sample, which is not known a priori.

For $2\pi/\theta$ mode reflectometers, an inverted CPC can be used in front of cavity radiation sources (46, 54) in order to reduce the non-Lambertian properties of such sources at larger angles. The HDR 100 hemiellipsoidal reflectometer from Surface Optics Corporation uses a reflective cone above a cavity radiation source (55).

D. Port and Auxiliary Optics Losses

Some radiation reflected from the sample in the $\theta/2\pi$ mode may be lost out of the entrance port when external optics are used or off of the focusing mirror or lens and its support structure when internal optics are used. This effect can result in an error in the measured reflectance. Similarly, for $2\pi/\theta$ mode reflectometers, the optics used to collect reflected radiation from the sample reduces the radiant power striking the sample and may affect the measured reflectance. If the sample is a Lambertian reflector and the illumination ($\theta/2\pi$) or viewing ($2\pi/\theta$) is near normal, then the fraction of radiation intercepting the sample optics is given by

$$F_o \approx \frac{A_o}{\pi \cdot d_o^2} \approx \frac{1}{4 \cdot (f/\#)^2}, \tag{8.6}$$

where A_o is the projected area of the limiting aperture of the sample optics, d_o is the distance between the sample and the limiting aperture, and $f/\#$ is the f- number of the sample optics. For $f/4$ optics and a Lambertian sample, 1.6% of the reflected radiation is lost through the sample optics.

For samples with an arbitrary BRDF, Coblentz (14) determined the size of this correction experimentally by placing a blackened disk near the beam port. The size of the beam optics loss can also be measured for $\theta/2\pi$ mode instruments by using a beamsplitter and a second detector (56). For $2\pi/\theta$

instruments having external optics, one can estimate the size of the beam port loss by placing a semitransparent conformal plug (i.e., a beam splitter) into the sampling port.

E. Interreflections

Interreflections between the detector and sample in the $\theta/2\pi$ mode or the source and sample in the $2\pi/\theta$ mode can introduce a systematic error that is dependent on the sample reflectance value as well as the sample BRDF. Additionally, the sizes of the sample, detector, and input spot in the $\theta/2\pi$ mode or the sample, source, and viewed spot in the $2\pi/\theta$ mode, and the magnification properties of the conic mirrors, are all factors that can affect the size of the error. For $2\pi/\theta$ instruments, interreflections can increase the source temperature. In the following analysis, it is assumed that the source temperature is stabilized (57).

For example, Sullivan and Allen (30) estimated the size of the interreflection error for a $\theta/2\pi$ hemiellipsoidal reflectometer by employing an averaging sphere at the detector focus. For an actual sample reflectance of 0.70, interreflections would cause the reflectance to be measured as 0.77.

Previously calibrated standards of similar reflectance and BRDF as the samples to be measured can be used to estimate the size of the interreflection uncertainty. However, this approach is limited by the a priori knowledge of the sample's reflectance and BRDF, as well as the availability of standards. The interreflection error can be directly minimized by using detectors ($\theta/2\pi$) or cavity radiation sources ($2\pi/\theta$) having a low reflectance.

In Section IV,E,1 we develop analog expressions for the measured reflectance, which include the effects of interreflection. Beckett and others have estimated the size of this correction analytically (15, 27, 56, 57) and with raytracing analyses (58). In Sections IV,E,2 and IV,E,3, we present methods developed by Beckett and Clarke for correcting of the interreflection error. In Section IV,E,4 we discuss the effects of reducing detector and source reflectance.

1. *Analytic Model*

The throughput T_s of a $\theta/2\pi$ collecting mirror reflectometer can be expressed as

$$\begin{aligned} T_s = {} & \rho_s \rho_{r,m}(1 - \rho'_D)(1 - F_o) \\ & + \rho_s \rho_{r,m}(1 - \rho'_D)(1 - F_o)^3 \rho_{r,m}^2 \rho'_s \rho'_D F_m F_s \\ & + \rho_s \rho_{r,m}(1 - \rho'_D)(1 - F_o)^5 \rho_{r,m}^4 \rho_s'^2 \rho_D'^2 F_m^2 F_s^2 + \cdots, \end{aligned} \tag{8.7}$$

where ρ_s is the sample directional/hemispherical reflectance, ρ'_s is the sample hemispherical/hemispherical reflectance, $\rho_{r,m}$ is the conic mirror specular reflectance, and ρ'_D is the detector hemispherical/hemispherical reflectance. An analogous expression can be applied to $2\pi/\theta$ instruments. In Eq. (8.7) we assume that the sample and detector are Lambertian reflectors, that the sample is fully illuminated by the beam, and that the image of the sample fully illuminates the detector after the first pass through the conic mirror. It is assumed that any radiation that overfills the sample is absorbed elsewhere in the system. The quantity F_m is the fraction of the radiation reflected from the detector that reenters the conic mirror above the detector. Collecting mirrors discussed previously (26, 27) that subtend less than the full hemisphere above the detector will have $F_m < 1$. The conic mirror above the sample is assumed to subtend 2π sr. F_o is the fraction of the radiation reflected from the sample that strikes the beam optics (Eq. 8.6). It is assumed that F_o can be used to describe the beam port losses from reflections at the detector. The quantity F_s is the fraction of the radiation reentering the conic mirror after reflecting off the detector that intercepts the sample. For a dual-paraboloidal mirror, $\rho_{r,m}$ must be replaced by $\rho^2_{r,m}$. Equation (8.7) can be reduced to

$$T_s = \frac{\rho_s \rho_{r,m}(1 - \rho'_D)(1 - F_o)}{1 - \rho^2_{r,m}\rho'_s\rho'_D F_m F_s(1 - F_o)^2}. \tag{8.8}$$

The denominator of Eq. (8.8) is an enhancement factor accounting for interreflection effects. For a hemiellipsoidal mirror with the sample and detector front faces positioned in the plane containing the major and minor axes of the hemiellipsoid, $F_m = 1$. When the sample is removed the throughput of the system for the reference measurement is $T_r = \rho_{r,m}(1 - \rho_D)$. The apparent sample reflectance obtained by ratioing the sample and reference measurement signals is given by the ratio of sample and reference throughputs:

$$\frac{T_s}{T_r} = \frac{\rho_s(1 - F_o)(1 - \rho'_D)/(1 - \rho_D)}{1 - \rho^2_{r,m}\rho'_s\rho'_D F_m F_s(1 - F_o)^2}. \tag{8.9}$$

Here, we neglect the interreflections between the reflectometer and the external optical system, which will depend on the details of the external system and need to be dealt with separately. Note that the factor by which the reflectance is enhanced due to multiple reflections can easily exceed 1.1 if the detector reflectance is greater than 20%. Clarke and Larkin (59) have suggested a separate correction for the interreflection error associated with the specular and diffuse components of samples with an arbitrary BRDF. Their technique involves measuring standards of known reflectance, such as pure barium sulfate powder and a first-surface aluminum mirror. For purely specular samples, the quantity $(1 - F_o)(1 - \rho'_D)$ in the numerator of Eq. (8.9) can be replaced with $1 - \rho_D$.

2. *Beckett's Method* ($\theta/2\pi$)

Beckett (15) observed that by interchanging the sample and detector so that the radiation beam is incident directly on the detector, one can obtain a detector signal that is proportional to

$$T_{r'} = \frac{(1 - \rho_D)}{1 - \rho_{r,m}^2 \rho'_s \rho'_D F_m F_s (1 - F_o)^2}. \tag{8.10}$$

Here, we have used the same assumptions that were used to derive Eq. (8.7). In Beckett's original derivation, the distinction between directional/hemispherical and hemispherical/hemispherical reflectance was ignored. The derivation of Eq. (8.10) also assumes that the detector is fully illuminated and that the sample exchange factor F_s is independent of which focus (sample or detector) is initially irradiated. This last assumption is not strictly correct because the spatial distribution of the irradiation on the detector will be different in the two cases. A raytracing analysis is needed to determine the size of the error related to this assumption. The ratio of the sample throughput result (Eq. 8.8) with Eq. (8.10) gives the apparent sample reflectance:

$$\frac{T_s}{T_{r'}} = \rho_s \rho_{r,m} (1 - F_o) \frac{(1 - \rho'_D)}{(1 - \rho_D)}. \tag{8.11}$$

The mirror reflectance can be independently measured relative to a reflectance standard with an absolute V-W apparatus (60) or with the conic mirror reflectometer itself. The measurement of F_o was discussed in Section V,D. An estimate of the hemispherical/hemispherical reflectance of the detector can be obtained by performing a series of measurements of the directional/hemispherical reflectance at various incidence angles. Beckett (15) did not suggest using the technique with $2\pi/\theta$ instruments; however, the application is straightforward.

3. *Clarke's Method* ($2\pi/\theta$)

Clarke and Larkin (59) have modified a hemispherical mirror reflectometer operated in the $2\pi/\theta$ mode so that the source can be viewed with and without the sample present. By ratioing these two measurements, they determine the factor by which the sample reflectance is enhanced due to interreflections. Clarke and Larkin then measure the sample reflected power and the source power passing through the sample aperture, using optics having the same étendue as that of the optics used to measure the source power when the sample was absent. From these four measurements, they determine the product of the sample reflectance and $1 - F_o$. Although the order of measurements is different, this technique is equivalent to the $2\pi/\theta$ analog of Beckett's (15) technique.

In order to illustrate the dependence on the variables defined previously, we slightly modify Clarke's method. It is assumed that the source and sample are Lambertian reflectors, that a viewing port in the conic mirror is symmetrically placed with respect to the sample and source, and that the sample area is defined by a perfectly black mask. If this sample area were fully illuminated, the image of the sample would just underfill the source aperture. Under these assumptions, the reflected sample power, as measured through the viewing port, is given by

$$P_s = \frac{P_{sc}\rho_{r,m}\rho_s(1 - F_o)F_s F_o}{1 - \rho_{r,m}^2 \rho_s' \rho_{sc}' F_s(1 - F_o)^2}, \tag{8.12}$$

where the source emitted power is P_{sc}, the source hemispherical/hemispherical reflectance is ρ_{sc}', and the sample hemispherical/hemispherical reflectance is ρ_s'. The source power exiting the viewing port when the sample is absent is equal to $P_{sc}F_o$. Thus, the ratio of the source power, as measured through the viewing port when the sample is present, to $P_{sc}F_o$ is given by

$$F = \frac{1}{1 - \rho_{r,m}^2 \rho_s' \rho_{sc}' F_s(1 - F_o)^2}, \tag{8.13}$$

This factor quantifies the enhanced brightness of the source due to interreflections. An identical factor present in Eq. (8.12) describes the enhanced brightness of the sample. If the source power is measured through the sample aperture with the same optics used to measure the reflected sample power, and with a conformal matched plug filling the view port, the result is

$$P_r = P_{sc}\rho_{r,m}F_s F_o. \tag{8.14}$$

Hence, the ratio of $P_s/(P_r F)$ gives $\rho_s(1 - F_o)$. The factor F_o can be determined from techniques discussed in Section V,D or by measuring the ratio of the power incident diffusely on the sample mask without a conformal matched plug in the viewing port to that with a plug.

4. *Low-Reflectance Detectors and Sources*

If one makes the corrections prescribed by Beckett or Clarke and simultaneously utilizes a low-reflectance detector ($\theta/2\pi$) or radiation source ($2\pi/\theta$), the resulting error due to interreflections can be greatly reduced. The normal incidence reflectance of gold–black coated detectors varies from near zero for wavelengths <2 μm to 2–6% for wavelengths at 20 μm, depending on the gold–black thickness. Thicker gold–black coatings reduce the reflectance at longer wavelengths but also increase the detector's thermal time constant.

The effective reflectance of the detector can be further reduced by using Paschen's (11) method. In particular, an inverted CPC, when positioned in front of a detector ($\theta/2\pi$) or source ($2\pi/\theta$) to reduce variations from ideal angular response as described in Section V,C, will also enhance the effective absorptance. Radiation reflected from the detector or source at angles larger than the CPC's critical angle is reflected back to the detector or source. If the detector or source is a Lambertian reflector and if the illumination of the CPC is Lambertian, then the effective reflectance of a CPC–detector combination is given by

$$\rho_{\text{net}} = \rho_{\text{D}} \cdot \rho_{\text{r, cpc}}^{2\langle n \rangle} \cdot \sin^2 \theta_{\text{c}} + \frac{(\rho_{\text{D}})^2 \cdot \rho_{\text{r, cpc}}^{2\langle n \rangle + \langle n' \rangle} \cdot (1 - \sin^2 \theta_{\text{c}}) \cdot \sin^2 \theta_{\text{c}}}{1 - \rho_{\text{D}} \cdot \rho_{\text{r, cpc}}^{\langle n + n' \rangle} \cdot (1 - \sin^2 \theta_{\text{c}})}, \tag{8.15}$$

where θ_c is the CPC critical angle, ρ_D is the directional/hemispherical reflectance of the detector (replace with ρ_{sc} for $2\pi/\theta$ instruments), $\rho_{r,\,cpc}$ is the regular reflectance of the CPC, $\langle n \rangle$ is the average number of reflections on the CPC for radiation entering the CPC's larger aperture at an angle $\theta < \theta_c$, $\langle n' \rangle$ is the average number of reflections on the CPC for radiation entering the CPC at an angle $\theta > \theta_c$, and $\sin^2 \theta$ is the fraction of power reflected from a Lambertian surface that is contained within an angle of θ from the surface normal. The denominator of the second term in Eq. (8.15) accounts for multiple reflection effects between the CPC and the detector/source. For a critical angle of 30°, a detector reflectance of $\rho_D = 0.05$, a CPC reflectance of 0.99, $\langle n \rangle = 1.5$, and $\langle n' \rangle = 2$, Eq. (8.15) predicts that the effective reflectance of the CPC–detector/source combination will be only 0.013. As noted previously, the use of an inverted CPC or hemispherical reflector requires a larger area detector, which can result in a reduced signal-to-noise ratio.

F. Sensitivity to Sample and Detector/Source Position

Conic mirror reflectometers can be significantly more sensitive to positioning errors of the sample and detector ($\theta/2\pi$) or source ($2\pi/\theta$) than integrating sphere reflectometers. In order to estimate the degree of sensitivity, a raytracing analysis of the three primary conic mirror designs was performed. For this study, the largest physical dimension of each conic mirror was assumed to be equal (i.e., $2R = 2a = 4f_s$), and the detector radius was fixed at $0.0427R$. For a hemisphere of radius 15.2 cm, this constraint fixes the detector radius at 13 mm, which is typical of large gold–black coated pyroelectric detectors. The circular sample was assumed to be a perfect diffuse reflector and the detector a blackbody ($\rho_D = 0$). A sufficiently large number of rays ($>50{,}000$) was used to ensure a standard uncertainty in the throughput of less than 0.25%. The diameter of the beam spot on the sample was set at 80% of the value needed [see Eqs. (8.2)–(8.4)] to avoid overfilling the detector. An $f/3.8$ beam port was positioned in the conic mirror for 8° incidence on the sample. For the hemisphere and hemiellipsoid, the eccentricities (ε, ε') were fixed at 0.1, and the sample radius (r_s) was determined from the constraint on the detector radius. For the dual-paraboloid, the sample radius and the eccentricity ($\varepsilon'' = r_s/2f_s$) were determined from the value of the detector radius. Under these conditions, the throughput of the three primary conic mirrors versus the sample position, as measured from the conic mirror baseplane, is shown in Fig. 7. The sample position is shown as a dimensionless variable with the normalization factor depending on the conic mirror.

The beam port loss is equal to 1 minus the ordinate value at $z = 0$. For the hemisphere and hemiellipsoid, the raytrace-derived value agrees well with the value predicted by Eq. (8.6). The slightly higher value observed for

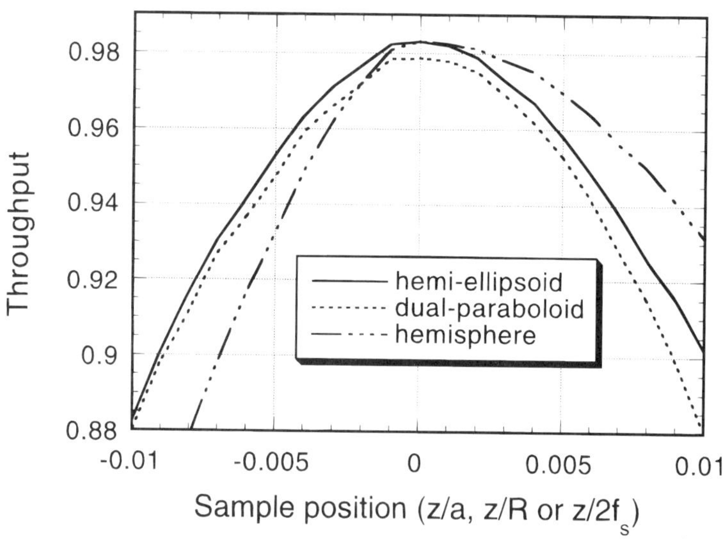

FIG. 7. Normalized throughput of a dual-paraboloid, hemiellipsoid, and hemispherical mirror reflectometer versus the sample position.

the dual-paraboloid may be due to the closer proximity of the port and the finite size of the beam spot on the sample. The dual-paraboloid's drop in throughput with sample z position is symmetric with respect to the sign of the position, whereas the hemisphere results exhibit significant asymmetry. The positioning tolerance required to ensure an error of $\leq 0.25\%$ is well within the capability of modern positioning equipment, assuming the conic mirror's largest physical dimension is at least 10 cm.

A more detailed study was conducted for a hemiellipsoid reflectometer with an eccentricity of 0.1 and a detector radius of $r_D/a = 0.0427$. Raytrace analyses were performed for various beam spot diameters on a Lambertian sample as a function of distance between the sample and the hemiellipsoid base plane. It was assumed that the base plane of the hemiellipsoid contained both semiaxes, that multiple reflections between the sample and detector could be ignored, and that an $f/3.8$ beam port was positioned for an average beam incidence angle of 8°. The beam spot diameters were undersized from the value predicted by the magnification formula given in Eq. (8.3). The results of this study are shown in Fig. 8. For the reflectometer and sample parameters studied, keeping the flux losses below 0.2 to 0.3% requires that the sample z coordinate be within the range $-.0015$

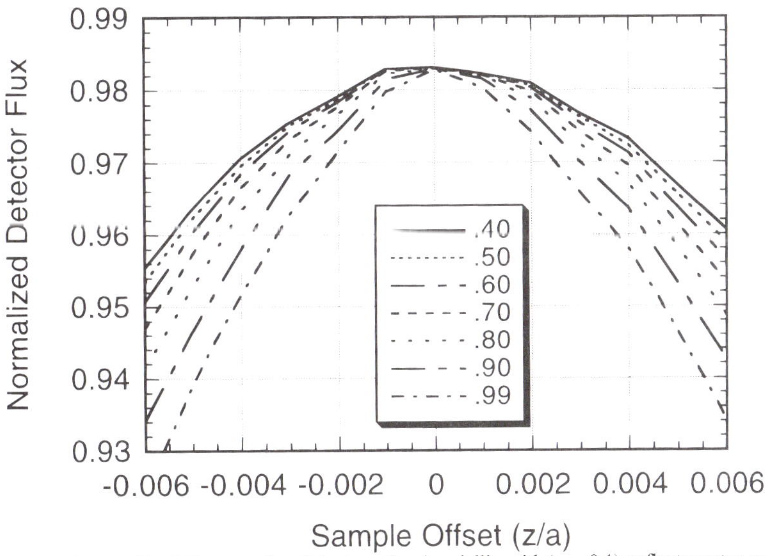

FIG. 8. Normalized flux on the detector of a hemiellipsoid ($\varepsilon = 0.1$) reflectometer versus sample position for various beam spot diameters on a Lambertian sample. The beam spot diameters are undersized by factors of 0.40–0.99 from the value defined by the ellipsoid magnification formula.

to $+.002a$ if the beam spot undersizing factor is ≤ 0.70. Note that the falloff in the detector flux is greater for lowering the sample because some scattered light is not collected by the hemiellipsoid. The beam port losses at $z = 0$ (1.7%) agree with the value predicted by Eq. (8.6). In general, increasing the eccentricity or the beam spot size will increase the sensitivity to sample positioning errors.

G. Shadowing Losses

When using an internal focusing mirror with a conic mirror, the supporting arm for the focusing mirror can introduce significant (>2%) shadowing errors. The losses from the optics have been discussed in Section V,D. Losses from the supporting arm can be reduced by minimizing the size of the arm and giving it an appropriate conic profile. For hemiellipsoidal instruments with the sample and detector normals coaxial with the semimajor axis (26, 27), the sample and its mount can block reflected radiation and cast a shadow on the detector. For instruments using two on-axis paraboloids or two on-axis ellipsoids, the detector and its mount could also block reflected radiation. These shadowing losses can be comparable in size to beam port losses and can easily exceed 1% of a Lambertian sample's reflectance if the reflectometer is not designed properly. Sample mount losses can be reduced by using small samples, locating the mount so that it does not intersect the specular beam, and minimizing the cross-sectional area of the mounts. The size of the losses for a Lambertian sample can be determined from a raytracing analysis or from radiation exchange factor tables.

H. Mirror Imperfections

Surface roughness and mirror slope errors can scatter or redirect radiation striking a conic mirror and potentially reduce the measured reflectance. The surface of a collecting mirror will always have some roughness due to microirregularities, polishing marks, dust, and/or single-point diamond-turning marks (61). If the root mean square (rms) surface roughness (σ) is much smaller than the radiation wavelength λ, then scalar scattering theory can be used to derive (62) the following relationship between the total integrated scatter (TIS) and rms surface roughness:

$$\mathrm{TIS} \equiv \frac{\rho_{\mathrm{d}}}{\rho_{\mathrm{r}}} \cong \left(\frac{4\pi\sigma}{\lambda}\right)^2, \tag{8.16}$$

where the total integrated scatter is defined as the ratio of the diffuse reflectance (i.e., directional/hemispherical-specular) to the directional/hemispherical reflectance. The smallest spatial wavelength on the mirror surface that scatters light can be determined from the diffraction grating equation. For normally incident light scattered into an angle of 90° from the mirror normal, the minimum spatial wavelength is equal to the radiation wavelength. Hence, when diamond turning a collecting mirror, it is best to have the period of the diamond-turning grooves less than the smallest wavelength of light that one plans to use with the mirror. Polishing the mirror after diamond turning can significantly reduce the rms surface roughness (63). Alloying materials in aluminium and other metals tend to segregate at grain boundaries as hard particles that polish more slowly and leave a residual source of scatter (61). The TIS versus rms surface roughness is shown in Fig. 9 for various wavelengths. These results demonstrate the major reason why collecting mirror reflectometers have found limited use in the visible.

Diamond-turning is an expensive process that is suitable for instruments developed at national standards labs. For commercial IR reflectometers, electroformed mirrors can be fabricated at a fraction of the cost of diamond-turned mirrors. Stresses in the electroforming mandril typically cause the mirror slope to deviate approximately 1–5 min of arc from the desired value within 13–25 mm of apertures. These deviations can cause detector ($\theta/2\pi$) overfilling errors when measuring Lambertian samples. Neu

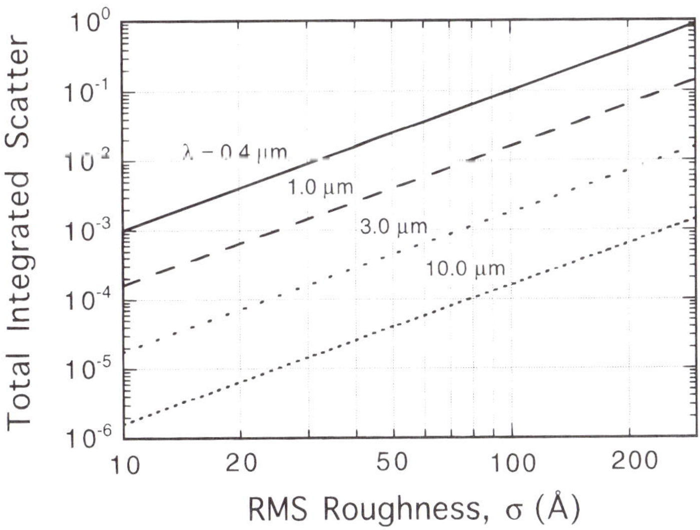

FIG. 9. Total integrated scatter (TIS) versus the RMS surface roughness for wavelengths of 0.4, 1, 3, and 10 μm.

et al. (64) discuss the focusing properties of a large electro-formed hemi-ellipsoidal mirror used in the $2\pi/\theta$ mode. They recommend expanding the ratio of the source diameter to the measured beam spot diameter defined by Eq. (8.3) by more than 50% to compensate for alignment effects and imperfections in the mirror.

VI. Summary and Conclusions

A comprehensive review of the published literature on conic mirror reflectometers has been conducted. Based on this review, the significant sources of measurement error present with conic mirror devices have been systematically analyzed, including important magnification effects. Recent developments in the design of hemispherical sources and detectors using nonimaging CPC concentrators have been reviewed. Extensive raytracing studies of the effect of sample positioning on the throughput of the primary conic mirrors have shown that for modestly sized mirrors (i.e., $R = a = 2f_s \approx 15$ cm), the positioning error is small ($<0.25\%$) if the sample offset from the conic mirror base plane can be controlled to ≤ 0.1–0.2 mm. However, the correction factor associated with interreflection effects can easily exceed 1.1 if the reflectance of the source ($2\pi/\theta$) or the detector ($\theta/2\pi$) is above 20%. Techniques for minimizing and correcting for the interreflection error are reviewed. All the reflectometer designs and equations discussed can be easily modified to perform and analyze transmittance measurements of diffusing samples.

Acknowledgments

The authors dedicate this chapter to the late Ternay Neu of Surface Optics Corporation. We thank Jack Hsia at NIST for his careful review of the manuscript and The Naval Warfare Assessment Division in Corona, California, for support of the work described.

References

1. Colorimetry: Official Recommendations of the CIE, Publication No. 15.2, p. 15. CIE, Vienna (1986).
2. Gier, J. T., Dunkle, R. V., and Bevans, J. T. *J. Opt. Soc. Am.* **44**, 558 (1954).
3. Reid, C. D., and McAlister, E. D. *J. Opt. Soc. Am.* **49**, 78 (1959).
4. Kubelka, P., and Munk, F. *Z. Tech. Phys.* **12**, 593 (1931).
5. Frei, R. W., and MacNeil, J. D. "Diffuse Reflectance Spectroscopy in Environmental Problem-Solving," p. 209. CRC Press, Cleveland (1973).

6. Budde, W. *J. Res. NBS* **80A**, 585 (1976).
7. Ooba, N., Erb, W. *et al.*, Absolute Methods for Reflectance Measurements, Publication No. 44 (TC-2.3). CIE, Vienna (1979).
8. International Lighting Vocabulary, Publication No. 17.4. CIE, Vienna (1987).
9. Clarke, F. J. J., and Parry, D. J. *Lighting Res. Technol.* **17**(1), 1 (1985).
10. Richter, W., and Erb, W. *Appl. Opt.* **26**, 4620 (1987).
11. Paschen, F. *Berlin Akad. d Wiss.*, April 27, 405 (1899).
12. Royds, T. *Phys. Zeit.* **11**, 316 (1910).
13. Royds, T. *Phil. Mag. J. Sci.* **21**, 167 (1911).
14. Coblentz, W. W. *Bull. Bur. Stds.* **9**, 283 (1913).
15. Beckett, H. E. *Proc. Phys. Soc. London* **43**, 227 (1931).
16. Sanderson, J. A. *J. Opt. Soc. Am.* **37**, 771 (1947).
17. Derksen, W. L., and Monahan, T. I. *J. Opt. Soc. Am.* **42**, 263 (1952).
18. Derksen, W. L., Monahan, T. I., and Lawes, A. J. *J. Opt. Soc. Am.* **47**, 995 (1957).
19. Birkebak, R. C., and Hartnett, J. P. *Trans. ASME* **80**, 373 (1958).
20. Kozyrev, B. P., and Vershinin, O. E. *Opt. Spectrosc.* **6**, 345 (1959).
21. Janssen, J. E., and Torborg, R. H. in "Measurement of Thermal Radiation Properties of Solids" (J. C. Richmond, Ed.), NASA SP-31, p. 169. U.S. Government Printing Office, Washington, DC (1963).
22. Kronstein, M., Kraushaar, R. J., and Deacle, R. E. *J. Opt. Soc. Am.* **53**, 458 (1963).
23. White, J. U. *J. Opt. Soc. Am.* **54**, 1332 (1964).
24. Keegan, H. J., and Weidner, V. R. *J. Opt. Soc. Am.* **56**, 523 (1966).
25. Brandenberg, W. M. *J. Opt. Soc. Am.* **54**, 1235 (1964).
26. Blevin, W. R., and Brown, W. J. *J. Sci. Instrum.* **42**, 385 (1965).
27. Dunn, S. T., Richmond, J. C., and Wiebelt, J. A. *J. Res. NBS* **70C**, 75 (1966).
28. Neu, J. T. Design, Fabrication and Performance of an Ellipsoidal Spectroreflectometer, NASA-CR-73193 (1968).
29. Heinisch, R. P., Bradac, F. J., and Perlick, D. B. *Appl. Opt.* **9**, 483 (1970).
30. Sullivan, J. D., and Allen, F. J. 'Tests of an Ellipsoidal Mirror Reflectometer,' Ballistic Res. Lab. Memo. Rep. No. 2551 (October 1975).
31. Wood, B. E., Pipes, J. G., Smith, A. M., and Roux, J. A. *Appl. Opt.* **15**, 940 (1976).
32. Dunkle, R. V., Gier, J. T., and Bevans, J. T. "Final Progress Report on the Snow Characteristics Project." Univ. Calif. Inst. Eng. Res., Berkeley (1955).
33. Dunkel, R. V. in "Surface Effects on Spacecraft Materials" (F. J. Clauss, Ed.), p. 117. Wiley, New York (1960).
34. Personal communication, Leonard M. Hanssen, NIST. The dual-ellipsoid design ($\theta/2\pi$) would use two prolate hemiellipsoids with a common focus and their base planes perpendicular to the major axis. The major axes of the hemiellipsoids, the sample normal and the detector normal, would be coaxial. The diffusely reflected radiation from the sample would then pass through the second focus of the sample ellipsoid before entering the detector ellipsoid. The incidence angle on the detector could be restricted by utilizing an off-axis ellipsoid or by increasing the detector ellipsoid's semimajor axis size and truncating the ellipsoid. Reciprocal mode ($2\pi/\theta$) designs also appear possible.
35. Neher, R. T., and Edwards, D. K. *Appl. Opt.* **4**, 775 (1965).
36. Clarke, F. J. J., and Larkin, J. A. *Infrared Phys.* **25**, 359 (1985).
37. Neu, J. T. in "Optical Scatter: Applications, Measurements and Theory II" (J. Stover, Ed.), Vol. 1995, p. 101. SPIE, Bellingham, WA (1994).
38. Willey, R. R. *Appl. Spectrosc.* **30**, 593 (1976).
39. Richter, W. *Appl. Spectrosc.* **37**, 32 (1983).
40. Oppenheim, U. P., Turner, M. G., and Wolfe, W. L. *IR Phys. Technol.* **35**, 873 (1994).

41. Hanssen, L. M. in "Spectrophotometry, Luminescence and Colour; Science and Compliance" (C. Burgess and D. G. Jones, Eds.), p. 115. Elsevier, Amsterdam (1995).
42. Labsphere, P.O. Box 70, North Sutton, NH 03260. The mention of manufacturers and model names is intended solely for the purpose of technical information useful to the reader and in no way should be construed as an endorsement of the named manufacturer or product.
43. Hanssen, L. M. *Appl. Opt.*, **35**, 3597 (1996).
44. Edwards, D. K. *Appl. Opt.* **5**, 175 (1966).
45. Griffiths, P. R., and de Haseth, J. A. "Fourier Transform Infrared Spectroscopy." Wiley, New York (1986).
46. Snail, K. A., and Hanssen, L. M. *Appl. Opt.*, submitted (November 1997).
47. Snail, K. A. *Appl. Opt.* **26**, 5326 (1987).
48. Dunn, S. T. NBS Tech. Note No. 279 (1965).
49. Hanssen, L. M., and Snail, K. A. unpublished data. The detector was obtained from Laser Probe in Utica, NY.
50. Personal communication, Marc Mermelstein, Naval Research Laboratory.
51. Edwards, D. K., Gier, J. T., Nelson, K. E., and Roddick, R. D. *Appl. Opt.* **51**, 1279 (1961).
52. Although the term intensity is used in conjunction with Fig. 5 in this reference, a related technical report indicates that the vertical scale is in units of spectral radiance.
53. Snail, K. A., and Carr, K. F. in "Infrared, Adaptive, and Synthetic Aperture Optical Systems" (R. Johnson, W. Wolfe and J. Fender, Eds.), Vol. 643. p. 75. SPIE, Bellingham, WA (1986).
54. Snail, K. A. U.S. Patent No. 4,988,205, dated January 29, 1991.
55. Personal communication, Ternay Neu, Surface Optics Corporation.
56. Snail, K. A. and Hanssen, L. M. *Appl. Opt.* **28**, 1793 (1989).
57. Clarke, F. J. J. in "Advances in Standards and Methodology in Spectrophotometry" (C. Burgess and K. D. Mielenz, Eds.), p. 235. Elsevier, Amsterdam (1987).
58. Snail, K. A. "Multiple reflection effects in hemi-ellipsoid reflectometers," presentation at the OSA Annual Meeting, October 20, 1987, Rochester, NY.
59. Clarke, F. J. J., and Larkin, J. A. in "Recent Developments and Applications of Infrared Analytical Instrumentation," Vol. 917, p. 7. SPIE, Bellingham, WA (1988).
60. Bennett, H. E., and Koehler, W. F. *J. Opt. Soc. Am.* **50**, 1 (1960).
61. Bennett, J. M., and Mattson, L., "Introduction to Surface Roughness and Scattering." Optical Society of America, Washington, DC (1989).
62. Bennett, H. E., and Porteus, J. O. *J. Opt. Soc. Am.* **51**, 123 (1961).
63. Parks, R. E., and Evans, C. J. *Prec. Eng.* **16**, 223 (1994).
64. Neu, J. T., Beecroft, M. T., and Schramm, R. in "Stray Radiation in Optical Systems III" (R. Breault, Ed.), Vol. 2260, p. 62. SPIE, Bellingham, WA (1994).

BIDIRECTIONAL REFLECTANCE DISTRIBUTION FUNCTION AS A MEASURE OF OPTICAL SCATTER

S. H. C. P. McCALL
Stellar Optics Research International Corporation (SORIC)

I. Introduction: Significance and Use of the Bidirectional Reflectance Distribution Function*

This chapter explains how to characterize the angular distribution of optical scatter from an opaque surface. In particular it focuses on measurement of the bidirectional reflectance distribution function (BRDF). BRDF is a convenient and well-accepted means of expressing optical scatter levels for many purposes (1, 2).

The angular distribution of scatter is a property of surfaces that may have direct consequences. Scatter from mirrors and other components in an optical system can be the limiting factor in resolution or signal-to-noise level. Scatter can be an important design parameter for telescopes. Scatter measurements are crucial to correct operation of ring laser gyros. Scatter from a painted surface such as on automobiles can influence sales appeal.

* Sections 1 and 2 are extracted, with permission, from ASTM standards: E1392-90, "Standard Practice for Angle Resolved Optical Scatter Measurements on Specular or Diffuse Surfaces" Copyright ASTM, 100 Barr Harbor Drive, West Consholtocken, PA 19428 (1).

ISBN 0-12-764070-3/$25.00.

The angular distribution of scatter from optically smooth surfaces can be used to calculate surface parameters or reveal surface characteristics. For example, the total scatter found by integrating the BRDF over the hemisphere can be related to surface roughness. The amount of scatter at a given scatter angle can be associated with a specific surface spatial frequency.

Additional data presentation formats described in Section 2.3 have advantages for certain applications. Surface parameters can be calculated from optical scatter data when assumptions are made about model relationships. The extrapolated parameters; total integrated scatter (TIS), roughness (σ), and power spectrum (PSD), are described in Section 2.4.

Optical scatter from an opaque surface results from surface topography, surface contamination, and subsurface effects. Scatter from small amounts of contamination can easily dominate the scatter from a smooth surface. Likewise, subsurface effects may play a more important scatter role than typically realized when surfaces are superpolished.

This chapter applies only to BRDF measurements on opaque samples. It does not apply to scatter from translucent or transparent materials for which there are subtle complications that affect measurement of translucent or transparent materials that are best addressed in standards (3, 4).

BRDF is simply the ratio of sample radiance (w/cm^2 sr) to sample irradiance (w/cm^2). It is convenient to incorporate radiance in the definition or the scatter would depend on instrument parameters (detector aperture and distance to the detector).

BDRF measurements can be made at any wavelength in the ultraviolet, visible, and infrared regions; however, the majority of instruments operate at 0.6328 μm because of convenience. Difficulty in obtaining appropriate sources, detectors, and low scatter optics complicates its practical application at wavelengths less than approximately 0.25 μm. Diffraction effects that start to become important for wavelengths greater than 15 μm complicate its practical application at longer wavelengths. Diffraction effects can be properly dealt with in scatter measurements (5), but they are not discussed in this chapter. If you want to know the amount of scatter at a specific wavelength, λ, and angle θ_s it is best to measure it at that specific λ and θ_s because angle and wavelength scaling is not exact or always possible. Thus, although such methods exist, this chapter does not provide (i) a method for wavelength scaling: a method to extrapolate from data for one wavelength to data for any other wavelength; or (ii) a method for angle scaling in which data taken at particular incident and scatter direction are not extrapolated to other directions.

Scatter measurements and instrumentation is a complex topic that cannot be wholly reviewed in this chapter. Although BRDF is defined in differential form it is measured with the incremental limitations imposed by real instrumentation. [For Section I, see (6).] All of the following produce noticeable deviations between the true BSDF and the measured BSDF: the finite detector aperture, scatter created within the instrument, calibration inaccuracies and practical equipment limitations such as noise, detector nonlinearity, and mechanical errors. In order to generate meaningful scatter specifications and fully utilize the data it is important to understand the deviations (6). The simple plane-of-incidence scatterometer outlined in Fig. 1 contains most of the components typically found in more sophisticated systems. These may be grouped into four categories: light source, sample mount, receiver (detector), and computer/electronics package, further described in Ref. (6). The ability to store, analyze, and display data in convenient graphical format is a key feature of the instrument. An example of BRDF data from a front-surface mirror is shown in Fig. 2. The horizontal axis is a log scale of degrees from specular, and the vertical axis is a log scale of the BRDF in sr^{-1}. The second plot, labeled "instrument signature," is a measure of light scattered by the instrument and is clearly a concern for BRDF interpretation in the near specular region. Reference (6) discusses the causes, measurement, and interpretation of instrument signature.

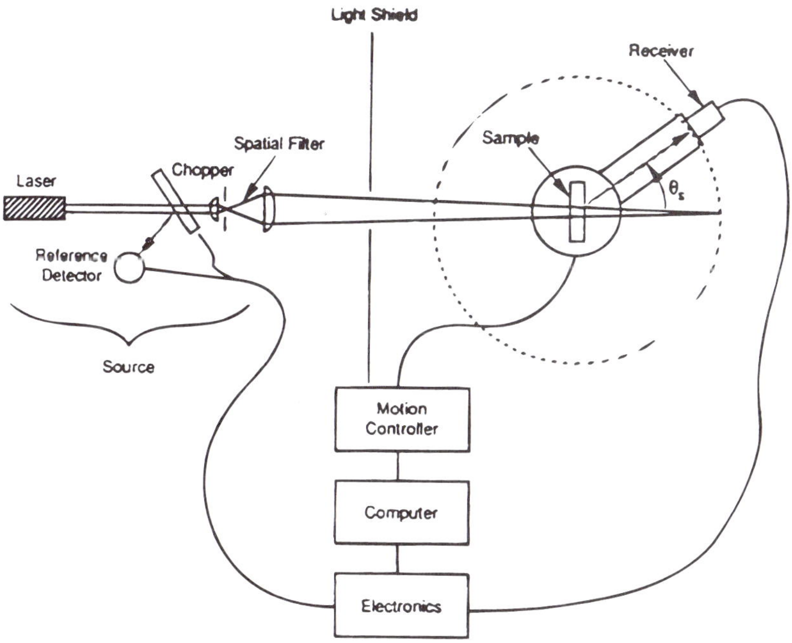

FIG. 1. Components of a typical BSDF scatterometer (6).

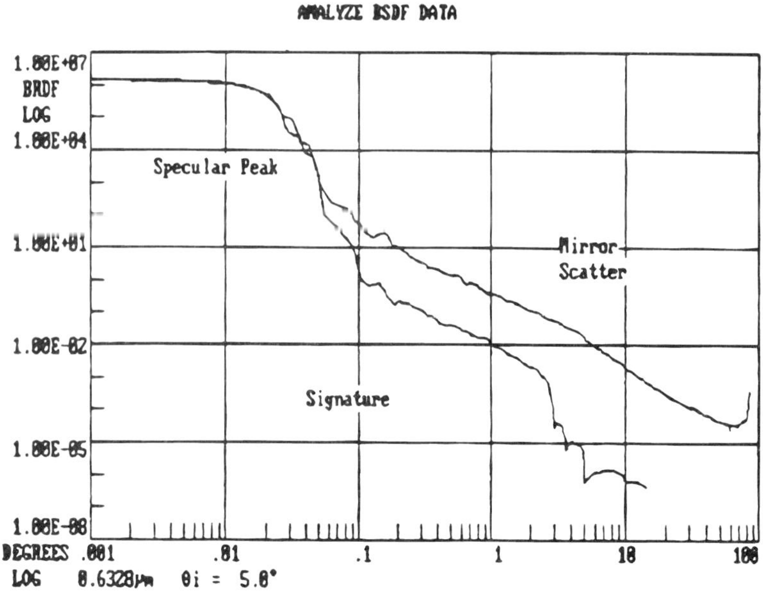

FIG. 2. The BRDF of a molybdenum mirror compared to instrument signature (6).

II. Terminology and Definition of BRDF

A. Definition of BRDF

The definitions here pertain to the American Society for Testing Materials (ASTM) "Angle Resolved Optical Scatter Measurements and Specular or Diffuse Samples" (ASTM E1392-90). Definitions of terms not included here will be found in definitions E284 (definitions of terms relating to appearance of materials) or ANSI Standard B46.1 (ANSI/ASME B46.1, "Surface Texture; Surface Roughness, Waviness, and Lay"). Additional graphic information will be found in Figs. A1–A3.

- *Scatter*: the radiant flux that has been redirected over a range of angles by interaction with the sample;
- *Sample coordinate system*: a coordinate system fixed to the sample and used to specify position on the sample surface for measurement. Note: The sample coordinate system is application and sample specific. The Cartesian coordinate system shown in Fig. A1 is recommended for flat samples. The origin is at the geometric center of the sample face with the Z axis normal to the sample. A fiducial mark must be shown at the

periphery of the sample and it will be assumed to mark the X axis unless otherwise indicated;

- *Plane of incidence*: (PLIN) the plane containing the sample normal and central ray of the incident flux;
- *Beam coordinate system, XB, YB, and ZB*: a Cartesian coordinate system with the origin on the central ray of the incident flux at the sample surface, the XB axis in the PLIN and the ZB axis normal to the surface as shown in Fig. A1. Note: The angle of incidence, scatter angle, and incident and scatter azimuth angles are defined with respect to the beam coordinate system;
- *Incident direction*: the central ray of the incident flux specified by θ_i and ϕ_i in the beam coordinate system;
- *Angle of incidence* (θ_i): angle between the central ray of the incident flux and the ZB axis;
- *Incident azimuth angle* (ϕ_i): the fixed 180° angle from the XB axis to the projection of the incident direction onto the XB–YB plane. Note: It is convenient to use a beam coordinate system (see Fig. A2) in which $\phi_i = 180°$ because this makes ϕ_s the correct angle to use directly in the familiar form of the grating equation. Conversion to a sample coordinate system is straight forward, provided the sample location and rotation are known;
- *Scatter direction*: the central ray of the scattered flux specified by θ_s and ϕ_s in the beam coordinate system;
- *Scatter angle* (θ_s): angle between the central ray of the scattered flux and the ZB axis;
- *Scatter azimuth angle* (ϕ_s): angle from XB axis to the projection of the scatter direction onto the XB–YB plane;
- *Specular direction*: the central ray of the reflected flux that lies in the PLIN with $\theta_i = \theta_s$ and $\phi_s = 0$;
- $\Delta\theta$: the angle between the specular direction and the scatter direction;
- $\Delta\beta$: the projection of $\Delta\theta$ onto the XB–YB plane; that is, the Δ angle measured in direction cosine space. For scatter in the PLIN, $\Delta\beta = \sin\theta_s - \sin\theta_i$. For scatter out of the PLIN, the calculation of $\Delta\beta$ becomes more complicated (see Section 2.3, $\Delta\beta$ Presentation);
- *Receiver solid angle* (Ω): the solid angle subtended by the receiver aperture stop from the sample origin;
- *Sample radiance* (L_e): a differential quantity that is the reflected radiant flux per unit projected receiver solid angle per unit sample area. Note: In practice L_e is an average calculated from scattered power, P_s, collected by the projected receiver solid angle, $\Omega \cos\theta_s$, from the illuminated area, A. The receiver aperture and distance from the sample determines Ω and the angular resolution of the instrument;

- *Sample irradiance* (E_e): the radiant flux incident on the sample surface per unit area. Note: In practice, E_e is an average calculated from the incident power, P_i, divided by the illuminated area, A. The incident flux should arrive from a single direction; however, the acceptable degree of collimation or amount of divergence is application specific and should be reported;
- *BRDF*: the sample radiance divided by the sample irradiance. The procedures given in this practice are correct only if the field of view determined by the receiver field stop is sufficiently large to include the entire illuminated area:

$$\mathrm{BRDF} = \frac{L_e}{E_e} = \frac{(P_s/\Omega A \cos\theta_s)}{(P_i/A)} = \frac{P_s}{P_i \Omega \cos_s} \ [\mathrm{sr}^{-1}]. \tag{1}$$

BRDF is a differential function dependent on the wavelength, incident direction, scatter direction, and polarization states of the incident and scattered flux. In practice it is calculated from the average radiance divided by the average irradiance. Notice that BRDF is referenced to the incident power and not the reflected power as some people might do for a specular sample. The $\cos\theta_s$ term takes into account foreshortening of A as θ_s increases so that we are using true radiance in our calculation. This keeps BRDF constant for a Lambertian (uniformly diffuse) surface. It the $\cos\theta_s$ term is not included, BRDF decreases with θ_s for a Lambertian surface; however, the advantage is that it would directly show the scattered power for a particular detector aperture. We must ensure that the detector can view the entire A and also that P_i includes the entire beam. Usually P_i is measured by removing the sample and swinging the detector into the straight through position. A reference detector is used to monitor incident beam power so that any variations after the initial V_i measurement can be taken into account;
- *Instrument signature*: the mean equivalent BRDF due to instrument limitations where there is no sample scatter present. Note: The limitation is normally stray from instrument components and out-of-plane aperture position errors from receiver positions near specular. At large scatter angles, the limitation is normally electronic background from the detector, but as θ_s approaches 90° the accuracy of θ_s becomes important because of the $1/\cos\theta_s$ term in BRDF. The signature can be measured by scanning a very low scatter reference sample, in which case the signature is adjusted by dividing by the reference sample reflectance. The signature is commonly measured by moving the receiver near the optical axis of the source and making an angle scan with no sample in the sample holder. It is necessary to furnish the instru-

ment signature when reporting BRDF data so that the user can decide at what θ_s the BRDF data are lost in the signature. Preferably, the signature is several decades below the sample data and can be ignored;

- *Noise equivalent BRDF* (*NEBRDF*): the root mean square (rms) noise fluctuation, or the standard deviation of the detector signal, expressed as equivalent BRDF. Note: Measurement precision is limited by the acceptable signal-to-noise ratio with respect to these fluctuations. It should be noted that although the detector noise is independent of θ_s, the NEBRDF will increase at large values of θ_s because of the $1/\cos\theta_s$ factor. Measurement precision can also be limited by other experimental parameters as discussed in Section 2.2.

B. Precision and Bias

1. *Precision*

The precision of BRDF is inconclusive based on the results of an interlaboratory round robin conducted in 1988 (7). This round robin was conducted at a single wavelength (0.6328 μm), angle of incidence (10°), polarization state (s incident), and with four specific sample surfaces. It was found that precision depends on the BRDF level and scatter angle as discussed in Ref. (8). Additional information on precision was accumulated in a 10.6-μm round robin conducted in 1989 (9).

A white diffuse sample with mean BRDF = 0.27/sr gave a fractional deviation (standard deviation of the 18 measurement sets divided by the mean BRDF) close to 17% at scatter angles from 15 to 70°. A black diffuse sample with mean BRDF = 0.01/sr gave fractional deviations from 24 to 39% depending on scatter angle. Specular mirrors gave fractional deviations from 31 to 134% depending on scatter angle. Variations were larger at large scatter angles in which detector noise levels of some instruments and errors in θ_s had a large effect. These variations are much larger than expected from a typical error analysis.

2. *Bias*

There is no bias inherent in BRDF. BRDF is a number derived from the ratio of physical parameters that can be specified in absolute units. However, individual laboratories may have measurement errors that lead to systematic offsets, such as an inaccurately measured solid angle. Other possible mechanisms are discussed in Ref. (8). It is not possible at this time to separate these systematic errors from bias; however, intralaboratory measurements on the same instrument typically repeat within 5% (10).

C. Additional Data Presentations

1. $\Delta\theta$ Presentation

It is common practice to plot BRDF with respect to the angle from the specular beam, $\Delta\theta$. If scatter is measured only in the PLIN, $\Delta\theta = \theta_s - \theta_i$. However, in the more general case for scatter out of the PLIN:

$$\Delta\theta = \cos^{-1}(\cos\theta_i \cos\theta_s + \sin\theta_i \sin\theta_s \cos\phi_s). \tag{2}$$

This is a useful angular reference for specular samples. However, when using this format, care must be taken that $\Delta\theta$ is not confused with θ_s in the calculation of BRDF. This presentation format is normally used only when $\Delta\theta$ passes through zero; that is, when scatter scan includes the specular beam. The terms "forward scatter" and "back scatter" refer to PLIN scatter directions for which $\Delta\theta$ is respectively positive or negative. Note that $\Delta\theta$ continues to increase as a negative angle when passing the surface normal because the sign of ϕ_s switches in Eq. (2).

2. $\Delta\beta$ Presentation

$\Delta\beta = \beta - \beta_0$, where $\beta = \sin\theta_s$ and $\beta_0 = \sin\theta_i$, is a method of expressing the angle between the specular and scatter directions in direction cosine space along the surface for scatter in the PLIN. This is very useful normalization when scatter results only from surface microroughness, and the grating equation:

$$\sin\theta_s = \sin\theta_i \pm n\lambda f, \tag{3}$$

where λ is the wavelength of the incident flux, f is the linear spatial frequency for the microroughness in the x or y direction, and n is the diffraction order.

Equation (3) can be used to relate θ_s to f. Only the first order ($n = 1$) is significant for roughness $\ll$ wavelength. BRDF can now be interpreted as the ability of each f to scatter light. If BRDF is plotted versus a $\Delta\beta$ scale it may be independent of θ_i and proportional to f. If the surface behaves in this way the BRDF is "shift invariant" (11). In the general case for scatter out of the PLIN the following two dimensional grating equations apply:

$$\cos\phi_s \sin\theta_s = \sin\theta_i \pm n\lambda f_x \tag{4}$$

$$\sin\theta_s \sin\phi_s = \pm n\lambda f_y. \tag{5}$$

The definition of $\Delta\beta$ must be expanded to include the projection of the scattered light in the x and y directions:

$$\Delta\beta^2 = \sin^2\theta_i + \sin^2\theta_s - 2\sin\theta_i \cdot \sin\theta_s \cdot \cos\phi_s. \tag{6}$$

3. *Reflectance Factor*

A measure of diffuse reflectance in common use is the reflectance factor R, that is the ratio of flux propagated from source to receiver in a reflectometer with a specimen to the flux propagated with a perfectly reflecting diffuser. Regarding a scatterometer as a very directional bidirectional reflectometer, the following relationship between R and BRDF is obtained:

$$R = \frac{\text{BRDF}}{\text{BRDF}_{\text{diffuser}}} = \frac{\text{BRDF}}{1/\pi} = \pi\text{BRDF}. \tag{7}$$

Additional information can be found in Ref. (3). Reflectance factor and specular reflectance share the same symbol, R, but they are not the same parameter.

D. Model-Dependent Calculated Parameters

1. *Total Integrated Scatter*

TIS can be calculated from BRDF by integrating BRDF over the hemisphere (12). Typically, a 5° total angle "hole" is left around the specular beam because specular light is not included in total integrated scatter (13).

For an isotropic surface we can measure in-plane BRDF at $\theta_i = 0$ and calculate the expected total integrated scatter by integrating over the angle limits specified in Ref. (13):

$$\text{TIS(calculated)} = 2\pi R^{-1} \int_{2.5^\circ}^{70^\circ} \cos\theta_s \,\text{BRDF} \sin\theta_s \, d\theta_s. \tag{8}$$

Sample specular reflectance, R, must be included because total integrated scatter is referenced to reflect and not incident power. The $\cos\theta_s$ term must be included because BRDF is defined in terms of the projected receiver aperture. This comparison between total integrated scatter and BRDF may not be exact because the total integrated scatter detector is less sensitive to light incident on the detector at large angles, and if low f (close to specular) scatter dominates, the 5° hole is critical. In addition, a TIS instrument is not polarization selective.

2. *Roughness*

The rms surface roughness, σ, is an often quoted number that can be obtained from direct profile measurements with stylus or optical profilometers. It can also be inferred from total integrated scatter when $\sigma \ll \lambda/4\pi$ for front surface scatter from a clean, smooth surface (14) as described in

Ref. (13):

$$\sigma = (\lambda/4\pi\sqrt{\text{TIS}}). \tag{9}$$

The user must confirm the usefulness of this σ calculation based on the particular measurement circumstances. It may have strong frequency limitations and not agree with surface roughness derived from optical or mechanical profile instruments [which can have different spatial frequency limits (15)].

3. *Power Spectrum*

The surface PSD function can be calculated from the BRDF through a scatter model. For example, the grating equation model discussed in Section 2.3.2 shows that high-frequency surface perturbations will scatter light far from specular and low-frequency perturbations will scatter close to specular. The PSD shows the amount of modulation versus f; that is, the square of the Fourier transform of the surface profile. Because it is a sample property, the same PSD should be obtained regardless of wavelength- and incident angle-dependent differences in the BRDF data.

Wavelength scaling is another check on system calibration. Smooth, clean, nonabsorbing front surface reflectors should yield the same PSD for different BRDF measurement wavelengths. If the instrument does not wavelength scale on appropriate samples, the BRDF measurement may be suspect. Polished molybdenum and aluminized glasses are two examples of mirrors that have been shown to wavelength scale from visible into the infrared. Many beryllium mirrors and silicon carbide mirrors have been shown to not wavelength scale because of anomalous scatter that arises from features other than surface roughness.

III. Interpreting BRDF Light Scatter Measurements*

In this Section, we discuss how light scatter measurements can be used to quantitatively determine the physical condition of a surface or bulk material. By studying typical data from various applications, it will be seen how scatter measurements offer a fast, sensitive quality check for a variety of industrial and laboratory applications, from detailed evaluations to simple pass/fail quality checks.

A. Where Does Scatter Come from?

Objects that are opaque and/or reflective scatter light from their outer surfaces. A transparent or partially transparent object may scatter light both from its outer surfaces and from its interior (or bulk).

* Sections 3 and 4 are duplicated from "Interpreting Light Scattering Measurements" (2) (Schmitt Measurement Systems, Inc., Portland OR, USA).

1. *Surface Defects*

A perfectly smooth mirror will reflect light according to the simple relationship angle of reflection = angle of incidence. This is termed specular reflection. If a narrow laser beam is directed at a perfectly smooth mirror, all the reflected light intensity will be in the narrow reflected beam. In the real world, surfaces are not perfectly smooth and have some types of defects, such as dust particles, cracks, residual film, microroughness, etc. These defects cause some incident light to be scattered or diffusely reflected off the surface in many directions. In the case of very rough surfaces, such as a sheet of sandpaper, the well-defined specular reflection disappears and all the reflected light is scattered.

2. *Bulk Defects*

When a beam of light passes through a transparent (clear) material, such as water or glass, it will be almost invisible from most angles. Very little light is scattered out of the direction of beam propagation. However, if the material has small particles or defects in it, some of the light will be scattered, making the beam visible from all directions. The simplest example of this is to consider a search light. It is virtually invisible in a clear night sky but clearly visible if it passes through smoke or a cloud. The reason it becomes visible is that the smoke particles or water droplets scatter the light in all directions.

B. How is Scatter Related to the Defects?

The size, shape, and distribution of the surface features or bulk defects determine how much light is scattered versus how much is reflected in a specular fashion. In addition, the distribution and size of these features determine the angular distribution of the scatter. Measurements of how the material scatters a well-characterized beam of light, such as a laser beam, can therefore be used to determine the size and distribution of the scattering features.

1. *Defining and Measuring Scatter*

Figure 3 shows how light may be scattered in both reflection and transmission. In both cases, the light may be scattered throughout a hemisphere, whereas the "specular" beam is confined to a single angle.

The terms most useful in defining or measuring scatter are collectively called BSDF. The two most commonly used are BRDF (for reflected scatter) and BTDF (for transmitted scatter). For simplicity, we concentrate on the more common BRDF surface measurements, but similar interpretations apply to BTDF measurements.

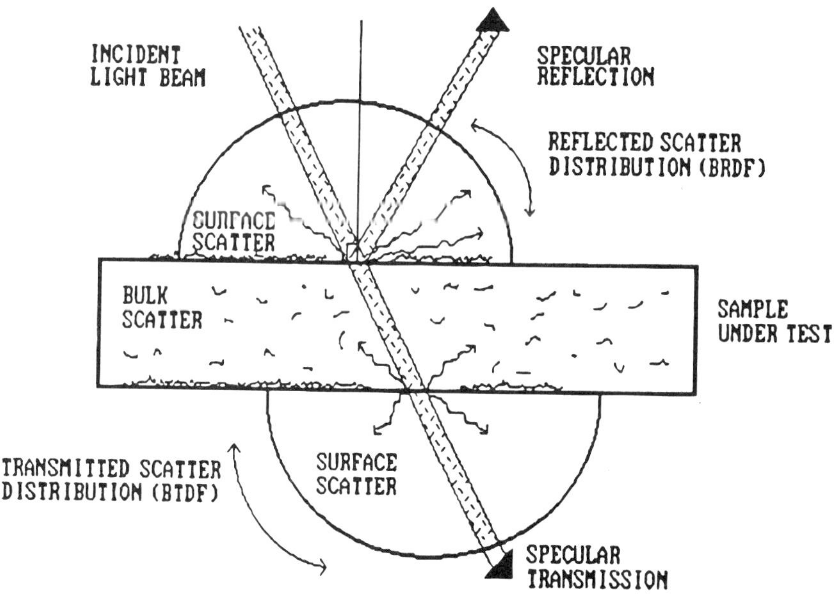

FIG. 3. Both surface and bulk imperfections contribute to both reflected and transmitted scatter (2).

The BRDF is a full description of the angular distribution of the reflected scattered light. It assumes a single-incident light beam at a fixed angle of incidence. The BRDF value at each scatter angle is the scattered light intensity at that angle divided by the incident intensity, and it includes various geometric factors (Fig. 4) BRDF can be measured by an array of detectors arranged in an arc or by sweeping a single detector through an arc.

The following are some important general rules to remember: The intensity of scatter increases with increased depth and density of the surface defects; the scatter angle (measured from specular) is dependent on the average width (or length) of the surface defects; and wider (longer) defects scatter at angles close to specular, whereas narrower (shorter) defects scatter at angles far from specular.

C. Interpreting Scatter Data

1. *BRDF Angle Plots*

In this section, we consider the most common type of measurement, BRDF angle scans. During measurements, such as with the CASI Scatterometer

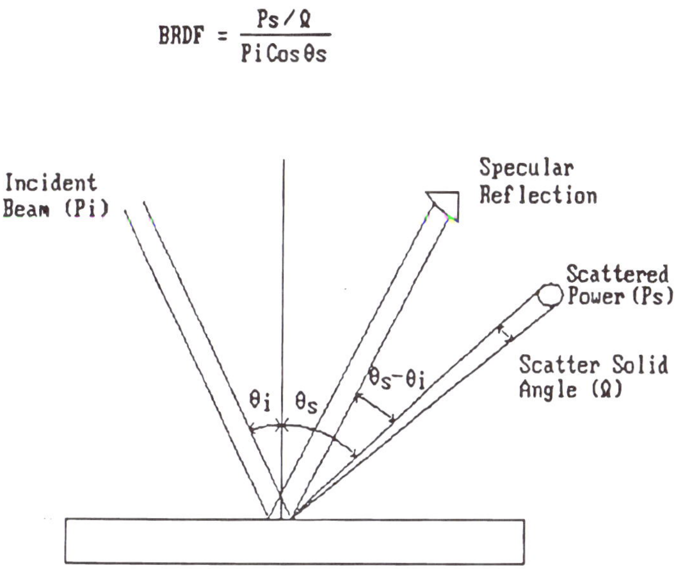

FIG. 4. Definition of BRDF (2).

(Schmitt Measurement Systems, Inc., Portland, OR, USA) the angle at which the laser strikes the sample (angle of incidence) is kept constant and the angle of the detector is varied to measure the scattered intensity at different angles. If the sample being tested has an optical use, the chosen angle of incidence may be the specific angle relating to its actual operation.

Surfaces Produced by Random Processes. The BRDF of a surface produced by random wear, such as by grinding or polishing, is simple to analyze in order to obtain valuable information about surface quality (smoothness or finish). The easiest way to understand such a plot is to examine some sample data.

Figure 5 shows the BRDF plot for two randomly polished mirrors. As explained previously, this is a plot of scattered intensity versus angle. The 0° point on the horizontal axis indicates the direction of the specular reflection. All other directions on the plot are relative to this direction ($\theta_s - \theta_i$ in Fig. 4). The vertical axis is the BRDF scatter (on a logarithmic scale). BRDF is typically plotted on a logarithmic scale because it changes over several orders of magnitude within a few degrees of the specular beam.

From Fig. 5, it can easily be seen that mirror 2 has a higher BRDF plot than mirror 1, indicating that mirror 2 scatters more light than mirror 1. Without any mathematics whatsoever, one can immediately tell that mirror 1 has a much smoother surface than mirror 2.

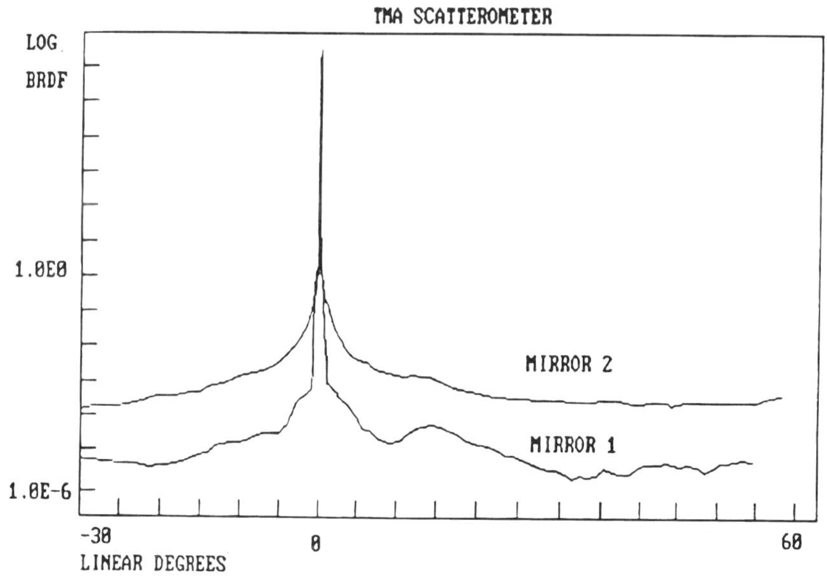

FIG. 5. BRDF plots for two mirrors (2).

BRDF is often displayed in a "log–log" plot. That is, both the BRDF and the angle of scatter (horizontal axis) are plotted on log scales. There are two very good reasons for this. First, it allows a more detailed study of the scatter distribution close to the specular direction and, for a polished surface, this is where most of the light is scattered. Second, for surfaces produced by a random process, a log–log plot generally approximates a straight line. The slope of this plot is characteristic of the process that is used to produce the surface under test. Fig. 6 shows log–log plots for the same two mirrors as in Fig. 5. These mirrors are both produced by the same process. Although the slopes of the log–log plots are very similar, mirror 2 clearly has a rougher surface finish. Conversely, Fig. 7 shows the BRDF plots for two mirrors produced by quite different processes. Mirror 1 has a higher scatter near the specular angle, whereas mirror 2 produces more scatter at higher angles. One can therefore deduce that mirror 1 has long surface characteristics compared to mirror 2, but the surface of mirror 2 has a greater number of short surface characteristics. Generally, for optical surfaces, higher scatter near specular results in a degradation of the resolution of an object or ability to clearly "image" objects very close to each other. Higher scatter at higher angles often produces flare or "noise" in an optical system.

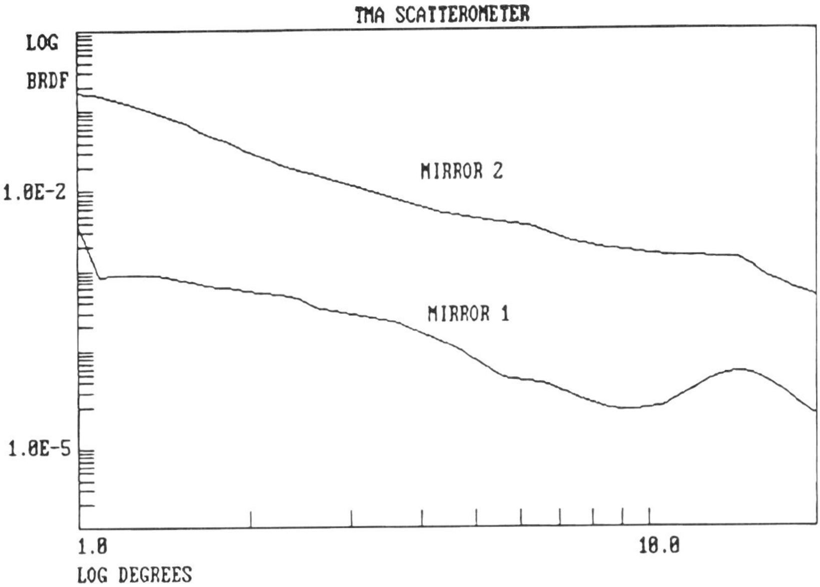

FIG. 6. Log-log BRDF plots for two mirrors produced by the same process (2).

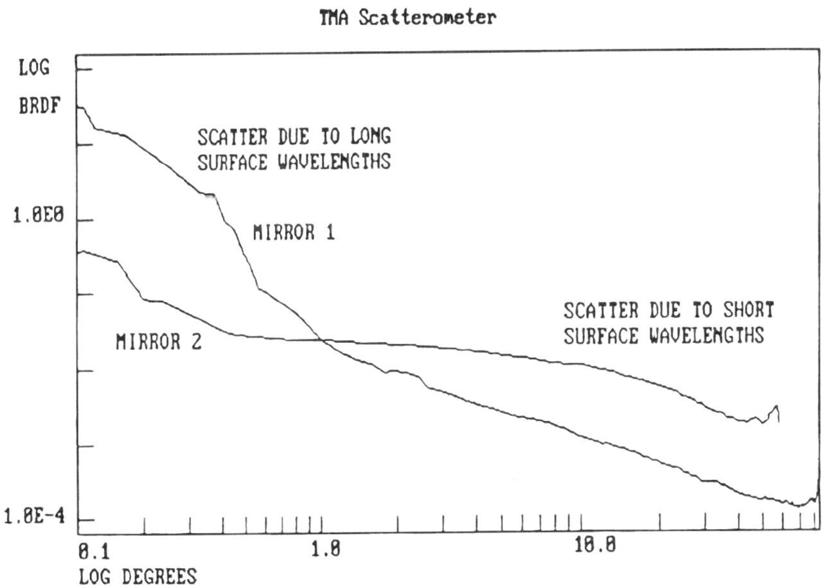

FIG. 7. Log-log BRDF plots for two mirrors produced by different processes (2).

This straight line plot is thus a very powerful diagnostic. If the log–log BRDF plot for a particular sample has an unusual slope or even a curvature, this could indicate that there are problems in the process used to produce the sample. If there are peaks in the curve, then the surface defects are not random.

Machined Surfaces/Periodic Processes. Machined surfaces are inherently different than random wear surfaces in that the surface defects tend to be formed in a regular or periodic manner. A simple example of this is a metal disc produced by a lathe-turning process that actually looks like a phonograph record when viewed under magnification. Periodic surface features such as tool marks cause light to be scattered into a few specific angles. This shows up on a BRDF plot as one or more sharp peaks or spikes. The number, height, and angular location of the peak(s) all give valuable information about the regular surface grooves or defects. Simply put, the depth (height) and density of the surface grooves or tool marks determine how high the peaks are in the BRDF plot, whereas the position of a peak on the plot indicates the width (surface wavelength) of the grooves causing the peak (Figs. 8 and 9).

The following is a simple mathematical analysis: The angle at which a peak occurs is given by

$$\sin(\theta_{scat}) = \sin(\theta_{inc}) + n\lambda/L, \tag{10}$$

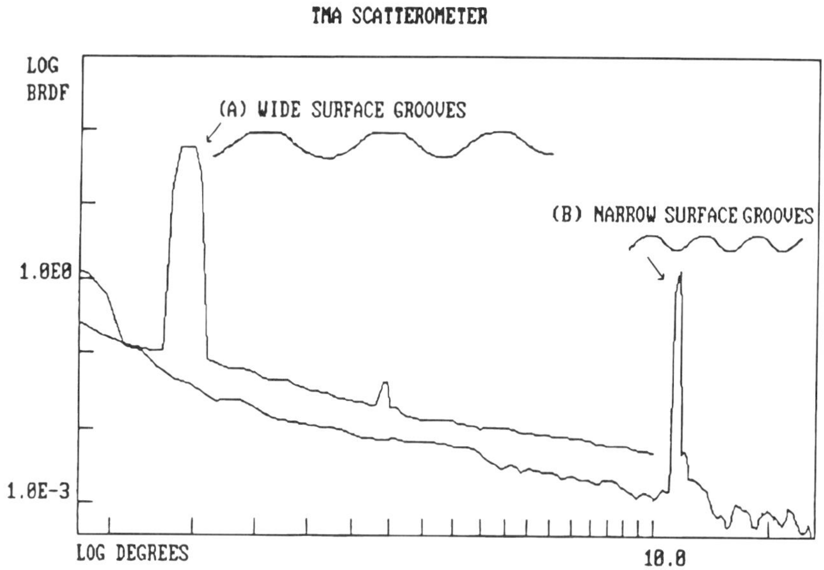

FIG. 8. The height and position of a peak are related to the depth and width of the surface grooves (2).

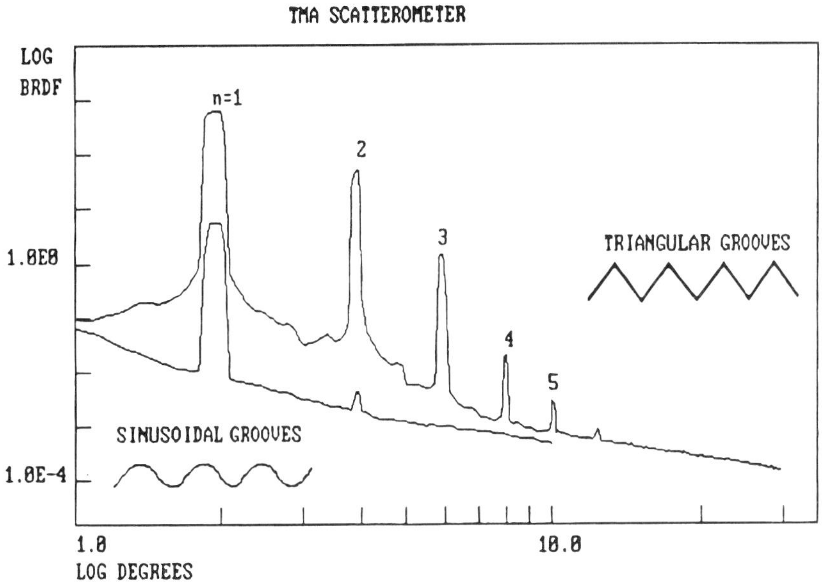

FIG. 9. Different shaped grooves can produce a series of peaks (2).

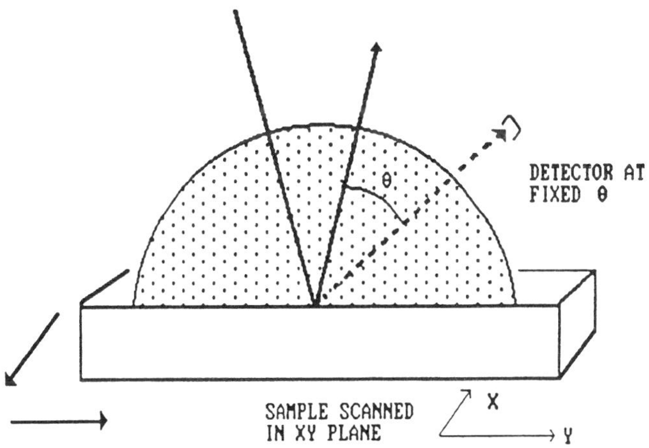

FIG. 10. In a raster scan the surface moves but both angles are fixed (2).

where L is the periodic spacing (length) between successive surface groves, λ is the wavelength used in the scatter test, θ_{inc} and θ_{scat} are the incident and scatter angles relative to surface normal, and n is a positive integer (1, 2, 3, $\cdots$). Peak position is determined by the periodic spacing as shown by the equation. The number and relative size of the peaks are determined by tool mark shape. Sinusoidal grooves give a dominant first order peak ($n = 1$) with only very small higher order scatter as long as the groove amplitude (depth) is small compared to a wavelength. Other shapes (square, triangle, cusp, etc.) change the relative size of the peaks (Fig. 9). A convenient parameter used to analyze surface roughness is the inverse scatter wavelength ($1/L$), which is known as the spatial frequency (f).

To summarize, we can tell the following from a BRDF plot containing sharp peaks:

1. The presence of sharp peaks on a BRDF or BTDF plot indicates regularly spaced defects or grooves on the surface or within a transparent material.
2. The height of the sharp peaks indicates how deep and dense the scratches or grooves are: Larger peaks indicate deeper surface scratches.
3. The spacing of the grooves (periodicity) can be simply derived from the angle at which the main peak(s) occurs using Eq. (10). The position of these peaks can be used to check that the machine generating the surface is still running at normal speed and feed rates.
4. The number of peaks in a related series and the relative intensity of these peaks are functions of the shape of the grooves or tool marks. Changes in this pattern are an excellent diagnostic for tool wear, chip drag, chatter, vibrations, and a number of other characteristic and potential problems in the machining process.
5. The height of the BRDF curve between the peaks indicates how much random roughness there is on the surface, as discussed previously.

2. *Raster Scans*

The BRDF plots previously described represent the most common types of scatter measurement. However, this type of measurement essentially measures the average surface properties over an area (or volume) defined by the size of the laser beam used in the scatterometer. Although this spot size can be varied, many applications benefit from information taken over a larger area.

In a raster area measurement, both the angle of incidence and the detector angle (scatter angle) are fixed. The sample is then moved in predetermined x–y steps (rastered) or rotated until the laser beam has sampled the entire surface (Fig. 10). As the sample moves, the scatter intensity is continuously recorded. The data are then presented as a map using multiple colors. Each color represents a certain range of scattered intensity (hence degree of roughness on nontransmitting materials). A raster scan is used for one or more of the following reasons:

1. To study a large surface or volume and better understand its overall, average, or statistical scatter and roughness.
2. To study the degree of homogeneity of such a surface.
3. To detect, locate, or count individual flaws.

In this type of measurement, both the angle of incidence and the detector angle (scatter angle) are fixed. The sample is then moved in predetermined x–y steps (or raster pattern) until the laser beam has sampled the entire surface (Fig. 10). As the sample moves, the scatter intensity is continuously recorded. The data are then presented as a map using multiple colors. Each color represents a certain range of scattered intensity (hence degree of roughness on nontransmitting materials). For example Fig. 11 show raster scans of two mirrors (a) #001 and (b) #002 measured at 0.514 μm. All

(a)

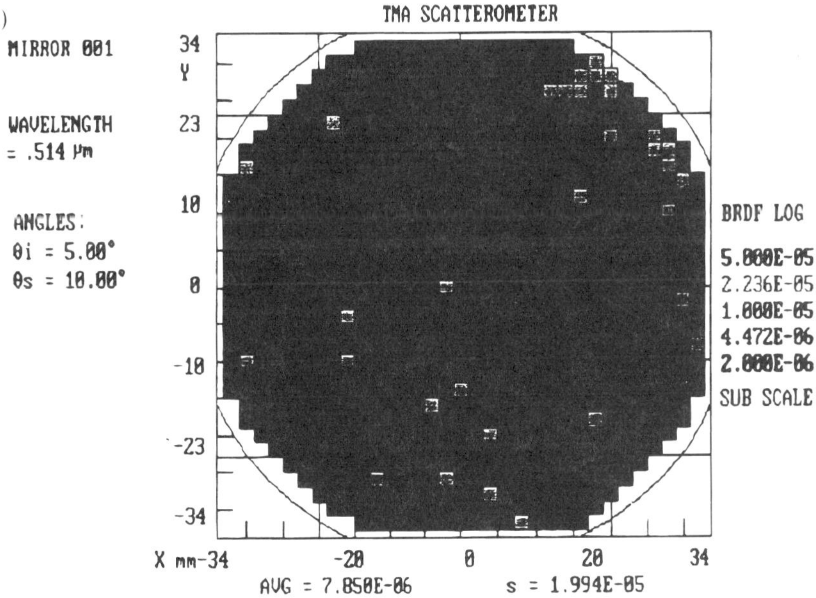

continued

(b)(*continued*)

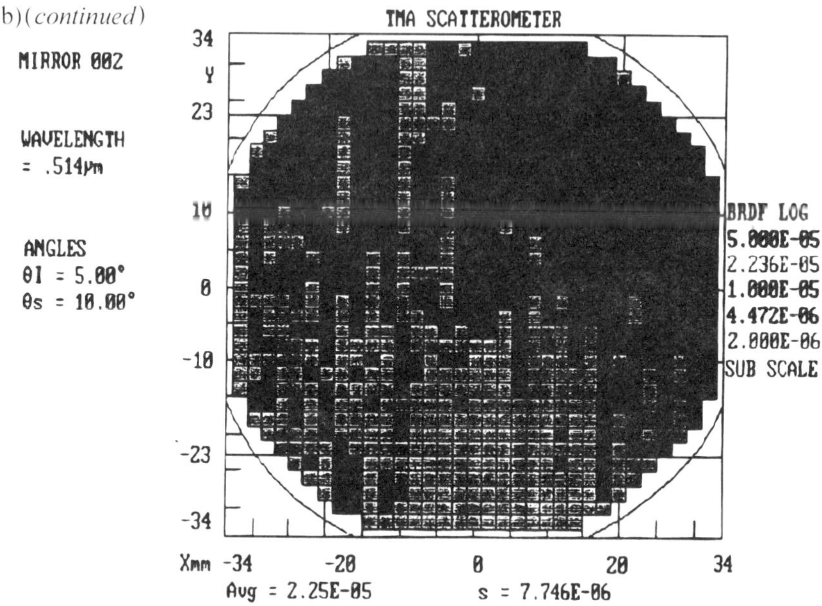

FIG. 11. (a) Raster scan of mirror 001. (b) Raster scan of mirror 002 (2).

areas shaded darkest have BRDF values below 2.00E-6, and all areas one shade lighter have BRDF values between 2.00E-6 and 4.472E-6. The areas shaded darkest are therefore smoother than those one shade lighter. For these scans, the detector was positioned at 10° from the specular angle. Notice that 001 has lower scatter than 002. 001 has less variation across its surface; it scatters more evenly. The process of creating 001 created smoother, higher quality regions but also left more high-scatter defect areas. Refinement of the 001 process could ultimately lead to an excellent surface relative to the 002 process.

Also printed on the plots are averaged and standard deviation values of the displayed area. In some applications using scatter as an on-line process control tool, these values can be a signal to correct the process or discard the sample.

3. *Calculation of Surface Roughness*

It has been shown that the measurement of scatter direction of prominent scatter peaks can be used to find the wavelength (or spatial frequency) of prominent periodic surface features. In addition, as random surface roughness increases the background scatter increases. Therefore, it should come as no surprise that in special cases the measured BRDF can be used to calculate roughness statistics. This is accomplished through the use of diffraction theory. A full explanation is beyond the scope of this document, but the results are fairly easy to understand, and making use of them can provide insight into the roughness characteristics of reflective samples.

For the special case in which reflective scatter comes from only small surface features (sometimes called microroughness), the BRDF is essentially proportional to a roughness characterization function known as the surface PSD function. Except for a change in units, and the fact that it is plotted against spatial frequency instead of scattering angle, a PSD plot looks a lot like the corresponding BRDF plot. Figures 13 and 14 show the BRDF and calculated PSD from a molybdenum mirror. The PSD has the very convenient feature that the rms surface roughness can be found by simply integrating the curve (finding the area under the curve) from some f_{min} to some f_{max}. The rms is equal to the square root of the calculated area under the curve. This is shown in Fig. 14 for the PSD of Fig. 13. Notice that if different frequency limits are chosen, different rms values will result. It is clear that the rms roughness of a surface depends on the defined bandwidth (f_{min} to f_{max}). This is a fact of life for all roughness measurements whether they

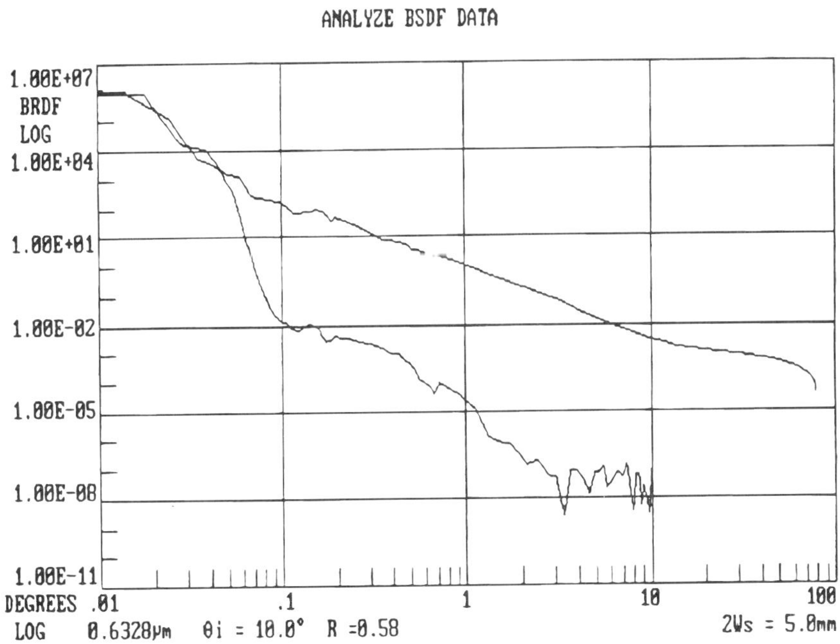

FIG. 12. BRDF from a Molybdenum mirror (2).

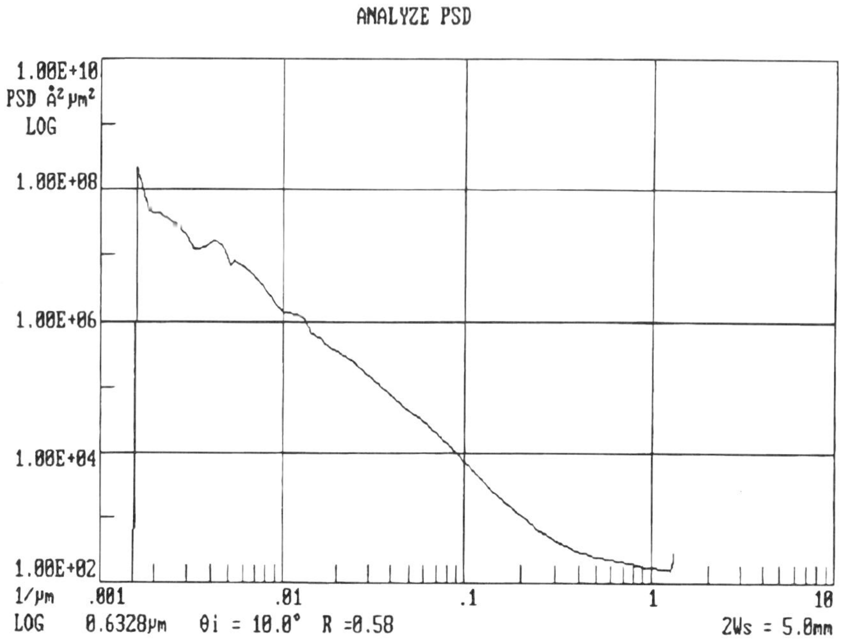

FIG. 13. Power Spectral Density (PSD) from a Molybdenum mirror (2).

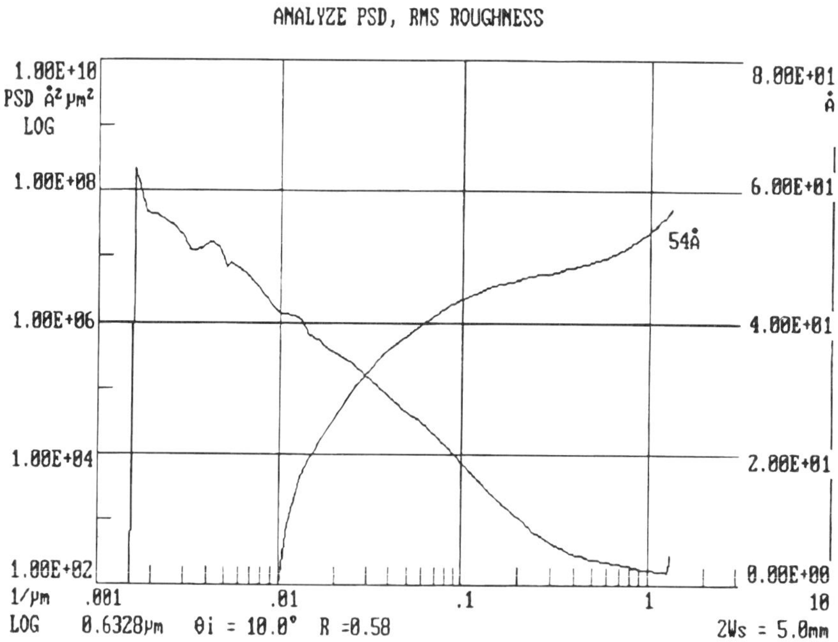

FIG. 14. Surface roughness calculation (rms) for the PSD of Fig. 11 (2).

are made by scatterometry or profilometry. Thus, in order for roughness measurements and specifications to be meaningful, the associated bandwidth must also be given.

As indicated previously, scatter can be used to accurately find roughness only if the scatter signal is caused exclusively by microroughness. Other sources of scatter, such as surface contamination, bulk defects, and very rough surface features, create scatter noise that prevents accurate roughness characterization. Thus, the surface must be a smooth (mirror-like), clean, front surface reflector for the technique to be used for accurate roughness calculations. However, scatter remains a sensitive indicator of surface features even when these conditions are not met. Even though a surface is too rough to meet the conditions for roughness calculation, it may still scatter more or less as the roughness changes and the measurement can be used to provide a fast, noncontact method to monitor changes in roughness.

IV. Summary

As seen in Section 3 excellent insight into surface quality, defects, and bulk purity can be obtained from scatter measurements. Also, astute analysis can result in simple, inexpensive in-process and final quality control procedures for production of a variety of surfaces, materials, and optics. As can be seen, much can be learned from visual examination of BRDF plots and raster scans. This is often sufficient for pass/fail tests of product quality. The data can also be analyzed in more detail using various formulas and computer programs. In addition, because the data are digital, they can be rapidly analyzed by computer in on-line process control applications. SORIC* and leaders in the stray light community have developed a BRDF database (16, 17) for commonly used surfaces and materials employed by scientists and engineers (18, 19, 20, 21, 22, 23).

Appendix

A-1. Relationship between the Sample (*X*, *Y*, and *Z*) and Beam (*XB*, *YB*, and *ZB*) Coordinate Systems

The *Z* and *ZB* axes are always the local normal to the sample face. Locations on the sample face are measured in the sample coordinate system. The incident and scatter directions are measured in the beam coordinate system. If the sample fiducial mark is not an *X* axis mark, the intended value must be indicated on the sample (Fig. A1).

* SORIC (Stellar Optics Research International Corporation) of Thornhill, Ontario, Canada (905-731-6088).

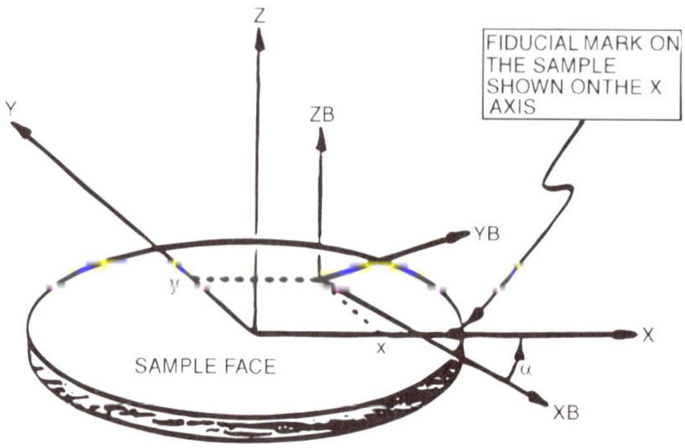

Note 1—The *x-y* zero position on the sample face is assumed to be the geometric center of the sample
Note 2—The fiducial mark can be on the edge or back of the sample

FIG. A1. Relationship between sample and beam coordinate systems (1).

A-2. ANGLE CONVENTIONS FOR THE INCIDENT AND SCATTERED LIGHT IN THE BEAM COORDINATE SYSTEM

The projection of the incident direction onto the sample face is the $-XB$ axis. Azimuth angles are measured from the XB axis. The incident azimuth angle, ϕ_i is always 180° so ϕ_s can be used directly in the common form of the grating equation (Fig. A2).

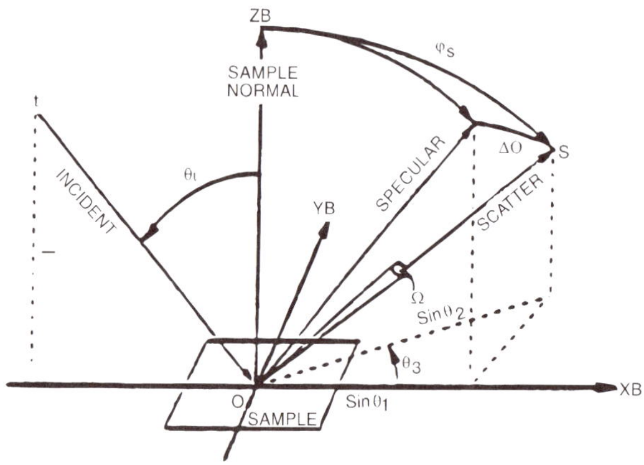

NOTE—The PLIN is the *I–O–ZB* plane The scatter plane is the *S–O–ZB* plane

FIG. A2. Angle conventions (1).

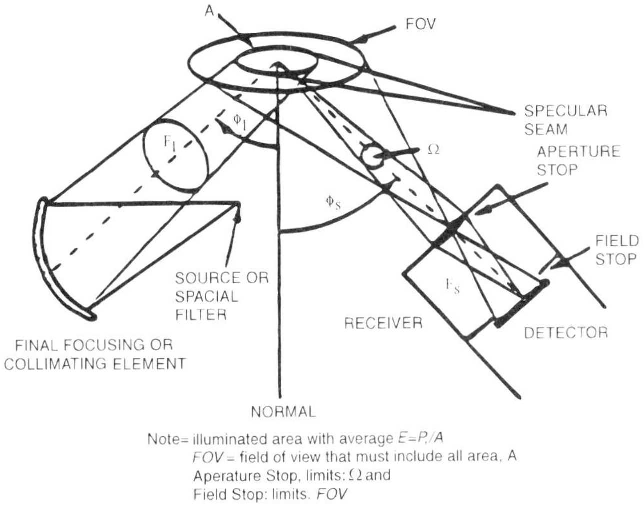

FIG. A3. Receiver geometry (1).

A-3. THE RECEIVER GEOMETRY

In many cases the field stop is set by the detector size; however, as the aperture stop approaches the field stop the risk of seeing unwanted stray light increases. Other receiver geometries may be used. They all have effective aperture and field stops and it is good operating practice to make them well defined (Fig. A3).

References

1. "Standard Practice for Angle Resolved Optical Scatter Measurements on Specular or Diffuse Surfaces", ASTM: E1392-90, 1990.
2. "Interpreting Light Scattering Measurements". Schmitt Measurement Systems, Inc., April, 1994. (Portland, OR, USA).
3. ASTM. E 167: Practice for goniophotometry of objects and materials.
4. ASTM. E 179: Practice for selection of geometric conditions for measurement of reflectance and transmittance.
5. Smith, S. M. (1988). Reflectance of AMES 24E, Inrablack and Martin Black. *Proc. SPIE* **967**, 251.
6. Stover, J. C. (1990). Optical scattering. Measurement and analysis. Copyright 1990 by SPIE.
7. Leonard, T. A., and Pantoliano, M. (1988). BRDF round robin. *Proc. SPIE* **967**, 226.
8. Leonard, T. A. (1988). The art of optical scatter measurement. Proceedings of the High power optical components conference, Boulder damage symposium. National Institute of Standards and Technology.

9. Leonard, T. A., Pantoliano, M., and James, R. (1989). Results of a CO2 BRDF round robin. *Proc. SPIE* **1165**.
10. Bjork, D. R., Rifkin, J., and Cady, F. M. (1989). BRDF error analysis. *Proc. SPIE* **1165**.
11. Harvey, J. E. (1977). Light scattering characteristics of optical surfaces, *Proc. SPIE* **107**, 41.
12. Stover, J. C., Hourmand, B., and Kahler, J. A. (1984). Comparison of roughness measurements by differential scatter and total integrated scatter. *Proc. SPIE* **511**.
13. ASTM. F 1048: Test method for measuring the effective surface roughness of optical components by total integrated scattering.
14. Stover, J. C. (1990). *Optical Scattering: Measurement and Analysis*, Chap. 4. McGraw-Hill, New York.
15. Bennett, J. M., and Mattson, L. (1989). *Introduction to Surface Roughness and Scattering*, p. 32. Optical Society of America, Washington, DC.
16. Breault, R. P. Stray light technology overview of the 1980 decade (and a peek into the future), No. 1401. SPIE.
17. Breault, R. P., Lange, S. R., and Fannin, B. B. Cleaning, coating and BRDF measurements of a two-meter low scatter mirror, No. 1404. SPIE.
18. McCall, S. H. C. P., Pompea, S. M., Breault, R. P., and Regens, N. (1992). *Reviews of Black Surfaces for Space and Ground-Based Optical Systems*, SPIE Proceedings, Vol. 1753, p. 158. Presented at the SPIE Conference, Stray Radiation in Optical Systems II, July 20–22, 1992, San Diego.
19. McCall, S. H. C. P., Breault, R. P., Piotrowski, A., Rodney, J. W., and Piotrowski, L. (1996). *A Black and White Surfaces and Materials Database*, SPIE Proceedings, Vol. 2864. Presented at the SPIE Annual Conference on Optical Science, Engineering and Instrumentation. Session: "Stray Light and System Optimization: Theory, Surface Spectral Characteristics, and Techniques," pp. 6–7 August, Denver, CO.
20. McCall, S. H. C. P., Sinclair, R., Pompea, S. M., and Breault, R. P. (1993). *Spectrally Selective Surfaces for Ground and Space-Based Instrumentation: Support for a Resource Base.* SPIE Proceedings, Vol. 1945, p. 497. Presented at the SPIE Conference: The International Symposium on Optical Engineering and Photonics in Aerospace and Remote Sensing, Session: "Space Astronomical Telescopes and Instruments II," April 12–16, Orlando, FL.
21. Pompea, S. M., and McCall, S. H. C. P. (1992). *Outline of Selection Processes for Black Baffle Surfaces in Optical Systems*, SPIE Proceedings, Vol. 1753, p. 92. Presented at the SPIE Conference, Stray Radiation in Optical Systems II, July 20–22, San Diego.
22. McCall, S. H. C. P., Pierre, J. E., Mabee, A., Tennyson, R., Morison, D., Kleiman, J., Iskanderova, Z., and Gudimenko, J. (1994). *The Atomic Oxygen Exposure Effects Module of the Database for the Properties of Black, White, Reflective and Transmissive Spectrally Selective Surfaces*, SPIE Proceedings, Vol. 2260. Presented at the SPIE Conference, Stray Radiation in Optical Systems II, July 24–29, San Diego.
23. McCall, S. H. C. P., Begrambekov, L., and Pierre, J. E. (1992). *A Review of Black and Reflective Spectrally Selective Surfaces Developed in the Former USSR*, SPIE Proceedings, Vol. 2260. Presented at the SPIE Conference, Stray Radiation in Optical Systems II, July 24–29, San Diego.

Additional Sources

1. Bartell, F. O., Dereniak, E. L., and Wolfe, W. L. (1980). The theory and measurement of bidirectional reflectance distribution function (BRDF) and bidirectional transmittance distribution function (BTDF). In *Radiation Scattering in Optical Systems*, Vol. 257, p. 154. SPIE.

2. Smith, S. M. The far infrared reflectance of optical black coatings, No. 759. SPIE.
3. Elson, J. M., and Bennett, H. E. Image degradation caused by direct scatter from optical components into the image plane, No. 1013. SPIE.
4. Greynolds, A. W. Relative micro-roughness scattering from the surfaces of a transmitting optical element, No. 1016. SPIE.
5. Anderson, D. S., and Frogner, E. A method for the evaluation of subsurface damage, No. 1090. SPIE.
6. Johnson, Jr., E. G. (1993). Simulating the scratch standards for optical surfaces: Theory. *Appl. Opt.* **22**(24), 4056.
7. Peters, P. J. Stray light control, evaluation, and suppression. No. 1381. SPIE.
8. Leonard, T. A. (1989). Results of a CO_2 BRDF round robin. *SPIE* **1165**, 444. Scatter in optical components.
9. Leonard, T. A. (1988). The art of optical scatter measurement. Paper prepared for the 20th symposium on optical materials for high power lasers (The Boulder Damage Symposium), Boulder, CO, October 26. NBS.
10. Leonard, T. A., and Pantoliano, M. (1988). BRDF round robin. *SPIE* **967**, 226.
11. Leonard, T. A. (1990). Standardization of optical scatter measurements. *SPIE* **1331**, 188.
12. Leonard, T. A., and Rudolph, P. (1993). BRDF round robin test of ASTM E1392, Paper No. 1995-30. Paper presented July 16 at the SPIE Symposium in San Diego.
13. Leonard, T. A. (1992). Scattered-light tests illuminate surface properties. *Laser Focus World*, p. 103.
14. Standard practise for angle resolved optical scatter measurement on specular or diffuse surfaces. ASTM Designation E 1392-90.
15. Bennett, J. M. (1992). Surface finish and its measurement, Part A and Part B (Vols. 1 and 2), Collected works in optics. Copyright 1992 by the Optical Society of America, Washington, DC.
16. Bennett, J. M., and Mattsson, L. (1989). Introduction to surface roughness and scattering. Copyright 1989 by the Optical Society of America, Washington DC.
17. Breault, R. P. Suppression of scattered light. Unpublished thesis.
18. Hsia, J. J., and Richmond, J. C. (1976). A high resolution laser bidirectional reflectometer with results on several optical coatings. *J. Res. NBS Phys. Chem.* **80A**, 189–205.
19. ASTM. E 284: Definitions of terms relating to appearance of materials.
20. ANSI/ASME. B46.1: Surface texture (surface roughness, waviness, and lay).

Section 3

PRACTICAL APPLICATION OF SPECTROSCOPIC MEASUREMENT

SELECTING A SPECTROSCOPIC METHOD BY INDUSTRIAL APPLICATION

EMIL W. CIURCZAK

I. The Process

Before deciding which method of monitoring and/or analyzing a product sample should be used, the chemical process itself must be evaluated. The engineer(s), production manager, and analyst must work together and decide which step or steps determine the quality of the product. The main problem with this approach comes from the evolution of different languages among these groups over time. It may well be decided that the traditional approach of submitting a finished sample to the quality control lab for acceptance or rejection is no longer valid. This approach only serves to prevent a substandard product from being sold but does nothing to "control" the quality of the product while it is being produced.

A typical example of a chemical reaction in need of control is the sulfation of a fatty alcohol to produce a surfactant (surface active agent.) In traditional batch-mode manufacturing, the chlorosulfonic acid is added to the alcohol, reacted, and then the mixture is made basic with sodium or ammonium hydroxide. The concentration of the sulfated product is assayed (usually by titration) and any necessary dilution made. Then the color, viscosity, percentage sulfate, pH, etc. are measured! This is a clear case in

ISBN 0-12-764070-3/$25.00.

which the reaction should be monitored by some rapid method.

In some cases, simple pH or temperature measurement gives the operator a handle on the reaction. In the previous example, too rapid a reaction gives rise to a sulfate ester. This bridged alcohol product does not easily neutralize immediately, but it eventually breaks down to give sulfuric acid and fatty alcohol. This lowers the pH, causing saponification of the sulfated product, which needs to be kept at a high pH. The released acid accelerates the degradation of more surfactant; thus, it is essentially self-catalyzing. [Another side effect of the lowered pH is that ethoxylated products break down to dimers of the ethylene oxide, 1,4-dioxane, a known carcinogen(1).]

This is a clear example of the need for real-time measurement. Possibly the temperature of the reaction may be sufficient, but the "bad actor," the sulfate ester, has a distinct infrared (IR) signature and may be produced, and thus detected, even when the temperature appears normal.

A. Characterizing the Process

One major consideration is the entry point of each raw material or precursor into the reaction stream. The importance of each chemical or precursor must be evaluated in light of the finished product. If a caustic solution is added until a particular pH is attained, then its precise assay need not be known. If, on the other hand, a preset amount is metered in at a certain rate, then the assay value is critical.

1. *Raw Materials*

"The product is only as good as the starting materials" may be a true statement, but only to a point. The terms necessary and sufficient are important to keep in mind. If unnecessary tests are run on raw materials, the cost of the final product is almost automatically increased with little or no gain in quality. If an organic material is to be used with no final cleanup step, then the highest purity may have to be used. If addition of a chemical is followed by a cleanup or recrystallization step, then technical-grade material may suffice. In this case, a battery of assays is a costly waste of time.

The initial tests must be defined in terms of use: will a simple identification suffice or is a full set of analytical tests required? The answer to this question yields more or less profit.

2. *Semifinished Products*

Often, a product is not manufactured in a complete "start-to-finish" process. The intermediate product(s) may be synthesized and stored for later use or the intermediate(s) may be supplied by another site or vendor. In either case, the chemistry must be defined enough to know the shelf-life or "use by" date. Also, if a material is out of date, which degradation products may be critical to the next step in the process must be determined.

If a material was tested as a "final" product before storage, then one or two tests may be used to confirm that it is still chemically useful. Often, color may be used for a quick check of purity, but more often an infrared/near-infrared (IR/NIR) identity is preferable.

3. *Finishers/Setting Agents*

This is a nebulous term including any chemical added to the product after synthesis for some real or imagined benefit. It can be as obvious as an oil coat or galvanizing iron nails or an agent designed to quench a polymerization. In these cases, the spectroscopic signature is fairly obvious and may be seen with a low-end instrument (e.g., a filter IR/NIR spectrometer.)

A polymeric material coated onto, for example, aluminum is easily measured by an attenuated total reflectance attachment. Vibrational spectroscopy, IR, or NIR may be used for this application. A possible postproduction reaction, polymerization, or coating each may be followed by a surface probe. The reaction requires only a moderately fast instrument, such as a rapid scanning IR or NIR. Monitoring a continuous process, e.g., a lamination of two or more layers, requires a rapidly scanning device. In this case, an interferometric IR/NIR or accoustooptic device is able to scan hundreds to thousands of sample points per minute.

B. Understanding the Process

The most important question demonstrating a clear understanding of the process, is stated as follows: "Is the measurement you are taking of a controllable parameter?" This is not a trivial question. If the pH is measured and there is no option to mediate with acid or base, then to what purpose is the measurement made? It is, at best, an interesting set of historical data you are piling on the shelf! [It may be used to describe a "good" run, but the purpose of this section is quality *control*.] Let us dissect the process and determine what is needed and what is fluff.

1. *What Chemistry Is Involved?*

This is a fair question, but it is not always easily answered. We often are aware of the blackboard equations, the starting materials, and the product. However, you must ask yourself if you really know what is going on "in the pot" at any given time. You cannot hope to control that which you do not comprehend. Often, the plant manager does not know that he/she does not know until real-time measurements are attempted.

Does the reaction proceed in a linear or convoluted manner? If there are numerous side reactions possible, then any number of them could lead to an out-of-spec product. Simple precautions, such as free radical scavengers or a different catalyst, may do wonders for yield. Looking for the presence of or monitoring the levels of multiple chemicals is the strength of vibrational spectroscopy combined with chemometric methods.

2. *Is There Cooling/Heating?*

The thermodynamics of the reaction can be an invaluable tool. Spectral changes along with temperature measurements are two of the easiest and telling pieces of data available to a production engineer. Careful monitoring of both can lead to exact heating/cooling requirements based on data and not merely on theory. As James Bond often states, "Things are a lot rougher in the field."

If the reaction is based on preliminary work in a flask, the heating of the material in a 10,000-gallon reactor may offer some surprises. Measuring temperature changes without following concomitant chemical changes gives only half the picture. It is imprudent to set up operational parameters for plant operators without clear data on what "too hot" or "too cold" really mean.

It would also be a good idea to monitor the chemistry at or near the heating or cooling interfaces. Monitoring only the "average chemistry" of the overall reaction may be hiding potential degradation along the heat exchangers. In vibrational spectroscopy, chemometrics are often needed to obviate the effects of temperature. It may be shown that the spectra taken to measure the chemistry of a reaction may be used to monitor the temperature as well (2).

3. *What Mechanical Actions Are There?*

Mixing. This consideration is most important when a rapid reaction is occurring, as in the example of chlorosulfonic acid and a fatty alcohol. In this case, heat, a gas and an unwanted product, may be generated. An IR/NIR probe, attached to a rapid-scanning instrument, at the point of mixing can be the tool used to adjust the rate of addition of materials and/or the rate of stirring or even help determine the type of stirring needed.

Pumping. Various pumps cause different turbulence. Without getting into chaos theory, the manner in which a material is pumped can greatly influence its mixing. Another factor, especially in viscous fluids, is that different impeller pumps may cause localized heating and/or cavitation. This effect may degrade the raw materials or cause variable mixing due to viscosity changes and aeration. A small IR/NIR monitor could be placed in such a way as to evaluate these potential problems.

II. The Product

At last, we arrive at the product! The analyst must have a clear idea of what is good and what is out of (meaningful!) specifications. Often, only the performance in its end use may determine the "goodness" of a product. Running tests for their own sake or giving the customer a large analysis sheet is a waste of time (read: money). The answer, "because I can," is not sufficient to perform any one particular test. Is it a meaningful piece of data or just an expensive number?

A. What Are Meaningful Parameters?

1. *Physical Properties*

The easiest (usually) tests to perform are physical types. These may be automated or may be run by operators with minimum training. As the analyst, you design and specify tests to be run. Ask yourself, "Could I release the product without this number?" Better still, "Is this test telling me specifics about the making of the product or the performance of a particular product for the customer?"

Viscosity. Viscosity is not as simple of a measurement as is normally thought. This is a common measurement and its actual complexity is rarely considered. Just defining whether a particular fluid is Newtonian or non-Newtonian is a possible master's project for a graduate student. It is often a difficult test to reproduce precisely because it is so complex (3). The values are often difficult to reproduce for a single sample, much less to determine accurately for different samples. However, the physical parameters of a fluid can be monitored by IR or NIR, which, as vibrational spectroscopic techniques, are extremely sensitive to matrix effects such as hydrogen bonding. Viscosity is often affected strongly by hydrogen bonding, thus its ability to be measured by vibrational spectroscopy.

Density. The same comments made for viscosity could be applied to density. Care must be taken on the part of the (minimally trained) operator to thermostat the sample, exclude bubbles, and obtain exact weights. This is another test that is not as simple as first perceived. The density of a fluid directly affects the transmission properties of light. A spectroscopic method would be simpler and more dependable than any physical method for density [possibly through refractive index changes (2)]. In this case, almost any type of light may be used, but certainly IR/NIR are leading contenders. The reason for a vibrational spectroscopic approach is that so much other information may be obtained simultaneously.

Color. It almost goes without saying that a visible spectrometer is better at color measurements than is the human eye. The first choice to measure color would appear to be a colorimeter. Some care must be taken in this choice, however, because a particular color may be produced by numerous species. It would be far better to use vibrational spectroscopy in combination with a colorimeter to assure the analyst that the apparently "proper" color measured is derived from the correct chemical species.

2. *Chemical Parameters*

Although the physical characteristics of a product may hint at problems, the chemistry is often the only sure way to determine goodness. Again, the level of chemical testing depends on the final use of the product. Is it taken internally by humans or used to clean bathtubs? [There is no such thing as one product being "more important" than another. If there is a market, there is a need.] The implication here is only that food and pharmaceuticals need be more carefully analyzed for safety reasons than products not coming into contact with humans.

Purity (*Assay*). There is one test run by almost every quality control lab to answer the following: How much of what I want is in the final product? The complexity of purity tests runs the gamut from a simple taste/smell test or specific gravity measurement to a gas or liquid chromatograph equipped with a mass spectrometer. In almost every case, an IR or NIR scan can give as much (or more) information as any complex test and may be performed on-line instead of in the lab.

By-Products. Not nearly as many companies care about the minor components found in their product as those that care about the assay of the "active." In industrial processes, it is often unimportant because further process reactions exclude the impurities: They are washed out, precipitated, distilled, etc., or they do not react in a subsequent step. In foodstuffs, cosmetics, and pharmaceuticals, it is not only important but, also mandatory by law to determine what possible impurities are present.

Functionalities (OH Number, Acid Number, and Iodine Number). Although not as conclusive as a direct assay, these circumstantial tests may be used *in toto* to define a product. The United States Pharmacopoeia uses this approach for "positive" identification of drug substances. Of course, they do admit infrared spectra as "alternative" information or for identification of pure substances. Pure substances have been fingerprinted for decades by infrared: so why not products, which often are only simple mixtures of substances?

B. What Are the Critical Parameters?

While considering all the tests that may be performed, I alluded to "necessary and sufficient" from time to time. By this, I mean which tests must be run and which give us information that is merely "nice to know." These tests are often defined by either the stability of the product or the end use of the customer.

The stability will determine the shelf-life and, therefore, batch size, date of manufacture ("just-in-time" or stockpile), need for turnover, etc. The final destination and use will determine the "fine" tests such as minor impurities, odor, and color. Is the product to be used in a perfume or toilet bowl cleaner?

1. *Which Ingredient/Parameter Controls/Defines the Process?*

One convenient method for following a process in real time is to choose one key parameter (viscosity, pH, etc.) or ingredient (polyamide, sugar, or alcohol) that defines the process. In most cases, it is irrelevant whether the material monitored is a starting material or a product/by-product. If the process is truly understood, the key component is usually known. If this is not the case, then some experimentation is needed on the part of the processor to determine the key active ingredient(s)/parameter(s).

2. *Can This Parameter Be Monitored?*

The following is a very important question: Can the "key" parameter(s) be measured by any existing methods or must the (instrumentation) customer innovate? Is the key parameter something an established, well-defined method can easily measure? In the case of on-line or real-time measurements, "easy" is relative. What is meant is that there must be a way to monitor the key ingredients (or process steps) 24 hr a day with maximum accuracy and minimum down time.

At this point, initial cost versus lifetime expenses often comes into play. A "cheap" answer looks good on the quarterly "bottom line," but if it needs constant maintenance, the cost is a ruse. Many plant managers and/or quality control managers decide at this point that spectroscopic measurements are the least labor-intensive of any type on the market. The probe is usually passive, not needing to be renewed or cleaned at frequent intervals, and the most difficult maintenance on the spectrometer normally consists of replacing the source lamp on a yearly basis.

Spectroscopic methods usually need the least frequent recalibration of virtually any analytical technique. No chemicals are needed and, with fiber optic probes and NEMA containment, spectrometers can be made quite safe to operate in hazardous atmospheres. The operator of the spectrometer may, indeed, even be located at a remote lab and still monitor the process in real time via computer and telecommunications networking.

III. The Measurement Procedure

A. What Are the Existing Solutions?

Is the customer aware of what competitors have done in the past? This information, positive or negative, will give him or her a direction to pursue or avoid. Often, an instrument manufacturer can "hint" at the correct way to proceed without actually betraying another customer's trust.

Now that the analyst has ascertained which species are to be monitored, the next logical question is "How?" The options for the specific hardware and software can run (in complexity and price) from an "off-the-shelf" setup to an entirely custom-built system, including in-house written software. Before a course of action is taken, several questions must be asked.

1. *Are Suitable Devices Already Sold?*

If the answer is yes, then half the battle has been won without a shot being fired. If a customer is conversant with other companies in the field, it is possible they might help with (nonproprietary) information.

More likely, however, the customer will have to attend trade shows and ask vendors and presenters many questions. Following the previously detailed steps in defining the problem, an analyst will be able to ask the right questions and understand the answers received. This is the step in which a selective auditory filter must be employed. If someone claims that a technique does not work, the analyst should have some doubts. This may be a person's way of covering the fact that he or she was not and familiar or skilled enough with the procedure to make it work!

If the person tells you that a particular piece of equipment does not work, lend a more credulous ear. In this case, he or she may have purchased the wrong instrument for the particular analysis. Again, the instrument might be functional—just not for the test for which it was purchased. The experience could save time and money when searching for the correct tool.

Ultraviolet, visible, near-infrared, and mid-range infrared spectroscopies all work as per Beer's law. This has been proven during the past 100 or so years in several million assays. Therefore, distinguish between a case in which the procedure works even if a particular instrument does not and one in which the procedure or technique does not work. [The best hitters in baseball do not get a hit 70% of their times at bat, but there are no cries for bats to be abandoned because they "don't work!"]

Also, even if a specific piece of hardware did not (appear to) work for the process for which it was purchased, the fault may have been in an incomplete training program, a misinformed sales representative, or one poorly built instrument. Have patience and faith in the science.

2. *Can the Instrument Supplier Help?*

In other words, is an instrument manufacturer willing to perform feasibility studies, lend equipment, or send an engineer and/or applications scientist to the plant location? Even if the specific application has not previously been performed, a reputable manufacturer will be more than able to assist you because he or she knows his or her equipment better than anyone else.

3. *Are There Any Available Commercially Built Devices?*

At some point, you must ask, "Can a device be built by the vendor or in-house technical staff if it is not currently available?" There is always the problem (referred to as "an insurmountable opportunity") that a prototype might cost 10–110 times what a production model might. Of course, the instrument might not be suitable for the application, even after it is designed and built!

This is the point at which the customer must ascertain the value of an on-line instrument. If it is really a major time/money saver, then it might be worth the effort to break ground. If it is for a minor improvement, then it might be better to wait for the instrument to become commercially available.

If the decision is to continue with development, there must be a complete openness between the customer and the instrument company. One cannot help the other if honesty is absent. [I have seen cases in which the instrument company promises more than it can possibly deliver and cases in which the buyer holds back critical information in a misguided fear of disclosure. It is in the instrument company's best interests to maintain a reputation for trust, thus, I cannot remember a case in which a reputable company ever disclosed any trade secrets to another company!]

Delays of years could occur if the instrument company is even slightly unsure of the exact purpose of the instrument. Here is where trade secrets become opened. This may be the major stumbling block for the project. The customer must be aware of this inevitable step and treat it as seriously as the monetary issues. The instrument manufacturer is also allowing the customer to work with equipment not yet protected by patents. Both must have clear (i.e., written) understandings of the limits of the "partnership."

Assuming that the equipment has been obtained (custom or standard), another important consideration is how to implement it. Often, the instrument company will have personnel who can help with its implementation. In many cases, additional personnel or a third party (consultant) must be added to the project. The eventual cost of this dedicated party is far less than continued failure due to inexperience. The learning curve can be slow and expensive when production is involved.

B. Feasibility/Calibration Studies

1. *Determine Needed Samples*

Good homework early in the project will have located the key ingredients. The actual instrumentation as well as the product will define the number and type of calibration samples needed. NIR spectrometers will need to have the species "introduced" in all possible matrices that might occur in a particular reaction/process. Unlike mid-range infrared (MIR), NIR seldom has clear, individual absorption peaks and may, initially, require more calibration samples.

In complex mixtures, powerful mathematical procedures, such as partial least squares (PLS) or principle component analysis (PCA), are used to quantify a particular species in a mixture. These procedures are based on the whole spectrum or a large portion thereof.

It is often possible to focus on a specific absorption band in the MIR and create a simple Beer's law equation for one or more species. Several wavelengths may be employed in a procedure called multiple linear regression, wherein a more complex version of Beer's law is developed. In very complex mixtures with severely overlapping absorption bands, PCA and

PLS are also used for MIR. Similar software is available for ultraviolet/visual is as well, but electronic bands are seldom rich (complex) enough to provide multiple component determinations. Numerous books are available on this topic (4–6) and may be consulted.

2. *Will the Work Be Performed by the Customer or Supplier?*

Depending on the specific case, up to several hundred or more than a thousand samples may be needed for calibration. The possibility exists that the customer may not have the personnel or expertise to perform all the analyses needed for the calibration. The instrument company, not usually equipped with a working analytical laboratory, may need a third-party lab to develop the initial prediction equation.

Careful communications at this point are needed. Many cases exist in which large numbers of chemical tests were run on inappropriate samples. This stage of the testing is often the roadblock that ends the project. It is usually best to allow the spectroscopist(s) working on the equation to have a strong influence on the samples that are taken for analysis.

3. *Gather Meaningful Samples*

The gathering of samples for a calibration set is not trivial (7). It must be determined at the beginning of the study whether synthetic samples may be used or actual production samples are needed. Often, subtle differences between production and synthetic samples exist. In this case, the time-consuming process of gathering samples from a production line, testing them, and running their spectra must be performed over a number of months. The majority of this work is performed routinely; the only added burden is the scanning of the samples, usually a small increment of time compared with the rest of the analysis.

Again, the instrument supplier should be able to guide the customer through the first trials. How many samples are needed, how often should they be taken, and how much sample is needed are questions best answered with the aid of an applications specialist. Keep in mind that a working referee method is needed to assay all these samples so that a prediction equation may be designed. The eventual accuracy of the prediction is dependent on the accuracy of the referee method.

4. *Building the Model Equation*

The act of turning spectra with chemical values into a meaningful equation is called "chemometrics" (5, 6, 8). As is said at the circus, "Do not try this at home, the sword-swallower is a trained professional!" With the powerful programs supplied by most instrument manufacturers, it is very tempting to try pushing buttons without experience. Do not!

Take a course, hire a consultant, pay the instrument company—just do not start pushing buttons. Anyone can eventually produce a straight calibration line, but it may not be meaningful. To quote Ann Landers, "Seek professional help!"

When a model equation has been put together and tested off-line, it is time to test it in the process environment. It is hoped that, by this time, the proper interface has been installed in the plant so that the instrument may be installed (remember "planning?") It is important to remember at this point that there is still a while to go before the instrumental readings may be fully trusted.

C. Monitor Actual Runs

When the process monitor is installed, regular production may resume. This is still the "shake-down" stage, and corrections will, in all likelihood, have to be made.

1. *Parallel Testing*

The primary method will need to be run on the samples being assayed by the spectroscopic method for a period of time. Because these may be taken from a point in the process at which samples were not taken previously, new sampling techniques may need to be developed. [No one said it was simple—just worth the effort . . . eventually.]

Off-Line. It may be decided that the parallel testing should be made with the device placed next to the process line. This would allow samples to be drawn from the stream and tested on the device and by the standard method with no difference between samples. When all the desired corrections have been made on the equation, then the system may be installed for in-line application.

The drawback to this approach is that, no matter how careful the engineer is, the samples are not really in the process stream. The analyst cannot fully trust the calibration to not need further adjustments when finally installed. A rough equation may be all that is needed at this point.

In-Line. Either at the start or soon after an at-line test period, the instrument must be installed. Again, there will need to be simultaneous testing by the standard method(s) during the evaluation period. In this case, the sampling will have to be rapid enough to simulate the point at which the spectrophotometric device has taken a reading. The sampling speed will be determined by the process, of course.

It is almost always the case that the preliminary equation will need to be "tuned." This will not be a reflection on the abilities of either the instrument manufacturer or the analysts. It is a fact of physics: Samples sitting in a beaker are not the same as flowing samples under pressure!

2. *Fine-Tuning*

During the in-line evaluation stage, there may be a need to upgrade the equation either through recalculation or, more likely, through more samples to enhance the ruggedness of the algorithms. When the instrument is agreeing with the standard method within predetermined limits (note that spectroscopic methods are never more accurate than the methods used to calibrate them; any statistic giving a lower error implies that the calibration equation is "overfitting" the data), it is time to release the product based on results from the equipment alone.

In the first year, the analyst will wish to double-check the process fairly often to assure himself/herself that the new instrument is doing its job. One valuable result of this close monitoring is the ability to plan a maintenance and calibration schedule.

3. *Evaluation with Time*

The results of the equipment should be double-checked at regularly scheduled intervals. This actually serves two purposes: The precision and accuracy of the automated method may be enhanced as well as corrected, and the analysts keep their ability to perform the referee tests in case of a breakdown of the instrument.

As the analytical personnel become familiar with the in-line system, subsequent systems become almost trivial to install. Although it provides little comfort at the time, the knowledge that the first system will always be the worst is a comfort *after* installation is complete.

IV. Summary and Conclusions

It is hoped that any attempt at in-line testing is preceded by an inventory of the chemical reactions in the process. Once the process is mapped, choosing the instrumentation is far simpler. It also must be remembered that the calibration of any spectroscopic method can take more time and effort than more common methods. The payback is in the low cost of running the method and the speed and accuracy of the results.

At a time when companies are reducing their workforces, competition and regulations are requiring more, not less, testing of the product. The fortunate coincidence is that there is a concomitant increase in the speed and accuracy of spectroscopic devices for process control. This trend will lead to more measurements of higher quality, allowing manufacturers to control their products in real time. Thus, higher quality, less expensive products will be the result of this revolution.

References

1. Ciurczak, E. W., and Robinson, J. R. (1980). *J. Soc. Cosmet. Chem.* **31**, 12.
2. Ciurczak. E. W., and Honigs, D. E. (1990). *The Measurement of the Refractive Index of Liquids via NIRS Utilizing a Single Fiber Optic Probe*, PittCon, New York.
3. Atkins, P. W. (1990). *Physical Chemistry, 4th ed.*, Freeman, New York.
4. Mark, H. (1991). *Principles and Practice of Spectroscopic Calibration* Wiley-Interscience, New York.
5. Mark, H. (1992). Data analysis: Multilinear regression and principle component analysis. In *Handbook of Near-Infrared Analysis* (D. A. Burns and E. W. Ciurczak, Eds.), pp. 107–158. Dekker, New York.
6. Bjorsvik. H-R., and Martens, H. (1992). Data analysis: Calibration of NIR instruments by PLS Regression. In *Handbook of Near-Infrared Analysis* (D. A. Burns and E. W. Ciurczak, Eds.), pp. 159–180. Dekker, New York.
7. Williams, P. C. (1992). Samples, sample preparation, and sample selection. In *Handbook of Near-Infrared Analysis* (D. A. Burns and E. W. Ciurczak, Eds.), pp. 281–316. Dekker, New York.
8. Workman, J. J., Jr. (1992). NIR spectroscopy calibration basics. In *Handbook of Near-Infrared Analysis* (D. A. Burns and E. W. Ciurczak, Eds.), pp. 247–280. Dekker, New York.

COLOR AND SOLAR TRANSMITTANCE MEASUREMENTS

J. MURRAY STEWART
Pilkington-Libbey Owens Ford Company

ISBN 0-12-764070-3/$25.00.

I. Introduction

Measuring the color of objects correctly is difficult. Instrument data at best are only one dimension of color perception. We know that we obtain all color information from energy in the spectral region between 0.38 and 0.76 μm. Spectrophotometers and colorimeters can easily obtain transmission and reflection data in this spectral region. However, these instruments can only measure objects with high precision that are compatible with their optical and physical geometries. Dedicated computer color programs easily calculate color in specified conventional ways. Spectrophotometers are also used to measure solar spectral properties of materials over a wider spectral region from 0.3 to 2.5 μm. These properties determine the solar heat filtration potential or solar exposure susceptibility of materials at the earth's surface.

This section is a cursory examination of the complicated subject of color measurement. The reference citations herein contain detailed analyses of the topics under discussion.

A. Color Measurements

In all applications of color measurement, the primary objective is to produce a color data history close to the originally perceived color situation (1–3). Practicing this idea is key to successful color measurements. Human color perception varies with individuals but is second nature to all of us in every conceivable situation. Most of this section is devoted to some of the practical problems and solutions of color measurements based on this circumstance.

B. Color Matching

People practice color relativity all the time. They may or may not be aware that they are comparing colors under conditions that may never reoccur (4). Standard viewing conditions have been established to discuss effectively a particular color matching event at some other time. The ground rules for color communication begin with the establishment of standard observers (5) and standard illuminants (6). This is absolutely necessary for comparing instrument data as well as for discussing visual appearance (7). A "standard observer" is a set of mathematical weight values representing the color matching stimuli of the cones in the average color normal observer's eyes. These weight values are part of the spectral calculation that is used to determine perceptual color information. If color matching is to be done visually by an individual, he or she must first be tested for color normalcy (8). This condition can be determined by the American Optical AO-HRR pseudoisochromatic plates and by the Farnsworth–Munsell 100 Hue Test. The observer must then use a light source that is radiometrically certified for color rendering use (9). He or she will then grade specimens by comparing them to known colors in a color order system. These systems, though imperfect (10), will allow him or her to assign names or numbers of similarly graded material color sets (11). An experienced observer with near-average color vision (12) may make a visual color match between a specimen and a labeled element of similar material in the color system under a defined lighting condition and at a specific viewing position sufficient for someone else to duplicate. However, these natural qualifiers change with time. Therefore, instrumental color matching, though temporally less precise (13), is easily quantified permanently using colorimeters and spectrophotometers. Spectrophotometers measure a specimen's spectral transmittance, reflectance, or absorptance at wavelength intervals. Colorimeters measure a specimen's transmittance, reflectance, or absorptance through expensive, near-standard observer and approximate illuminant filters (14). Spectrophotometers may soon become the only instruments available to measure color because they are more precise, and with their associated computers, which are now equivalent in size, they may soon cost less than colorimeters.

C. Color Communication

Throughout history, literature, abounds with references to color and man's struggle to communicate accurate impressions of it (15, 16). Now we are attempting to wed modern instruments, "artificial eyes," with mathematical algorithms that are equivalent to color spaces (17) approximating human perception (18). Even before the computer age, scientists were investigating color communication mathematically (19, 20). Since 1913 the International Commission on Illumination (CIE) has worked to bring worldwide lighting concepts into common focus (21). CIE publications Nos. 15 (22) and 15.2 (23) describe international colorimetry standards. Internationally accepted terms in lighting are defined in CIE publication No. 17 (24), but a more extensive reference of color and appearance terms is available from the American Society for Testing and Materials (ASTM) (25). Pamphlets published by some of today's color instrument manufacturer's define and illustrate these terms superbly in color pictures (3, 26–28).

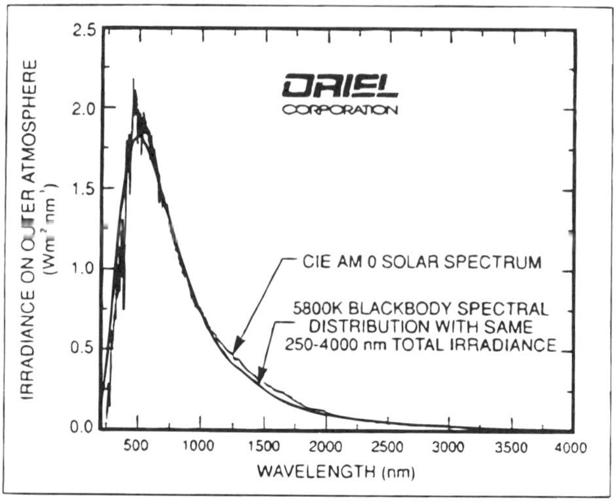

FIG. 1. The spectrum of radiation from the sum is similar to that from a 5800K blackbody (29).

Usually, we assume daylight illumination on or through an object when we discuss its color. A correlated color temperature of 6500°K closely approximates "average" overcast daylight color at the earth's surface (29). However, this temperature is too hot to physically produce as a standard in laboratories so the mathematical equivalent is factored in by computer when making daylight color calculations (Fig. 1).

II. Current Color Spectrophotometer, Principles and Geometries

Operators of color measuring instruments must have a general background in color theory and instrument theory (1, 2, 30). Most instrument companies include a general outline of both with their instrument instructions (31). One excellent book on color theory is *The Physics and Chemistry of Color* by Kurt Nassau (32). A good example of a general text on essential instrument theory is Varian's *Optimum Parameters for Spectrophotometry* (33). Some optical companies also include excellent background information on calibration filters, instrument components, and instrument theory in their catalogs (29, 34–36).

Next in importance to acquiring a general background in color and instrument theory is selecting a suitable instrument. Measuring a specimen's spectral transmittance or reflectance properly requires careful selection of instrument geometry to faithfully reproduce the desired color aspect of the specimen. There are a number of unique instrument geometries available for this purpose.

Conventional instrument geometries incorporate a source, prism or grating to disperse the light from the source, a detector, and necessary accessories such as lenses, mirrors, sample holders, and movable slits. If the spectral range of the instrument extends beyond the visible into the near-infrared and ultraviolet, dual sources, detectors, and ancillary electronics are required. The Perkin–Elmer Lambda 19, diagramed in Fig. 2, is an example of this technology (37).

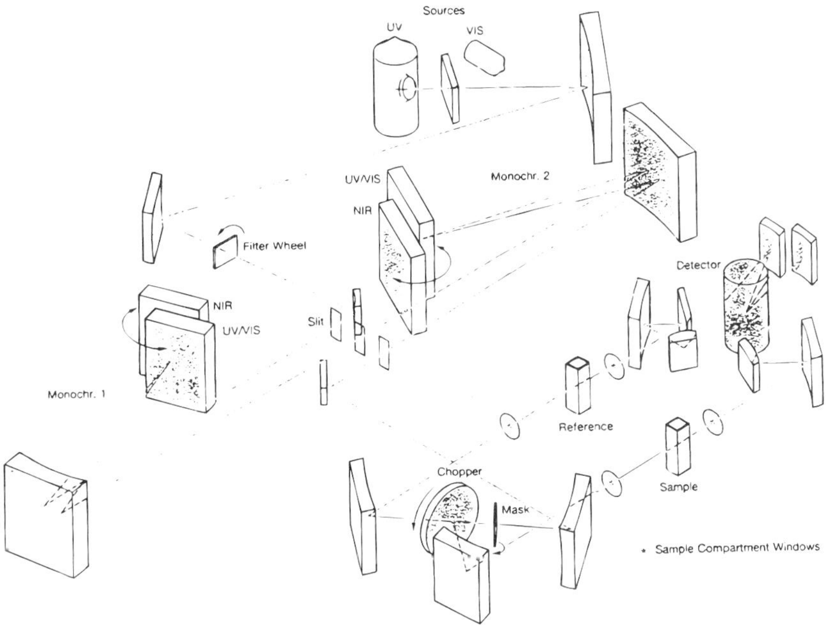

FIG. 2. Optical performance as a basic feature for analytical accuracy (37). A high-performance optical system is the basis for analytical accuracy in Lambda 19 spectrometers. The engineering task was to design an energy-optimized optical system to achieve specifications at the technical limits. The result is an optical system in λ Lambda 19 spectrometers that incorporates large collimating mirrors and gratings for maximum energy throughput. The wavelength range is from 175 to 3200 nm: synchronized slit mechanisms are microcomputer controlled for optimized and accurate setting of spectral bandwidth. Holographically ruled gratings with 1440 liters/mm in the UV/VIS and 360 liters/mm in the NIR guarantee high linearity, resolution, and wavelength accuracy. The Lambda 19 features a "pencil beam" which has a uniform geometry over the entire 120-mm width of the standard sample compartment. A common beam mask can limit the beam to 5 mm in height if small samples have to be analyzed. The signal-to-noise ratio achieved with this pencil beam design provides outstanding spectral quality for accurate results in many applications.

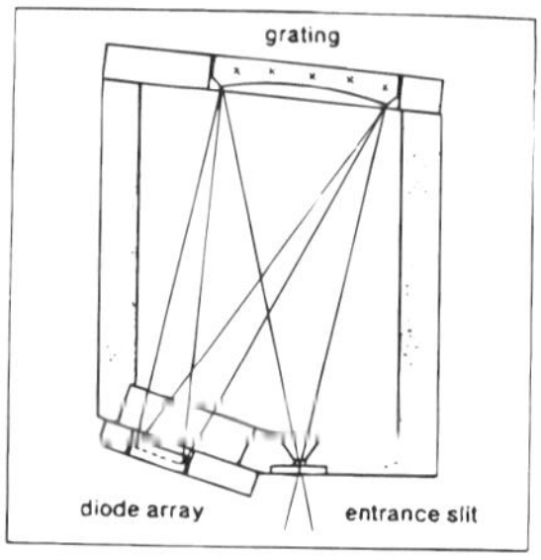

FIG. 3. Optical setup (38).

Diode array technology is another leading candidate for measuring color, as well as other visible properties, of specimens. Polychromatic light reflected or transmitted from a specimen is dispersed by a grating and focused on an array of detectors, each detecting a specific wavelength interval (38) (Fig. 3). Instrument examples of this are Zeiss OFT 311, OFR 311, and MCS 2 × 511 (Fig. 4).

A spinning grating monochromator by Rees Instruments, Ltd. (39) is another successful and unique solution to color measurement technology (Fig. 5).

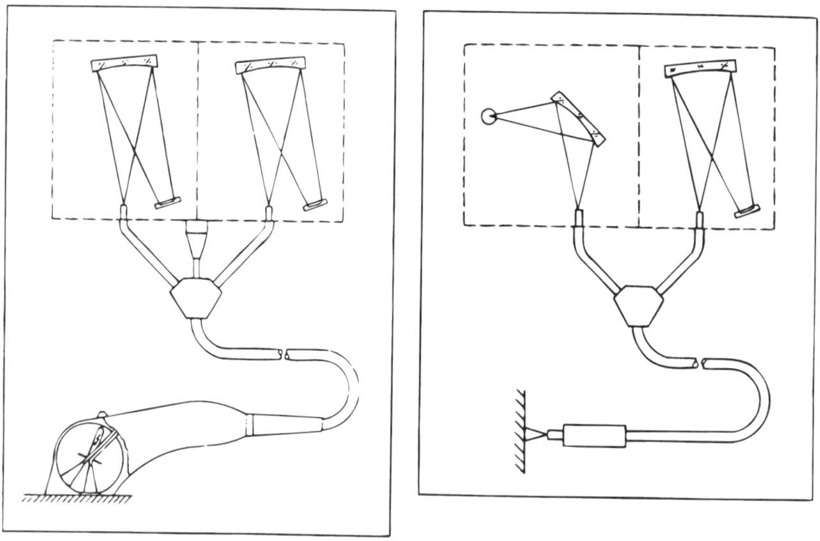

FIG. 4. (*Left*) Schematic of the MCS 2 × 512 VIS color-measuring instrument, consisting of two identical spectrometer modules and the movable measuring head with integrating sphere and flash tube. (*Right*) Measuring setup for the thickness measurement of films and layers, consisting of light source, diode array spectrometer, and fiber bundles (38).

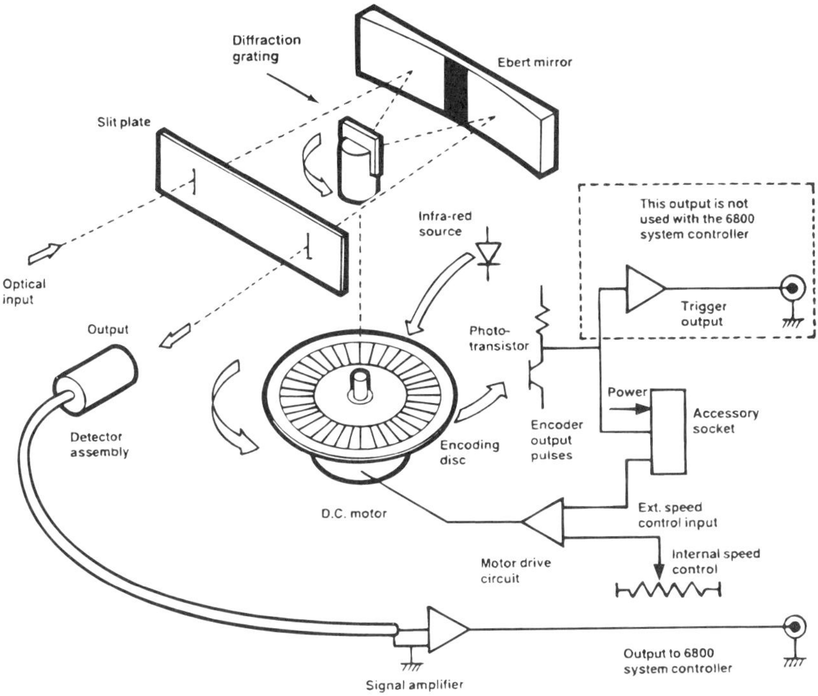

FIG. 5. Schematic diagram of the Monolight monochromator (39).

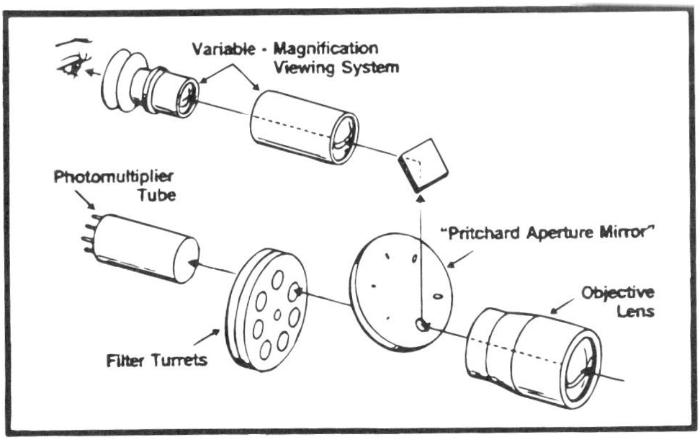

FIG. 6. Pritchard optical system (40).

The Pritchard optical system allows the operator to simultaneously view and measure the brightness and color of small target samples through a variable magnification and variable aperture photometer (40) (Fig. 6). Each of these instruments measures a specimen's color in one limited way. These data, in turn, are only one dimension of the specimen's appearance (41) (See Fig. 7).

A. Relation to Human Sight

Color measurement cannot be divorced from our simultaneous perception of other specimen characteristics (43) such as gloss and haze. Instrument measurements will also be affected by these other specimen attributes. It is

OBJECT CLASS (Examples)	IDEALIZED DISTRIBUTION OF LIGHT	COLOR ASSOCIATED WITH THE DOMINANT DISTRIBUTION OF LIGHT	APPEARANCE ATTRIBUTES ASSOCIATED WITH OTHER DISTRIBUTIONS OF LIGHT
OPAQUE NONMETALS (Painted Panel, Ceramic Tile, Heavy Fabric, Pad of Paper)	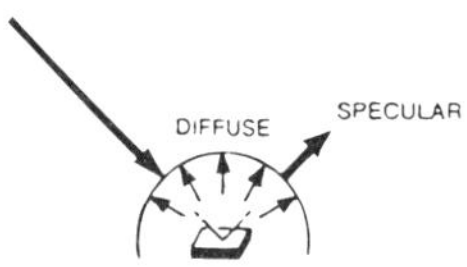	COLOR BY DIFFUSE REFLECTION Analyzed by: Spectrophotometric Curve or CIE X, Y, Z; or Y, x, y; or L, a, b, or Special Scales or Color Difference from Standard	GLOSS BY SPECULAR REFLECTION Analyzed by: Specular Gloss, Distinctness-of-Image Gloss, Luster or Contrast Gloss, Sheen, Surface Texture, Reflection Haze, or Directionality
OPAQUE METALS (Auto Bumper, Brass Doorplate)	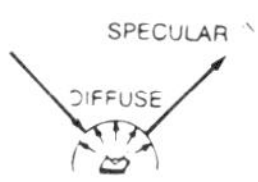	COLOR BY SPECULAR REFLECTION Analyzed by: Color Scales, Specular Gloss, Distinctness-Of-Image Gloss, Luster, Directionality, or Surface Texture	HAZE BY DIFFUSE REFLECTION Analyzed by: Reflection Haze
TRANSLUCENT (Plastic Light Enclosure)	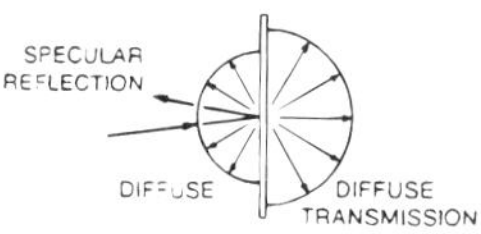	COLOR BY DIFFUSE TRANSMISSION Analyzed by Color Scales	GLOSS AND COLOR BY DIFFUSE REFLECTION Analyzed by Gloss and Color Scales
TRANSPARENT (Vegetable Oil, Sunglass Lens)	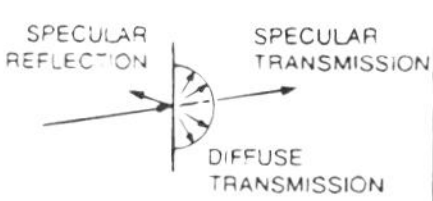	COLOR BY SPECULAR TRANSMISSION Analyzed by Color Scales	GLOSS BY SPECULAR REFLECTION; HAZE OR TURBIDITY BY DIFFUSE TRANSMISSION Analyzed by Transmission Gloss and Haze or Turbidity Scales

FIG. 7. Idealized geometric light distribution and appearance attributes for four classes of objects (42).

difficult to define and to measure each separately. Perceived gloss, for example, is a complicated subject (44). We use mental associations with colors to help us evaluate objects (45). Studies as fundamental as those on human visual discrimination (46) are necessary to correlate our instrument measurements with our perception of color.

B. Instrument Geometries

Figure 8, which shows instrument types (42), provides helpful ways to determine the type of instrument needed for a particular measurement. Detailed text standards on most subjects involving perception, color, and object appearance in general are available from ASTM (47). Also, Wyszecki and

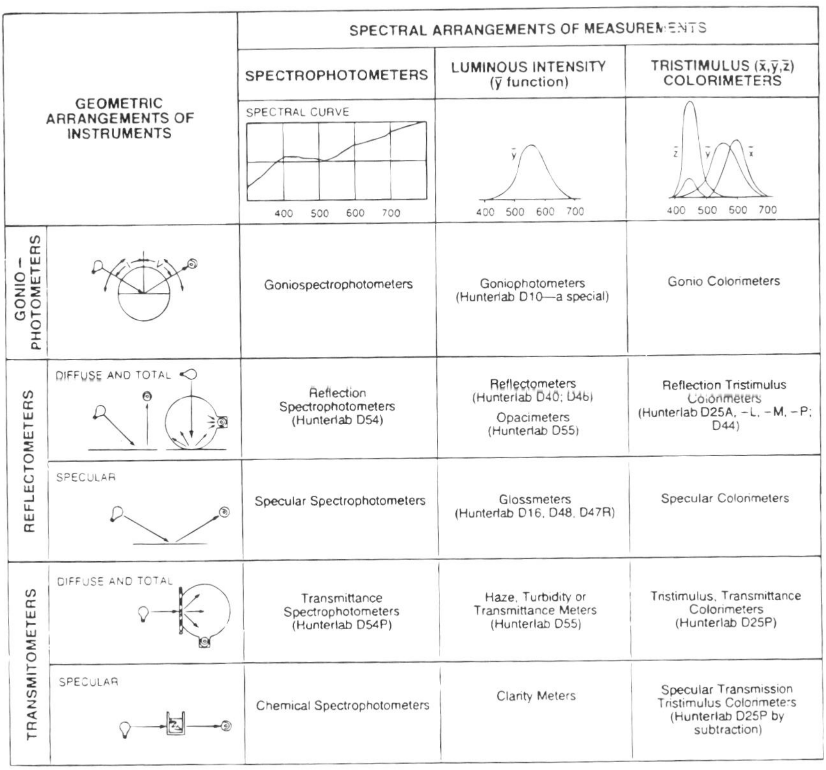

FIG. 8. Instrument classification by geometric and spectral light characteristics (42).

Stiles (48) have published a comprhensive authoritative textbook on these subjects.

C. Instrument Accessories

An integrating sphere attachment to the measuring instrument is usually required because so many color measurements are made on specimens that diffuse light. With the addition of this attachment comes the responsibility of knowing much more about instrument geometries (49), limitations, and new requisite measuring color standards (50, 51). Integrating spheres are coated with a very highly reflecting and diffuse white coating. One of the best is polytetrafluorethylene (PTFE) powder, whose reflectance has been studied extensively (52). It has nearly replaced MgO and Ba_2SO_4 for this application. Pressed into block forms with carbon black or phosphors, PTFE might serve as a reflectance calibration standard for gray scale and fluorescent materials (53).

III. Instrument Applications

Spectrophotometers can operate as separate units with small dedicated computers to integrate and transform measured spectral data into color system information. Networking of a number of specialized instruments through one central processing unit offers a number of advantages for the laboratory. Calibration, data acquisition, and printouts can all be standardized. Learning operating routines requires less time, and common run mistakes and systematic errors are easier to recognize. The end user of the color information can best understand the results from different instruments when they are presented in a common format.

A. Instrument Calibrations

Calibrations involve three interactive elements: instrument operating methods (standard practices), physical standards, and calculation conventions. Industry-accepted calibration methods are obtainable from ISO, CIE, IES, ANSI, ASTM, and other standards writing and publishing institutions as well as from papers published by national standards laboratories such as the following:

National Physical Laboratory, England
Deutsches Amt für Messwesen und Warenprüfung, Germany
Electrotechnical Laboratory, Japan
Mendeleev Institute of Metrology, Russia
National Institute of Science and Technology, United States
National Research Council, Canada
National Standards Laboratory, Australia
Pysikalisch-Technische Bundesanstalt, Germany

Two examples of these practices are ASTM E925, *Practice for the Periodic Calibration of Narrow Band-Pass Spectrophotometers*, and NIST Technical Note 594-3, *Photometric Calibration Procedures*. Most ASTM practices that deal with spectral calibration procedures combine instrument theory with good practice procedures. ASTM E1341-91, *Standard Practice for Obtaining Spectroradiometric Data from Radiant Sources for Colorimetry* (54), for example, devotes about one-fourth of its text to advice on verification of instrument linearity, wavelength scale, spectral bandwidth, stray light, and overall system condition. National standardizing laboratories also measure physical specimens, against their primary standards, for distribution as transfer standards to industrial laboratories. These physical standards are termed standard reference materials (SRMs). Industrial instruments are calibrated with these transfer standards by adjusting them to read the same values, calculated by an agreed convention from measured spectral data, such as those certified by the national standardizing laboratory.

Instruments should be calibrated according to the instrument manufacturer's instructions in operating manuals. Particular instrument procedures may be required to accommodate general instructions found in national standard practices (55) such as those mentioned previously. Most instrument manuals are written explicitly enough, however, to be used even by novice instrument operators who have some background in spectrophotometry (56). The subsequent calculations performed by the dedicated computer or mainframe are usually transparent to the operator. These calculations transform the raw spectral data into conventional color information that can be compared to similar historical color data. Good general color standards do not change with technological changes. The CIE 1931 standard (2°) observer is used to weigh specimen spectral data to the color matching function of the average observer. This standard of tabulated data is still in common use in many industries and is likely to remain a viable standard (57).

Some absolute instrument calibrations do not rely on national references. For example, absolute reflection of opaque, front surface mirrors can be determined using a Strong V-W instrument attachment (58). The absolute reflectance of integrating sphere diffuse coatings can also be calculated from instrument measurements (59).

B. Standards: National, Company, and Plant

Liquid absorbance standards provide a theoretically perfect way of checking the linearity of the photometric scale of an instrument because absorbance is directly proportional to concentration or path length. However, technical imperfections in instruments, cuvets, and filter solutions limit the accuracy of absolute photometric determinations (60). Liquid standards should be prepared for each calibration session. Solid samples should be checked and recertified periodically by an accredited standards testing laboratory. Color standards that change with temperature or light exposure such as the National Bureau of Standards red glass filter No. 2101 that is doped with selenium (50) must be used cautiously as recommended by the standards issuing laboratory.

Most industrial products must conform in quality to national material standards set by industry agreements. A primary source of instructional "text" standards is in the *ASTM Annual Book of Standards*. This is actually a 70-volume set of books (for 1994). Each book is a section on certain types of materials (61). ASTM also sells companion books on standards of specific materials (62). Members of ASTM, who author the standards, include people from industry (manufacturers), the government (regulators), and others who are interested in the quality of a product or service (consumers). Spectral methods and standard practices about color are published annually and will be published in 1994 in Section 16 (63). Previously, color and appearance subjects were encompassed in Section 14, titled "General Methods and Instrumentation."

Industries set their own standards for certifying their products based on procedures and requirements in ISO, CIE, ASTM, and other national and international standards. Companies buy physical standards from optical supply houses (64) or SRMs from national laboratories to calibrate their spectrophotometers. Optics filters are sold with certificates certifying their spectral properties. Certificates from national standardizing laboratories hold the highest credibility and are based on international round-robin comparisons (65, 66). Standards from optical supply companies are often identical in material, shape, and size to SRMs from national laboratories. These standards are certified by instruments calibrated using national laboratory standards. The added uncertainty introduced by the second measuring instrument lowers the certified spectral credibility, value, and acquisition time. Usually, if traceability from a company's standards to the national standard is certified by a reputable standards supplier (67), the spectral precision will suffice for accreditation by consumers and regulatory agencies (68). Standard filters may be solid, liquid, or gas that either scatter or absorb energy in a certifiable way (69). Usually, these primary company standards are used at a central research laboratory of the organization to calibrate its best spectrophotometers. Representative product samples measured here are distributed to the manufacturing centers as transfer standards to be used to calibrate on-line quality control product monitors.

C. Sample Types: Physical Considerations

A technique for measuring specimens on spectrophotometers cannot be generalized because there are many sample types. Physical characteristics of samples will determine the optimum instrument to use and the method of spectral analysis necessary to produce the best color data. Most specimens will fall into one of the following six categories and require solutions to these related questions before they are measured:

1. Transparent: Is the sample's transmittance or reflectance to be measured? Is the measurement to be absolute or relative to another specimen? If a relative measurement is needed, are the two samples physically and optically similar? If reflectance is to be measured are one or both sides filmed? Reflectance will be measured from which side of the sample? How will the side opposite to the reflectance-measured side be treated during the measurement? Will a light trap prevent transmitted light or multiple reflections from behind the sample from also being measured? Will the specimen's backside be backed with index oil and black glass to prevent second-surface reflections? How thick is the sample? Can it be measured in an area whose surfaces are free of dirt and scratches? Are the surfaces parallel or will the sample introduce a lens effect into the sample-measuring beam? Is there a curvature to the sample in the measurement area (70)? Is the sample rigid or must the sample be mounted in a custom jig for support? Will the sample's size and weight allow it to fit into the spectrophotometer properly? Is the sample sufficiently transparent for accurate measurements?
2. Semitransparent: In addition to the questions for transparent samples, more questions need to be answered for translucent specimens (71). Will all of the light that is scattered by the sample be collected by the optics and included in the measurement? Will the amount of scattered light vary with wavelength, mounting angle (72), or beam geometry?

3. Opaque: Is the sample surface smooth or rough? Is opacity due to the sample material or to an opaque coating on one or both sides? Is the reflectance specular, diffuse, or both? What reflectance geometry is necessary to measure the sample? Must a reflectance accessory and reflectance standards be used? Which physical standards are needed for calibration and are they similar to the sample to be measured? Will all the reflected light from the sample be collected and measured? Is there a preferred (flop) angle of reflectance and is this accounted for in the measurement (73)? Is the product to be used for high visibility at night, i.e., is it a retroreflecting material? Road signs, highway markers, and jogging apparel are special materials that should be evaluated on-site at night under specified distance and lighting conditions. Measurement conditions must be stated very precisely if brightness or color difference measurements will be made to determine durability.
4. Films: Is the film thick or thin. Is it on a substrate and, if so, is the film or substrate opaque? Is the film hard or soft? Is there one or more (stacked) films? Are they uniform over the filmed area and do they have "pinholes" in them? If there is a film stack, is interference between films at various wavelengths accounted for in the measurement or calculations (74)? These and other considerations are important in establishing film specifications (75).
5. Inhomogenous: Can the sample be measured in an area that is unaffected by the inhomogeneity? Can the sample beam be reduced in cross section enough to miss major color anomalies? Will reducing the beam diameter reduce the energy to a point that the signal-to-noise ratio is unacceptable? Can the specimen, for example, a broadweave cloth, be measured over a large enough surface area to average out local color differences?
6. Fluorescent/thermochromic/photochromic: Is the sample radiating or absorbing as well as transmitting and reflecting (76) due to the measuring instrument's source radiation? If yes, what are the excitation wavelengths and what wavelengths are affected?

D. Spectral Measurement Ranges

Energy from the sun reaches the earth over a spectral range from 280 to 2500 nm. This spectral range may be subdivided into three intervals according to a special function of each. Properties of specimens in the ultraviolet (UV) range from 280 nm to approximately 400 nm determine if these materials offer effective UV protection. The spectral region from approximately 400 to 750 nm is considered when a specimen's brightness, color, gloss, and other aspects of appearance are to be assessed. Instrument geometries of color systems are critical factors in the production of data that correspond to human perception (77). Even under near-perfect viewing conditions controversy and investigation continues on the relationship of brightness to luminance (78).

The usual illumination of an object for color discrimination is daylight. It excites a photopic response of the eye cones. At lower levels of illuminance color discrimination falls off. Intermediate light levels excite both the rods and the cones in the eye and at low light levels only rod stimulation occurs. Spectral weights at visible wavelengths of average Mesozoic and Scoptic vision have been standardized and adopted internationally (66). Beyond the red end of the visible spectrum, the near-infrared (NIR) from 750 to 2500 nm is usually associated with heat perception. Actually, approximately one-half of the solar energy is IR heat; the other half comes from solar energy in the visible region. Most heat absorbing glasses filter out more IR energy than visible energy by absorption because they contain from 0.5 to 1% iron by weight. Iron absorbs heavily from 500 to 1500 nm (79). The IR absorption of an iron-containing glass increases proportionally to the amount of iron and to the amount of it in its reduced FeO state. It is responsible for the green tint in inexpensive glasses. Fe_2O_3 only mildly absorbs energy in the UV at approximately 0.38 μm. A glass specimen's integrated IR spectra with respect to the solar energy incident on it gives us a more sensitive way to ascertain heat filtration than if the whole total solar spectrum is considered with respect to the glass.

Spectral measurements in the NIR region are used for spectral analysis of trapped water in samples because water has many close absorption bands at approximately 1900 nm. Ambient moisture bands from the air must be factored out by calculation or by using a dry air purge system during the spectral run.

Automobile windshields are glass sandwiches held together with polyvinyl buteral (PVB) plastic. Spectral analysis of the windshield in the NIR reveals glass absorption spectra overlaying plastic (methylene) absorption spectra at approximately 1700 nm and water absorbance at approximately 1925 nm. Subtracting the glass absorption from the plastic's and applying the linear relationship between plastic thickness and absorption allows the thickness of the plastic to be calculated from spectral data. Moisture content is determined by the same procedure. The ratio of moisture to the plastic thickness yields the percentage moisture in the plastic, which is a measure of PVB to glass bond strength. This spectral method can be performed without damaging the windshield that is being measured. Instruments are commercially available to specifically measure and calculate this property of windshields at fabricating plants following windshield assembly operations. Plastic companies, such as DuPont and Monsanto, determine the relationship between plastic thickness and spectral absorbtion by gravimetric experiments.

E. Static and Dynamic Measurements: Exposure Tests

The most precise way to calculate a spectral curve that has sharp absorbance peaks in it is to scan to predetermined wavelengths, set the slit width and bandwidth at optimum settings, measure the spectral property (%*T* or %*R*), and repeat this process at each wavelength that is needed to calculate the color of the object. Usually, this is too time consuming to be practical, so scanning during measurement is required. Multiple scans must be made at various speeds and with varying slit widths until little or no change in the spectral curves occurs. Faster scan speeds will not seriously affect the color calculations of transmittance or absorbance spectra if the colorant concentration is low and the specimen is thin because the spectral curves will lack sharp absorbtion peaks.

Spectral properties of some products may change during measurement due to prolonged exposure to the measuring beam. Some scanning spectrophotometers are designed to minimize exposure at excitation wavelengths measuring light-sensitive specimens with monochromatic rather than polychromatic (white) light. These instruments are sold with "reverse" optics and cost considerably more than conventional scanning spectrophotometers. The measured results should be the same when nonsensitive specimens are measured by either optical arrangement according to the Helmholtz reciprocity relation in optics. Some glasses doped with photochromic-sensitive salts and all fluorescent paints are sensitive to light and will require monochromatic mode illumination during spectral measurements on scanning spectrophotometers. There are alternative solutions to this measurement exposure problem. Diode array spectrophotometers measure the spectral properties of specimens by exposing them to a momentary burst of energy from the source. Some colorimeters also use this flash exposure technology with very good measurement results on many types of products.

F. Sample Fingerprints

Chemical elements and compounds are characterized by their selective absorbtion bands. Organic materials and rare earth elements usually have steep and narrow spectral absorbtion bands that are used as wavelength calibration standards. Inorganic colorant materials, on the other hand, generally have relatively shallow and broad spectral bands that are easier to measure correctly (79). Spectral absorbtion data from samples taken from different production times can be integrated and transformed into a color space and the vector displacements related to bad or good color changes.

G. Quality and Quantity Measurements

There is always a trade-off between quality and quantity measurements. Each measuring situation requires a number of compromise considerations concerning accuracy, precision, and repeatability. If the samples to be measured are known to be uniform in quality, or have been chosen as the best example of a product, then more time and attention should be given to making high-precision measurements on a few of them. This generally involves frequent calibrations between sample runs at more wavelengths in each mode of measurement on high-quality spectrophotometers. These instruments should be certified for accuracy, precision, and repeatability by the field service engineer who inspects them at least annually. Quantity measurements, at a lower level of precision, are usually made for statistical-quality analysis on average product specimens. Faster, cheaper, simpler, and more rugged measuring instruments are used for these measurements. When multiple spectral runs can be superimposed on the same screen to monitor changes in one or more colorants, integrated color data often are not required to correct colorant concentrations.

H. Quality Control Considerations

Quality control (QC) specimens are usually not ideal samples. If large numbers are to be measured for statistical certification, the measuring procedure and equipment must be routinely checked. Operators must be able to continuously distinguish system drift changes from slow product changes. One aid to reducing the number of monitored parameters is to use a jig to mount samples in the same position relative to the sample beam for measurement. QC instruments should not require a darkroom for their use nor should they be seriously impaired by stray light during use. Many instruments today use high-intensity flashlight sources or chopped beams to avoid both of these measurement constraints.

IV. Measurement Techniques

Three areas of expertise are involved in measurement techniques. First is specifying the specimen color parameters necessary to satisfy the customer (80). Second is taking actions to ensure maximum precision from measuring instruments. A U.S. ASTM standard that addresses these issues is provided in Ref. (81). The third area involves sample–instrument interactions by making the best use of measuring equipment to obtain the best color information.

A. Sample Preparations

Specimen color of gases is not usually measured. Gas sample holders, available from spectral supply houses, have transparent windows at both ends of a cylinder up to 10 cm long that can be mounted in the sample beam of some spectrophotometers. A 1-cm cuvet is the principal instrument accessory for holding liquid samples. Both holders can introduce significant errors into spectral measurements (60). Turbidity, air bubbles, and evaporation must also be avoided in order to obtain acceptable spectral results from liquid specimens. Cleaning and surface shape problems are inherent with solid samples. A good detergent washing with a clean, soft towel followed by an isopropyl alcohol rinse and air drying is the usual cleaning method used on smooth solid samples before spectral analysis. It may be necessary to clean delicate samples simply with a camel's hair brush or a Staticmaster (Nuclear Products Co.) brush. Size, shape, and surface(s) texture must be considered before making spectral measurements. Instrument considerations include sample beam incident angle, reflected light back into the instrument, sample orientation if polarization is present in the sample, and edge scattering effects if the sample edges are close to the measuring beam. The specimen's edge brightness should be reduced to a minimum when the sample has many small surfaces, such as jewels (82), or if the sample is being measured at angles at which the sample beam can extend over a considerable area to the edge of the specimen.

Diffuse samples are of two types. Either the sample or its surface(s) scatter light. If the surface is smooth, gloss measurements can be made at angles (83). Transmittance and reflectance of translucent materials, though not measured directly, can be calculated from spectral measurements (71). Scattering is a variable component of the total light envelope. It can be either wide- or narrow-angle scattering. Narrow-angle scattering is generally perceived as small blue–white halos of light from a sample's surface. Large-angle scattering is perceived as contributing a matte or translucent character to the specimen. An integrating sphere must be used to gather all the light reflected or transmitted from either type of sample. The larger the integrating sphere, the more light from the second surface of a solid sample will be returned to the sphere and measured in the reflectance mode (84). If the sample scatters light but the surfaces are smooth, reflectance measurements usually will not be a problem. Diffuse transmittance measurements require an integrating sphere to collect all the light. If measurements are made on diffuse samples with instruments that lack an integrating sphere, tests with scattering neutral density screens may help to determine how much light is lost by scattering. Large samples that cannot be moved or have been installed elsewhere can be photographed in color and the photograph's colors measured and compared with a photograph of similar material exposed under similar viewing conditions elsewhere. If large color tolerances are acceptable to the customer for a particular product, this method may be acceptable as the basis of a warrantee agreement.

B. Sample Positions

Some sample types require very precise alignment in measuring instruments in order to obtain meaningful spectral results (72, 85, 86). Reflectance measurements are about twice as sensitive to position errors in the measuring beam as are transmission measurements. The operator should realize that spectrophotometers are specifically built to measure only a few specimen types precisely. Other specimen types may require instrument accessories and will usually involve a compromise in measured data accuracy. If a sample is exceptionally thick, such as tank periscope windows, there may be no right way to position it in the sample beam of an instrument that was not designed to measure it. If the sample is oversized and the instrument operates only when the sample is fully inside the enclosed sample compartment, black photographic material may suffice to extend the sample compartment. Small diffuse front surface reflecting samples can be measured at angles to the incident beam inside an integrating sphere. Very large spectrally reflecting filmed window samples can be run on some spectrophotometers using a Herrick sample beam attachment. A very precise sample holder is required to maintain the sample stationary and normal to the extended sample beam during the reflectance measurement scan.

C. Spatial Frequency: Gradients

When specimens have nonuniform surfaces, repositioning them will usually result in different measurements. Recording exactly what position was measured, what viewing conditions (instrument geometry) were used, and what limitations should be expected if the measurement should be repeated should be considered part of the run data. Sometimes the nature of a color gradient is the point of a measurement—for example, colors of variable thickness films, some paint drawdowns with substandard hiding power, and automotive windshield fade-off color bands. Such samples should be masked and measured in contiguous positions and the spectral data plotted as a curve or a family of curves to describe the gradient condition in a most useful way.

D. Polarization Orientation

If a sample's transmittance or reflectance varies in different orientations normal to the sample beam, the measured position and the degree of sample beam polarization must be accounted for along with the sample data. Depolarizing attachments are available for some instruments. These attachments eliminate the need to selectively orientate the polarizing sample in the sample beam. However, they attenuate the energy available for measurement and so may increase the signal-to-noise ratio beyond acceptable levels. Samples with films applies in a preferred direction onto substrates, glass samples under stress or measured at angles, napped cloth materials, and some metallic paints will fall into this category.

E. Sample Holders

If an instrument is to be used to make many solid sample transmission measurements at normal incidence to the sample beam, a machined and optically aligned sample jig should be used. This will avoid many questions about sample positioning of similar specimens. Liquid cuvets usually come in matched pairs. These pairs are supposed to have the same path length, to have parallel sides, and to be made of the same material. They should be checked in a dual-beam instrument scanning the full instrument wavelength range. Both cells should be filled with distilled water and run first in the sample and reference beams and then reversed and run again. If the two cells are evenly matched the curves will fall on top of one another; if not, as is often the case, the curves can be used to determine the correction factors. Gas samples should be placed in gas cells that are as long as physically possible to fit into the spectrophotometer's sample compartment.

F. Instrument Slit Widths and Scan Rates

Scanning spectrophotometers also designed to operate at optimum slit widths and scan rates. The two are closely related and are the first considerations one must make when considering which combination will yield the best spectral data from a specific type of specimen (33). Instrument sources vary in intensity with wavelength. Lens/mirror systems absorb and scatter different amounts of energy at different wavelengths. Detectors are not uniformly sensitive to energy at different wavelengths nor are they uniformly sensitive over their detection surface areas. These are the instrument variables that are incorporated into the slit width and scan rate configurations by the instrument manufacturer. Additionally, specimen properties, such as absorption bandwidths, optical density, and whether the sample is to be run against a standard other than air with enough relative energy to identify spectral differences between them and background noise, must be considered when determining the proper slit widths and scan rates before a specimen run. Maximum spectral resolution is a goal achieved at minimum slit widths but at a considerable loss of energy. The energy that is allowed to pass through the monochromator slits is usually a linear function of wavelength at each slit edge. The total energy is the area under a triangle that must vary with wavelength to provide optimum spectral data results (Fig. 9).

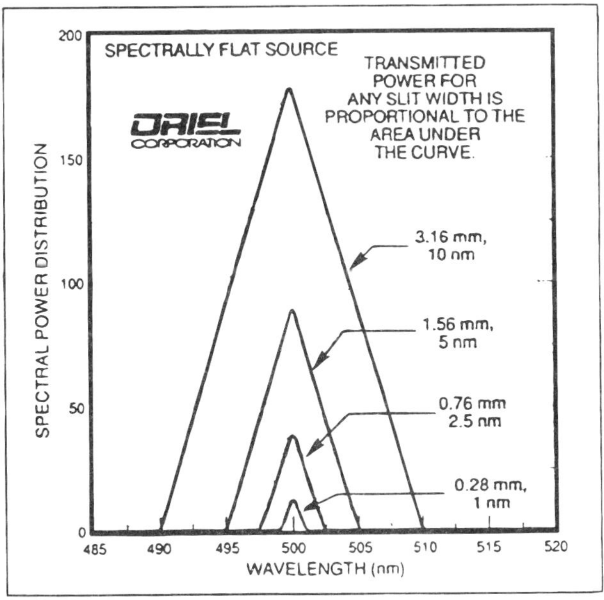

FIG. 9. Spectral power distribution for 77,200 1/4 m Monochromator with various slits. This is for a spectrally flat source uniformly filling the input. Note the shape is triangular and not the rectangular profile one might intuitively expect (29).

G. Scan Modes: Subtraction and Overlay

Specimen spectral data can be obtained relative to air or to a standard. The specimen's absolute spectral properties can then be calculated from these data (87). In double-beam instruments there is an advantage to running a specimen relative to a standard of similar material. Energy levels between the two beams remain approximately equal over much of the spectral scan. All the variable parameters discussed previously occur simultaneously in both sample and reference beams to the sample and standard. Single-beam instruments can store a reference spectra during an initial scan and can also store subsequent spectral data on a sample. These two sets of data can be overlaid in curve form to ascertain changes in specific chemical concentrations if the instrument remained stable and calibrated between runs. Data sets can be subtracted and displayed in tabular format when small changes at specific wavelengths are of interest. Transformed into a color system, visible data files can be subtracted to indicate shifts in color relative to established color tolerances (88).

H. Accessories: Inspection, Run Alignment, and Calibration

Mounting specimens can be a real challenge when they must be measured in the exact position after returning from a prolonged exposure test. Make a special jig for each project of this nature. Describe the sample positioning using the jig and mark run orientations on the samples if possible. One should be able to replace the sample in the same area, on the same side, and at the same angle as it was measured before degenerative exposure. This is most important if the sample is nonuniform across its surface in appearance either before or after exposure.

A dentist's 1-in. diameter tooth mirror is a useful inspection tool when mounting small samples so that they will be centered in a sample beam of almost the same size. Sample beams at sample mounting positions can be outlined on single sheets of paper and then transferred to templates for making special jigs to mount odd-sized specimens in the sample beam. A single sheet of Polaroid plastic rotated in the sample beam with and without the sample is useful in predicting polarization effects from the instrument and sample. Set the wavelength at 550 nm (high energy from the source, sensitive detector wavelength, and low signal-to-noise ratio at a low slit width) for checking instrumental reactions to polarization and for visual inspections.

Physical standards, certification certificates, and yearly calibration data for all instruments should be kept together with instrument manuals. This is a key action of the standards traceability process. Solid standards should be kept in hard containers in positions that will not allow their measuring surfaces to touch anything but air. They should be handled by their edges only and inspected for oxide accumulation before use. Original standard certifications and all subsequent recertifications from standards testing laboratories should be kept in a chronological file. Instrument calibration data and curves should also be kept in a chronological file along with copies of any procedure standards that apply to the instrument or specimens to be certified. Standards data should include instrument type and serial number, operator's name, run date, spectral calibration and background scans, standards scans and other associated data that may be required by customers during inspection of the laboratory.

V. Mathematical Considerations

The luminous illuminant A transmittance ($\%T_A$) of a specimen can be measured directly on an optical bench if the bench is equipped with (i) a light source radiometrically calibrated to the correlated color temperature of approximately 2856°K, (ii) a constant electric current and voltage to operate the lamp at the certified color temperature, (iii) photopic filter that will correct the detector to the approximate photopic $V(\lambda)$ brightness response of the average human eye, (iv) blocking cutoff filters to exclude energy from the source that would be detected outside of the visible region, and (v) a diffusing filter to ensure equal distribution of the source energy across the surface of the detector. Good measurement approximations ($\sim \pm 0.5\%$T abs) to the theoretically correct specimen value are possible with careful adherence to these requirements but often this is hard to do (89). However, a set of neutral density filters or a set of standard products, such as those that are measured on the optical bench, are needed to verify a specimen's measured values by the optical bench. These verification sets must be certified from spectral measurements because these values are calculated using the mathematically perfect energy distribution for illuminant A and the mathematically perfect photometric standard observer for photopic vision $V(\lambda)$. The brightness color matching function $Y(\lambda)$ is equivalent to the photopic relative luminous efficiency of radiation $V(\lambda)$. This and the other two weight sets [$X(\lambda)$ and $Z(\lambda)$] of the tristimulus weighting functions are published in ASTM Method E308 (90). The average spectral sensitivities of the cones in the eye can be appreciated when these stimulus values are plotted over the visible wavelength region (Fig. 10; see Example 1).

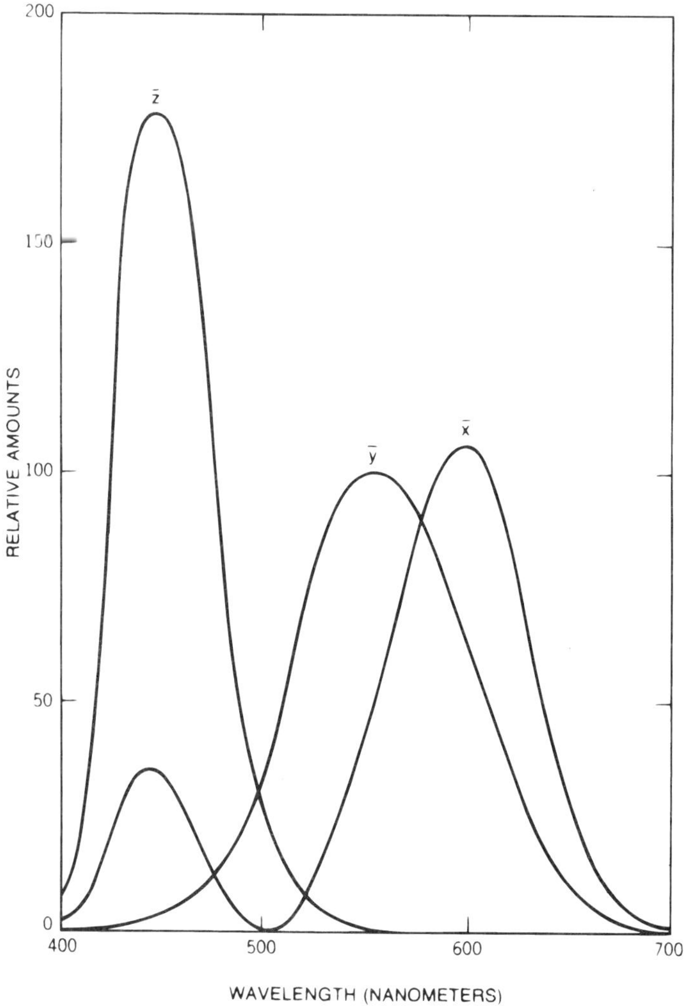

FIG. 10. E1931 CIE Standard Observer (42).

Color is calculated from specimen spectral data multiplied by these tristimulus weight sets and by the spectral distribution of the lamination. Today's spectrophotometers and their dedicated computers are fast and their standardized algorithms are usually transparent to the user. The operator may forget that there are still changes taking place in the standard methods for calculating color (91–94). The standard data and algorithms in the computers of spectrophotometers are seldom if ever updated. It is important to have instrument calculation algorithms and standard data files (Table 4, see page 371) on file if color calculation "errors" occur between different instrument computers. These systematic errors can lead to even larger discrepancies when the data are plotted in color spaces that are not uniform (95, 96).

It is therefore necessary for the operator to understand what math is being performed by the color calculating computer. A specimen's color is calculated from its measured spectral transmittance or reflectance property multiplied by a standard colorimetric observer and by a standard-color light source. The calculation algorithm for multiplication is usually the conventional rectangular method that is simply stated in equation form in procedure standards and used in tabular form. Modern, slightly more sophisticated calculation algorithms, such as Simpson's rule and trapezoidal integration, are more accurate and just as easy for the computers to calculate. Trapezoidal integration, for example, will yield much closer results than rectangular integration when different wavelength intervals are used to calculate areas under spectral curves. In this method of integration one-half of each wavelength interval is calculated into the next contiguous wavelength interval. Spectral differences between the rectangular and trapezoidal methods are greatest when either the energy or the spectral curve of the specimen change slope quickly. Calculated tristimulus data can be reduced to many color conventions. The following wavelength-abbreviated calculation illustrates the mathematical (rectangular) process to determine color appearance of an object from measured spectral data.

Example 1: Compute the color of a blue paint as viewed at a 2° acceptance cone angle by a standard observer on an average overcast day (illuminant C).

TABLE 1

Wavelength (nm) and hue	Energy of source Daylight (C)	Response of eye cones—2° observer		
		Red (X)	Green (Y)	Blue (Z)
400 Violet	6.3	0.01	0.00	0.07
450 Blue	12.4	0.34	0.04	1.77
500 Green	11.2	0.00	0.32	0.27
550 Yellow/green	10.5	0.43	1.00	0.01
600 Orange	9.0	1.06	0.63	0.00
650 Red	8.8	0.28	0.11	0.00
700 Dark red	7.6	0.01	0.00	0.00

TABLE 2

Wavelength (nm) and hue	Normalized tristimulus values of observer to daylight $(C)\times(X)$/Sum *Red*	$(C)\times(Y)$/Sum *Green*	$(C)\times(Z)$/Sum *Blue*
400 Violet	0.00	0.00	0.02
450 Blue	0.20	0.02	1.04
500 Green	0.00	0.17	0.14
550 Yellow/green	0.22	0.49	0.00
600 Orange	0.45	0.27	0.00
650 Red	0.12	0.04	0.00
700 Dark red	0.00	0.00	0.00
Sums	0.99	1.00	1.21

TABLE 3

Wavelength (nm) and hue	% Reflectance of paint (%R)	Daylight sensations of paint on eyes (%R)$\times(C)\times(X)$ *Red*	(%R)$\times(C)\times(Y)$ *Green*	(%R)$\times(C)\times(Z)$ *Blue*
400 Violet	85	0.36	0.01	1.72
450 Blue	76	14.95	1.69	78.80
500 Green	52	0.13	8.88	7.48
550 Yellow/green	45	9.68	22.22	0.19
600 Orange	30	13.49	8.01	0.01
650 Red	27	3.19	1.20	0.00
700 Dark red	25	0.10	0.04	0.00
Sums		41.9	42.1	88.2

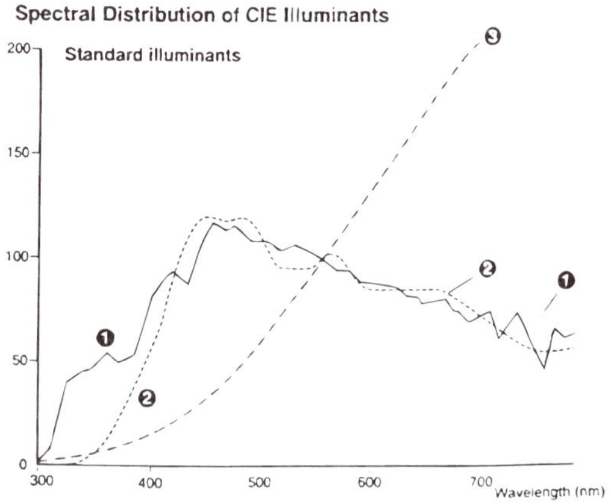

FIG. 11. Spectral distribution of CIE illuminants (26). (1) Standard illuminant D_{65}: Average daylight (including ultraviolet wavelength region) with a correlated color temperature of 6504K; should be used for measuring specimens which will be illuminated by daylight

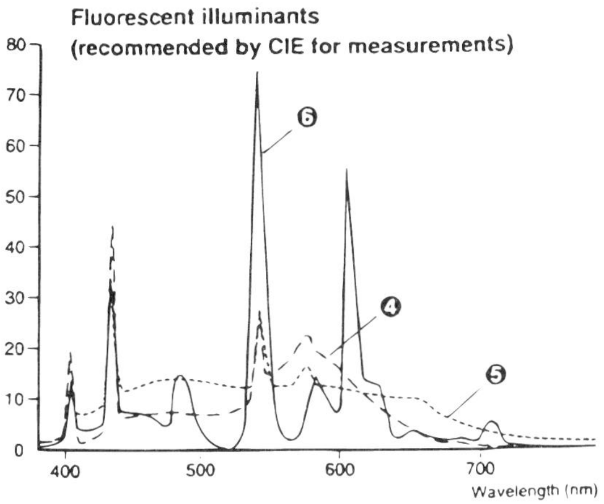

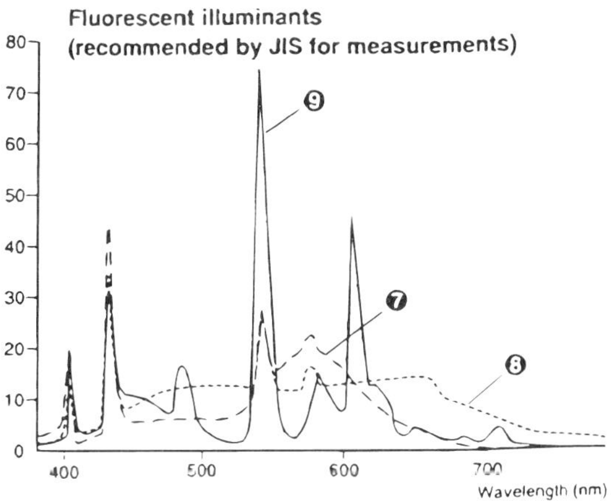

FIG. 11. (continued)

including ultraviolet radiation. (2) Standard illuminant C: Average daylight (not including ultraviolet wavelength region) with a correlated color temperature of 6774K; should be used for measuring specimens which will be illuminated by daylight in the visible wavelength range but not including ultraviolet radiation. (3) Standard illuminant A: Incandescent light with a correlated color temperature of 2856K; should be used for measuring specimens which will be illuminated by incandescent lamps. (4), F2, cool white; (5) F7, daylight; (6) F11, three narrow-band cool white; (7) F6, cool white; (8) F8, daylight white; and (9) F10, Three narrow-band daylight white (26).

Conclusion: The paint looks blue under these lighting and viewing conditions because there is twice as much blue sensation as there is either red or green sensation in the eye.

The color of an object will also vary with the color of the light source. Other standard illuminants may radically change the color appearance of an object because the area under their curves usually varies considerably with wavelength (Fig. 11).

Figure 12 compares an apple's red color under daylight limitation to its color under a standard light bulb at night. This is a pictoral representation of the calculation procedure in Tables 1–4. Notice the area-under-the-product curves vary in proportion to the color change.

Examples 2–4, using the previous tables calculation procedure, will show the differences in color appearance of an object when alternate standard observers and standard sources are used. These will serve to emphasize the importance of choosing the required combination of observer and source to compute the resultant color from the measured spectral data.

Table 4 presents alternative weight values from those in Table 1. They will be used in the recalculation of the blue paint color in Examples 2–4.

Example 2: Change Sources of Example 1 [daylight (illuminant C) to light bulb (illuminant A)]

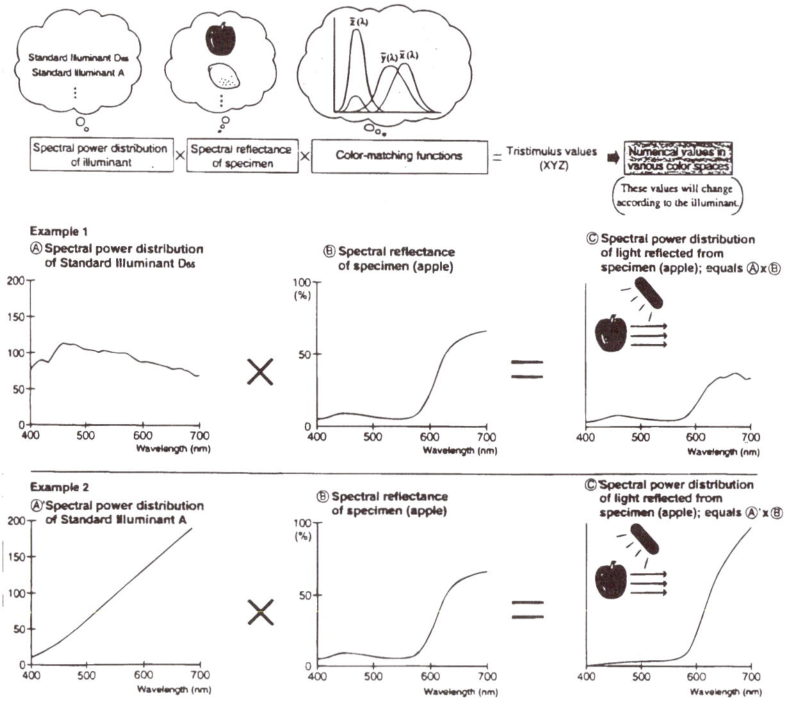

FIG. 12. Apple color variations (26).

TABLE 4

Wavelength (nm) and hue	Energy of source Light bulb (*A*)	Response of eye cones—10° observer		
		Red (*X*)	Green (*Y*)	Blue (*Z*)
400 Violet	1.5	0.02	0.00	0.09
450 Blue	3.3	0.37	0.09	1.99
500 Green	6.0	0.00	0.46	0.22
550 Yellow/green	9.3	0.53	0.99	0.00
600 Orange	12.9	1.12	0.66	0.00
650 Red	16.5	0.27	0.11	0.00
700 Dark red	19.8	0.01	0.00	0.00

- Substitute column 2 of Table 4 into column 2 of Table 1
- Recalculation yields Table 3 sums: *red* = *38.1*, *green* = *38.5*, *blue* = *25.5*

Example 3: Change sources and observers of Example 1 (daylight to light bulb and 2 to 10°)

- Substitute columns 2–5 of Table 4 into columns 2–5 of Table 1
- Recalculation yields Table 3 sums: *red* = *38.8*, *green* = *39.2*, *blue* = *25.8*

Example 4: Change observers of Example 1 (2 to 10° acceptance angle of observation)

- Substitute columns 3–5 of Table 4 into columns 3–5 of Table 1
- Recalculation yields Table 3 sums: *red* = *41.6*, *green* = *43.5*, *blue* = *87.1*

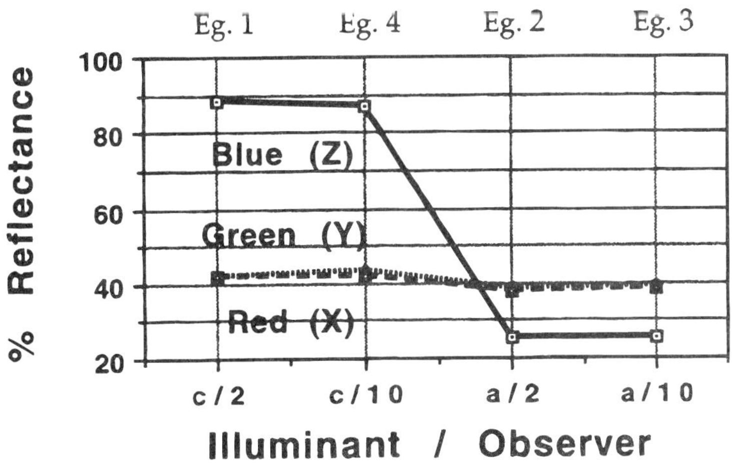

The graph shows the tristimulus values, the amount of stimulus to the eye cones, that these viewing conditions provide. Two general observations follow from the graph of the calculated tristimulus values. First, the change of observers has an almost negligible effect on the evaluation of the paints color. Second, the change of illuminants has considerable effect on the paint color.

These values are usually transformed into a color system that is easier to relate to conceptually. L,a,b color systems space opaque reflective colors of paints, papers, and textiles like a standard observer would when he or she compares colors of these objects. Note that the L,a,b, values are derived from the X,Y,Z values, which vary with observer and illuminant (Fig. 13)

The arithmetic color calculations illustrated in the previous examples are simple for standard illuminants and standard observers but the math and science behind the development of the weight table are far from simple or complete. The relationship between radiometry and photometry (97) is the first important step in the development of color standards. The color of light sources in relationship to color differences and color matching must be incorporated into any color order system (98). The weight sets for color matching at various size wavelength intervals has changed slightly but

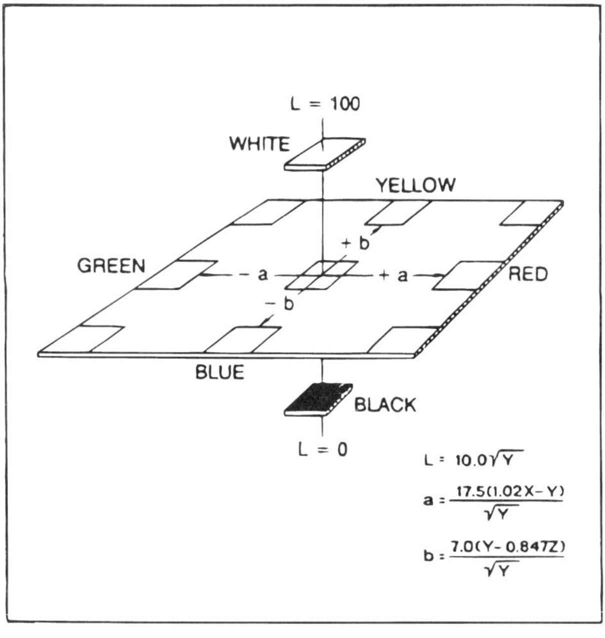

FIG. 13. Hunter L,a,b color coordinate system (42).

noticeable over the years (91, 99). Currently, the internationally accepted standard values for all these variables are stated in CIE publication No. 15.2 (23).

A. Spectral Ranges

Because color is a function of many gradient functions, there are no natural finite limits to color vision. The blue limit of the visible is taken to be somewhere between 360 and 400 nm for mathematical calculations. The red limit of vision likewise has been arbitrarily defined to be between 700 and 830 nm. In most practical calculations of color or color differences there is little or nothing to be gained by carrying calculations beyond 400 and 700 nm. Exceptions are some red, blue, or purple specimens or sources with high spectral responses in the dark color limit areas (82). Whereas the exact weight factors in these limit are built into the data tables and algorithms, the limited instrument response here mey overshadow any advantages from the extended calculations.

B. Calculation Methods

The physical laws of substance reactions to changes in thickness, concentration, and reflection losses at surfaces due to index changes are well known. With extreme care, measurements will come close to theoretical calculations. Usually, however, measurements must be made under less than perfect conditions. Spectral models help to guide spectral investigations about new products that are associated with these changes. Statistical measurements, on the other hand, give a better idea not only of changes due to theoretical variables but also to production limitations and customer acceptance levels of imperfect products (100).

The tables incorporated into color calculating routines by spectrophotometer manufacturers can be found in the ASTM publication *Standard Test Method for Computing the Color of Objects by Using the CIE System* (90). This ASTM standard is a compilation of calculated product tables for various illuminants with the two standard observers for color matching. Both standard observers are used in various industries to specify color. The operator must choose which observer is to be incorporated into the color calculations for the measured specimens. The overall difference in the color of an object observed by the 10° CIE 1964 Supplementary Observer or the 2° CIE 1931 Observer is usually small but noticeable. This is due to an uneven distribution of cone receptors in the retina of the standard eye. At a solid acceptance (cone) angle of 2°, light from the object activates the cones around the fovea, which are relatively close together and allow maximum color discrimination. A 10° cone angle of viewing a colored specimen, though not as color sensitive because the cones are not packed so closely over that portion of the retina, is a better average of the way color is viewed in everyday circumstances. Use the 2° observer to compare data to old color data but use the 10° observer to calculate new color data. Colors change when standard sources are varied; therefore, the operator must choose a standard source to calculate color that is similar to that under which the color of a specimen is to be evaluated. Color changes due to changes in source colors alone is the basis of a CIE 1972 special metamerism index (48, p. 169). The color change of an object is the most complicated and common issue involved in colorimetry (101). It is known that samples will not appear metameric unless their spectral curves cross at least three times in the visible region (102). There is much uncertainty in the color science community as to how to define and measure metamerism (8, 9, 12, 103, 104).

C. Perception vs Model Calculations

Perception is easy—perhaps so easy we have a hard time finding words or formulas to describe the complicated appearance of objects. Many excellent scientific articles on the subject of metamerism appear each year in color literature (105) but even commonly accepted color tolerance models to which they refer are still under fire (106) (see Fig. 11).

The difficult task of expressing our everyday experiences of color matching often cannot be adequately defined in a color space. Extended light sources, flicker, stroboscopic effects, variable contrast backgrounds and surrounds, matte surfaces, and polished top layers such as a leaf (107) and others cannot be properly defined mathematically. Calculations that involve such parameters are only approximations of the situations they represent. Try to imagine how dull a scene would be if you only perceived instrument measured color information.

VI. Pitfalls: User Techniques

Selecting a standard illuminant, a standard observer, and the proper expression of results is essential to obtaining color data that correspond to color perception. However, because instruments are simple to operate, many operators are untrained in instrument and color theory. They make bad parameter choices or instrument adjustments and "measure" specimen colors incorrectly (30). If the operator can define the color task, he or she will probably choose the correct variables and solve his or her color measuring problems satisfactorily (108). This assumes that the calibration and measuring procedures have been executed properly.

If the operator does not know the instrumental reason for calibration, he or she probably will not make the necessary calibrations properly or when required. An operator must be able to detect systematic errors in wavelength, bandwidth, detector linearity, nonstandard geometry, and polarization that affect accuracy. He or she must also be able to detect random errors, such as drift, electronic noise, and improper sample preparations, that affect repeatability (109).

Instrument operators must always scrutinize color samples for unusual spectral properties. Thermochromism, for example, is a specimen property that will cause its color to change with temperature. The temperature must be recorded along with the color measurements to be meaningful to future users of the color information. Figure 14 demonstrates graphically this condition in a colored glass long bandpass filter.

Automation implies simplification of instrument operations. This in turn usually implies restrictions on sample types that can be measured accurately. The spectrophotometer operator must be able to properly reconfigure such a system to obtain good color measurements if sample types change. Solid multilayered filmed samples can change color by transmittance or reflectance with small changes of surface viewing. When similar material standards, with certified spectral values, are not available for calibration, the instrument and measurement geometry must be described along with the measurements. Failure to do so will inevitably lead to dissimilar results when the specimen's color is remeasured. Many color spec-

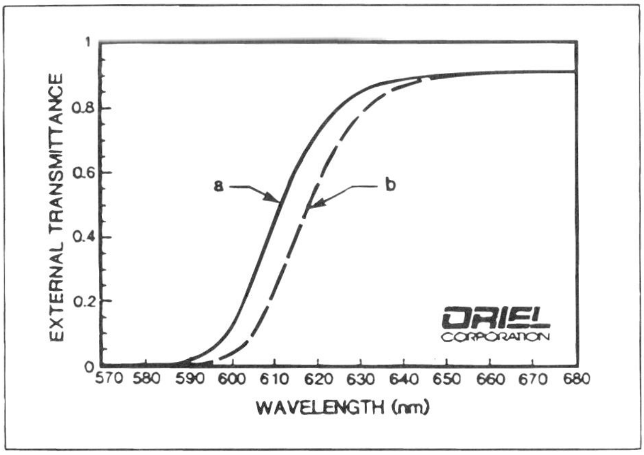

FIG. 14. 51312 long pass filter at (a) 20°C and (b) 60°C (64).

trophotometers are specifically designed for measuring colored liquids in cuvettes. The optical geometry of the measuring beam in the sample compartment usually cannot be altered to accommodate other sample types. If a homemade sample holder is to be substituted for the standard instrument holder, the instrument maker should be consulted about its effect on the resulting measurements and asked about recommendations for calibrations and other changes to the operating procedures.

Instruments and their associated computers operate best in a nonhostile environment. Controlled laboratory conditions are recommended. Temperature should be maintained near 70°F and the relative humidity should be close to 50%. This will reduce the chances of dry air static charges damaging computer operations and prevent rust forming on internal unprotected instrument parts when the air is moist. The laboratory air should be continuously circulated to filter out dust. Fortunately, these conditions are also generally favorable for laboratory workers. However, these set points are for optimum instrument conditions. The lab may seem too hot and muggy in the winter and too cool and dry in the summer to those who are in the lab for short periods of time. Even minor adjustments in the room conditions for "creature comfort" may create a temporarily destabilized atmosphere that could result in accelerated rust or static attacks on the instruments. A constant voltage supply and a stable, preferably vibration-free instrument platform should be installation preconditions for research-grade color instruments. One should consider locating a color laboratory with research-grade instruments on the ground floor of multilevel buildings to reduce vibrations during measurements. Probably the worst user pitfall is forgetting to clean the air filters of ventilation fans for instrument light sources. Even in a "clean" filtered laboratory area, cool air must circulate through instrument filters to cool the source and to keep the optical system dust free. When air filters (which unfortunately are often not easily seen) clog with dust, the heat from the source may destroy its cooling fan and the instrument. Be aware of any burning smell and inspect by touch for overheating, especially when the instruments must be on for extended periods. Make filter cleaning a regular chore of lab attendants.

A. Sample Cleaning, Marking, and Orientation Mistakes

Some cleaning agents are harmful to people and to specimens. MSDS sheets and cleaning instructions should always be read and followed in the laboratory. For example, pure soft metal films without overcoats (front surface Al or Au reflectance standards) are samples that can be destroyed by any type of cleaning agent other than ionized air. A small amount of dust that sticks conspicuously to the surface of a specimen will affect the measurement far less than will permanent damage incurred from even gentle cleaning of such materials. Some samples can be irreparably damaged by improper labeling. Diamond or carbide scribes are commonly used to identify hard smooth surface specimens. These scratches will scatter light from the sample beam if they are in the measurement area. Sample markings from felt-tip pens will contribute unwanted errors to color measurements. Erasable labels can disappear and therefore should be avoided as a rule, especially if the specimens are to be measured after further testing. Filmed samples can be damaged by contact with the glue from their protective envelopes. Most spectrophotometer measurements are made at near-normal sample orientations to the measuring beam. If polarization is suspected in the sample or instrument, rotate the specimen until the photometric level is maximum when the wavelength is set at approximately 550 nm. When the specimen is measured, record its orientation along with the color data and instrument make. At near-normal incidence to the measuring beam, total luminous transmittance values will be the same when measured from either side. This is not the case for most filmed products or for samples with different surface textures on opposing surfaces. The side facing the measuring beam or integrating sphere sample port must be noted along with the spectral data.

VII. Limitations: Instruments and Techniques

All color measuring instruments should not calculate color using the same source-observer weight tables. Significant color measurement errors can arise when this axiom is not observed by instrument manufacturers (110). Integration tables have been recalculated since most instruments were programmed to make color calculations. The table values are now indicative of a triangular slit bias rather than a rectangular one. The changes in the tables, though not great, will result in calculations from most instruments in closer agreement now. This is because the energy distribution at most slits is triangular. Other instrument errors without values determined by the manufacturers may be photopic, unwanted UV, IR, cosine or surrounding field responses, nonlinearity and fatigue of the detector, readout errors, modulation and polarization, range change temperature coefficient, and other calibration errors. Equations for each of these errors exist (111) if the operator wishes to check out his or her instrument.

Color appearance of an object will vary with the color, size, and position of the light source but the color of the object is a fixed property of the material (112). Our eyes, like a measuring instrument's detector, receive different stimuli under different lighting conditions but our experience allows our perception of the object to remain essentially unchanged in common viewing circumstances. Measured color data are crudely manipulated to approximate one perception circumstance in a mathematical color space (113). These data give us little information on perceived color constancy that our minds create by distorting the viewed impression as light sources and backgrounds change. For example, many cloths measure a different color during the day [illuminant D65] and at night [illuminant A], but we are not usually aware of the change except by changes in contrast with the background. Daytime or nighttime viewing of an object invokes a mental experience correction to the observer's perception; similar corrections are not made by instruments (114). Perception of haze (forward scattered light) can vary from measured magnitudes depending on instrument geometry and the angle of scatter. The circumstances are so varied that a convention was adopted more than 40 years ago to describe it. Instruments have changed over the years and now only some are suitable for this measurement (115).

A detector should ideally have high sensitivity, high stability, uniform sensitivity, and linearity over several orders of magnitude in incidence flux. "No single detector has all four attributes" (8). The best detector in any particular instrument is therefore a compromise of these characteristics. Light sources vary considerably in spectral output.

A. Area, Size, and Background Contrast

Instruments are at a very great disadvantage to humans when total visual performance is expected. Our visual performance depends on luminance adaptation, object size and shape, observation time, subject areas, and background contrasts and color (116). These parameters determine visual thresholds and glare tolerances that cannot be measured on spectrophotometers. Shadows contribute significantly to our visual perception but cannot be measured spectrally (117). Instruments attempt to measure gloss as we see surface reflections from smooth surfaces but are not always successful because gloss is another complicated appearance attribute (118). Absolute color is measured against white calibration standards and relative color differences of specimens are measured as contrasts following white background measurements. Schönfelder's law (1933), however, would indicate that better color spacing is achieved between objects viewed against a black background than a white one (119).

Color differences of installed products can limit color measurements to on-site locations. Specimen size on-site is often a major problem of color measurement. Two adjacent windows in a building, for example, may not appear to be the same color after installation and therefore must be measured on-site to determine if the difference is great enough to warrant replacement. Two possible solutions to on-site color measurements are (1) to measure the reflectance of a white wall illuminated by a controlled light source through each of the two windows and then calculate the color difference, or (ii) to photograph both windows under the same conditions and analyze the color differences from the pictures in a color laboratory.

Some situations defy measurement because they cannot be adapted to known instrument geometries. For example, repeatable transmittance measurements cannot be made of runway strip lights through multiple-pane windows (at steep viewing angles) in airport towers. The measured color of a second-surface mirror, though properly measured, may fall within an agreed color tolerance but be unacceptable as installed because it was not measured on-site. This situation may occur, for example, when mirrors are installed in a bathroom and multiple reflections amplify the color differences of the mirror substrates under typical house yellow light bulb illumination at night. The observer's reflection may appear green and is understandably unacceptable. Stresses and strains of tempered car glass back windows create visible color patterns at times but the color cannot be measured by conventional means.

B. Measurement Data vs Perception Information

Background is a minor factor when retroreflecting materials are observed at night. Measurement of color of retroreflectors is still being perfected (86). One of the hardest problems to solve is how to measure the color of a CRT display because conventional methods of measurement do not correlate well with perceived color (120). There are many other aspects of perceived color that cannot be measured directly, such as the color of the sea, shading colors of objects at sundown, and colored patterns in a butterfly's wing. There is much work to be done to invent machines that can measure and relate the same object color in an intelligible way to everyone.

C. Dynamic Measurements Scan vs Reaction Times

Chemical reactions can be monitored effectively by measuring color changes. Physical reactions such as adding multiple pigments to a clear base can also be monitored by measuring color changes. These reactions cause dynamic selective absorption across the visible spectrum. Scanning spectrophotometers can be set at a sensitive wavelength that changes fastest during the reaction. It can be monitored as a function of either a decrease in transmittance or an increase in absorbance. Although reaction rates can be critically monitored in this way, overall color changes and positions of isosbestic points, for calculating the total concentration of an equilibrium mixture, cannot be determined. Diode array spectroscopy can determine both. This technology probably will replace the older technologies of colorimeters and scanning spectrophotometers because dynamic and static color tasks are performed by diode arrays without limitations or modifications to the equipment.

D. Atmospheric Conditions; Temperature, Humidity, and Purges

Spectrophotometers are made with many temperature- and humidity-sensitive components. A hostile environment for a research-grade instrument may be defined as one in which the operating conditions are outside $72^\circ \pm 5^\circ$F or $50 \pm 10\%$ relative humidity or a minimum particulate count in the air. Less expensive, sometimes portable models are usually less sensitive to atmospheric conditions because they are not made to operate at high-precision levels that are affected by these changes. Portable models can be subjected to very hostile environments in field use. These conditions should be noted when applicable. If near-IR spectral regions are to be monitored, either a purge system or a dual-beam instrument or both should be used to factor out atmospheric moisture in spectral data over wavelength regions 1350–1400 and 1800–1950 nm. Some specimens may suffer from desiccation during the purge and measurement intervals. They can change shape (curl or shrivel), altering the sample beam direction and creating erroneous data. Check the sample condition before and after a scan when a purge system is used.

E. Multiple Glazing Measurements and Calculations: Filmed

When a specimen is composed of a stack of objects, instrument beam geometry is more critical to obtaining good spectral data. Sometimes the individual layers of a specimen can be measured accurately and mathematically joined in the unit calculation. For example, each component of a double- or triple-pane window, filmed or unfilmed, can be measured by reflectance and transmittance then combined by equations to determine heat gain levels of sunlight through the windows (121). Multiple ply glass laminates for bullet-resisting applications may be too thick to measure directly in an instrument's sample compartment but the individual ply spectral properties can be combined mathematically at each wavelength and then integrated to determine the specimen's color or total visible transmission using the equation: $\%T_{\text{sum}} = \%Tm \cdot (T_1/Tm) \cdot (T_2/Tm) \cdot (T_3/Tm) \cdots$.

Instrument limitations suggest that representative specimens be sent from a prospective instrument buyer to instrument manufacturers for analysis. It is important to know what and how much change can be expected from data obtained from old and new instruments.

VIII. Advances Needed

The foregoing discussion indicates many incompatibilities between human and instrument color evaluation. These situations give direction to the advances that are needed in this field. Improvements are coming and at an ever faster rate as we learn more about how we evaluate color and solar-heated situations. Some of the technological improvements the color community is striving for are the following:

- Quicker and easier networking of spectrometers and computers within company laboratories. This will cut the learning curves of operators and should improve communications between the technical and management areas of research organizations.
- Increased macroprogramming allowing tailored instrument routines to build large sample databases around less fundamental, higher level color parameters. Industrial custom color-tolerance spaces and multi-transformed spectral data calculations to improve visualization of R&D trends are developing rapidly.
- As miniaturization and precision continue to take center stage in instrument development, color measurements are taking place more in the field and less in specialized laboratories for color evaluations. In the past colorists concentrated on problems of color measurements that fell within the restrictions of laboratory instruments. Most of our color evaluations do not fit into laboratory situations, and field measurements, if accurate, will correspond closer to our color perceptions on-site.
- Faster instruments (diode arrays) can evaluate dynamic color changes in snapshot fashion that approaches virtual reality. Such measuring techniques will transform color evaluation from simple static situations to a more natural macrounderstanding of color and other appearance factors that are simultaneously involved in our perceptions of objects.
- It has always been harder to define color of most observations scientifically than to translate color descriptions into other human languages. Both procedures are improving rapidly. As primary concepts of color become incorporated into macroconcepts that approach our natural perception of the world, communication improves because more words are added to languages to describe these ideas. Laws and standards are written and enforced by mutual consent of many nations at lower operating levels following strict traceability procedures (122). ISO 9000, for example, sets out a detailed comprehensive methodology for all operations of industry. Other international standards organizations continue to write new procedures on new

subjects with the cooperation of more participating nations than in the past. These standards are, in turn, the basis of more standards on specific topics such as color and solar measurements.

- Physical standards certified by national standards laboratories should be specimens of industrial products for those industries. Generic standard reference materials suffice to certify an instrument for general use but these standards do not necessarily suffice to certify values obtained from products that are dissimilar from those standards. Representative industrial specimens should be submitted and measured as transfer standards by national testing laboratories. When they are returned to a company, they should be recognized and used as SRMs. This reduces the traceability path and thus increases the credibility of measurements on similar materials.
- Ongoing downsizing along with higher precision and simplification of measurements are instrument development objectives. The overall customer objective is to be able to transfer expertise, encapsulated in convenient and simple equipment, out into the field where color problems need analyzing on-site. Instruments will be developed that will be better suited to measure more types of colored products with a higher degree of precision and reproducibility.
- Development of color detectors and sources is satisfactory. Beyond the visible range lamps and sensors may not be adequate. They do not have wider responses to rapidly measure and analyze solar properties of materials as efficiently as color instruments measure color. More development of solar measuring instruments is needed for UV filtration potentials of glass products, UV degradation resistance of other products, and overall solar comfort heat indexes of glass and plastic glazing materials. These data are important to many customers because they impact activities such as plant growing in hot houses, comfortable air conditioning in buildings and cars, and many aerospace applications.
- As computers become faster, more computer time will be allocated for error-checking routines that monitor instrument and sampling procedures; for example, self-calibration using internal standards, monitoring of sample alignment in the sample beam, warnings of parts needing replacement, and standby status instrument programs that will allow instruments to remain on for stable operation when needed but otherwise unattended.
- Color formulas and computer algorithms will continue to be refined as instrument technologies achieve accommodating levels of development. Multifaceted color appearance measurement tasks in particular need to be developed for specific products.

References

1. IESNA Color Committee, *Color and Illumination*, DG-1-90 ed., p. 44. New York (1990).
2. General Electric Company, L. B. D., *Light and Color*, TP-119 ed., p. 31. Author, Cleveland, OH (1974).
3. Minolta Camera Co., Ltd., *Precise Color Communication*, Vol. PCC 001E-E4. Author, Osaka, Japan (1990).
4. Thornton, B., Comprehending color, in *Lighting Design Appl.*, 3 (1992).
5. International Standards Organization and Commission International de L'Eclairage, *CIE Standard Colorimetric Observer*. Author (1991).
6. International Standards Organization and Commission International de L'Eclairage, *CIE Standard Colorimetric Illuminants*, ISO 10527:1991(E). Author (1991).
7. Hale, W. N., Jr., Visual assessment of appearance, in *ASTM Standard.*, 46–49 (1987).
8. Photometry of Light Source Subcommittee of the IES Testing Procedures Committee, IES practical guide to colorimetry of light sources, *J. Illuminating Eng. Soc.* **18**(2), 122–127 (1989).
9. Thornton, B., Comprehending color, in *Lighting Design Appl.*, 12 (1993a).
10. Caivano, J. L., Cesia: A system of visual signs complementing color, *Color Res. Appl.* **16**(4), 258–280 (1991).
11. Jacobson, E., *Basic Color*, 1st ed., pp. 123–178. Poole Bros., Chicago (1948).
12. Allen, E., Some new advances in the study of metamerism, in *Color Eng.*, 35–51 (1969).
13. Nimeroff, I., *Colorimetry*. U.S. Department of Commerce, Washington, DC (1968).
14. Nielsen, O., The importance of spectral match in photometry, in *Lighting Design Appl.*, 38–42 (1987).
15. Birren, F., Color and Human response, *Color Res. Appl.* **8**(2), 75–81 (1983).
16. Kurth, R. H., Colors in glass, in *Color Eng.*, 37–41 (1970).
17. Billmeyer, F. W., Jr., and H. K. Hammond, III, ASTM standards on color-difference measurements, *Color Res. Appl.* **15**(4), 206–209 (1990).
18. Robertson, A. R., Colour Order systems: An introductory review, *Color Res. Appl.* **9**(4), 234–240 (1984).
19. NBS, *Contributions to Color Science*, NBO Standards, (Ed.). U.S. Government Printing Office; Washington, DC (1979).
20. Hardy, A. C., *Handbook of Colorimetry*, 5th ed. (1970), T.C.M. Laboratory, (Ed.), Technology Press, Cambridge, MA (1936).
21. USNC, *United States National Committee of CIE*. Commission International de L'Eclairage (1989).
22. CIE, *Colorimetry*, Publ. No. 15 (E-1.3.1). Author (1971).
23. CIE, *Colorimetry, Second Edition*, Publ. No. 15.2. Author (1986).
24. CIE, *International Lighting Vocabulary*, Publ. No. 17. Author (1970).
25. ASTM, *E-284 Standard Terminology of Appearance*, 1992 ed., Appearance of Materials, E-12, Vol. 06.01, *Annual Book of Standards*. Author, Philadelphia (1992).
26. Minolta Camera Co., Ltd., *Precise Color Communication*, J303(E)-A1 ed. Author, Osaka, Japan (1993)
27. X-Rite, Inc., *Understanding Color Communication*, 8-90 ed., Vol. 918-801, p. 19. Author, Grandville, MI (1990).
28. Minolta Camera Co., Ltd., *Chroma Meters*. Author, Osaka, Japan (1991).
29. Oriel Corp., *ORIEL Corp. Light Sources, Monochromators & Spectrographs, Detectors & Detection Systems, Fiber Optics* (the Oriel three volume catalog) (E. Esposito and N. Fernandes, Eds.), Vol. 2. Author, Stratford, CT (1994).

30. Billmeyer, F. W., Jr., Instrumental color measurement, in *ASTM Standard. News*, 32–34 (1987).
31. Gardner, B. Y. K. (Ed.), *Spectrogard II Color System*, Version 1.2, Vol. OPMN0002. Silver Springs, MD (1990) published by B. Y. K. Gardner.
32. Nassau, K., *The Physics and Chemistry of Color*. (Wiley series in pure and applied optics). Wiley-Interscience, New York (1983).
33. Varian Instrument Division, *Optimum Parameters for Spectrophotometry* (1973).
34. Melles Griot, Optics Guide 3, in *Catalogue* (1985).
35. Laurin, T. C., *et al.*, The photonics design and applications handbook, in *The International Handbook and Reference to Photonics* (T. C. Laurin, Ed.). Laurin, Pittsfield, MA (1992).
36. Klinger Scientific Corp., *High Accuracy Optical Components* (1978).
37. Perkin–Elmer, *Lambda 19 UV*/VIS/NIR Spectrometer (1992).
38. Machler, M., and Schlemmer, H., MCS multichannel spectrometer, *Zeiss Information* **30**(100E), 16–19 (1989).
39. Monolight, *Optical Spectrum Analyser*, Rees Instruments, Ltd. (1992).
40. Photo Research, *PR-1980A Pritchard Photometer*, Kollmorgen Div (1991).
41. Rennilson, J. J., and Hale, W. N. Jr. (Eds.), *Review and Evaluation of Appearance: Methods and Techniques*, ASTM Special Technical Publication. ASTM, Philadelphia (1984).
42. HunterLab, *Analyzing Appearance by Measurements*, Hunter Associates Laboratory, Inc. (~1970).
43. Hutchings, J., Factors affecting total appearance, in *ISCC News*, 13 (1986).
44. Seve, R., Problems connected with the concept of gloss, *Color Res. Appl.* **18**(4), 241–252 (1993).
45. Danger, E. P., *The Color Handbook: How to Use Color in Commerce and Industry*. Gower, Hamshire, UK (1987).
46. Lipetz, L. E., Universal human visual pigment curves from psychophysical data, *Color Res. Appl.* **13**(5), 276–288 (1988).
47. ASTM, *ASTM Standards on Color and Appearance Measurement*, 3rd ed., A406. Author, Philadelphia (1991).
48. Wyszecki, G., and Stiles, W. S., *Color Science: 2nd ed. Concepts and Methods, Quantitive Data and Formulae*. Wiley-Interscience, New York (1982).
49. Goebel, D. G., Generalized integrating-sphere theory, *Appl. Optics* **6**(1), 125–128 (1967).
50. Keegan, H. J., Schleter, J. C. and Judd, D. B., Glass filters for checking performance of spectrophotometer-integrator systems of color measurement, *J. Res. Nat. Bureau Standards* **66A**(3), 203–221 (1962).
51. Eckerle, K. L., and Venable, W. H., Jr., 1976 remeasurement of NBS spectrophotometer-integrator filters, *Color Res. Appl.* **2**(3), 137–141 (1977).
52. Weidner, V. R., and Hsia, J. J., Reflection properties of pressed polytetrafluoroethylene powder, *J. Optical Soc. Am.* **71**(7), 856–861 (1981).
53. Weidner, V. R., Hsia, J. J., and Eckerle, K. L., Exploratory research in reflectance and fluorescence standards at the National Bureau of Standards, *Optics News* **12**(11), 18–20 (1986).
54. ASTM, *E1341 Standard Practice for Obtaining Spectroradiometric Data from Radiant Sources for Colorimetry*, Appearance of Materials, E-12, Vol. 06.01, p. 5, *Annual Book of Standards*. Author, Philadelphia (1991).
55. ASTM, *E805 Standard Practice for Identification of Instrumental Methods of Color of Color-Difference Measurement of Materials*, 1994 ed.; Appearance of Materials, E-12, Vol. 06.01, *Annual Book of Standards*. Author, Philadelphia (1987).

56. Diano Corp., *Match-Scan Wavelength Calibration*, EI000221 ed. Author, Woburn, MA (1972).
57. Fairchild, M. D., The CIE 1931 standard colorimetric observer: Mandatory retirement at age 65?, *Color Res. Appl.* **18**(2), 129–133 (1993).
58. Strong, J., *et al.*, *Procedures in Experimental Physics*, pp. 375–376. Prentice Hall, Englewood Cliffs; NJ (1938).
59. ASTM, *E306 Standard Method for Absolute Calibration of Reflectance Standards*, 1994 ed.; Appearance of Materials, E-12, discontinued 1989. Author, Philadelphia (1976).
60. Burke, R. W., and Mavrodineanu, R., *Accuracy in Analytical Spectrophotometry*. NBS (1983).
61. ASTM, *1994 Annual Book of ASTM Standards Catalogue*, p. 45. Author, Philadelphia (1994).
62. ASTM, ASTM Standards, in *Catalogue*, p. 50 (1993).
63. ASTM, *1994 Annual Book of ASTM Standards*; Paint-Tests and Analysis of Physical Optical Properties; Appearance; Durability of Nonmetalic Materials, E-12, Vol. 06.01. Author, Philadelphia (1994).
64. Oriel Corp., *Optics & Filters* (the Oriel three volume catalog) (E. Esposito and N. Fernandes, Eds.), Vol. 3. Author, Stratford, CT (1990).
65. Eckerle, K. L., *et al.*, International intercomparison of regular transmittance scales, *Metrologia* **27**, 33–38 (1990).
66. CIE, *Principles of Light Measurements*, Publ. No. 18 (E-1.2). Author, (1970).
67. ASTM, International directory of testing laboratories, in *Standardization News*, 80 (1993).
68. ANSI, *American National Standard for Safety Glazing Materials for Glazing Motor Vehicles Operating on Land Highways*, Safety Code Z26.1–1983. (1983).
69. Alexander, W. B., Absorption filter selection: How to, in *Optical Spectra*, 29–33 (1975).
70. Grunler, B., and Treffs, H., Influence of the radius of curvature of spherical glass samples on IR reflectance spectra, *Silikattechnik* **40**(1), 7–9 (1989).
71. Johnston, R. M., *et al.*, Measurement of the transmittance and reflectance of translucent, in *Coloring of Plastics VII*. Cherry Hill, NJ, Proceedings SPE (Society of Plastic Surgeons) (1973).
72. ASTM, *E1392 Standard Practice for Angle Resolved Optical Scatter Measurements on Specular or Diffuse Surfaces*; Appearance of Materials, E-12, Vol. 06.01, *Annual Book of Standards*. Author, Philadelphia (1990).
73. Fairchild, M. D., *et al.*, Absolute reflectance factor calibration for goniospectrophotometry, *Color Res. Appl.* **15**(6), 311–320 (1990).
74. Griffiths, P. R., and Haseth, J. A. D., *Fourier Transform Infrared Spectrometry* (P. J. Elving and J. D. Winfordner, Eds.), Vol. 83. Wiley, New York (1986).
75. Barr Associates, I., *Some General Considerations for Setting Specifications* (1990).
76. ASTM, *E991 Standard Practice for Color Measurement of Fluorescent Specimens*; Appearance of Materials, E-12, Vol. 06.01, p. 3, *Annual Book of Standards*. Author, Philadelphia (1990).
77. Burns, M., You can check color appearance by spectrophotometer, in *Industrial Research and Development*, 126–130 (1980).
78. Alman, D. H., Breton, M. E., and Barbour, J., New results on the brightness matching of heterochromatic stimuli, *J. Illuminating Eng. Soc.* **12**(4), 268–274 (1983).
79. Bamford, C. R., Colour generation and control in glass, in *Glass Science and Technology*, 2, Elsevier, New York (1977).
80. Venable, W. H., Take a moment to specify appearance measurements, in *Appearance*. Williamsburg, VA, ISCC (Inter-Society Color Council) (1987).
81. ASTM, *E1164 Standard Practice for Obtaining Spectrophotometric Data for Object-Color*

Evaluation; Appearance of Materials, E-12, Vol. 06.01, *Annual Book of Standards.* Author, Philadelphia (1991).

82. Boguth, W., and Schweiz, R., Microscopic colour measurement on mineral sections and precious stones, *Zeiss Information* **30**(100E), 12–15 (1989).
83. ASTM, *E430 Standard Test Methods for Measurement of Gloss of High-Gloss Surfaces by Goniophotometry*; Appearance of Materials, E-12, Vol. 06.01, *Annual Book of Standards.* Author, Philadelphia (1991).
84. Clarke, F. J. J., and Compton, J. A., Correction methods for integrating-sphere measurement of hemispherical reflectance, *Color Res. Appl.* **11**(4) (1986).
85. ASTM, *E810 Standard Test Method for Coefficient of Retroreflection of Retroreflective Sheeting*; Appearance of Materials, E-12, Vol. 06.01, *Annual Book of Standards.* Author, Philadelphia (1991).
86. Rennilson, J. J., Retroreflection, in *ASTM Standardization News*, 42–44 (1987).
87. Springsteen, A., *A Guide to Reflectance Spectroscopy.* Vol. Tech. Guide, p. 41. Labsphere, North Sutton, NH (1992).
88. X-Rite Inc., *Understanding Color Tolerancing*, Vol. L10-024. Author, Grandville, MI (1992).
89. Spears, G. R., High speed photometry and colorimetry of lamps & luminaires, in *Lighting Design Appl.*, 38–42 (1987).
90. ASTM, *E308 Standard Test Method for Computing the Colors of Objects by Using the CIE System*; Appearance of Materials, E-12, Vol. 06.01, *Annual Book of Standards.* Author, Philadelphia (1990).
91. Fairman, H. S., The calculation of weight factors for tristimulus integration, *Color Res. Appl.* **10**(4), 199–203 (1985).
92. Liu, Y., Berns, R. S., and Shu, Y., Optimization algorithm for designing colored glass filters to simulate CIE Illuminant D65, *Color Res. Appl.* **16**(2), 89–96 (1991).
93. David, H. A., CIE technical committee 1–29, industrial color-difference evaluation progress report, *Color Res. Appl.* **18**(2), 137–139 (1993).
94. Billmeyer, F. W., Jr., Redetermination of CIE standard source C, *Color Res. Appl.* **8**(2), 90–96 (1983).
95. Thornton, B., Comprehending color, in *Lighting Design Appl.*, 8 (1993b).
96. Miller, K., Call to colors, in *Photonics Spectra*, 75–82 (1985).
97. Jones, O. C., and Preston, J. S., *Photometric Standards and the Unit of Light.* National Physical Laboratory (1969).
98. Davis, R., Gibson, K. S., and Haupt, G. W., Spectral energy distribution of the International Commission of Illumination Light Sources A, B, and C, *J. Res. Natl. Bureau Standards* **50**(1), 31–37 (1953).
99. Foster, W. H., Jr., *et al.*, Weights for calculation of tristimulus values from sixteen reflectance values, in *Color Eng.*, 35–47 (1970).
100. ASTM, *D4356 Standard Practice for Establishing Consistant Test Method Tolerances*, Quality and Statistics, E-11, Vol. 14.02, *Annual Book of Standards.* Author, Philadelphia (1989).
101. Thornton, B., Comprehending color, in *Lighting Design Appl.* (1993c).
102. ASTM, *D4086 Standard Practice for Visual Evaluation of Metamerism*; Appearance of Materials, E-12, Vol. 06.01, *Annual Book of Standards.* Author, Philadelphia (1986).
103. Billmeyer, F. W., Jr., Color forum, *Color Res. Appl.* **16**(5), 342–343 (1991).
104. Thornton, B., Comprehending color, in *Lighting Design Appl.*, 10 (1994).
105. Fairman, H. S., Recommended terminology for matrix R and metamerism, *Color Res. Appl.* **16**(5), 337–341 (1991).
106. Boynton, R. M., Nagy, A. L., and Olson, C. X., A flaw in equations for predicting chro-

matic differences, *Color Res. Appl.* **8**(2) (1983).

107. Worthey, J. A., Light source area, shading, and glare, *J. Illuminating Eng. Soc.* **20**(2), 29–36 (1991).
108. Grahm, L. A., How to solve your color problem—DEFINE IT!, in *Color Eng.*, 30–33 (1966).
109. Berns, R. S., Empirical modeling of systematic spectrophotometric errors, *Color Res. Appl.* **13**(4), 243–256 (1988).
110. Venable, W. H., Accurate tristimulus values from spectral data, *Color Res. Appl.* **14**(5), 260–267 (1989).
111. Light Measurement Talk (LMT), *How to Select Photometer*, Application note AN5-688 (1988).
112. Corth, R., Human visual perception, in *Lighting Design Appl.*, 20–24 (1987).
113. Berns, R. S., and Rose, B. J., Color reproduction: Good news and bad news, in *ISCC News*, 5 (1991).
114. Phillips, E., Legal reflections, in *Lighting Design Appl.*, 10–11 (1993).
115. Billmeyer, F. W., Jr., and Chen, Y., On the measurement of haze, *Color Res. Appl.* **10**(4), 219–224 (1985).
116. Highway Signs Subcommittee of the Roadway Lighting Committee, Recommended practice for roadway sign lighting, *J. Illuminating Eng. Soc.* **12**(3), 141–145 (1983).
117. Erhardt, L., Views on the visual environment, in *Lighting Design Appl.*, 16–17 (1993).
118. Hammond, H. K., III, Gloss, in *ASTM Standardization News*, 36–40 (1987).
119. Judd, D. B., Ideal color space, in *Color Eng.*, 37–53 (1970).
120. Taylor, J. M., Just another pretty picture? Hardly!, in *ASTM Standardization News*, 50–52 (1987).
121. ISO, *Glass in Building—Determination of Light Transmittance, Solar Direct Transmittance, Total Solar Energy Transmittance and Ultraviolet Transmittance, and Related Glazing Factors.* International Standards Organization, Switzerland (1990).
122. Billmeyer, F. W., Jr., A national standardization program for spectrophotometry, *Color Res. Appl.* **8**(3), 182–185 (1983).

OPTICAL SPECTROSCOPY OF NEW MATERIALS

SUSAN WHITE
Thermal Protection Materials and Systems
NASA Ames Research Center

I. Background

Composites are formed from two or more insoluble materials to produce a material having superior mechanical, thermal, or other required properties for the intended use. The classic composite consists of a filler in a ductile matrix. The filler or structural component consists of fibers, particulates, or, less frequently, flakes. The filler strengthens the composite in tension, impedes dislocations in order to delay the onset of failure through cracking, or enhances the electrical conductivity, hardness, or another desirable property. The matrix holds the filler components in place and typically contributes strength in compression, transfers mechanical loads of the filler phase, transfers heat by conduction, or provides environmental protection from oxidation or moisture (Schwartz, 1984; Epstein, 1969).

A remarkable variety of composites have been produced by nature, operating through evolution or geologic processes, and manufactured composites in the form of straw-laden bricks have been used since antiquity. However, the explosive growth and development of composites is a modern phenomenon. Composites are currently used for a rapidly expanding number of diverse applications including aircraft structures, rocket nozzles, metal matrix heat sinks for integrating circuits, spacecraft and satellite structural materials, ablative surfaces, lightweight automobile components, cutting tools, optical-grade mirrors, and sports equipment such as skis, tennis rackets, and bicycles.

Composites are categorized as either homogenous or gradient materials, which are further divided into isotropic or anisotropic materials. In a

ISBN 0-12-764070-3/$25.00.

homogeneous composite, the fibers or particles in a composite are evenly distributed throughout the material. In a gradient material, the filler component is unevenly distributed in the matrix. Laminates are the most widely used gradient composites, in which the laminar or layered structure is used to enhance the strength on the upper and lower surface of a load-bearing part. The laminae themselves can be composite materials. In isotropic composites, particulates or randomly oriented fibers or flakes are used as filler. In an anisotropic composite, preferentially oriented fibers are the filler of choice for reinforcement in the critical direction(s). Fibers can be oriented parallel (1D), or in a parallel–perpendicular orientation (2D), either using a woven fabric or by layups in alternating parallel directions. A 3D weave of fibers produces a nonrandom, highly reinforced material with strength in the three principal axes. In fiber–fiber composites, one fiber generally contributes mechanical strength, and the second fiber contributes refractory or other desirable properties to the end product.

Composites range from dispersion-hardened alloys to advanced fiber-reinforced composites. Advanced fiber-reinforced composites are high-performance materials with very high strength-to-weight ratios. Fibers are used either in chopped form or in filament-wound composites. Fiber-reinforced composites typically use glass, graphite, boron, metal, or aramid fibers or filaments in a matrix of polymer, ceramic, or metal. The organic fibers are generally lightweight, flexible, have elastic, but have a limited temperature range. The inorganic fibers are strong, rigid, brittle, and function well at elevated temperatures. Whiskers are used to maximize the strength-to-weight ratio of a fiber-reinforced composite. Whiskers are extremely strong, ductile single-crystal metal or oxide fibers, having diameters in the range of 0.1–2 μm, which are grown nearly defect-free. The strength of a crystal without defects can approach the theoretical strength limit of the constituent material because defects precipitate failure of a material under tension by functioning as initiating cracks. Whiskers respond elastically to extremely high levels of stress before fracturing or yielding, so a matrix can transfer very high loads when filled with whiskers.

In high-strength materials, it is critical to obtain good bonding between the fibers and the matrix and to minimize the presence of voids, both during production and when the composites are under loads. A coupling agent or an interphase is used to mechanically or chemically enhance the bond between the fibers and matrix.

II. Current State of the Art

Optical measurements on composite and porous materials present unique challenges because the presence of multiple interfaces between the different phases or component materials strongly affects the transport of the optical signal of the radiation beam through the sample. Radiation is scattered at every interface between any two materials having different refractive indices. This changes the direction of the incident radiation beam and increases the overall reflectance of a material.

The hemispherical total and spectral reflectance, emittance, transmittance, and absorbance are measured to calculate energy balances for high-temperature or high-heat flux applications, including ablative materials, reusable reentry heat shields, and rocket nozzles. To measure the optical properties for energy balances, integrating spheres are widely used. For this purpose it is necessary to capture both the diffuse or scattered component and the specular or normal component of the reflected or transmitted energy from a material (Edwards *et al.*, 1969; Messerschmidt, 1985; Clarke and Larkin, 1985; Duclos, 1985; Roos *et al.*, 1988). The specular component is calculated by taking the difference between two measurements, in which the specular component is either included or removed by using either a diffuse reflector or an absorber at the specular reflectance angle in the integrating sphere. Roos *et al.* (1988) formulated a method to analyze the reflectance measurements made in integrating spheres, to select the appropriate reference material for the ultraviolet/visible (UV/VIS) range, and to solve the anomalies associated with non-Lambertian samples; that is, samples that do not reflect light in the ideal pattern of a specular component and a pure diffuse component. Clarke and Larkin (1988) developed a reflectometer/transmissometer device to measure the diffuse and normal components of the IR transmittance or reflectance using hemispherical reflectors in different configurations.

Rustagi and Bhawalkar (1990) reviewed the optical phenomena associated with the presence of submicron-sized particles. Flytzanis *et al.* (1986) and Niklasson (1989) discussed the optical nonlinearities and optical and electrical properties of composites.

Potter *et al.* (1992) reviewed the literature on the microstructure of semiconductors in a composite structure with an insulating matrix. These composites offer insight into the electronic band structure and the optical properties of the materials. Simmons *et al.* (1991) discussed the electronic band structure of composites formed from semiconductors and glass components.

III. Current Applications

Optical spectroscopy is used to characterize the following properties of composites (NMR, TEM, SEM, Ramen, and Mossbauer spectroscopy are also widely used to analyze composites):

- Measure the thermal radiative properties of spacecraft thermal protection materials: Covington and Sawko, Stewart, and White have measured the reflectance and transmittance for a variety of spacecraft thermal protection materials and calculated the appropriate blackbody-weighted emittance and absorbance for the predicted conditions. In addition, spectroscopy has been used to detect changes in spacecraft surface materials resulting from long-term exposure to the low earth orbit environment.
- Investigate the burn rate of rocket propellants containing metal complexes: Stoner *et al.* (1992) investigated the burn rate of modified rocket propellants containing metal complexes using Fourier transform IR (FT-IR) spectroscopy by heating the composite at a rate of 100–250°C/sec.
- Evaluate the bonding between the matrix and the fibers: Nishijima *et al.* (1984) correlated the transmittance of a laser beam through an epoxy matrix composite with the strain amplitude dependence, which was in turn used to estimate the fracture of the interface between the components of the composite.
- Maintain quality control and to monitor polymerization, prepregs, and precursors: A prepreg is a moldable paper, cloth, or a mat impregnated with a partly cured resin. It can be laid up in the final form and cured by heating under pressure. Lauver and Vannucci (1979) spectroscopically monitored the chemical precursors of a resin and correlated the aging of the ester/acid solution and the ester impurities with the mechanical properties of graphite fiber-reinforced composites periodically made from the resins. May *et al.* (1976) proposed IR techniques to evaluate the chemical properties of epoxy prepregs. Baraton *et al.* (1989) used FT-IR for surface characterization of an aerogel precursor to cordierite. Ashley *et al.* (1992) investigated aerogel composites for radio luminescent light/power sources.
- Measure surface contamination: Blair and Ward (1991) demonstrated their method for measuring surface contamination by using FT-IR combined with microscopy to locate a single cellulose acetate fiber on a gold-coated surface. In their method, the FT-IR spectrometer is coupled with a motorized microscope stage to locate and identify surface contaminants as small as 10 μm in diameter. It is not necessary to know the optimum characteristic absorption frequency or the chemical composition of the contaminant in advance to use this method, although it is advantageous. Blair and Ward used a Nicolet 7199 FT-IR coupled to a Spectra Tech infrared microscope, with a motorized stage moving in 1-μm increments. Pearson (1990) used IR spectroscopy to measure reflectance in order to detect and identify

bonding surface contamination. Using polarized oblique incident light maximized the sensitivity of this method.

- Characterize the void formation: The formation and propagation of voids in composite materials must be minimized because, like cracks in a bulk material, they lead to mechanical failure of the part in use. Manson and Sefaris (1989) developed a void characterization technique using FT-IR and other techniques to investigate the initiation and propagation of voids during processing of advanced semicrystalline thermoplastic composites.
- Characterize the surface of the fiber component or a metallic inclusion: Sellitti *et al.* (1988) used FT-IR attenuated total reflection spectroscopy to characterize the surface of graphitized carbon fibers and identify the functional groups, and reactions between the fibers and an epoxy resin coating were investigated. Devaty (1989) evaluated the IR absorption of composites with metallic particles.
- Characterize the orientations and distributions of the fibers: An opaque or highly reflective matrix can totally mask the presence of the filler phase. However, total internal reflection in an optically transparent fiber phase can produce an optical waveguide effect, resulting in a finite transmittance even through an opaque matrix. Rayment and Majumdar (1978) exploited this effect to estimate the angular distribution of glass fibers in concrete and plaster matrices and to detect anisotropy and local variations in fiber orientation. A light guide/photodiode mechanically scanned the back surface of the sample to quantify the transmitted signal. The authors propose this method to replace the painstaking photographic cross-section method to estimate fiber distribution.
- Measure the properties of optical devices: Transparent or reflective composites that are used directly in optical applications such as optical waveguides and mirrors require measurement of the relevant optical properties, as described by Helms *et al.* (1989) and by Mohn and Vukobratovich (1988), and evaluation of the optical uniformity of the material, the transmission, index of refraction, and optical surface stability.
- Evaluate the state of internal stress in plastics: This application was discussed by Epstein (1969).

IV. Pitfalls of Techniques for Basic Users

Commercially available spectrophotometers are designed to be user-friendly, but pitfalls still await the user; thus, caution is required to perform accurate measurements.

In general, the optical beam is scattered at each interface between the two phases of the material, the matrix and the filler, just as water droplets in the air scatter visible light to give clouds a white or gray appearance and to produce rainbows. Direct measurements made on thin slab of an optically thin composite material of the apparent absorption include both the absorption of the filler and the matrix phases, as well as the outscattering of the beam at the interfaces. The apparent absorption is greater than the sum of the absorption of each component material. In the worst case, the surrounding matrix can be opaque or highly reflective in the wavelength range of interest and will totally mask the presence of the filler phase to spectroscopic measurements.

Spurious effects can result from the composite structure itself—in particular when it is highly structured or anisotropic. Like woven materials, composite materials frequently contain fibers with a strongly preferred orientation. Preferentially oriented fibers have the potential to cause two different problems: polarization and direct illumination of the detector. Polarization by the sample interacts with partially polarized light from the instrument mirrors and gratings. Directional scattering or reflection by the material can directly illuminate the detector in an integrating sphere or increase the apparent absorption in a transmittance configuration. Anomalies can result from these effects, e.g., an apparent spectral reflectance of greater than 100% or an error as high as 50% of the reflectance. This alerts the user abruptly to the existence of a problem. The more subtle problem occurs when the measurement is physically reasonable but still incorrect.

Some degree of partial polarization is introduced into the spectrophotometer by the source, mirrors, lenses, prisms, gratings, and the detector. This partial polarization is highly wavelength dependent. Robertson (1972) analyzed the impact of polarization on reflectance measurements. In order to measure the transmission of polarized light through anisotropic samples without using the labor-intensive and signal-reducing process of inserting polarizers into the path of the incident beam, Zwinkels and Dodd (1989) developed a method to exploit the known polarization of the incident beam in the Perkin–Elmer Lambda-9 spectrophotometer in the range from 250 to 950 nm. Anderson (1990) performed a mathematical analysis of the polarization effects in integrating spheres, of the sphere walls, and the effects of the presence of a nonideal diffuse target. To detect polarization effects, the experimenter can rotate any preferentially oriented specimen in the plane of the sample stage through 30, 45, or 90°, depending on the fiber or flake orientation, and compare the spectral measurements against the original measurements. Inserting polarizers into the path of the incident beam to polarize the incoming light can compensate for a given orientation of a structured sample, but this method is labor-intensive, particularly if the user accounts for the partial polarization of the beam being wavelength dependent, and polarizers will reduce the signal throughput for the measurements.

Spectrophotometers are designed for and calibrated with ideal diffuse and specular standards. The material sample is assumed to reflect or transmit radiation with a perfectly diffuse and a perfectly specular component. In an integrating sphere, when preferential scattering or reflectance occurs in a solid angle that includes the detector, the direct illumination of the detector will lead to spurious results. Hanssen and Snail (1987) used a ruled gold surface to quantify the importance of limiting the detector's field of view. To prevent the direct illumination of the detector, integrating spheres can be equipped with internal baffles. Roos *et al.* (1988) formulated a method for measuring the reflectance of preferentially oriented or structured samples, such as woven fabrics, brushed copper, or rolled aluminum, to compensate for the phenomenon of preferential reflection of the incident beam. Their measurements show, for example, that a rolled metal surface reflects light in a narrow disk rather than in the ideal specular or diffuse components.

Analytical work has been published on the optical behavior of composite materials. Edwards *et al.* (1962) analyzed the spectral reflectance and transmittance of imperfectly diffuse samples in an integrating sphere and provided calculations of the measurement error resulting from the nonideal material. Scattering in composites has been predicted in recent works by Varadan (1991), White *et al.*, and Lee *et al.* Individually, each fiber scatters the incident radiation in a forward-facing cone centered around the incident direction.

V. Limitations of Current Instrumentation and Techniques

Because advanced composites are used under extreme environmental conditions, such as at high-temperature or high-heat fluxes, room temperature measurements may be inadequate to predict the performance under operating conditions. Heated and cooled stages and sample compartments with controlled atmospheres are commercially available. The optical bench mirrors must be protected from inadvertent deposition of chemical species from the specimen by sample compartment windows. The windows themselves may become contaminated. An off-the-shelf integrating sphere with this capability is not available at the current time. Hanssen and Snail (1987) have developed a diffuse gold integrating sphere for IR spectroscopy in which the sample can be heated to 250°C. Off-the-shelf bidirectional reflectance accessories with controlled atmosphere and a sample stage that can be heated from cryogenic temperatures to 500°C are commercially available.

Problems independent of the sample being a composite material are discussed in the appropriate sections in this book and in the manufacturer's literature.

VI. Advances Needed in the Field

Advances in spectroscopy are needed to evaluate the interface between the matrix and the fiber, plate, or particulate filler in composite materials and to improve nondestructive testing and process monitoring.

Currently, large margins of safety and expensive individual component testing are required for high-performance composites (Schwartz, 1984) because simple, low-cost nondestructive testing methods are insufficiently developed. Better methods of nondestructive testing are needed both for quality control during production of composites and for periodic inspection of composites that are frequently used under harsh operating conditions. Burleigh (1989) has summarized and developed nondestructive methods for testing composites using thermography, which could be extended for other spectroscopic methods. Further improvements in process monitoring are needed to evaluate the state of cure of the resin matrix, to quantify the moisture absorption of epoxy–resin composites that sharply degrade its high-temperature properties, and to evaluate the orientation of fibers in fiber composites.

The bonding reaction between the different materials in a composite is particularly critical for structural composites, but it remains little understood. It depends on wettability, chemical compatibility, adsorption characteristics, as well as differential stresses. Coupling agents, silanes, and chrome complexes are used to increase bond strength and prevent water adsorption. This is a fertile area for optical spectroscopy, which can be explored using Blair and Ward's (1991) technique coupling microscopic analysis with a moving stage.

References

Best References in the Field

Clarke, J. J., and Larkin, J. A. *High Temperatures—High Pressures* **17**, 89 (1985).

Potter, B. G., Clausen, E. M., Simmons, C. J., and Simmons, J. H. *Ultrastruct. Process. Adv. Mater.* **4**, 491 (1992).

Roos, A., and Ribbing, C. G. *Appl. Optics* **27**, 3833 (1988).

Roos, A., Ribbing, C. G., and M. Bergkvist, *Appl. Optics* **27**, 3828 (1988).

Schwartz, M. M. "Composite Materials Handbook," McGraw-Hill, New York (1984).

Zwinkels, J. C., and Dodd, C. X. *Appl. Optics* **28**, 2381 (1989).

General References

Anderson, R. *Appl. Optics* **29**, 4235 (1990).

Ashley, C. S., Reed, S. T., Brinker, C. J., Walko, R. J., Ellefson, R. E., and Gill, J. T. In "Chemical Proceedings of Advanced Materials" (L. L. Hench and J. K. West, eds.), p. 989, Wiley, New York (1992).

Baraton, M. I., Merle-Mejean, T., Quintard, P., and Lorenzelli, V. *J. Phys. Chem.* **94**, 5930 (1990); *J. Phys. Colloq.* **C4**, 239 (1989).

Blair, D. S., and Ward, K. J. *SPIE Appl. Spectrosc. Mater. Sci.* **1437**, 76 (1991).

Burleigh, D. D. *SPIE 780 Thermosense IX* 250 (1987); *SPIE 1094 Thermosense XI* 175 (1989).

Devaty, R. P. *Physica A* **157**, 301 (1989).

Duclos, J. *Instrumentation Res.* 105 (1985).

Edwards, D. K., Gier, J. T., Nelson, K. E., and Roddick, R. D. *Appl. Optics* 1279 (1962).

Epstein, G. In "Handbook of Fiberglass and Advanced Plastics Composites" (G. Lubin, ed.), p. 661, Van Nostrand Reinhold, New York (1969).

Flytzanis, C., Hache, F., Ricard, D., and Roussignol, P. *Springer, Proc. Phys.* **13**, 331 (1986).

Hanssen, L. M., and Snail, K. A. *SPIE* **807**, 148 (1987).

Helms, R. G. *et al., SPIE* **1114**, 426 (1989).

Lauver, R. W., and Vannucci, R. D. NASA-TM-79068 (1979).

Manson, J. E., and Sefaris, J. C. *Sci. Eng. Composites* **1**, 75 (1989).

May, C. A., Helminiak, T. E., and Newey, H. A. In "Bicentennial of Materials," p. 274, SAMPE, Azusa, CA (1976).

Messerschmidt, R. G. *Appl. Spectrosc.* **39**, 737 (1985).

Mohn, W. R., and Vukobratovich, D. *Sampe J.* **24**, 26 (1988).

Niklasson, G. A. *Physica A* **157**, 482 (1989).

Nishijima, S., Matsushita, K., Okada, T., Okamoto, T., and Hagihara, T. In "Nonmetallic Materials and Composites at Low Temperatures," p. 143, Plenum Press, New York (1984).

Pearson, L. H. In "Review of Progress in Quantitative Nondestructive Evaluation," p. 2017, Plenum Press, New York (1990).

Rayment, D. L., and Majumdar, A. J. *J. Mater. Sci.* **13**, 817 (1978).

Robertson, A. R. *Appl. Optics* **11**, 1936 (1972).

Rustagi, K. C., and Bhawalkar, D. D. *Ferroelectrics* **102**, 367 (1990).

Sellitti, C., Koenig, J. L., and Ishida, H. In "Interfaces in Polymer, Ceramic, and Metal Matrix Composites," p. 163, Elsevier, New York (1988).

Simmons, J. H., Potter, B. G., Kumar, P., and Neidt, T. M. *Ceram. Trans.* **19**, 333 (1991).

Stoner, C. E., and Brill, T. B. *Combustion Flame* **83**, 302 (1991).

Stoner, C. E., Haggerty, B. S., Rheingold, A. L., and Brill, T. B. *Propellants, Explosives Pyrotechnics* **17**, 82 (1992).

Varadan, V. K. *SPIE* **1558** (1991).

SPECTROSCOPY OF CERAMICS

JAMES F. CORDARO
QivaStar, Inc.

I. Introduction

Ceramics constitute a very broad category of materials. Hence, the purpose of this chapter is to define and to limit the scope of our discussion to a manageable set of topics but still retain the character of a survey. A discussion on ceramic materials in a spectroscopy handbook should be prefaced by an answer to the question, "What constitutes a ceramic?" Ceramics are broadly considered to be inorganic, nonmetallic solids such as metal oxides, very often as polycrystalline solids or in powder form. Oxide glasses are also a part of ceramic materials; however, glasses will not be considered here because these are covered in a separate chapter. Specifically, ceramics will be defined as crystalline inorganic, nonmetallic materials that have undergone high-temperature processing. Naturally occurring solids such as unprocessed minerals are excluded. Synthetic single crystals, such as synthetic sapphire or synthetic rutile, will be included in our discussion.

A discussion of the optical properties of ceramics would be incomplete

ISBN 0-12-764070-3/$25.00.

without a discussion of light scattering. The scattering of incident light is what produces the major optical effects in many ceramics. Overall, optical properties of materials are largely determined by their structural form, e.g., powder or single crystal. For example, the transparency and reflectivity of ceramics are determined, most significantly, by light scattering. Although the overall color is determined by the selective absorption of various wavelengths, light scattering effects can influence the shade visible to the eye. However, the color we observe from an object does not imply uniqueness in absorption because the physiology of the eye is also an important factor. In our discussion on light scattering, we specifically mean elastic scattering of light by ceramics, i.e., the wavelength of the incident monochromatic light remains (practically speaking) unchanged after scattering. We shall not consider the effects of inelastic scattering in which the wavelength changes such as in Raman scattering, which is a separate subject.

II. Scattered Light and Its Relationship to Reflectance of Ceramics

A. General Discussion

Several treatises on light scattering and optical reflectance are readily available for extended information (1–6). The discussions in Bohren and Huffman (4) are particularly lucid. For purposes of our discussion, consider a beam of light as it enters a chamber of length h within which are suspended small particles of matter. The transmitted irradiance, I_t, of the light will be diminished (i.e., attenuated) as it exits the chamber according to the Beer–Lambert law $I_t = I_i \exp[-a_e h]$, where I_i is the incident irradiance. Irradiance (sometimes called intensity) is measured in units of W cm^{-2}. The irradiance of light should not be confused with radiance, which is irradiance per unit solid angle. Radiance is expressed in units of W cm^{-2} sr^{-1}. a_e is the attenuation coefficient, usually expressed in units of cm^{-1}.

Attenuation of light consists of the sum of absorption and scattering, or $a_e = a + s$; s is known as the scattering coefficient, in units of cm^{-1}, and a is the absorption coefficient, also in cm^{-1}. By light scattering, it is meant that the irradiance of the light has diminished due to the deviation of the light from the direct beam, i.e., the light does not completely disappear: It is merely redirected from the original path. By contrast, for true absorption, the light does disappear as the energy of the photons is consumed either by electronic or atomic excitation processes within the solid. Virtually all the light energy consumed is eventually converted to heat as phonon excitation, regardless of how it is consumed. For the suspended particles described previously, if the true absorption processes are small, the dominant mechanism by which the irradiance of the beam is diminished is that of light scattering and then $I_t \cong I_i \exp(-sh)$.

When a beam of light is incident upon a polycrystalline ceramic body, much of the light penetrates the surface and undergoes both scattering and absorption. The characteristic penetration depth is estimated from a_e^{-1}, which is the depth at which the irradiance of the light diminishes by a factor of e^{-1} or about 0.37. The scattered light eventually returns to the surface of the body, where it exits from the surface in all directions. The directionality of this scattered light exiting the surface falls into two categories: specularly reflected light and diffusely reflected light. These will be discussed in Section III, A.

There are numerous mechanisms by which light can be scattered; however, we can divide light scattering into three regimes based on two length scales; the wavelength, λ, of the light and the diameter, d, of the particle that does the scattering. From these length scales, a unitless size parameter x is determined (2), where $x = \pi d/\lambda$. The three regimes are (1) $x \ll 1$, (2) $x \cong 1$, and (3) $x \gg 1$. They are known as the Rayleigh regime, the Mie (or resonance) regime, and geometrical optics, respectively.

Scattering by very small particles ($x \ll 1$), e.g., particles whose diameters are approximately $\lambda/10$ or less, is commonly known as Rayleigh scattering. An historical account of Rayleigh scattering can be found in the article by Young (7). The irradiance of Rayleigh scattered light, I_s, follows the well-known λ^{-4} dependence. For unpolarized light, the scattered irradiance for a particle that is small compared to the wavelength is described (4) by

$$I_s = I_i \frac{8\pi^4 n_1 d^6}{\lambda^4 r^2} \left| \frac{m^2 - 1}{m^2 + 2} \right|^2 (1 - \cos^2 \theta_s), \tag{1}$$

where θ_s is the scattering angle ($\theta_s = 0°$ corresponds to the forward direction and $\theta_s = 180°$ corresponds to the backward direction), d is particle diameter, r is radial distance from the particle, m is the relative index of refraction (particle/medium), and n_1 is the index of the medium. Although the Rayleigh effect is most notable for scattering by gas molecules in the atmosphere and by suspensions of small particles, it has been demonstrated as occurring in specially prepared inorganic glasses (8). Light scattering in the visible by ceramic particles (e.g., for opacifiers in glazes) is maximized for particle sizes near the Mie regime.

The Mie (or resonance) regime is an intermediate ($x \cong 1$) scattering regime. Precisely, Mie theory refers to light scattering by spherical particles of any size; however, more common usage has come to mean scattering by particles of approximately the same diameter as the wavelength. For details of the theory of Mie scattering, the reader is referred to Van de Hulst (2) or

Refs (1) and (4). Mie scattering (in the common usage) produces more complex light scattering effects than does Rayleigh scattering. The scattering coefficient, s, for particles tends to maximize around the Mie regime.

Now consider scattering by particles that are larger than the wavelength of the light but still close to the resonance regime ($10 < x < 10^2$). For particles of large diameter compared to λ, the light will scatter following the laws of diffraction. This could be considered to be the major light scattering effect for consolidated (i.e., sintered) ceramic bodies. It is predominant at interfaces that occur along various grain boundaries in ceramic bodies and where large-diameter porosity is present. This can also occur with large-diameter powders and with second phases. For the extreme ($x \gg 1$) case in which "particle" diameter is very large, for example, $x \geq 10^3$, the light will scatter according to the laws of geometrical optics. Given common grain sizes of 5–50 μm, ceramics are more likely to be closer to the case of diffractive optics than to the case of geometrical optics. However, geometrical optics can still be useful for insight.

B. Light Scattering by Grain Boundaries

Light scattering from grain boundaries is not significant when the crystals (i.e., grains) have the same phase and when the index of refraction within a crystal is isotropic. For solids in the cubic crystal system, n is crystallographically isotropic, and grain boundaries for cubic crystal system materials would not be expected to contribute significantly to the scattering of light.

The light scattering from grain boundaries will occur under two conditions: (1) when the ceramic is composed of a mixture of grains or of crystals having different phases and (2) when the ceramic is composed of birefringent (i.e., doubly refractive) crystals of the same phase. If the grains are substantially larger than the wavelength of the light, the laws of geometrical optics become a useful tool in order to illustrate the behavior of light at the grain boundary interfaces.

The absorption coefficient, a, is related to a unitless parameter that is also used as a measure of the absorption, known as the index of absorption, k. The absorption coefficient can be derived in terms of k as

$$a = 2\omega k/c, \tag{2}$$

where ω is the angular frequency ($\omega = 2\pi\nu$) and c is the speed of light in vacuum. k is used to specify the complex index of refraction $n^* = n + ik$ for a solid. A complete discussion of the relationships between the optical constants n and k is well presented by Coelho (9).

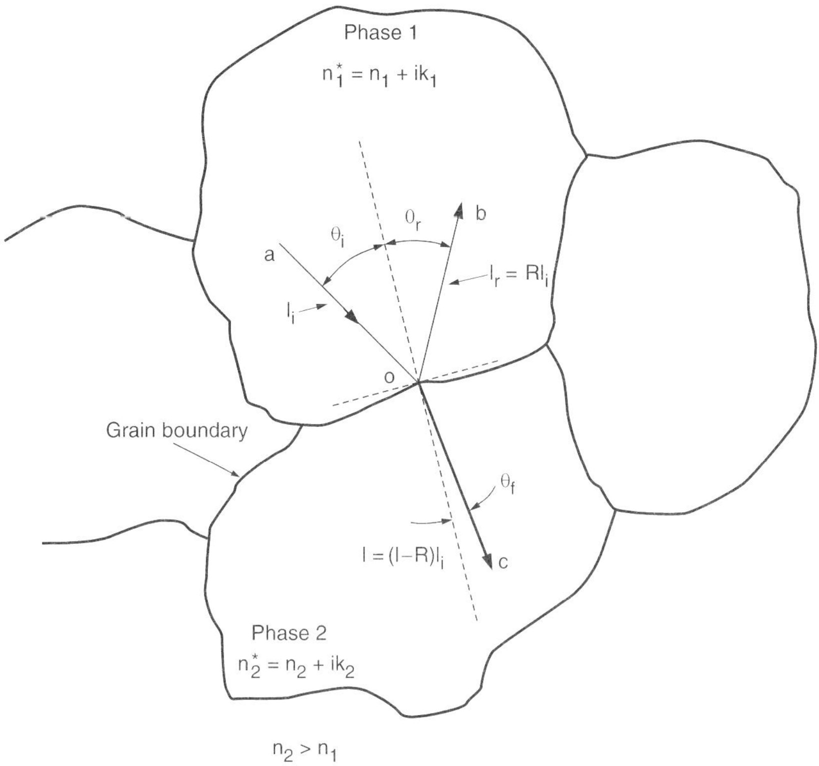

FIG. 1. Propagation of light across a grain boundary or other interface consisting of materials of two different phases.

Figure 1 is a diagram of a light ray *ao* that is incident upon a grain boundary interface. The material is composed of grains of phase 1, having complex index of refraction $n_1^* = n_1 + ik_1$, and grains of phase 2, having complex index of refraction $n_2^* = n_2 + ik_2$. The real parts of the complex indices of refraction differ such that $n_2 > n_1$. The light ray *ao* undergoes both reflection and refraction at the interface. If the grains are not highly absorbing (i.e., k_1 and $k_2 \ll n_1$ and n_2), then a small fraction of the light is reflected at the interface along path *aob*. The reflected light follows the behavior that the angle of incidence, θ_i, is equal to the angle of reflection, θ_r. The greater irradiance of the light is refracted along path *aoc*, such that ray *oc* is bent toward the normal to the interface. The angle of refraction θ_f follows Snell's law, which is given by $n_1 \sin \theta_i = n_2 \sin \theta_f$. After the light ray undergoes refraction at several boundaries, the cumulative effect is that the light is scattered away from its original path. Such scattering from grain boundaries is not the major reason a ceramic appears opaque, unless the indices of refraction differ significantly (i.e., the relative index of refraction, m, is large). Discussions of geometrical optics can be found in Refs. (10)–(12).

With the exception of those solids in the cubic crystal system, most ceramic crystals possess some degree of optical anisotropy. The index of refraction is usually defined in terms of two indices using plane-polarized light: $n_{\perp}$, where the **E** field is perpendicular to the optic axis, and $n_{\parallel}$, where the **E** field is parallel to the optic axis. The term optic axis refers to an axis of optical and crystallographic symmetry. A consequence of optical aniosotropy is that when a ray of unpolarized light enters a birefringent crystal, the beam will be split into two rays, the ordinary ray (o ray) and the extraordinary ray (e ray).

Light scattering from the interfaces of birefringent grains can be significant. In a birefringent crystal, the indices of refraction along different crystallographic directions can differ by as much as 0.2 or 0.3. A random orientation of grains in a ceramic means that as a light ray traverses a grain boundary, it is likely to pass from one grain, in which the indices of refraction are fixed in orientation relative to the incident path, to another grain, in which the light ray will encounter a different orientation for the indices of refraction, relative to the incident path.

Such a situation of birefringent grains of the same material is illustrated in Fig. 2, in which the index of refraction for the solid has been indicated to be different along the a and the c crystallographic axes. The grains are randomly oriented and the light ray undergoes reflection (path aob) and refraction (path aoc) at the interface. An example of such a solid would be ceramic TiO_2 as polycrystalline rutile. Synthetic rutile has a tetragonal crystal structure and a very strong optical anisotropy. The indices of refraction along the a and the c axes differ significantly. Perpendicular to the c axis, the sodium D line (589.3 nm) index of refraction $n_{\perp} = 2.909$ and parallel to the c axis $n_{\parallel} = 2.613$. Such a difference Δn of approximately 0.3 is a substantial one.

C. Light Scattering by Pores

Residual porosity is the major contribution to light scattering within a sintered ceramic body. The effect will appear similar to that of Fig. 1, in which the first phase (phase 1) now constitutes the pore that contains various gasses. In this case, $n_1 = 1.00$ and the resulting difference in the indices of refraction can be approximately 0.5 or greater. Such an enormous difference means that only small amounts of residual porosity are required to create a large amount of light scattering. Small amounts of residual porosity effec-

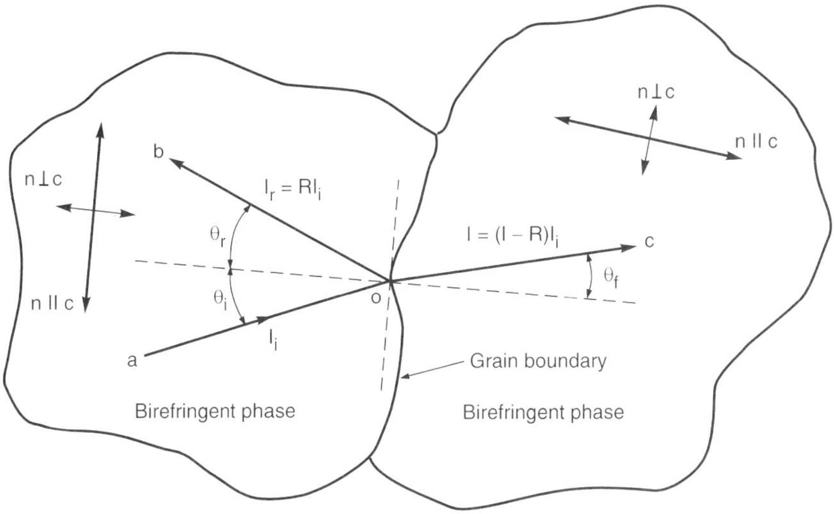

FIG. 2. Propagation of light across a grain boundary consisting of two grains of the same phase but having an optical anisotropy (birefringence).

tively render sintered ceramics opaque to the transmission of light. In this context, the term opaque means a very high capacity of the material to diffuse light.

When steps are taken during fabrication to substantially reduce the amount of residual porosity, the ceramic undergoes a transformation from opaque to translucent or even to nearly transparent. Such a transition can be observed in special polycrystalline alumina ceramics. Normally, sintered alumina bodies are observed to be opaque to somewhat translucent. However, an alumina material is produced under the trade name of Lucalox (General Electric Co.) in which there is low residual porosity. The ceramic is almost transparent to light and is used in the production of lamp envelopes for exterior lighting.

III. Reflected Light from Opaque Ceramics

A. GENERAL DISCUSSION

Because ceramics are, in general, not optically transparent as a result of either porosity or strong absorption, reflectance spectroscopy (1, 3, 5, 6) provides a means to study the optical properties of such materials. The spectral qualities of the reflected light from the surface of a polycrystalline ceramic depend on both light scattering and true absorption. In our discussion of reflection, it is first necessary to elaborate on the directionality of the reflected light.

With no contribution to light scattering, a beam of light incident on a surface will be reflected at an angle equal to the angle of incidence. Such reflection is considered to be specular. When light scattering is present, a diffuse component to the reflected light arises from multiple scattering of the incident beam. The angle of reflection is no longer equal to the angle of incidence. The light undergoes reflection at angles other than the angle of incidence (i.e., is scattered) in both forward and backward directions. With increasingly diffuse reflection, the specular irradiance is reduced. Such a surface appears to have a matte finish. For a perfectly Lambertian diffuser, the specular component of the irradiance is a minimum and the angular distribution of the scattered irradiance would be proportional to the square of the cosine of the scattering angle, $I_s \propto \cos^2 \theta_s$. A perfectly Lambertian diffuser is a limiting case, as is pure specular reflection, and in practice a material is never perfectly Lambertian.

The two sources of the diffusely reflected light are the surface roughness (3) and subsurface scattering features. The surface roughness consists of the microscopic hills and valleys of the air–material interface. The surface roughness contributes at a maximum to the diffusely reflected light when the wavelength of the light has approximately the same dimensions as the surface features, for example, the root mean square roughness of the surface scratches. A practical treatise on the effects of surface roughness can be found in the work of Bennett and Mattson (3). For wavelengths much larger or much smaller than the dimensions of the roughness, the amount of scattering is decreased. For ceramics, subsurface features such as pores and grain boundaries significantly scatter light that penetrates the surface.

Consider a ceramic specimen in the form of a plate that has front and back surfaces. For research purposes, a reflectance measurement would be made from a specimen in which the amount of light reflected from the back surface, and lost from transmission through the specimen, is negligible compared to light scattered from the interior and the front surface. In other words, the specimen would appear infinitely thick to light that penetrates the surface. With such a semiinfinite specimen, the total reflection coefficient (from the combined diffuse and specular components) is designated as R_∞.

Because R_∞ contains the effects of both true absorption and light scattering, quantitative information on true absorption alone is difficult to obtain from a reflectance measurement. A parameter related to absorption, but only qualitatively the same, is the relative absorptance. The relative absorptance, A_r, is obtained from R_∞ by

$$A_r = \log_{10}(1/R_\infty). \tag{3}$$

A_r can be used to provide qualitative information on absorption mechanisms.

Determination of the absorption coefficient a from R_∞ could be done using the Kubelka–Munk equations (5,6). In order to do this, an independent measurement of the scattering coefficient must be made. Procedures for the measurement of s are described in reflectance spectroscopy references by Wendlandt and Hecht (5) and by Kortum (6). In these references we find that a and s are related to R_∞ by

$$a/s = (1 - R_\infty)^2/2R_\infty. \tag{4}$$

Determination of the scattering coefficient over a range of wavelengths is tedious because s is a function of the wavelength. At one particular wavelength, s can be estimated from several measurements of the reflected light from very thin layers of material. There are various assumptions inherent within the Kubelka–Munk equations and the reader should consider the discussion in Hapke (1) on the limitations of the equations.

B. Instrumentation

Commercial instruments, such as the Lambda 19 (Perkin–Elmer Corp.), can be used for the measurement of the ultraviolet (UV), visible (VIS), and near-infrared (NIR) optical properties of ceramics. Such an instrument, when used in conjunction with the RSA PE 90 (Labsphere, Inc.) integrating sphere, can be used to measure the spectral reflectance of opaque ceramics. Commercial instruments are excellent for quality control and industrial and analytical analyses of ceramics.

A noncommercial apparatus will be described in this section. The apparatus is constructed of separate instruments arranged on an optical bench. The advantages over a commercial unit are that it provides a more pedagogic approach to the subject and has a higher flexibility in configuration for unusual experiments. The disadvantages are that it occupies more floor space and is much more time consuming to operate.

What is described is a ratiometric, double-beam apparatus that uses an integrating sphere (13, 14). It functions in the UV–VIS–NIR wavelengths from approximately 240 nm to 2.5 μm. Figure 3 shows a diagram of the apparatus in the configuration for the shorter wavelength (240 nm up to 700 nm) range. Two separate light sources are used: a 300-W xenon (Xe) arc lamp to cover the UV–VIS and a 100-W quartz tungsten halogen lamp to cover the VIS–NIR. A flowing water filter is a necessity to filter the

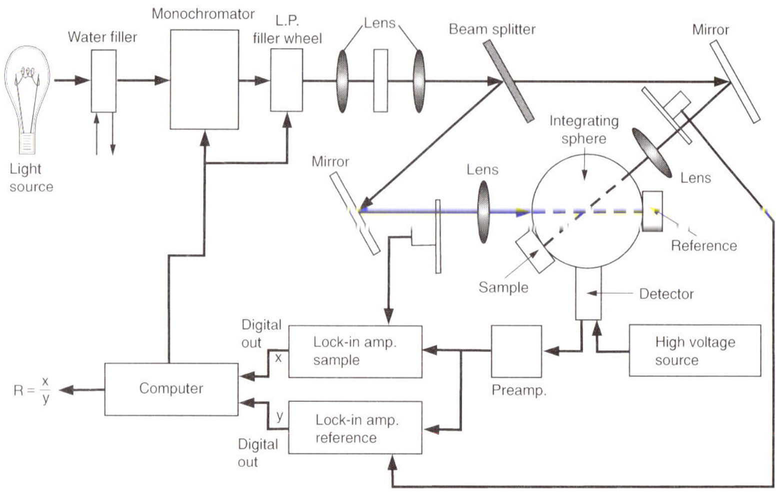

FIG. 3. Double-beam, ratiometric spectrophotometer for total reflectance measurements.

output of the Xe lamp in order to remove most of the IR. The unfiltered output of a 300-W Xe lamp could damage a monochromator and diffraction gratings. Three hundred watts appears to be excessive power for UV–VIS spectroscopy; however, the UV–VIS irradiance becomes very useful for the study of low-reflectivity black ceramics.

The white light is passed through a monochromator having variable slits in order to control the bandwidth. A monochromator with multiple diffraction gratings is necessary to cover the entire wavelength range (the one shown in Fig. 3 is a three-grating, Czerny–Turner configuration). A computer should be able to control grating changes. A filter wheel with long-pass filters is mounted after the monochromator. The filters are necessary in order to remove light due to second order diffraction. Filter changing should be able to be computer controlled. A variable aperture is used to control the spot size of the light on the specimen. Typical spot sizes with appropriate lenses are 10 mm down to 3 mm. Because the light exiting a monochromator is partially polarized, a broadband, common beam depolarizer can be placed after the aperture.

The light is split into two separate beams; one beam for the ceramic specimen and another beam for a reference material. A broadband beam splitter should be used. One option is called a "polka-dot" splitter, which consists of mirror dots on a fused-silica plate whereby the split ratio becomes 50:50 (relatively independent of wavelength) for beam cross sections above a certain minimum. The beams pass through two rotating wheel choppers that operate at different frequencies. A frequency ratio of 4:7 is typically used. After the choppers, the beams are directed through ports on an integrating sphere and onto the ceramic specimen and reference material.

The integrating sphere is 150 mm in diameter and is coated on the interior with a barium sulfate-based material. The coating has an excellent reflectance over the UV–VIS–NIR but is rather fragile and can be easily damaged. The sphere collects the specimen-reflected light, which is then detected by a single photomultiplier tube (PMT) mounted by a port on the bottom. The PMT detector is used for the UV–VIS range. A PbS cell would be substituted with an appropriate amplifier in order to cover the NIR. The gain on the PMT is controlled by the voltage on a stable, high-voltage power supply. Current from the PMT is converted to a voltage signal by a current-sensitive preamplifier (also known as a transimpedance amplifier). The preamplifier also has a gain adjustment. The signal from the preamplifier is then distributed to two single-phase lock-in amplifiers.

The lock-in amplifiers essentially function as one-component Fourier analyzers. In order to function, the lock-ins require reference signals, which are the signals from the beam choppers. The lock-ins separate the light signals from the same PMT because of their different chopping frequencies and reject background light. In addition, there are separate gain adjustments and noise-filtering options on the lock-ins. By using a single detector for both the specimen and the reference beams, we can compensate for the effects of drift in detector sensitivity. The signals from both lock-in amplifiers are output as digital information to a computer.

Prior to the measurement of an unknown, the optics are calibrated as a function of wavelength and the calibration curve is stored in the computer. Calibration is carried out by placing a reference material (standard) at both the sample port (port a) and at the reference port (port b) of the integrating sphere. The lock-in amplifier voltage signals V_a and V_b corresponding to the a and b ports are divided within the computer at each wavelength to form a calibration curve, R_{cal}:

$$R_{cal} = V_{a,\,stand}/V_{b,\,stand}. \tag{5}$$

This calibration curve takes into account factors such as the varying efficiencies of the diffraction gratings as a function of wavelength, the wavelength dependence of the light source irradiance, wavelength dependence of detector sensitivity, and unequal beam losses.

To run a spectrum, the standard in port a is replaced with the unknown. The unknown signal, $V_{a,\,un}$, and the reference signal, $V_{b,\,stand}$, are then divided into the ratio $V_{a,\,un}/V_{b,\,stand}$ within the computer. This ratio is then divided by the calibration curve, R_{cal}, stored in the computer memory. The reflectance of the specimen is

$$R(\lambda) = (V_{a,\,un}/V_{b,\,stand})(1/R_{cal})R_{stand}, \tag{6}$$

where R_{stand} is an absolute reflectance calibration curve for the reference material, if one is available.

Regarding spectral noise, in addition to using the filtering time constants on the lock-in amplifiers, noise reduction in the calibration curve and the unknown spectra can also be accomplished by using measurement averaging on the computer. This is a very effective method. Multiple measurements of M number are made at the same wavelength and then averaged to become one reflectance data point, where $R_{ave}(\lambda) = (1/M)\ (R_1(\lambda) + R_2(\lambda) + R_3(\lambda) + \cdots + R_M(\lambda)$. $M = 5$–8 appears to be effective in reducing noise without significantly slowing down the time required for recording a spectrum.

Typical reference materials have included freshly formed MgO powder, polytetrafluoroethylene powder, and barium sulfate powder. Much more convenient reference materials that are now available are Spectralon (Labsphere, Inc.) reflectance standards.

The total reflectance measurement on the specimen is carried out at an 8° angle of the incident light. This is one commonly used angle of incidence for these measurements. The reflectance of a surface is a function of angle and is discussed in Chapter 9.

C. Sintered Ceramics

Consolidated or sintered ceramic bodies can be easily measured if a flat surface is available. In the form of a pellet, the ceramic can be mounted to a holder, such as an aluminum alloy plate, and then attached to the port of the integrating sphere. Some caution should be exercised regarding the depth of penetration of the light into the pellet. The light will penetrate a sintered pellet much farther than a ceramic powder due to the smaller amount of light scattering in a pellet. If a significant fraction of light penetrates entirely through the pellet to the holder, then reflectance becomes dependent on the geometry (thickness) of the specimen rather than on intrinsic material properties.

The reflectance spectra for highly absorbing or "black" ceramics contain more noise than do other spectra due to the much lower signal-to-noise ratio. In black ceramics, the higher power Xe lamp (300 W) tends to compensate for the small amount of reflected light. Several other techniques can be utilized in order to improve the signal-to-noise ratio. One is to open the slits (widen the bandpass) of the monochromator in order to let more light through to the specimen; however, this has the disadvantage in that widening the bandpass reduces the wavelength resolution of the spectra. Another technique is to take higher averages using the computer: $M = 16$ averages work well for black material but higher averages help little to improve signal-to-noise (S/N) ratio. S/N scales with M statistically independent samples as $M^{-1/2}$.

A reasonable practice is to match the spectral characteristics of the reference material to that of the specimen being measured, such that there is not a very large difference in reflected light between the specimen and the reference. If a black ceramic is being measured, then it is generally helpful to use a black reference material. Carbon black powder is a possible reference. For other various neutral (gray) ceramics and for colored ceramics, the Spectralon materials are also available as calibrated references having a variety of spectral characteristics.

D. Ceramic Powders

The reflectance of ceramic powders can be measured relatively easily. The large difference in index of refraction between powder particle, $n \cong 1.5$–2.5, and the space between the particles, $n = 1.00$, produces a large amount of light scattering. Hence, in general, only a thin layer of powder is necessary to achieve the condition of reflection from a semiinfinite surface.

Figure 4 shows a cross section of a powder specimen holder for use with an integrating sphere for total reflectance measurement. The specimen holder is made from aluminum alloy and mounts to the exterior surface of an integrating sphere. The powder is packed into a recessed cup on the surface of the holder. The thickness of the powder is 3 or 4 mm. This thickness is generally sufficient to provide the conditions for a semiinfinite surface.

Light incident upon the surface of the powder undergoes both true absorption and multiple reflections. The light is reflected and refracted at the particle–air interfaces and the incident light diffuses into the powder. Some of the light penetrates only a thin surface layer until it is either completely absorbed or exits the surface as reflected light. Another portion of the light penetrates deeper into the powder until being absorbed or directed back toward the surface. The characteristic penetration depth is equal to a_e^{-1}, which originates from the Beer–Lambert law. Assuming that the specimen is a typical "white" ceramic powder, where $s \gg a$ and $s \cong 400\ \text{cm}^{-1}$, then the characteristic penetration depth $h_c \cong 0.025$ mm—a small distance. For such a scattering coefficient, a powder thickness of approximately $5 \times h_c$ is, in principle, sufficient so that the incident light does not signifi-

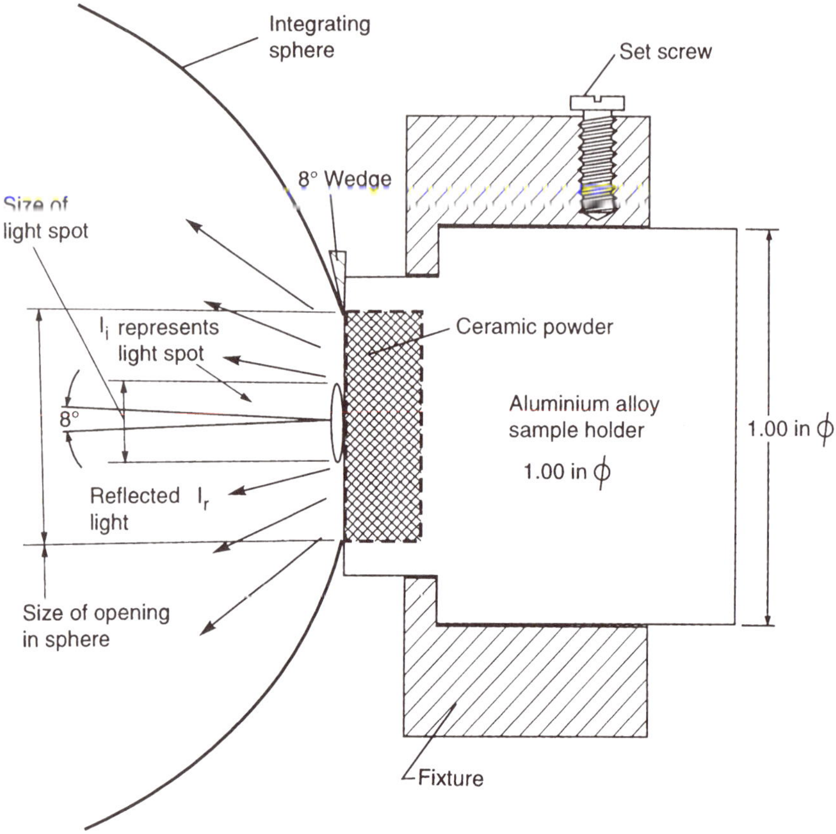

FIG. 4. Sample holder for ceramic powders illustrated as mounted to an integrating sphere.

cantly "see" the back of the specimen holder. For example, the thickness of a layer of TiO_2-based paint, which can hide a surface, may be on the order of 5 mils, or about 0.15 mm. However, a general thickness of 3 or 4 mm for powders is recommended as a rule of thumb.

The reflected irradiance of the powder does change with particle packing density; however, the change is not substantial (> 1%) unless the powder is highly compressed. Packing by hand using a flat surface, for example, is sufficient. It is important that the placement of the specimen holder on the integrating sphere be uniform from specimen to specimen. The machined front of the holder should correspond in placement to the diameter of the

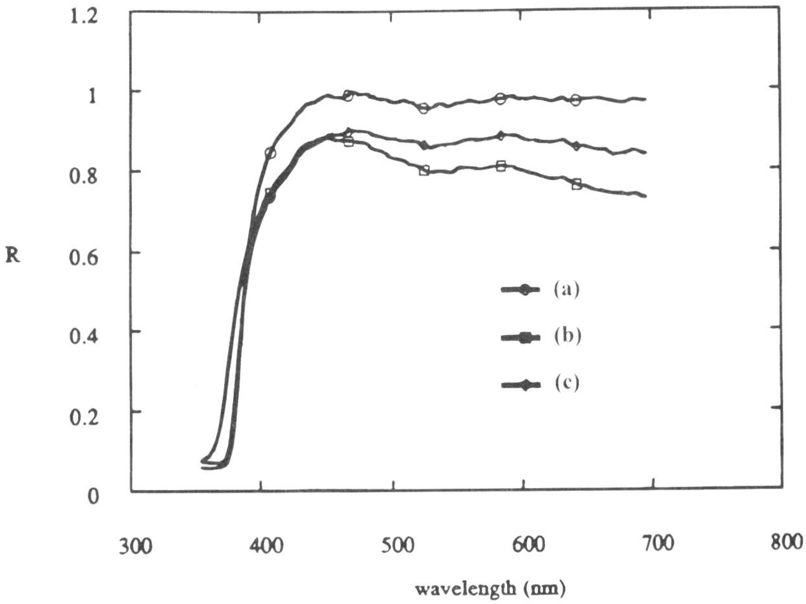

FIG. 5. Total reflectance spectra (diffuse and specular components) for three commercial zinc oxide ceramic powders.

integrating sphere as closely as possible. Figure 5 shows a few examples of total reflectance spectra for ceramic powders.

IV. Transmitted Light through Ceramic Crystals

Internal transmission of light through a crystal, assuming that s is negligibly small and that corrections have been made for reflection losses from external surfaces, follows the Beer–Lambert relation $I_t \cong I_i \exp(-ah)$ mentioned in Section II, A. Optical absorption in the visible generally occurs by the excitation of electrons on impurity atoms or on native defects. If the interactions between impurities are negligible, e.g., for low concentrations or where quantum–mechanical exchange interactions do not occur, then a is linearly proportional to the impurity concentration N:

$$a = \sigma N, \tag{7}$$

where σ is the absorption cross section of the impurity, generally given in units of cm^2. σ is a function of many factors: the nature of the impurity, the impurity–host material interactions, and wavelength. The quantum–

mechanical descriptions of σ are beyond the scope of the discussion for this section; however, they can be found in numerous solid-state physics and quantum mechanics treatises and in Ref. (15). In the chemical literature, cross sections are often referred to as molar extinction coefficients, ε, where ε is expressed in units of liters per mole per cm and the concentration N_m of the absorbing species is in moles per liter:

$$a = \varepsilon N_m. \tag{8}$$

The irradiance of light that emerges after passing through a solid is often recorded as percentage transmission $(\%T) = 100T$, where the transmittance, T, varies between 0 and 1 and is given by

$$T = I_t/I_i. \tag{9}$$

Another measure of the irradiance of light diminishing in a solid is the absorbance, A, which is recorded in units of optical density (OD), where OD is defined as

$$\mathrm{OD} = \log_{10}(I_i/I_t). \tag{10}$$

An absorbance of 1 corresponds to 0.10 transmittancc, or 10%T, whereas an absorbance of 2 corresponds to 0.01 transmittance, or 1.0%T.

A distinction should be made between total transmission and internal transmission because a fraction of the light passing through a plate is lost by reflection at the external surfaces of a slab of material. The irradiance of the reflected light is a complex function of the angle of incidence, θ_i, and polarization and is presented in Refs. (4) and (10) and derived in the work by Heavens (12). For unpolarized light at normal ($\theta_i = 0$) incidence, R is obtained from

$$R = (n_2^* - n_1^*)^2/(n_2^* + n_1^*)^2, \tag{11}$$

where n_2^* and n_1^* are the complex indices of refraction. If substance 1 is air, i.e., $n_1 = 1.00$ and $k_1 = 0$, then Eq. (11) reduces to

$$R = [(n_2 - 1)^2 + k_2]/[(n_2 + 1)^2 + k_2]. \tag{12}$$

With relatively transparent ceramics, in general $k \ll n$ and Eq. (12) becomes

$$R \cong (n_2 - 1)^2/(n_2 + 1)^2. \tag{13}$$

For a ceramic having an index of refraction of approximately 1.5 at the wavelength of measurement, one surface will reflect approximately 4% of the light at normal incidence. For a plate having a front and back surface, approximately 7 or 8% of the light is lost by reflection. If $n_2 = 2$, then approximately 20 or 21% is lost by reflection.

Considering multiple reflections at front and back surfaces, the irradiance of unpolarized light transmitted at normal incidence ($\theta_i = 0$) through a plate of thickness, h, is determined from a geometrical progression and is given (4) by

$$I_t = I_i[((1 - R)^2 \exp(-ah))/(1 - R^2 \exp(-2ah))] \tag{14}$$

If $R^2 \exp(-2ah) \ll 1$, which would normally be the case, then

$$I_t \cong I_i(1 - R)^2 \exp(-ah). \tag{15}$$

For extended discussion of the three-layer problem, the reader is referred to Heavens (12). Accurate determinations of absorbance require that spectrophotometer measurements be corrected to represent the actual internal transmission. Hence, $dn(\lambda)/d\lambda$, which is the dispersion in index of refraction, must be known over the wavelength range of interest in order to calculate R. $R(\lambda)$ can be measured using either an integrating sphere or a spectroscopic ellipsometer. As previously discussed, birefringent crystals are polarization sensitive. Because light exiting a grating monochromator is partially polarized, a common beam depolarizer can be used after the exit slit during measurements in order to depolarize the light prior to reaching a crystal.

V. Mechanisms of Optical Absorption in Ceramics

A. General Discussion

Most ceramic materials are electrically insulating, having energy band gaps of 3 eV or greater; hence, in single crystal form the materials are relatively transparent. Although there are many other ceramics, including those that show metallic conduction, this discussion will focus largely on those materials that fit the category of transparent in single-crystal form. Mechanisms of optical absorption (16–19) are so varied that it becomes difficult to categorize them; however, for the purposes of this discussion we will first make the broadest distinctions in terms of relative wavelength: ultraviolet, visible, and infrared absorption.

Light can interact with matter in only one of two ways: the electric field of the light wave can couple with an electric moment or the magnetic field of the light wave can couple with a magnetic moment. For almost all cases, the light couples with an electric dipole moment in the solid because the magnetic moments are comparatively weak. The coupling of the light wave creates elementary excitations in the solid and the light is absorbed in energy quanta, $h\nu$, corresponding to those excitations. Some of the more common excitations are the following: UV and near-UV (190–400 nm), electron band to band (extended state) transitions and exciton formation; VIS (400–700 nm), localized electronic excitations either from (i) electrons bound to impurities or (ii) electrons or holes bound to native defects (i.e., color centers); near-IR and IR (700 nm–100 μm), coupling to lattice vibrations (i.e., phonons) and coupling to conduction electrons (plasmon excitations). With the exception of lattice vibration excitations, all the absorption mechanisms involve some form of electronic excitation.

The most noticeable feature concerning the spectroscopy of solids in the visible range is the broadness of the absorption bands. The optical absorption for gasses occurs at specific wavelengths and the spectra show discrete absorption lines, whereas it is found that absorption in solids occurs in bands. The large width is often referred to as inhomogeneous broadening. It results in part from the coupling of phonons with the electronic excitation processes. In addition, the local environment of the absorbing species is slightly different from atom to atom, i.e., no crystal is completely perfect. Because of phonon broadening effects, electronic absorption spectra become sharper as the temperature of the specimen is lowered toward liquid helium temperatures (4.2°K). As the temperature decreases, so does the phonon population density.

The color we observe in transmission from a ceramic crystal does not correspond to the wavelengths of the light that are being absorbed. For example, the Co^{2+} ion in ceramics is a very strongly absorbing species, or chromophore. If we observe Co^{2+}-doped ZnO through an external white light source, the color we observe is blue. However, the optical absorption is occurring in the red. We only observe that light which is not absorbed by the chromophore. The energy of the absorbed light is eventually converted to heat. Most of the transitions from the excited state to the ground state are radiationless, giving up the energy as phonons rather than photons. In certain circumstances, however, photoluminescence in ceramics can be observed as some of the absorbed light energy is reradiated as photons.

B. Ultraviolet Wavelengths

The ultraviolet absorption edge arises due to band to band electronic transitions. Consequently, the onset of this fundamental UV absorption occurs at wavelengths that correspond to energies near the band gap energy, E_g. These electronic transitions correspond to excitations of electrons between extended states. The quantum states spatially extend throughout the crystal due to the overlap of wavefunctions from certain atomic orbitals. In a ceramic, e.g., a metal oxide, ground state electrons reside in valence bands mainly formed from the overlap of largely electron-occupied O 2p orbitals. This valence band is separated in energy by the band gap from a conduction band. In the typical metal oxide, the conduction band is generally formed from largely unoccupied metal cation orbitals.

The absorption coefficient is very large for band to band transitions, becoming on the order of 10^4 cm^{-1}. Consequently, this UV absorption functions as a cutoff (or "edge") at which light of energy larger than E_g is mostly absorbed and the ceramic becomes opaque. Hence, good transparency in the UV requires a solid having a large band gap. Sapphire (single-crystal Al_2O_3), synthetic fluorite (CaF_2), and UV-grade fused silica all have $E_g > 5$ eV and are used as UV-transmitting lenses and windows.

Another type of UV optical absorption in ceramics is that of exciton formation. There are many different types of excitons and for a summary the reader is referred to the work of Hayes and Stoneham (20). For this discussion, an exciton can be considered as a bound electron–hole pair formed when a photon, having an energy slightly less than the band gap energy, is absorbed. Exciton peaks in absorption spectra can be observed in many ceramics on the shoulder of the band to band UV absorption edge. The energy of an exciton peak provides a good estimate of the optical band gap energy of the ceramic.

C. Near-Infrared and Infrared Wavelengths

Absorption in the infrared is mainly due to the coupling of the light wave with optically active atomic vibrations. However, it is worthwhile to make this distinction finer by distinguishing between coupling of the light with (i) molecular vibrations and (ii) lattice vibrations.

The coupling of light with molecular vibrations forms the basis for traditional IR spectroscopy. The consequent absorptions result in many of the minima (in transmission) observed from typical IR spectra. The light wave interacts with various vibrational modes (bending, rotational, etc.) of molecular groups that are not coupled with others of the same species. Such vibrational modes can in the broadest sense be termed "localized" in order to emphasize that the vibrations are not transmitted long distances throughout the crystal.

In crystalline ceramics, coupling of the light also occurs with lattice vibrations. The difference from localized vibrations is that such non-localized vibrations, or phonons, propagate throughout the lattice of the entire crystal. There are two basic categories of lattice vibrations: optical and acoustic. Each category in a typical crystal composed of more than one chemical species of atom has both transverse and longitudinal branches. The light wave couples only with the optical phonons because it is these phonons that provide the appropriate electric dipole moments. The acoustic branches, e.g., longitudinal acoustic phonons, are optically inactive but do function to transmit compression and shear (i.e., sound) waves through the crystal.

Absorption due to lattice vibrations is responsible for the IR absorption edge in ceramic crystals. The resonance absorption wavelength, or Restrahlen wavelength, corresponds to long wavelength vibrations and is a function of the crystal structure, the masses of the atoms, and the atomic bond strength. For details, the reader is referred to nearly any book on solid-state physics. The resonance wavelength shifts to longer wavelengths for both heavier ionic masses and/or weaker atomic bonding. The farther into the IR is the Restrahlen wavelength, the wider will be the range for the IR wavelength "transparency" of the crystal. For example, ZnS has a Restrahlen wavelength of 35 μm, whereas MgO, which has lighter atomic masses and stronger bonding, has a shorter Restrahlen wavelength of 25 μm. Zinc sulfide is often used as a material for infrared windows.

A different type of IR absorption mechanism involving electronic transitions rather than phonons does occur in ceramics. This is absorption due to the excitation of conduction (sometimes called "free") electrons. Such excitations are known as plasmons. For many ceramics (and also semiconductors) the plasmon frequency happens to lie in the infrared because the concentration of conduction electrons is appropriate [see Eq. (16)] for that range of wavelengths. Very simply, the plasmon frequency, ω_p, can be derived as

$$\omega_p = (n_e e^2/\varepsilon_0 m^*)^{1/2}, \tag{16}$$

where n_e is the conduction electron concentration and m^* is the electron effective mass. For example, for ZnO of 10^{17} cm^{-3} conduction electron concentration, the plasmon frequency lies at approximately 16 or 17 μm. For metals, the conduction electron concentration is much higher, approximately, 10^{22} cm^{-3}, and hence ω_p lies in the ultraviolet region of the spectrum. This can be considered one of the reasons smooth metal surfaces appear reflective to the eye; that is, the frequency of visible light is less than the plasmon frequency for a metal and the propagation of the light in the metal is forbidden.

D. Visible Wavelengths

The vast majority of the optical absorption in the visible is due to localized electronic transitions; electronic excitations on impurity atoms and on native defects.

Native disorder, e.g., cation and anion vacancies and interstitial atoms, can form defects referred to as color centers (21, 22). Color centers can be induced in a ceramic by radiation damage, by chemical additivity (e.g., heating ZnO crystal in Zn vapor), by chemical reduction (e.g., heating TiO_2 crystal in an H_2—containing atmosphere), or by mechanical damage as from severe grinding. These color centers are structural (atomic) defects that have trapped electrons or holes and absorb light through localized electronic transitions. The most common color center is the F center (F for farbe—German meaning color). In an alkali halide, the F center consists of a halogen vacancy with one trapped electron. In a metal oxide, the F center consists of an oxygen ion vacancy with two trapped electrons. A strong absorption band arises from the excitation of the electron from a ground state *s*-like orbital to a first excited state *p*-like orbital, along with other bands due to different transitions. For a comprehensive discussion of color centers the reader is referred to the treatise by Fowler (22).

All color centers share one common feature; that is, they are inherently unstable and hence are not used for commercial purposes to give ceramics color. Color centers can be "bleached" or removed from the material both thermally (heating the ceramic) or by exposure to certain wavelengths of light (photobleaching).

The most common (and most commercially important) method of coloration of ceramics is by the substitution of transition element impurity atoms for host atoms. This is mainly the 3d transition elements but also includes rare earth impurities. The iron group transition elements possess partially filled 3d orbitals. In a uniform, spherically symmetric potential field, the 3d orbitals of an iron group cation are energetically degenerate, meaning that each of the five d orbitals have nearly the same energy. However, the d orbitals have an angular dependency, and therefore when the ion is situated on a lattice site in a crystal, the energy of the orbitals is split by the surrounding crystal field. The orbitals that are directed toward the ligands are raised in energy relative to the orbitals that are directed between the ligands.

For example, for the Co^{2+} chromophore in a tetrahedral crystal field, the t_{2g} orbitals (d_{xy}, d_{yz}, d_{zx}) are raised in energy relative to the e_g orbitals ($d_{x^2-y^2}$, d_{z^2}). The amount of splitting is referred to as the crystal field splitting energy. Because the Co^{2+} ion has the [Ar]$3d^7$ electronic configuration, the t_{2g} orbitals in the ground state are partially filled and electronic transitions between the t_{2g} and e_g can occur, thereby absorbing light. The major absorption band (actually a triplet) occurs in the red region of the visible spectrum, making such a Co^{2+}-doped solid appear blue in transmitted light. For a comprehensive treatise on the crystal field and such absorption mechanisms, the reader is referred to the work of Wood (23). Other factors, such as the Jahn–Teller effect, additionally split the energy levels, further complicating the spectra.

The wavelength of the light absorbed by a 3d transition ion is not a function of the particular transition element alone. Other factors, such as temperature, the oxidation state of the ion, the chemical nature of the particular host solid, and the nature of the ligand field, all have a strong influence on the absorption.

Rare earth element impurities are also used as commercial colorants in ceramics. In this case, these ions possess partially filled f orbitals. Similar effects occur as with the 3d transition elements; however, the electronic transitions occur between the energetically split f orbitals. Some ions commonly found to induce color are Ce^{3+}, Pr^{3+}, and Nd^{3+}.

Appendix

Symbols Used in Text and Their SI or Other Commonly Used Units

Symbol	Name	SI unit	Common usage
A	Absorbance		OD, optical density
A_r	Relative absorptance	Unitless	
a	Absorption coefficient	m^{-1}	cm^{-1}
a_e	Attenuation coefficient	m^{-1}	cm^{-1}
d	Particle diameter	m	nm or μm
e	Elementary charge	C	
h	Thickness of slab	m	cm
h_c	Characteristic penetration depth	m	cm
I_i	Incident irradiance	$J\ m^{-2}\ sec^{-1}$	$W\ cm^{-2}$
I_r	Reflected irradiance	$J\ m^{-2}\ sec^{-1}$	$W\ cm^{-2}$
I_s	Scattered irradiance	$J\ m^{-2}\ sec^{-1}$	$W\ cm^{-2}$
k	Index of absorption	Unitless	
M	Number of measurements	Unitless	
m	Relative index of refraction	Unitless	
m^*	Electron effective mass	kg	
N	Concentration	m^{-3}	cm^{-3}
N_m	Molar concentration	mol $liter^{-1}$	
n	Index of refraction	Unitless	
n^*	Complex index of refraction	Unitless	
n_e	Conduction electron concentration	m^{-3}	cm^{-3}
R	Reflectance of unpolarized light	Unitless	
R_∞	Reflectance of a semiinfinite surface	Unitless	
r	Radial distance from scattering particle	m	
s	Scattering coefficient	m^{-1}	cm^{-1}
T	Transmittance	Unitless	
x	Scattering size parameter	Unitless	

(continued)

TABLE (*continued*)

α	Polarizability	F m^2	cm^3 (Gaussian)
ε	Molar extinction coefficient		liter mol^{-1} cm^{-1}
ε_0	Permittivity of vacuum	F m^{-1}	
θ_i	Angle of incidence		Degrees
θ_f	Angle of refraction		Degrees
θ_r	Angle of reflectance		Degrees
θ_s	Scattering angle		Degrees
λ	Wavelength	m	nm
ν	Linear frequency	sec^{-1} (Hz)	
σ	Absorption cross section	m^2	cm^2
ω	Angular frequency	sec^{-1} (Hz)	
ω_p	Plasmon frequency	sec^{-1} (Hz)	

References

1. B. Hapke, "Theory of Reflectance and Emittance Spectroscopy," Cambridge Univ. Press, Cambridge, UK (1993).
2. H. C. van de Hulst, "Light Scattering by Small Particles," Dover, New York (1981).
3. J. M. Bennett and L. Mattson, "Introduction to Surface Roughness and Scattering," Optical Society of America, Washington, DC (1989).
4. C. F. Bohren and D. R. Huffman, "Absorption and Scattering of Light by Small Particles," Wiley, New York (1983).
5. W. W. Wendlandt and H. G. Hecht, "Reflectance Spectroscopy," Interscience, New York (1966).
6. G. Kortum, "Reflectance Spectroscopy; Principles, Methods, Applications," Springer-Verlag, New York (1969).
7. A. T. Young, *Phys. Today* *35*, 42 (1982).
8. J. F. Cordaro, *Am. J. Phys.* *56*, 948 (1988).
9. R. Coelho, "Physics of Dielectrics for the Engineer," Elsevier, Amsterdam (1979).
10. F. A. Jenkins and H. E. White, "Fundamentals of Optics," McGraw-Hill, New York (1976).
11. M. Born and E. Wolf, "Principles of Optics," Pergamon, Oxford, UK (1980).
12. O. S. Heavens, "Optical Properties of Thin Solid Films," Dover, New York (1991).
13. "Making Ratio Measurements—The Right Way," technical document IAN 51, Ithaco, Inc., Ithaca, NY.
14. A. Springsteen, "A Guide to Reflectance Spectroscopy," technical document RSG/92, Labsphere Corp., North Sutton, NH.
15. J. C. Slater, "Quantum Theory of Atomic Structure," McGraw-Hill, New York (1960).
16. B. Henderson and G. F. Imbusch, "Optical Spectroscopy of Inorganic Solids," Clarendon, Oxford, UK (1989).
17. S. Nudelman and S. S. Mitra (Eds.), "Optical Properties of Solids," Plenum, New York (1969).
18. F. Wooten, "Optical Properties of Solids," Academic Press, New York (1972).
19. J. N. Hodgson, "Optical Absorption and Dispersion in Solids," Chapman & Hall, London (1970).

20. W. Hayes and A. M. Stoneham, "Defect and Defect Processes in Nonmetallic Solids," Wiley, New York (1985).
21. P. D. Townsend and J. C. Kelly, "Colour Centers and Imperfections in Insulators and Semiconductors," Crane, Russak, & Co., New York (1972).
22. W. Fowler, "Physics of Color Centers," Academic Press, New York (1968).
23. D. L. Wood, Spectra of ions in crystals, in "Optical Properties of Solids" (S. Nudelman and S. S. Mitra, Eds.), p. 571, Plenum, New York (1969).

SPECTROSCOPY USING FLOWING SYSTEMS FOR CHEMICAL, PHARMACEUTICAL, AND BIOLOGICAL APPLICATIONS

EMIL W. CIURCZAK

In this chapter, I use the word fluids (i.e., flowing systems) to include gases and liquids in the physical chemistry definition (1), but in some cases one may consider a powder a "fluid." Although liquids make up the bulk of industrial fluid applications, gases are also important. In any analysis of a flowing system, the more that is known about the physical and chemical parameters, the more likely it may be controlled. Reference (2) provides some good insights into the instrumentation, but this chapter will be devoted to the philosophy of "where and why" we monitor flowing systems.

I. Defining the Sample/System

A. What Is the Matrix?

In order to choose the correct method of analysis, it must be clearly understood what effects the matrix will have on the measurement. In an ideal

ISBN 0-12-764070-3/$25.00.

Beer's law calibration, the matrix is nonabsorbing (and nonscattering) and does not interact with the analyte (3). This case is so rare (in industrial systems) as to not be worth considering. In fact, if it describes the system under consideration, this section is irrelevant. If, however, the system is like most, the matrix will be a major consideration in how the analysis is performed.

1. *Is the Analyte a Major or Minor Component?*

One of the most important considerations is the concentration of the component being measured. The larger the concentration, the more options for the analyst. When considering vibrational spectroscopic analyzers, a major component will have numerous wavelengths at which it may be analyzed. Minor components require the analyst to seek wavelengths at which they have major absorbances and, almost invariably, use multiple wavelength correlation techniques such as partial least squares (PLS) or principle component analysis.

2. *Is the Solution Colorless, Colored, or Cloudy?*

In addition to the analyte to be determined, the depth of penetration of the light beam will have a profound effect on the wavelength of light and path length selected for analytical use. Whether the method employs transmission or reflectance will often depend on the light absorbing or scattering properties of the fluid being analyzed.

B. What Is the Path (Length)?

This is not unlike choosing a car because of its headroom. The point of sampling has to be chosen with an equal amount of room for movement within and without the sampling point (e.g., pipeline). Other considerations for flowing systems include the inner diameter of the transport pipe and the ideal size of instrumentation that will fit outside the pipe.

1. *Is the Sampling Point in a Major Pipeline or a Sidestream?*

If the process may be represented by a 1-cm inner diameter tube, then measurements become relatively simple because either reflectance or transmittance may be performed. If, however, the measurements must be made in a large stream, not only path length but also physical parameters begin to seriously affect the measurement.

2. *Is the Sample under Pressure?*

Although liquids do not compress to any great extent, gases certainly do. This compression is realized as either a concentration or pseudo path-length increase. In a liquid, the hydraulic pressure on the surface of the measurement device is a primary concern. Most, but not all, commercial fiber optic probes are strengthened to withstand a pressure of at least 1000 psi or approximately 68 atm. If a high pressure is encountered, care must be taken to test any probe for pressure and leakage resistance.

When gases are pressurized, there is always a chance of the spectral signal changing spectra proportional with an increase in density. The analyst might even consider working in molality *in lieu* of molarity or g/ml. Molality does not change with the expansion or contraction of the solution.

C. Mitigating Parameters

In the "real world," samples are not well behaved. Pressure has already been mentioned, but it is merely one parameter that causes the analyst to be most careful.

1. *Viscosity*

Fluctuations in viscosity are real and common facts of chemical processes. As a polymer increases in chain length, the viscosity will surely increase—and not always in a manner resembling linearity. In the example of sugar solutions (sugar syrup) in a fermentation process, as the sugar is consumed by metabolic respiration, the viscosity will decrease. These changes affect, at a minimum, the flow characteristics, density, and pressure of the solution. These types of changes must be anticipated for an analytical method to be effective.

2. *Bubbles or Concentration Gradients*

Although at first glance the existence of entrapped gases (bubbles) and the presence of concentration gradients may not seem related, the effects of these on process analytical results are quite similar. Sudden jumps of the baseline and peak shape changes may be a result of either of these two phenomena. Debubblers (3) exist to remove entrapped gases for sidestream measurements. For main line measurements, these effects may be averaged out or modeled into a calibration equation using chemometrics (chemometrics techniques are described in other chapters of this book).

Concentration gradients may be handled in two ways: (i) slow down the measurement interval that effectively averages the gradient over time, or (ii) employ extremely rapid techniques that follow the gradient over short increments. The approach taken will depend on the application and whether a large or small change is important to process control (and, of course, which analytical equipment the manufacturer has on hand).

3. *Corrosive or Explosive Tendencies*

In theses cases, NEMA containers or, better still, remotely located electronics are called for. The obvious choice of measurement requires fiber optic probes with the light source located as far away from the process as possible. For those engaged in nuclear work, radioactive nuclides are almost always measured from a respectable distance (4).

II. Choosing the Appropriate Measuring Instrumentation

Because this book is dedicated to vibrational spectroscopy, the choices are more clear than in an open-ended text. The parameters, strong and weak points of each methodology will be discussed without prejudice. This is important because, in some cases, the best measurement choice for one process will be the absolute worst choice for another process. The analogy of a mismatched system and process may be made to a Corvette being used to haul cement for a construction project. The mismatch does not lower the intrinsic value of either the process or the measuring system, but it is a wrong marriage of two perfectly good instrument systems.

A. Infrared, Near-Infrared, or Raman

These three techniques represent some of the strongest and best analytical tools available today. Of the three, mid-range infrared (MIR) is the oldest commercial instrumentation available. (The mid-range IR extends between 4000 and 600 cm^{-1} or 2500 and 17,000 nm.) Commercial IR instrumentation developed rapidly during World War II principally to analyze synthetic rubber; MIR is outstanding as a structural elucidation tool. In recent years, rugged, rapid scanning instruments have been developed for process applications. Use of chemometrics allows for rapid, multicomponent analyses using the IR region of the spectrum.

The weakness of any MIR instrument has nothing to do with the quality of instrument engineering but rather the laws of physics: Infrared radiation is readily absorbed by nearly *everything*! Everything, including the matrix of most samples, "soaks up" MIR radiation quite readily. Extinction coefficients are rather large and light sources used in the MIR are merely heat sources and are relatively weak. The result is a loss of most MIR light energy impinging upon a sample. For most analyses, the MIR is limited to surface or thin film analyses. This is fine if samples are homogeneous. If there is a difference between the surface of the pipe or tank and the bulk of the system, there are some problems.

An attenuated total reflection (ATR) cell consists of an IR transparent material into which the IR radiation enters at a predetermined angle. The light strikes a surface at an acute enough angle such that the refractive index between the crystal and the sample causes the light to reflect back into the crystal. A long thin crystal can effect many of these reflections. At each reflective interface, a small portion of the light interacts with the sample, giving rise to a spectrum. Cells that may be placed in-line, through a reactor wall, or at the base of a reactor are commercially available.

Having intrinsically lower extinction coefficients, near-infrared (NIR) spectroscopy is increasingly being used in process measurements. (Near-infrared covers the range from 780 to 2500 nm, or the beginning of the MIR.) Because few peaks are single in NIR and the position of the peaks is often subject to hydrogen-bonding shifts, using the energy term, cm^{-1} (wavenumber), is not necessary. The term allows structural interpretation in the MIR but is inconvenient in the NIR. Also, many NIR instruments derived, initially, from ultraviolet/visible (UV/VIS) equipment, which uses nanometers for wavelength. NIR is especially useful for analyses of aqueous samples. Because of the lower absorptivities (extinction coefficients), longer path lengths may be employed, leading to a more representative sampling of the system. In nonaqueous systems, path lengths of 10–200 cm are not unheard of. At the lower end of the NIR spectrum, between 800 and 1100 nm, path lengths in the 2- to 10-cm range are used with aqueous systems.

The interfaces between NIR equipment and the process line are usually fiber optic bundles or single fibers. A simple transmission pair of fiber optic cables or a fiber bundle probe with a mirrored end (to reflect the light beam back to the illuminating probe) are the most common interfaces. A Swage-Lok fitting is often used to insert the probe into the path of flowing systems.

Because NIR is composed of overlapping overtones and combinations of bands originating in the MIR, a chemometric algorithm is needed for all but the simplest chemical systems. This is hardly a problem with the latest generation of fast processor-based computers and chemometric software.

Raman spectroscopy is fast becoming a valuable process measuring tool. Several companies have introduced Fourier transform (FT) Raman and dispersive Raman instruments designed for process control. This is important because laboratory instruments seldom are robust enough for the process environment. Because Raman is not truly an absorption type of spectroscopy, matrix effects are seldom as severe as they are with NIR or MIR. Water may be used as a solvent with no loss in signal or resolution. Glass, even tinted glasses, does not interfere with the Raman spectra; thus, simple glass view ports may be used *in lieu* of the sapphire ports for NIR or special materials used for IR.

Several of the early problems associated with Raman have been overcome. Fluorescence emission from the sample was a problem when visible light sources were used. Now, NIR sources such as the Nd:YAG (neodymium:yttrium aluminum garnet) laser diminish fluorescence and provide variable power levels for measurement optimization. The major difficulty with Raman is the lack of experienced interpretive spectroscopists and detailed texts. Even NIR seems commonplace when compared with the number of Raman applications. Only time will solve these latter problems as more applications occur and more are published.

B. Level of Sophistication

1. *Discrete Wavelengths*

In many reactions, it is often necessary to monitor only one or two discrete wavelengths. Common ingredients or functional groups, such as water, carbon dioxide, carbonyl, amine, or hydroxyl, may be followed with a minimum number of wavelengths (and effort). Numerous companies supply simple instruments that perform single measurements of from 2 to 40 wavelengths only. If 1 or 2 wavelengths are called for, sophisticated instruments are too expensive and complicated for the task.

Filters. A time-honored method of measuring a single wavelength is with interference filters. Fixed or movable types are available in both NIR and MIR. There are multifilter instruments that have improved during the past 25 years that give fast, reliable measurements. The major shortcoming of dedicated filter instruments is that they are not easily converted from one consistent measurement to another in a production setting. Nonetheless, if a process may be monitored by one inexpensive device, it is surely an extravagance to purchase a full-spectrum instrument. Note that rapid formulation changes may require multiple filter wheel configurations.

One clear advantage of the filter instrument is that there is (usually) a higher throughput of light energy than in one using a monochromator. The simplicity of the filter instrument often translates into lower development and maintenance costs.

Diodes. There are several companies that offer light emitting diode-containing instrumentation, often using wavelengths at the lower end of the NIR spectrum. MIR diodes are not commonplace or inexpensive enough to be widely used as yet. The strongest advantage of diodes is their durability. In most cases, a perfect match of wavelength needed and diode available does not exist. The diode nearest to the desired wavelength is fitted with an interference filter to produce a transmission profile for the desired wavelength(s).

For a long-running, dedicated instrument, diodes provide outstanding light sources. As with filter instruments, they may lack versatility, but this is seldom a problem with most routine analyses.

2. *Full Spectrum*

For more complex mixtures, multiple constituent analyses, or to satisfy a regulatory agency, a full-spectrum technique may be required. Numerous instrument manufacturers have either hardened their lab instruments or developed process instruments in recent years for both MIR and NIR applications. The biggest differences are in the types of monochromators employed in these instruments.

Scanning Gratings. The majority of scanning grating instruments used in production are in the NIR range. With quieter detectors (high signal-to-noise ratio) more intense light sources, and weaker absorptivities, NIR instruments have not required the throughput and multiplex advantages of interferometric instruments.

Models available are divided into pre- and postsample wavelength dispersion. Some manufacturers send white light through the sample, then impinge it onto a scanning grating and then onto the detector. Several others disperse the light with a grating, sending consecutive monochromatic wavelengths through the sample and then back to a detector. The advantage of the former is a higher light throughput, allowing longer fiber lengths, whereas the advantage of the latter is the ability to have a detector placed next to the sample, obviating the return fiber and thus saving energy and the cost of the fiber.

Both configurations are dependable but are among the slowest techniques available. Although a single spectrum may take only 1 sec to scan, the manufacturers recommend up to 32 or 64 scans be coadded for smooth, low-noise spectra. Thus, in a very fast (i.e., gas) process, these instruments may miss data. However, although the process is not exceedingly fast, these are rugged instruments to have in place. There are some moving parts that will eventually wear out, but their mean time between failure statistics are quite good. The sources must, of course, be isolated from fumes that might ignite, but NEMA enclosures are usually included in the purchase price.

Fourier Transform or Interferometric Types. The biggest leap in MIR technology occurred approximately 20 years ago. The Michelson interferometer was incorporated into commercial infrared spectrometers. (Newer instruments have more rugged corner cubes and end-swivel mirrors, but these are just improvements on the basic idea.) The speed of this type of instrument is several orders of magnitude greater than that previously described for grating (dispersive)-based types. This improvement in measurement speed and precision has made MIR an extremely useful technique, even for quantitative analysis.

The spectral improvements made in the FT-IR were not required for NIR spectrometers. However, the speed of FT instruments allows NIR spectra to be collected at ever faster rates with greater frequency domain precision. The FT approach has also made Raman a practical process tool due to the rapid data collection rate.

Diode Array. Many advantages of a fixed photodiode array spectrometer are immediately apparent. There are no moving parts, the multiple (more than 1000 in some cases) wavelength data channels obtained is instantaneous, and the data are quite reproducible. The primary spectral range for diode array elements is the short-wavelength NIR and the UV/VIS region (i.e., 200–1080 nm for silicon). Continuous measurement applications, such as gasoline blending and other clear liquids, lend themselves to (postdispersive) diode array systems. Because of the geometry of these instruments, they have not been applied to reflectance applications until recently.

Acoustooptic Tunable Filters. These are relatively new (commercially) on the process control scene. The physics of these devices are described in numerous articles (2, 5). As far as the operational performance of these systems is concerned, the best term is "fast." Although, for practical purposes, the speed of the system is comparable to diode array and FT instruments, it does have several unique features.

Because the wavelengths are produced by a radio frequency-driven transducer, random selection of wavelengths is possible. That is, unlike the grating or FT instruments, an analyst does not have to scan an entire spectrum—merely the desired wavelengths or specific sections of the spectrum. This is the feature that makes the acoustooptic tunable filter (AOTF) faster than the FT instruments for selected applications using partial scanning measurements.

III. Applications

A. Chemical

The largest number of applications, by sheer volume, would appear to be in the field of chemical synthesis. By this, I mean noningested chemicals. Drugs, foods, and food additives are listed in separate categories.

1. *Polymers*

Synthetic polymers are a fact of modern life. Bakelite, a phenolformaldehyde resin, was introduced in 1907. Styrene was used in synthetic rubbers in the early 1900s, whereas polystyrene was commercially produced in Germany in about 1930 and in the United States in 1937. Polymers are used in nearly every facet of production and are ubiquitous in homes. Although the chemistry of polymerization is well defined, often analyses lag behind in sophistication.

Even though polymer analysis by vibrational spectroscopy has been common for decades (6–9), the control of the process seems to have been overlooked. Of course, there have never been instruments as sophisticated as those existing today.

Whether condensation or addition processes are incorporated, the system may be characterized as "flowing." The monomers may start out as gasses or liquids, but the final product(s) is a viscous liquid or solid. These both "flow" at some point in the synthesis. Often, it is easier to measure the disappearance of the starting material, but following the appearance of the final product will be the best way to ensure a "good" product.

Because polymers are difficult to characterize as "good" or "bad" under normal conditions, a multilevel testing procedure is helpful. Viscosity, end-group analyses, percentage branching, cross-linking, degree of crystallinity, density, and average molecular weight are just some of the parameters tested in polymeric substances. Virtually all these may be determined by vibrational spectroscopy. In other words, not only may the process be followed for "completeness," but also the product may be totally evaluated in real time. In some cases, the production parameters may be adjusted "on-the-fly" to salvage an otherwise poor product. Because most tests are now run "postproduction," this alone is an improvement over current methods.

2. *Organic Synthesis*

For 100 years, organic chemists have performed reactions in the same manner: Place the starting materials into a vessel, react, sample, and analyze. Process chemistry is largely the same as it was a century ago! Mix! Cook! Sample! With so many ways to follow the reactions, this approach is, indeed, primitive.

Even reactions taking place at nearly the speed of enzyme-driven reactions may be followed by diode array or AOTF instruments. Certainly, slower, more conventional reactions are easily followed by nearly any instrument previously described.

With chemometrics, both the fates of the starting materials and the condition (purity) of the product may easily be followed simultaneously. Using some of the *caveats* mentioned earlier, there will, most likely, be several instruments that may be used for any particular application.

3. *Blending*

This process may sound trivial at first. When the physical chemistry involved is studied, it is seen that this is not a simple process. For example, 500 ml of water brought to 1000 ml with ethanol will not produce the same material as 500 ml of ethanol brought to 1000 ml with water. This example may be trivial, but not all mixing processes are. There are examples in which two different ratios of the same solvents may give either an endothermic or exothermic reaction (e.g., tetrahydrofuran and water, depending on which is in large excess). Whether a system heats or cools spontaneously is hardly a trivial matter.

In either case, the spectral as well as physical properties continuously change. Using refractive index or another simple device to follow mixing is never as precise as a spectral method using chemometrics.

B. Pharmaceutical/Biomedical

Although a great deal of the final products from the pharmaceutical industry are solid, the measurement of liquids is not a trivial portion of the analyses performed in the industry. Even when the final product is a solid, many intermediate steps take place in solution. Monitoring *via* spectroscopy is quite an important tool.

1. *Fermentation*

Fermentation is not an easy example with which to start. In most broths, the combination of nutrients and materials being produced may cause a hazy to opaque system. In cases such as this, the depth of penetration (of MIR radiation) along an ATR's surface is not enough to be bothered by particulates. The reaction may be followed by materials that are soluble and, thus, by definition are homogeneously distributed through the system.

If a greater depth is needed, NIR may be used. The more intense sources and lower extinction coefficients allow deeper penetration into any mixture. If the turbidity is severe enough, diffuse transmittance or reflectance are quite commonly used. Raman is another approach for highly turbid solutions because it is not as susceptible to measurement variations due to scattering. Also, although few Raman applications have been published in this field, it should seriously be considered for a measuring technique.

2. *Synthesis*

Virtually all the active substances in pharmaceutical products are synthesized in solution. The reactions are mostly followed by wet methods or chromatography. This is yet another case in which an ATR (MIR), fiber optic probe (MIR/NIR/Raman), or window (NIR/Raman) into the reaction vessel would allow for real-time analyses.

3. *"Dilution" of Live Vaccines*

Vaccines for a large number of illnesses contain active cultures: For example, measles, mumps, smallpox (essentially extinct now), and influenza. When the culture or fermentation is complete, the vaccine is diluted to the correct "titre" with liquids ranging from distilled water to inorganic suspensions. Often, several levels of vaccine are produced—all by dilution. The initial volume may be less than 100 liters. The dilution is a continuous flow and difficult to monitor in real time by any conventional method. In addition to the slow nature of biological tests, there is always an exposure hazard to the analysts sampling the vaccine.

Vibrational spectroscopic methods provide a rapid, noninvasive technique for monitoring a potentially hazardous material. A detector with a rapid feedback loop to the valves or pumps involved could allow for control of the system. Although small in volume, the value of each batch is quite high. In addition, there are usually a number of customers waiting for vaccines, so a ruined batch costs more than just money.

4. *Dissolution Studies*

Not only must a dosage form contain the correct amount of drug but also it must be able to release it into the system of the patient in a reproducible manner. One quality test used to measure this is called "dissolution." A tablet or capsule is placed in a measured volume of solvent (usually water) thermostatted to 37°C. The solution is stirred *as per* the method standard operating procedure and an aliquot or aliquots are taken for assay. The amount of drug to be released within a specified time is stated in the specifications. The assay method most often uses high-performance liquid chromatography or possibly just UV spectroscopy.

Time may be saved by measuring the solution *in situ* using a dipping fiber optic probe, by pumping through a flowcell, or by taking the aliquot to the laboratory and analyzing it in a cuvet *via* NIR. Some work has been presented on this topic (10). The NIR, MIR, or even Raman have the advantage of being more specific than UV/VIS (especially for multicomponent formulations) and are certainly faster than chromatography.

C. Food

This subject has been covered in so many publications that it will be just briefly discussed here. The topic was most popular in the first decade or two of NIR instrumentation because most of the early practitioners were agricultural scientists.

1. *Dairy Products*

Dairy products provide a natural use for a system to measure flowing product. Milk, cream, and soft cheese products have all been measured by NIR (11–14). Butter fat, moisture, oil, and protein are all easily determined spectroscopically in seconds. The most surprising fact herein might be the analysis of processed cheese spread *via* transmittance, where one might be tempted to try reflectance. It must be remembered that although visible light does not penetrate the sample, NIR might because of low extinction coefficients. Empirical data are your best test case.

2. *Vegetable Oil*

Vegetable oil is a relatively simple mixture of fatty acids and lends itself to MIR/NIR analyses. The unsaturation is easily seen in all the vibrational spectroscopic techniques discussed thus far. Because the iodine number may be easily seen, the hydrogenation of vegetable oil to margarine should also be an easy analysis to perform in real time. Multiple simultaneous determinations of iodine value, protein, moisture, and color are relatively routine in nature. Simple PLS or MLR algorithms make this an efficient method for obtaining these values in an in-process measurement.

3. *Grains*

It is not necessary for a material to be a fluid for it to flow. Wheat, rice, corn, barley, and soybeans most certainly flow to and from grain silos. It is possible to analyze these products for moisture, protein, oil, and hardness while they are flowing at a surprisingly high rate of speed (2).

IV. Conclusion

From the study of available literature, it would appear that nearly any substance that flows is able to be assayed on the fly. Rapid, rugged, and accurate vibrational spectroscopic analyzers are currently available. The plethora of chemometric algorithms available through the manufacturers and third-party vendors allows even the novice (with appropriate assistance) to go on-line with almost any assay now performed in the laboratory.

References

1. P. W. Atkins, *Physical Chemistry, 4th ed.* Freeman, New York (1990).
2. G. J. Kemeny, Process analysis, in *Handbook of Near-Infrared Analysis* (D. A. Burns and E. W. Ciurczak, Eds.), pp. 53–106. Dekker, New York (1992).
3. H. H. Willard, L. L. Merritt, Jr., J. A. Dean, and F. A. Settle, *Instrumental Methods of Analysis, 6th ed.*, Wadsworth, Belmont, CA (1981).
4. B. Buchannan and P. O'Rourke, *Spectroscopy* **9**(3), 12 (1994).
5. E. W. Ciurczak, A novel spectrophotometric system based on an acousto-optic tunable filter, *Spectroscopy* **5**(1), 10 (1990).
6. A. Elliot, *Infrared Spectra and Structure of Organic Long-Chain Polymers, 6th ed.* St. Martain's, New York (1969).
7. D. O. Hummel, *Infrared Spectra of Polymers: in the Medium and Long Wavelength Region.* Wiley, New York (1966).
8. J. L. Koenig, Raman scattering in synthetic polymers, *American Physics Society Division of High Polymer Physics*, Dallas, TX, March, 1970.
9. I. Koessler, *Infrared Absorption Spectroscopy, Encyclopedia of Polymer Science and Technology* (H. F. Mark, N. G. Gaylord, and N. M. Bikales, Eds.), Vol. 7. Wiley, New York (1967).
10. E. W. Ciurczak, The use of NIRS for following the dissolution of multicomponent pharmaceutical dosage forms, *Proceedings of the 3rd. International Conference on NIRS*, Brussels, Belgium, June, 1991.
11. A. M. C. Davies and A. Grant, *Int. J. Food Sci. Technol.* **22**, 191 (1987).
12. T. C. A. McGann, *Irish J. Food Sci. Technol.* **2**, 141 (1978).
13. D. A. Biggs, *J. Assoc. Off. Anal. Chem.* **55**, 488 (1972).
14. A. G. Coventry and M. J. Hudson, *Cereal Foods World* **29**(11), 715 (1984).

TEXTILE APPLICATION OF MOLECULAR SPECTROSCOPY

SUBHAS GHOSH
Institute of Textile Technology

I. Introduction

Most traditional textile testing procedures use wet chemistry methods that require a very high level of analytical precision to detect small changes in the materials and also require a long time to perform the tests. Furthermore, some of these chemical reagents are considered environmentally unsafe. Also, they are costly to dispose after use. The simplified near-infrared (NIR) procedures, which require almost no sample preparation take only a few minutes to perform a test, and eliminate the use of hazardous chemicals, created great interest in the textile manufacturing community. The initial success of the NIR technique in determining synthetic fiber thermal history led to the development of various applications to characterize textile raw materials, in-process stocks, and finished products.

Mid-IR analysis has been used in the textile industry for the past few years in some specific applications that are worth mentioning along with the near-IR applications. Some background information on related textile products and processes will be provided to establish the need for these measurements.

ISBN 0-12-764070-3/$25.00.

II. Textile Test Specimens

Textile test specimens vary widely in both physical and chemical characteristics; however, most samples can be characterized without any special preparation. Usual samples include fibers, yarns, and fabrics that can be presented directly to most commercial spectrophotometers. It is important to realize that variability within the textile sample is fairly high. A statistical sampling scheme is necessary to achieve a fair representation of a production lot. In the case of fabric samples, several layers of fabric or an appropriate background, such as ceramic or Teflon tile, will be necessary to create diffuse reflectance or transreflectance signals. Textile yarns and fabrics have bidirectional orientations and, hence, either a rotational sample cup or at least three rotations of the stationary cup are necessary to compensate for the differences owing to the sample orientation. Another difficulty with the textile fabrics is that they are usually dyed with various dark or pastel colors. Black and gray samples are more difficult to measure; however, using appropriate procedure dyed fabrics are analyzed quite frequently. It is not uncommon to use separate calibrations for (i) white or pastel, (ii) medium shade, and (iii) dark color samples.

III. Applications in Greige Manufacturing Processes

All processes involved in converting textile fibers and continuous filaments into fabrics are usually called greige manufacturing in the textile industry. Various processing steps, such as opening and cleaning, carding, drawing, and spinning, are involved in converting fibers into yarns. Again, yarns can be converted into fabric by different methods, such as weaving, knitting, and nonwovens. All these processes are generally referred to as greige manufacturing processes. Chemically, most textile samples are complex, predominantly showing overtones and combination bands of various C–H and O–H vibrations. Complexity with color, size, and physical configuration of the samples make the measurement difficult using reflectance spectroscopy, whereas the main advantage of the near-IR method is the simplicity of the sample preparation.

A. Fiber Blend Analysis

This is by far the most common application of the near-IR analysis in textiles. It is a common practice to mix two or more types of fibers to impart some desirable properties to the fabric. Some of the common blends are polyester/cotton, polyester/wool, polyester/Rayon, etc. These fibers are chemically very different from one another and hence it is not difficult to separate them using the near-IR method. The important consideration is that textile manufacturing needs an accuracy of measurement of less than 1%. Near-IR spectra of cotton, polyester, wool, acrylic, and polypropylene are shown in Fig. 1. Because both cotton and Rayon are cellulosic fibers, it is difficult to separate these two fiber types by the near-IR method.

Sample preparation for this application varies according to the previous history of the materials. These are two types of blending, e.g., intimate blend and draw blend. Intimate blending provides a homogeneous mixture of constituent fibers and hence the measurement is straightforward. The draw blending, however, creates an inhomogeneous mixture of fibers that necessitates additional sample preparation. A preblending of fibers is necessary using a laboratory blender that provides adequate sample preparation prior to the measurement. Normal sample cups provided by the instrument companies are adequate. If a remote fiber optic system is used the probe can be placed directly over the specimen. Even though the number of samples needed to obtain a significant average should be determined statistically, four readings per lot are usually sufficient.

The most common reference method is the gravimetric procedure in which one of the components is dissolved in a solvent and the weight of the

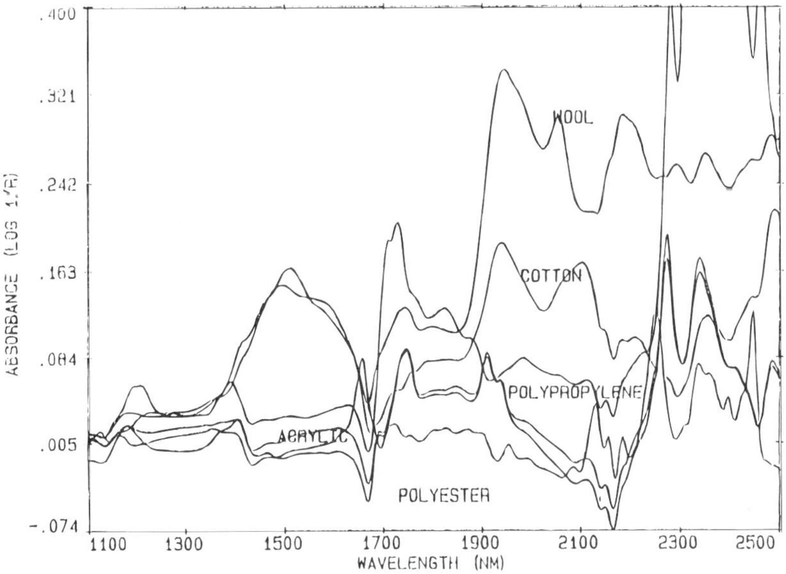

FIG. 1. Near-IR spectra of different textile fibers.

other component is expressed as a percentage of the total weight of the sample. As it appears from the spectra of various fibers (Fig. 1), band selection to separate these fibers is not a difficult task. Both the multilinear regression or a partial least square method could be used to develop a prediction model. The most common blend ratios are 50/50, 65/35, 40/60, etc. A strong prediction model can be developed without any complexity at this high concentration. The teaching set should contain 45–50 samples covering a range of blends to be determined. A knowledge of standard error of estimate of the gravimetric method is helpful to estimate the accuracy to be expected by the near-IR technique.

B. Cotton Fiber Maturity

Maturity of cotton fiber is one of the most important properties that influences the aesthetic value of the cotton fabrics. Immature fibers cause undesirable uneven dyeing and white specks such as those defects in dyed goods. A brief introduction to the cotton fiber morphology may be helpful to better understand the term maturity. A schematic description of the cotton fiber structure is depicted in Fig. 2 (1). The fiber can be described in four parts: a cuticle, a primary wall, a secondary wall, and the lumen. A cuticle is the wavy layer surrounding the fiber. Beneath the cuticle is the primary wall, which is at least 50% cellulose in the form of microfibrils. The second-

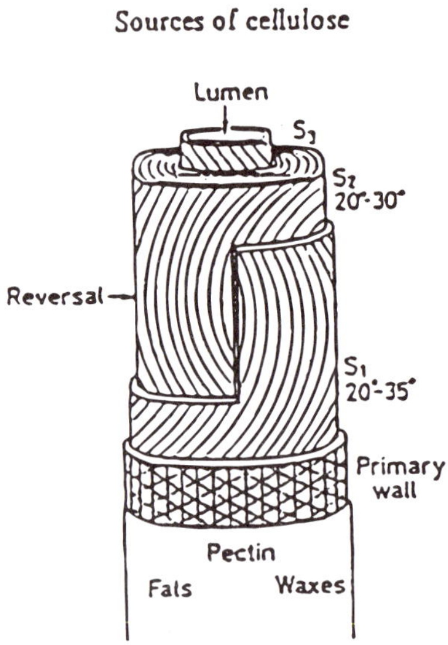

Fig. 2. A schematic diagram of a cotton fiber structure (1).

Fiber Wall Parameters

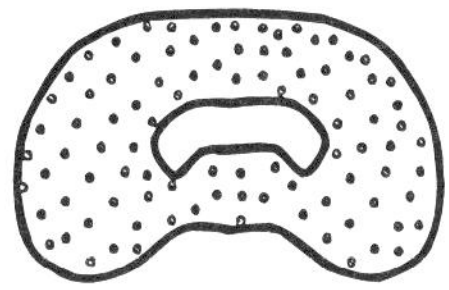

Maturity = Circularity (Theta)

$$\text{Theta} = \frac{4\,\pi\,A}{p^2}$$

FIG. 3. Diagram showing maturity calculation using the reference method (12).

ary wall is responsible for 95% of the fiber weight and is composed of two layers, S_1 and S_2. After the fiber completes its longitudinal growth, the S_1 layer of the cellulose is created from microfibrils arranged in a blended structure of alternate tight and loose packing. The S_2 layer consists of many lamellae, which have a spiraling fibrillar configuration. The lumen is the

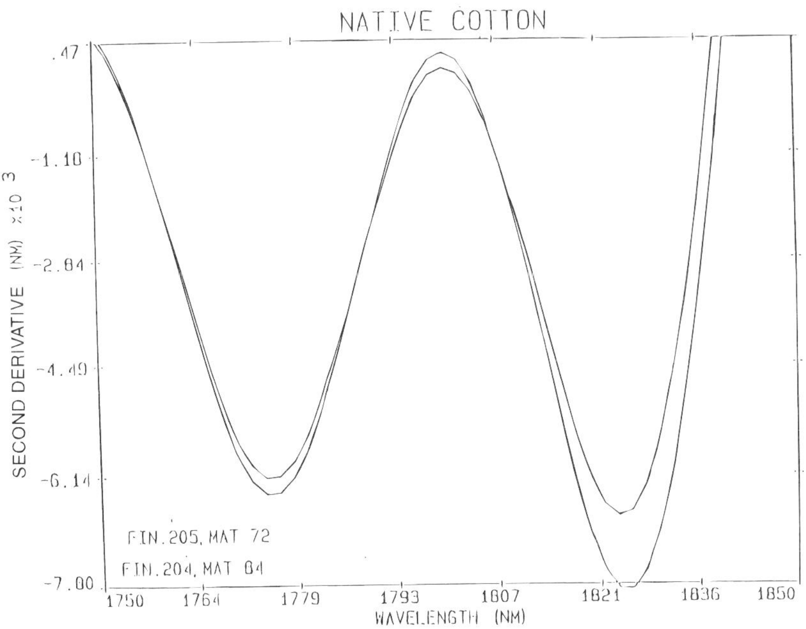

FIG. 4. Second derivative spectra of cotton fibers having the same fineness, but different maturity.

central channel of the fiber. The maturity is defined as the thickening of the secondary cellulose wall (S_2).

In the reference method (developed by Thibodeaux) 500 or more fiber cross sections are analyzed using a computerized image analyzer. As depicted in Fig. 3 (2), fiber maturity (wall thickness) is calculated from the wall area and perimeter values. The reference method, although accurate, takes a long time to characterize a sample and hence its practical use is limited.

It is important to realize that fiber fineness has a significant effect on the cotton spectra and thus fineness parameter should be included in the method development. Second-derivative spectra of cotton samples having different maturity but similar fineness are shown in Figs. 4 and 5, which show two regions near 1900 and 1820 that are more sensitive to maturity changes. Because of the variable nature of cotton fiber the teaching set should contain samples from different growth regions. A multivariate regression equation was constructed by correlating intensity values at specific wavelengths with the maturity values obtained from the reference image analysis method. Maturity values for a set of cotton from different

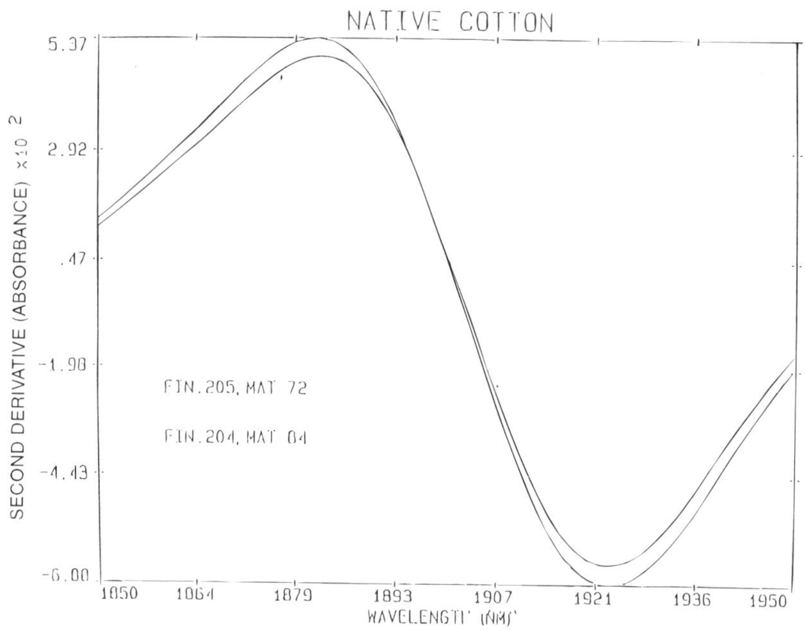

FIG. 5. Second derivative spectra of cotton fibers having the same fineness, but different maturity.

TABLE 1
MATURITY OF COTTON FROM DIFFERENT GROWTH REGIONS

Cotton type	Microscopic maturity (θ)	Near-IR maturity
Pima	0.648	78.0
Israeli HQ	0.602	77.1
Russian PV 2	0.610	75.1
Sudan X62 B	0.545	70.2
Pakistani	0.516	73.9
Sudan GB	0.495	70.4
Indian, Suvin	0.493	68.1
Turkish	0.473	71.9
Sudan DG 6B	0.303	64.2
	$r = 0.90$	

growth regions in the world are shown in Table 1, which shows a very good agreement between the reference and the near-IR method.

In a manufacturing location, usually in excess of 2000 samples are tested to engineer raw material for yarn manufacturing; thus, further automation of the near-IR technique would be helpful. Figures 6 and 7 show photogra-

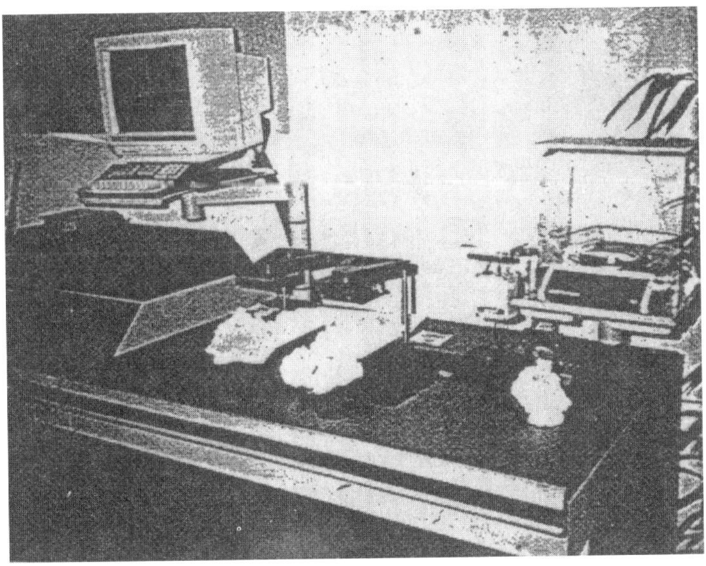

FIG. 6. Automated fiber testing line including near-IR device by Zellweger Uster, Inc.

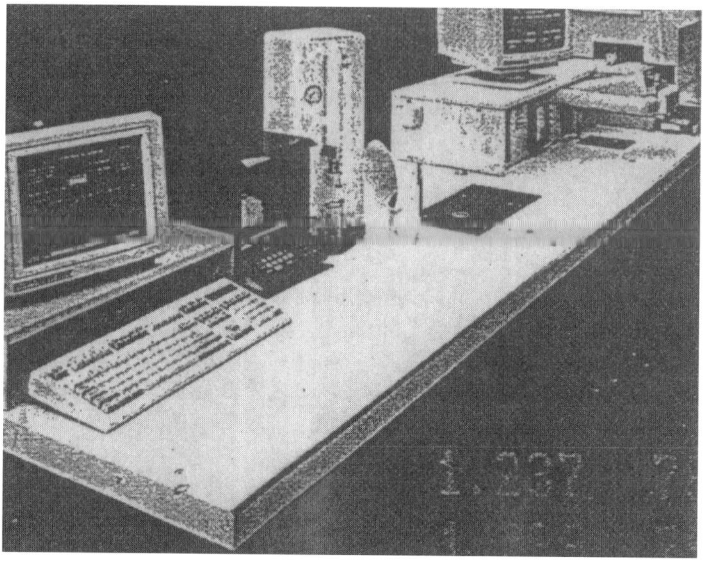

FIG. 7. Automated fiber testing line including near-IR device by Motion Control, Inc.

phs of the automated near-IR instruments along with other fiber-characterizing devices that allow a cotton sample to be tested for fineness, maturity, color, length, strength, and elongation in less than 1 min.

IV. Thermal History of Synthetic Fibers

Heat history of synthetic fibers is one of the important properties that largely control the textile properties, such as tenacity, elongation, dyeability, and aesthetics, because it affects the fine structure and morphology of the fiber. This application was developed by the authors in cooperation with James E. Rodgers (Monsanto Chemical Company). Inconsistent heat setting of the thermoplastic synthetic fibers causes undesirable defects in fabrics. When synthetic fibers, such as polyester and nylon, are subjected to thermal treatment, molecular order increases, which enhances the degree of crystallinity and crystalline perfection. Crystallinity of nylon and polyester have been successfully measured by polymer scientists using the mid-IR spectroscopy method. Absorbance spectra and second-derivative absorbance spectra of nylon samples annealed at different temperatures are shown in Figs. 8 and 9. Absorbance changes due to the thermal differences can be

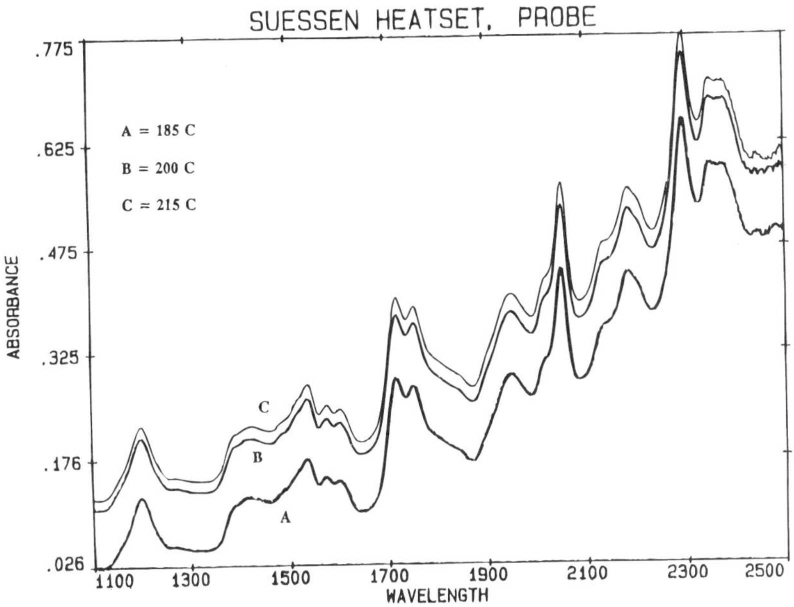

FIG. 8. Absorbance spectra of nylon yarns having different thermal histories.

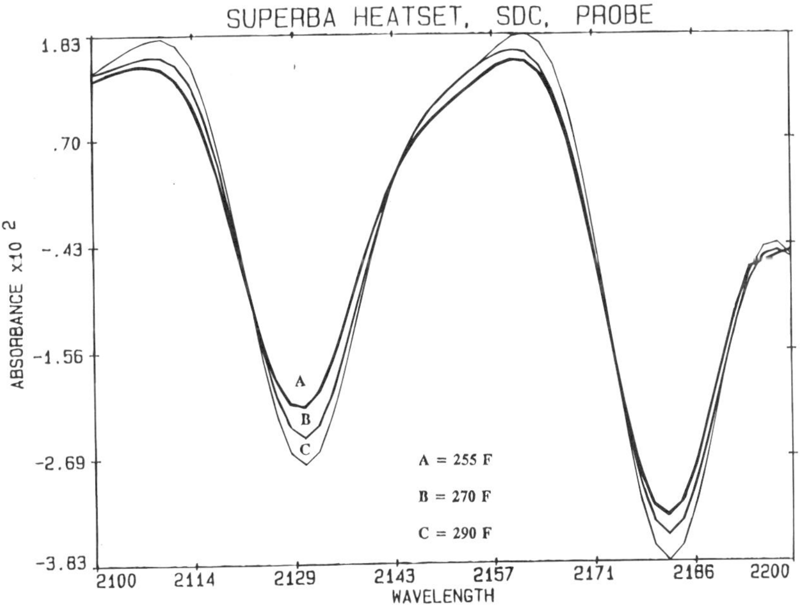

FIG. 9. Second derivative absorbance spectra of nylon yarns having different thermal histories.

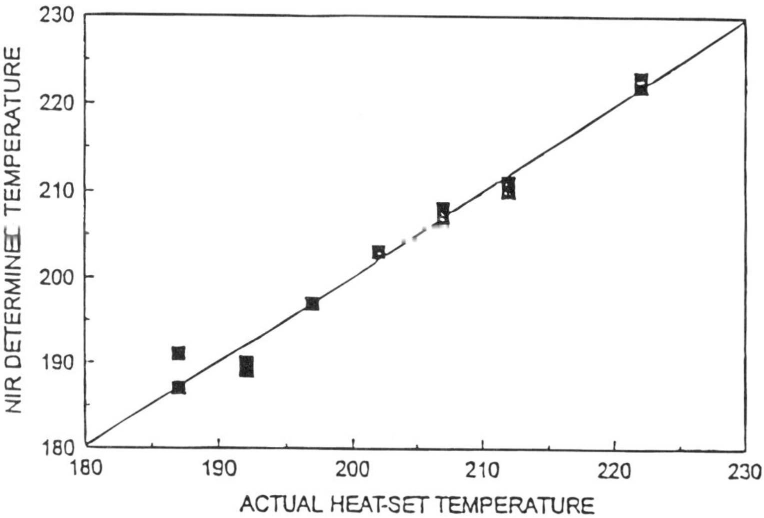

FIG. 10. Prediction of thermal history of nylon yarns using the near-IR method.

noted in the regions near 2130 and 2180 nm that are usually assigned to C=O stretch/C–H stretch combination, and amide III/N–H deformation, or N–H deformation/N–H stretch combination vibrations. Itoh (3) has studied nylon crystallinity using IR spectroscopy and reported that the amide–association in nylon chains is the main factor controlling configuration and packing of the molecular chains in an ordered structure. Because thermal treatment increases the molecular order and thus induces crystallinity, it is expected to affect the functional groups (C=O, N–H) involved in the protein-like linkages (C=O—N–H) in the nylon chains. A partial least squares (PLS) prediction model was developed from the teaching set of nylon fibers having different thermal histories. The prediction equation was used to determine heat set temperature of a set of nylon carpet yarns. An excellent prediction was obtained as seen in Fig. 10.

V. Fabric Finishing Application: Durable Press Resin on Fabrics

Fabrics containing cellulosic fibers are often treated with a resin to provide durable press properties by improving appearance retention. The most common durable press resins are dimethyloldihydroxyethyleneurea (DMDHEU) and a methylated derivative of DMDHEU that is known as DMeDHEU. They react disfunctionally with two cellulose molecules to form a covalent bond. This cross-linking prevents wrinkling of fabrics by increasing resistance to the deforming stresses applied to the fibers.

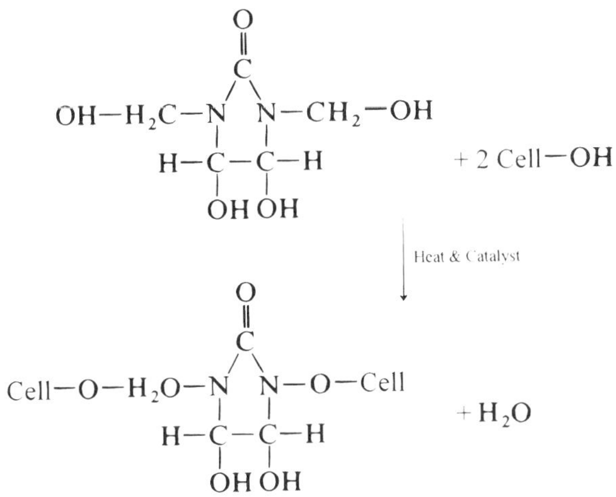

A Kjeldahl nitrogen analysis is normally used as a reference method to develop a teaching set. A problem with using the Kjeldahl analysis as a reference method for this application is that often fibers are treated with other finishing agents that also contain N_2. It is therefore necessary to know the complete finishing history of the fabrics before the measurements can be made. Spectra of the DMDHEU-treated fabrics are shown in Figs. 11–14, which show the regions that are sensitive to the changes in resin content (4). Several bands associated with the O–H, C–H and C=O vibrations are evident in both the resin and the polyester/cotton fabrics as seen in Figs. 11–14. A few common bands are expected because of the structural similarities between the resin and the fabrics. Three wavelengths can be used, such as 1452, 1722, and 1892 nm, for developing the multivariate prediction equation. These bands could also be included in a PLS model. The band near 1892 nm is related to the C=O stretching second overtone (4). The 1892-nm band is probably a better location than the wavelength at the peak maxima of the band because the maximum slope change between 1880 and 1920 nm can be detected at 1892 nm, which clearly separates the fabric treated with different levels of DMDHEU. The 1452-nm band is related to the O–H stretching first overtone (4). A broad band with shoulder is normally observed between 1400 and 1500 nm due to the O–H groups that are H-bonded to various degrees in the resin. However, the presence of cellulose as well as polyethylene terephthalete (PET) show an

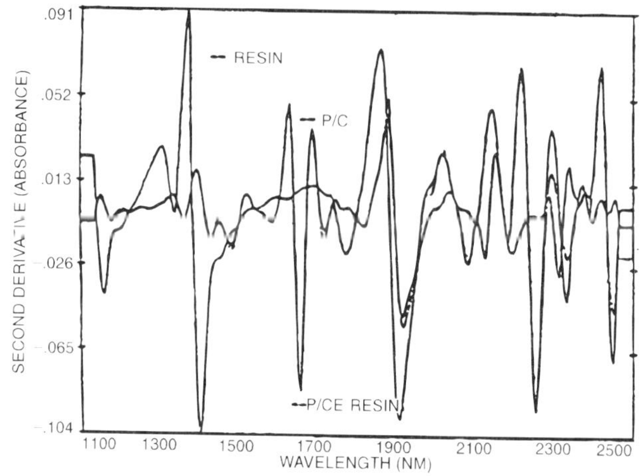

FIG. 11. Second derivative spectra of DMDHEU and polyester/cotton fabric.

O–H band in this region arising from the glycol portion of the PET structure and/or from the primary alcohol group in cellulose which further complicate the overall spectra of the resin treated fabric.

The absorption bands arising from the $-CH_2$ group when attached to tertiary nitrogen, such as $N-CH_2OH/N-CH_2-O-CH_3$, are shifted to a

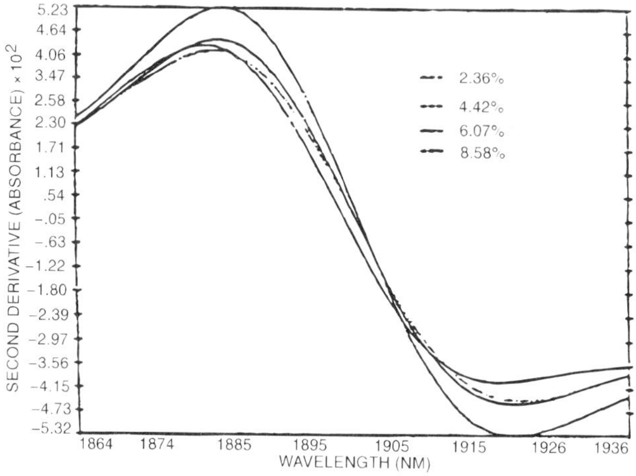

FIG. 12. Second derivative spectra of polyester/cotton fabric containing different amounts of resin.

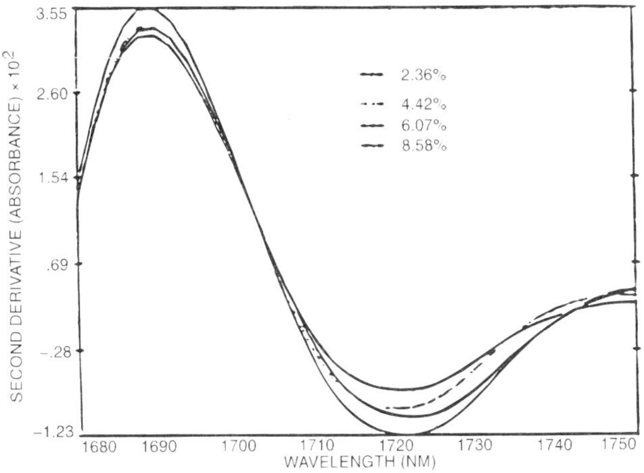

FIG. 13. Second derivative spectra of polyester/cotton fabric containing different amounts of resin.

lower value in comparison to C–CH_2OH linkages in cellulose (4). The C=O group in the resin structure polarizes into $C^+–O^-$ because oxygen is more electronegative than carbon and attracts the bonding electron more vigorously; hence, the higher nuclear change on the oxygen provides a greater attractive force than carbon (5). As a result, the C=O bond is

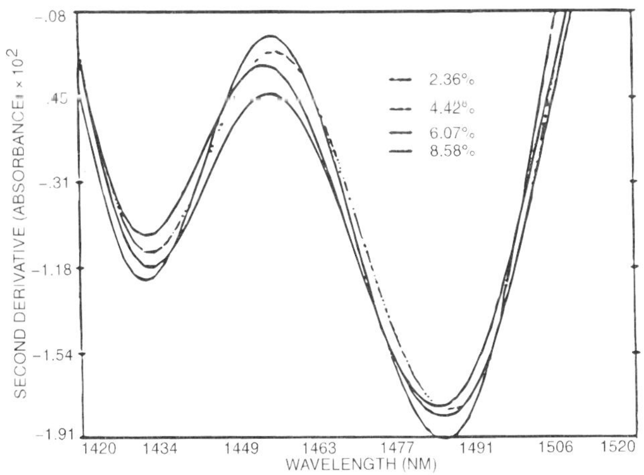

FIG. 14. Second derivative spectra of polyester/cotton fabric containing different amounts of resin.

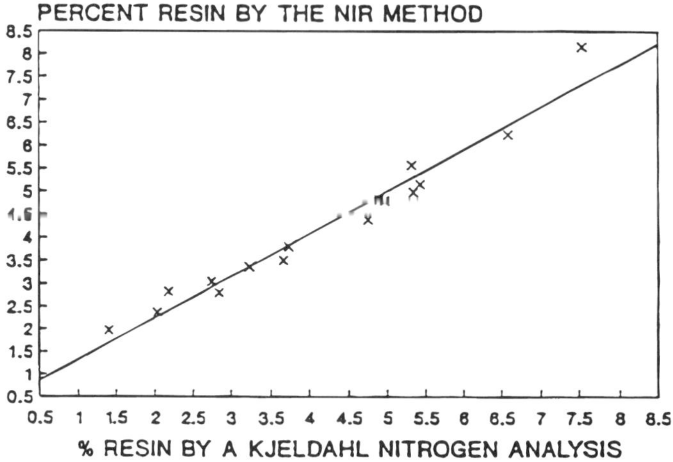

FIG. 15. Graph showing the comparison between Kjeldahl and near-IR methods of determining the amount of resin of polyester/cotton fibers.

ionized in the direction of C^+–O^-. The C^+ end of C^+–O^- draws electron density from nitrogen, which in turn influences electron flow from the carbon center of the CH_2 groups, thus affecting the absorption of both C=O and C–H bands present in the resin structure. A simple prediction model was developed using the PLS algorithm that included the bands discussed earlier and was then used to predict a set of resin-treated polyester/cotton fabrics. The agreement between the Kjeldahl and the near-IR method was very strong, with a standard error of estimate of 0·2%, as seen in Fig. 15.

In this chapter, a few applications of the near-IR spectroscopy in textiles have been discussed. Several other applications have also been developed such as sugar on cotton, size add-on warp threads, degree of mercerization, oil on fibers, etc. The near-IR method provides an excellent tool to the textile manufacturer for fast characterization of material in the high-speed processes, thus allowing a real-time product information.

VI. Application of Infrared Spectroscopy in Textiles

Infrared spectroscopy (mid-IR range) has been used extensively in characterizing textile materials for both qualitative and quantitative analyses. These tests include detection of contaminants, amount of lubricants on yarn and fabrics, and information relating to the polymer structure and morphology. In this section, the use of mid-IR analysis to obtain useful information relating to fiber morphology will be discussed using polypropylene as an example.

Consumption of polypropylene yarns and tapes (for geotextiles) has shown a steady growth in various applications such as home furnishing, automotive carpets, geotextiles structures, etc. In order to produce successful geotextile structures and other useful fabrics, polypropylene tape strength must be maintained high. It is well known that strength of the polymeric fiber, which is largely controlled by its molecular orientation and crystallinity and consequently these two fiber or tape properties, should be optimized.

A. IR Dichroic Orientation

Samuels (6) and deVries (8) used IR dichroic ratio of uniaxially oriented polypropylene fiber/tape to determine molecular orientation. The dichroic ratio is defined as the ratio of the absorbances of an absorption band (at a specific wavelength) when the electric vector of the plane-polarized infrared beam at normal incidence is first parallel and then perpendicular to a given reference direction (9). The dichroic ratio, D, is a function of two characteristic orientation angles, θ and α (6). θ is the angle that polymer chains make with the direction of drawing (longitudinal axis of the tape) and α is the angle that the transition moment makes with the polymer chain axis (10). Molecular vibrations involving atoms create electronically polarized bonds producing an oscillating dipole that is referred to as the transition moment. Fraser (8) derived a general equation relating θ and α to the dichroic ratio of an uniaxially oriented polymer. He demonstrated that for an oriented fiber with perfectly aligned chains, the dichroic ratio is a function of transition moment α as follows:

$$D_0 = 2 \cot^2 \alpha,$$

where D_0 is the dichroic ratio of perfectly aligned chains. In a more practical level, when molecular chains make an average angle θ to the direction of drawing or longitudinal tape axis orientation function f is expressed (6) as follows:

$$f = (D - 1)(D_0 + 2)/(D + 2)(D_0 - 1) = (3\,\overline{\cos^2 \theta} - 1)/2,$$

where D is the measured dichroic ratio $(d_{\parallel}/d_{\perp})$ at a given band. For an infrared band that absorbs in the crystalline region only, f is then expected to be equivalent to f_c, Herman's orientation function (11).

The 1256 cm^{-1} band however, arises from both crystalline and amorphous absorptions and, hence, infrared dichroism at 1256 cm^{-1} should show a strong correlation with average molecular orientation function, f_{av}.

Samuels (6) expressed average orientation function (f_{av}) as average orientation of each phase (i.e., crystalline and amorphous) after weighing them by the amount of each phase present, which led to the following equation:

$$f_{av} \equiv \beta f_c + (1 - \beta) f_{am},$$

where, crystalline fraction of the material, β, is usually obtained from polymer density, f_c crystalline orientation function is obtained from X ray diffraction measurements, and amorphous orientation (f_{am}) can be extracted from crystalline and total orientation (f_{av}). Usually, f_{av} is obtained from the birefringence method for tape or fiber samples. Samuels (6) developed a plot f_{av} versus $(D - 1)/(D + 2)$ for the 1256 cm^{-1} band for oriented isotactic polypropylene samples. A good linear relationship with a zero intercept has been published as illustrated in Fig. 16.

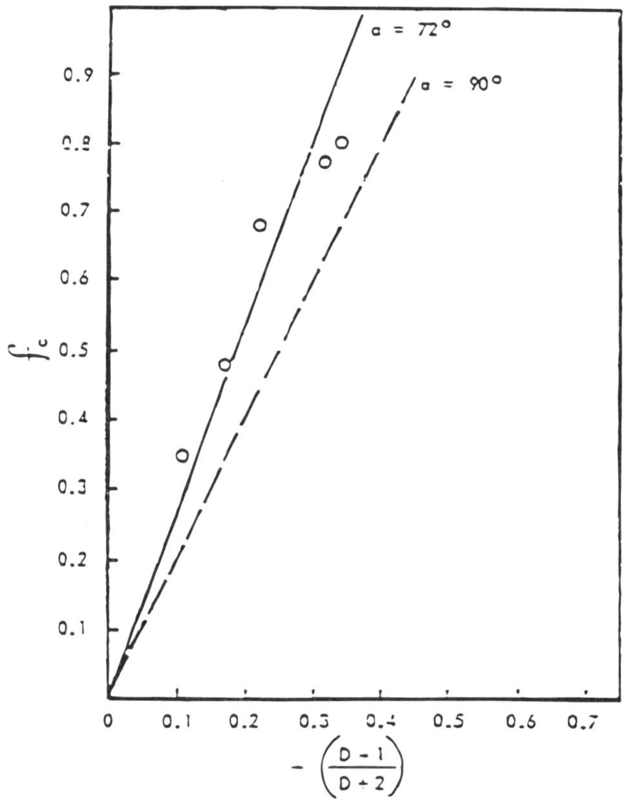

FIG. 16. Relationship between orientation function of polypropylene and IR dichroism at 1220 cm^{-1} band (6). Samuels, R. J., *J. Polymer Sci. A*, **3** (1965), Wiley, NY. Reprinted by permission of John Wiley & Sons, Inc.

From the foregoing discussion, it is evident that IR dichroic ratio of polypropylene is related to molecular orientation of polypropylene tape or fiber when the 1256 cm^{-1} band is chosen, whereas crystalline orientation can be obtained using the 1220 cm^{-1} band. The polypropylene tape orientation can be determined using IR transmission spectra. Tape mounting arrangement on a Perkin–Elmer 881 IR spectrophotometer is illustrated in Fig. 17. A microspecimen holder can be used for sample presentation that enables 2 or 3-mm wide single tape to be presented to the instrument. A Willis 4 × beam attenuator is used to minimize the loss of intensity in the transmissive mode.

The ratio of absorbances at the 1256 and 1220 cm^{-1} bands is measured when the electric vector of the plane-polarized beam at normal incidence is parallel to and perpendicular to an arbitrary reference direction. The reference direction is usually chosen as the direction of stretch (tape drawing) and, hence, tape longitudinal axis, at which the band intensity is maximum. The dichroic ratio of an absorption band is determined by placing the polarizer at the light source. Orientation function (f_c or f_{av}) is calculated using the Samuels' (6) relationship:

$$f_{av} \text{ or } f_c = \frac{D-1}{D+2}; \quad D = d_{\parallel}/d_{\perp},$$

where D is the dichroic ratio, $d_{\parallel}$ is the intensity of parallel polarized light, and $d_{\perp}$ is the intensity of perpendicular polarized light.

Orientation functions derived from dichroic ratio of the isotactic polypropylene films stretched at different draw ratios are shown in Table 2.

FIG. 17. Illustrates tape mounting on a Perkin-Elmer 881 IR spectrophotometer.

TABLE 2
ORIENTATION FUNCTIONS OF POLYPROPYLENE TAPES AS DETERMINED BY IR DICHROIC RATIO

		Average orientation		
Sample	$d1256_{11}$	$d1256_{1}$	$D = d_{\parallel}/d_{\perp}$	$f_{av} = D - 1/D + 2$
GWE 4:1	17.30	0.96	18.03	0.850
GWE 6:1	17.75	0.90	19.72	0.862
GWE 8:1	19.50	0.90	21.60	0.873
Repsol 4:1	15.00	0.90	16.67	0.839
Repsol 6:1	15.30	0.87	17.64	0.847
Repsol 7:1	17.25	0.90	19.16	0.858
Repsol 8:1	17.50	0.90	19.44	0.860
		Crystalline orientation		
Sample	$d1220_{1}$	$d1220_{11}$	$D = d_{\perp}/d_{\parallel}$	$f_c = D - 1/D + 2$
GWE 4:1	5.55	0.59	9.41	0.737
GWE 6:1	7.32	0.50	14.64	0.820
GWE 8:1	7.56	0.50	15.12	6.825
Repsol 4:1	4.74	0.75	6.32	0.639
Repsol 6:1	5.20	0.60	8.67	0.719
Repsol 7:1	6.09	0.50	12.18	0.788
Repsol 8:1	6.24	0.45	13.87	0.811

B. ISOTACTICITY/CRYSTALLINITY BY IR SPECTROSCOPY

Most polypropylene used in various textile products is produced from isotactic polymers. Isotacticity of polypropylene has significant influence on tape or fiber property. Only iso- and syndiotactic polymer have sufficient molecular order to create polycrystalline structure and hence acceptable physical and tensile properties for oriented tapes. Isotactic polymer chain is characterized by the arrangement of the functional group on one side of the plane.

IR crystallinity and isotacticity of polypropylene are not always clearly distinguishable parameters (12); hence, it is often considered a composite measurement of the two. IR spectra of both isotactic and atactic polypropylene are illustrated in Fig. 18 as published by Luongo (12). However, changes in infrared band intensities at 995 cm^{-1} which relate to the ordering of helical, isotactic chains will most likely reflect variations in crystallinity of tapes or filaments produced from the same polypropylene material.

The amorphous polypropylene material does not show well-defined absorption and broad bands with shoulders are observed that impede accu-

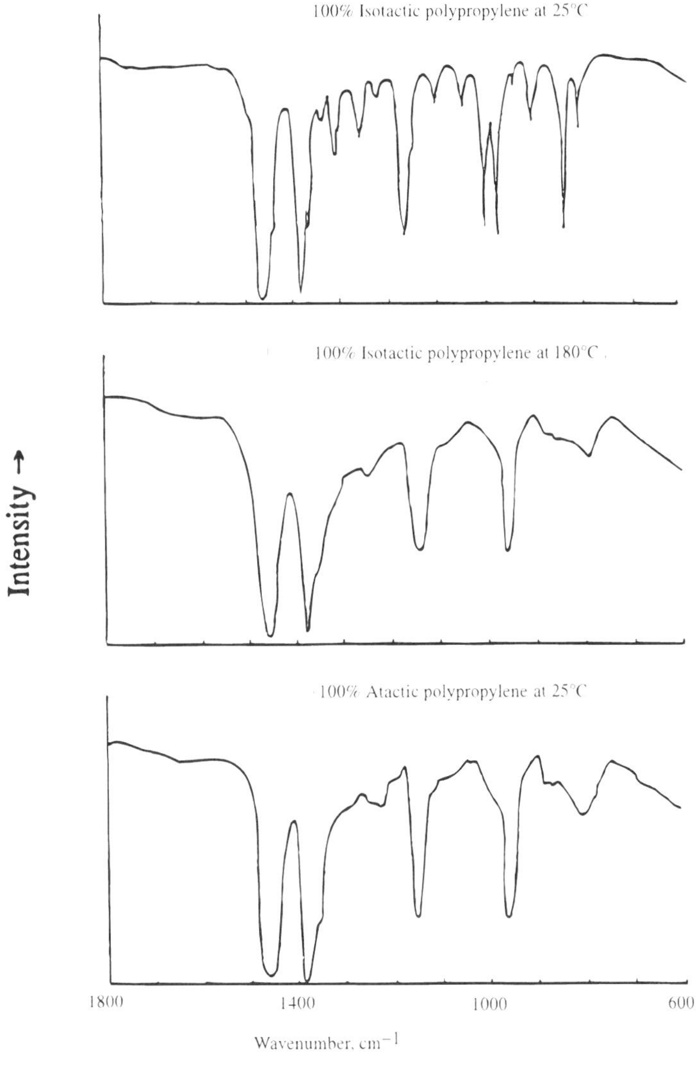

FIG. 18. IR transmission spectra of isotactic and atactic polypropylene. (12). Infrared study of polypropylene. *J. Appl. Polymer Sci.* **3**(9) (1960). Wiley, NY. Reprinted by permission of John Wiley & Sons, Inc.

rate quantitative measurements. The measurement of amorphous content has been attempted (10) using the background absorption underlying the crystalline bands in the 750–920-cm^{-1} region. It is more common practice to determine crystalline content of polypropylene using the intensity ratio at 995 and 974 cm^{-1} from the transmission spectra, where the 995-cm^{-1}

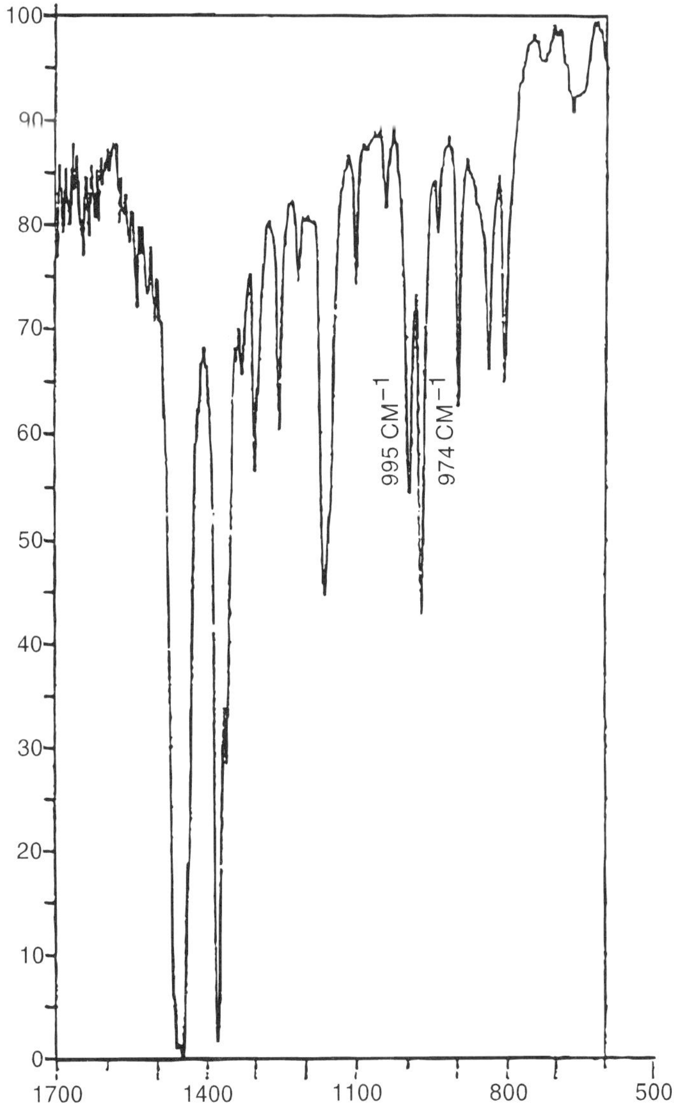

FIG. 19. IR transmission spectra of polypropylene showing the crystalline band at 995 cm^{-1}.

TABLE 3
IR CRYSTALLINITY VALUES

Polymer type	Draw ratio	Crystallinity fraction
ICI GWEZP PP	4:1	0.733
	6:1	0.740
	8:1	0.794
Repsol Isplen PP	4:1	0.759
	6:1	0.796
	8:1	0.802

band is highly sensitive to crystallinity changes in comparison to the 974-cm^{-1} band as shown in Fig. 19. It is recommended that at least eight scans averaging should be performed to achieve high-precision measurements. The changes in polymer crystallinity owing to the different draw ratios in tape manufacturing are shown in Table 3.

References

1. Nevell, T. P., and Zeronian, S. H., *Cellulose Chemistry and Its Applications*, pp. 15–180. Wiley, New York (1985).
2. Thibodeaux, D. P., and Price, J. B., Reference method for determining maturity of cotton fibers, Melliand Textilberlechte, 4/1989, E95-96.
3. Itoh, T., Changes with temperatures in crystal structures of nylon 6, 66 and 610. *Jpn J. Appl. Phys.* **15**(12), (1976).
4. Ghosh, S., and Brodmann, G. L., On-line measurement of durable press resin on fabrics using the NIR spectroscopy method. *Textile Chem. Colorist* **25**(4), 11–14 (1993).
5. Streitwieser, A., and Heathcock, C. H., *Introduction to Organic Chemistry*, 2nd ed., p. 357. MacMillan, New York (1981).
6. Samuels, R. J., *J. Polymer Sci. A*, **3** (1965) Wiley, New York.
7. deVries, A. J., International Union of Pure and Applied Chemistry, Macromolecular Division, Rhóne-Poulenc Research Center, Aubervilliers, France (1981).
8. Fraser, R. D. B., *Analytical Method of Protein Chemistry* (Alexander and Block, Eds.), Vol. 2. Pergamon, London (1960).
9. Gailey, J. A., *Anal. Chem.* **33**(13) (1961).
10. Miyazawa, T., *Vibrational Analysis of the Infrared Spectra of Steroregular Polypropylene, The Stero Chemistry of Maeromolecules* (A. D. Kelley, Ed.), Vol 3. Dekker, New York (1967).
11. Hermans, J. J., Hermans, P. H., Vernon, D., and Weidinger, A., *Rec. Frav. Chem. Pays-Bas* **65** (1946).
12. Luongo, J. P., Infrared study of polypropylene. *J. Appl. Polymer Sci.* **3**(9) (1960) Wiley, New York.

SOLAR MEASUREMENTS

J. MURRAY STEWART
Pilkington-Libbey Owens Ford Company

I. Introduction

The sun's energy at the earth's surface varies continuously in time and space. The solar constant derived from ground-based measurements has varied from 1323 to 1428 W m^{-2} (1). Outside the atmosphere (air mass = 0) the sun's energy is more predictable (American Society for Testing and Materials [ASTM, E490]). At the annual mean solar distance from the sun, we define a solar constant of radiation for the earth. This is the measured amount of solar energy flux incident normally on a unit area in a unit of time ($1.373 \pm 0.008 \times 10^6$ erg/sec cm^2). Typical atmospheric absorbers (water, ozone, CO_2, dust, and industrial pollutants) scatter and absorb

ISBN 0-12-764070-3/$25.00.

energy at selective wavelengths over the solar range from 0.3 to 2.5 μm. Scattered skylight plus the direct sunlight is referred to as the global radiation. It has a different spectral energy distribution than direct sunlight at the earth's surface, adding energy to blue and ultraviolet (UV) wavelengths (2, 3). The absorption effects of average amounts of atmospheric absorbers, superimposed on the solar spectral energy distribution above the earth's atmosphere, simulate solar energy at the earth's surface. A number of proposed standard solar irradiance curves (W/m^2 μm) are used to characterize solar properties of materials (4). When the sun is overhead, solar energy penetrates unit thickness of air (air mass = 1). The absorption increases as the path length increases. Both vary approximately as the reciprocal sine of the sun's altitude angle. When the sun is 30° above the horizon, for example, the solar energy must pass through twice as much air (air mass = 2) as it would if the sun were at the zenith position. A small correction must be applied to this simple relationship to correct for atmospheric refraction, especially if the sun is lower than 30° from the horizon (2) (Fig. 1).

Atmospheric density decreases disproportionately with altitude. Most of the atmosphere is below an altitude of 57 miles and 20% of it is under 1 mile above sea level. All these attributes of solar energy are constantly changing. A constant value must be chosen for each in order to compare the solar spectral properties of materials. Effective communications about an object's solar properties must be limited to its spectral response under specified solar energy conventions and under standardized measuring conditions.

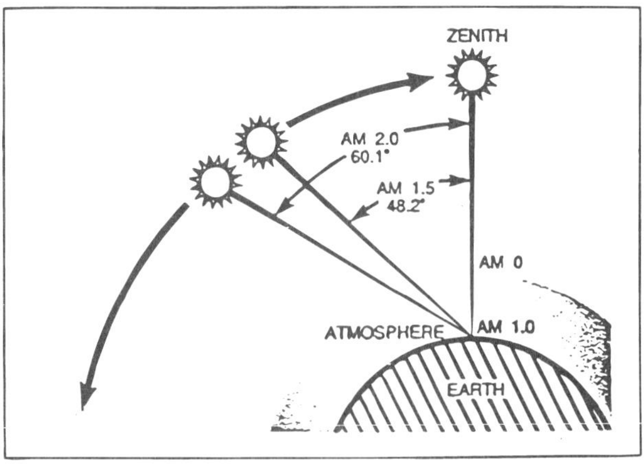

FIG. 1. Air masses describe the path length through the atmosphere (25).

Pyroheliometers, pyranometers, thermocouples, bolometers, solar arrays, and spectroheliographs measure solar energy directly (5). They are radiometers that convert solar heat into electrical current (6). Their measurement fluctuations are generally so large over time, due primarily to moving clouds, that average values are the only meaningful quantities of solar energy associated with an object's solar properties at sea level. This chapter will not deal with direct solar measurements of this type. Discussion will be limited to laboratory spectrophotometric measurements of specimens and to the indirect calculations of their solar spectral properties (7). For example, flat glass products filter sunlight to reduce heat inside cars and buildings. Spectrophotometric transmittance and reflectance data can be used to calculate a specimen's protective power from conventional solar energy (8). Most of the terms used in this article are defined in ASTM *Standard Definitions of Terms and Symbols Relating to Molecular Spectroscopy* (9).

II. Spectrophotometer: Principles and Geometries

Specimens must be measured correctly before spectral properties can be determined by calculation conventions. Scanning spectrophotometers require a number of important optimizing operations during calibration and sample runs (10). These operations ensure wavelength accuracy and a correct spectral bandpass to achieve good photometric accuracy (11). Spectral bandpass is a variable because detectors are not uniformly sensitive to all measured wavelengths nor do sources radiate with uniform intensity at these wavelengths. Detectors in particular (Fig. 2), and also many sources, cannot operate over the entire solar wavelength region (300–2500 nm). Spectrophotometers that operate over this range usually must employ two sources and two detectors to obtain the signal-to-noise ratio for accurate photometric measurements. An example of this type of instrument is diagrammed in Fig. 3.

Detector sensitivity generally falls off quickly as wavelength scans progress away from their optimum wavelengths. Minimum photometric sensitivity occurs at the crossover wavelength of two detectors covering the solar wavelength region. Detectors are chosen to maximize sensitivity at this point between the red and infrared spectral regions as illustrated in Fig. 4.

A. Instrument Geometries

Most high-accuracy spectrophotometers incorporate an integrating sphere attachment to collect diffuse reflections or transmissions from specimens.

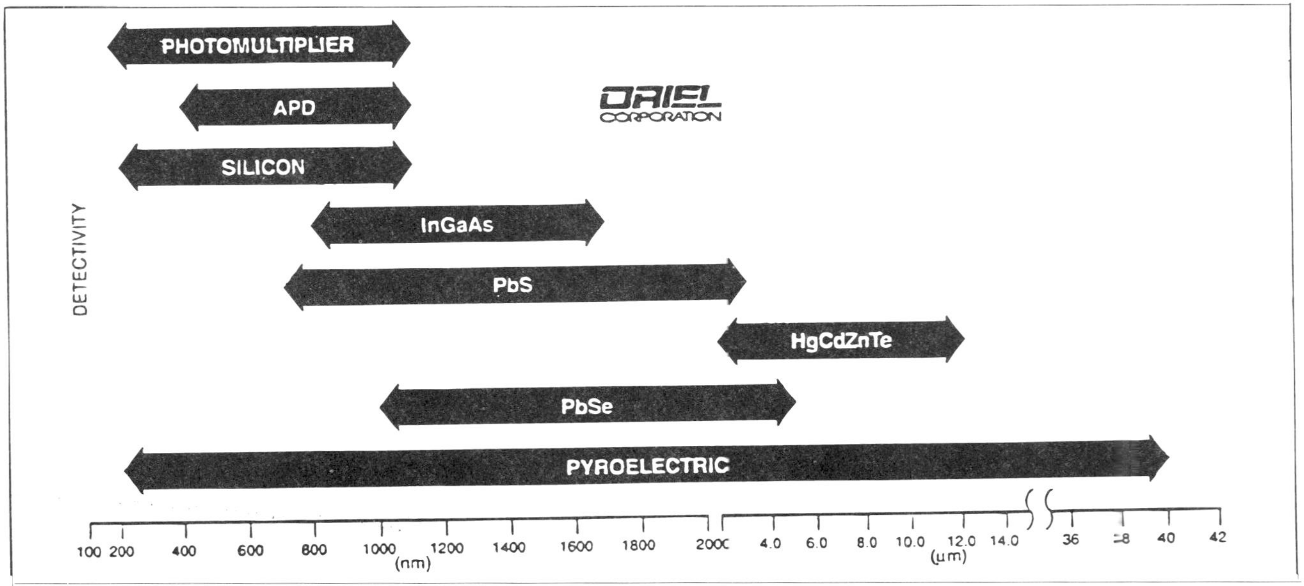

FIG. 2. Usable wavelength ranges of Merlin detectors (25).

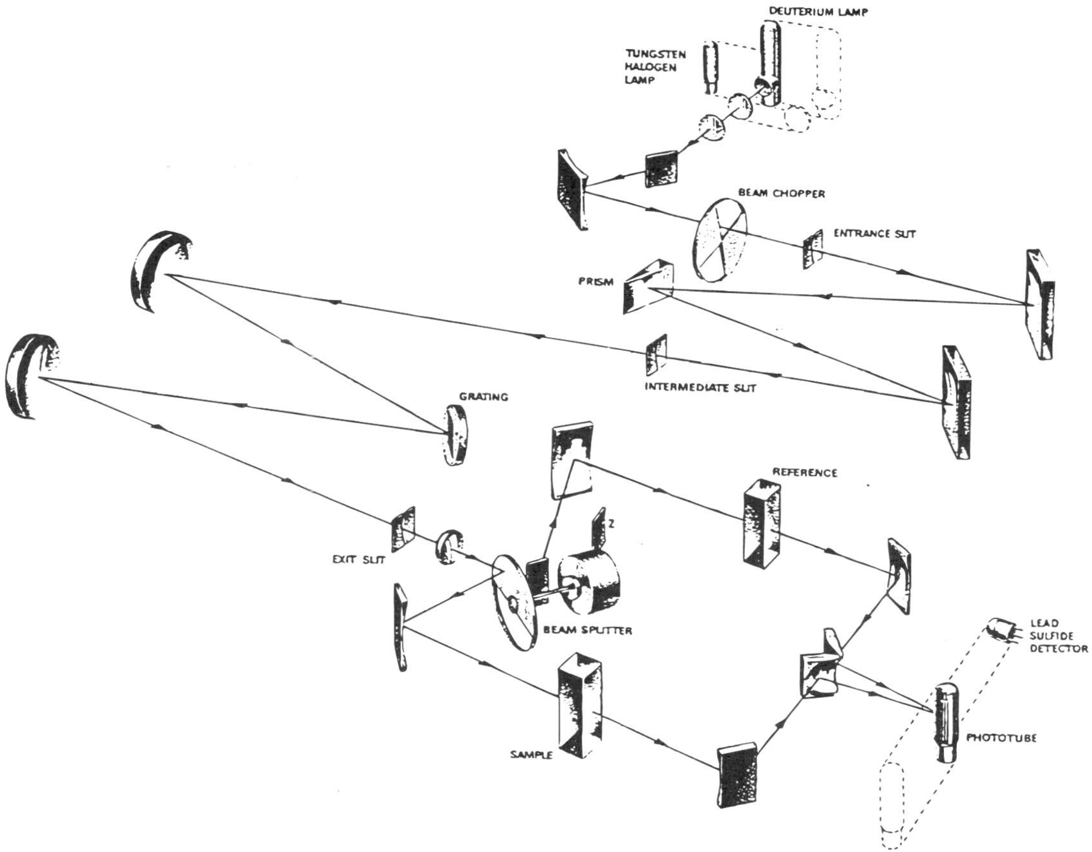

FIG. 3. Cary 17 optical diagram (from Cary 17 Recording Spectrophotometer Bulletin No. 117, 1970).

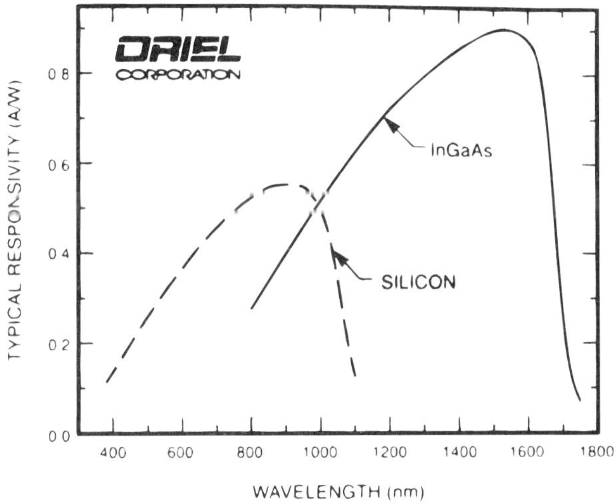

FIG. 4. Typical responsivity of Si and InGaAs detectors (25).

Double-beam instruments may compare the spectral response of a specimen in the sample beam to the diffuse white coating on the sphere wall. See also the discussion on instrument limitations. If the sample is primarily a specular reflector or transmitter the reference beam should contain a similar type

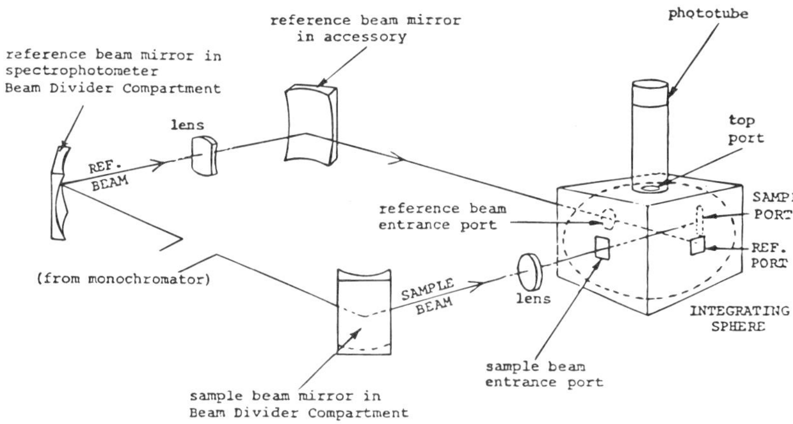

FIG. 5. Type I illumination system with integrating sphere (from Cary Model 1711 diffuse reflectance accessory manual).

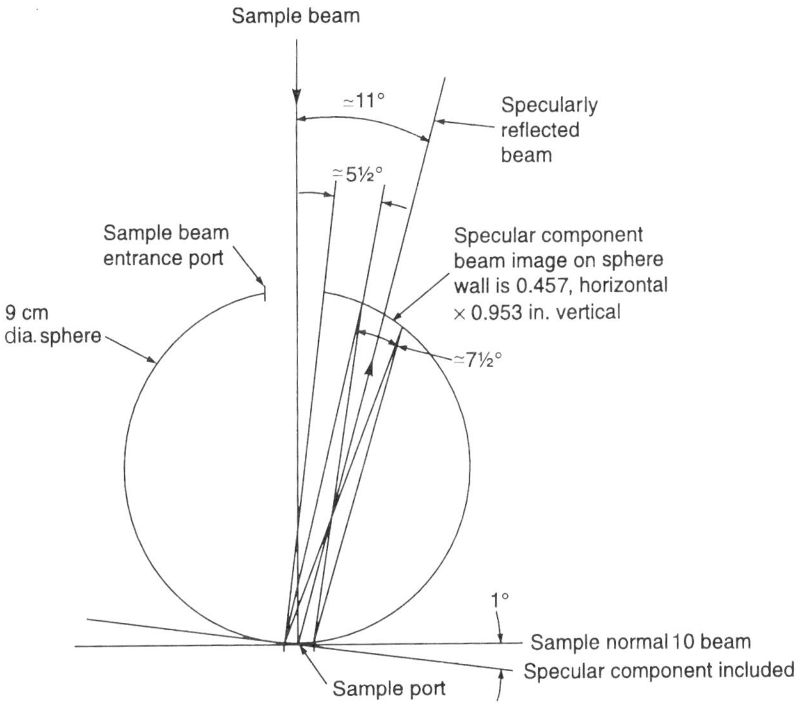

FIG. 6. Sample beam geometry when specular component is included in measurement (from Cary Mode 1711 diffuse reflectance accessory manual).

of standard. Figure 5 provides an example of an instrument geometry for this application.

Sphere geometries vary considerably to accommodate different instruments and specimen types. Figure 6 shows a typical example of a sphere's optical design considerations required to include or exclude a specimen's specular reflectance component. Convenient cross-tables of applications and instruments are available from some instrument companies (12, 13). Useful guides to help one choose and use the right instrument geometry to measure transmittance or reflectance of specific specimen types are available from ASTM (14). If spectral measurements must be made at angles to a specimen's surface, a standard practice is available from ASTM (15).

III. Instrument Applications

A. Instrument Calibrations: Standards

Absolute reflectance measurements can be made on front-surface opaque mirrors using the strong "V-W" reflectance attachment (16) diagrammed in Fig. 7.

An absolute method of checking a spectrophotometer's photometric linearity exists. It combines the use of Bouguer's law and the superposition of optical fields (17). A simple example of this technique from Hawes' paper explains the method clearly. Two neutral density filters are measured separately and then together. If the transmittance of one is 57.19 and the other is 48.66, the transmittance of the two together, separated by an air space, would be 57.19·0.4866 = 27.83. Other filters can be measured separately and combined in multiple stacks to check various photometric levels.

Many spectral calibrations on primary instrument are carried out using glass standards issued from a national standards issuing laboratory. Unfilmed glass standards are very stable reference materials but are checked by national laboratories periodically anyway (18). Glasses doped with rare earth oxides, such as didymium and holmium oxide, are stable transmittance standards that can be used to calibrate wavelength scales (19). Holium oxide crystals desolved in perchloric acid can be used as a liquid wavelength calibration standard in a cuvet (19). In powder form, these same rare earth oxides can be pressed into solid diffuse reflecting near-infrared (NIR) wavelength standards (20). The National Institute of

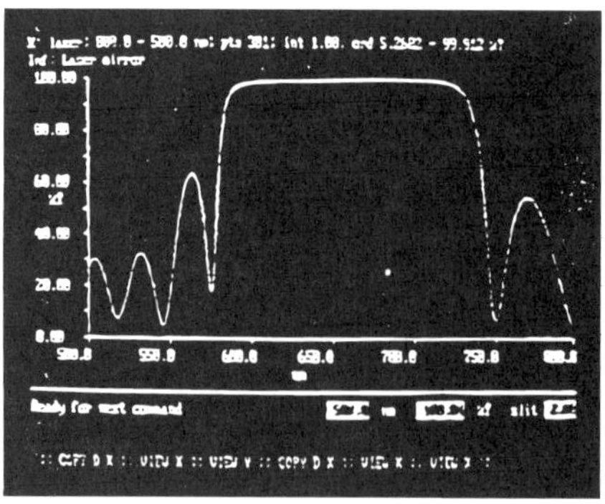

Fig. 7. The reflectance accessory according to Strong is a tool to measure absolute values from high-reflectance mirrors. Results are obtained in R^2 that can be easily converted into percentage reflectivity. The spectrum shows the reflectivity of a laser mirror for a HeNe laser (12).

Standards and Technology [NIST; formally the National Bureau of Standards (NBS)] occasionally issues special publications describing their standards testing programs (21). These pamphlets are instructive and informative about calibration materials and about measurement issues and techniques. Specimens sent to these national laboratories for measurement are returned with certificates of their spectral values as well as a valuable description of the measuring process (22, 23). Lists of available solar spectral standards are obtainable from a number of sources, such as NIST's standard reference material (SRM) annual catalog, the Photonics Directory, the Lawrence Berkeley Laboratory (24), and standards supply companies (25). Most solar spectral reflectance standards are calibrated opaque, specular metal mirrors with spectral curves like those in Fig. 8. These pure metals are soft and the mirror standards can be destroyed even with gentle handling or cleaning if they do not have a protective overcoat. Second-surface laminated SRM mirrors are available from NIST to avoid this problem.

These standards are also more appropriate as reference materials when they are used to measure second-surface specular samples with an integrating sphere attachment because they have similar geometric dimensions. Gold and rhodium do not tarnish but are the most expensive standards. Rhodium mirrors are extremely hard and spectrally "flat" at a fairly high level of reflectivity over most of the solar spectrum (26). Silver and aluminum tarnish and scratch easily. They make good temporary transfer and working standards but must be recalibrated as they age. Polished solid copper coupons or Cu-filmed glass are the cheapest standard specular reflectance materials recommended for calibrating instruments in the NIR and IR wavelength ranges. After a spectrophotometer has been calibrated with one of these transfer standards from a national testing laboratory, one should measure a representative product specimen. These will become

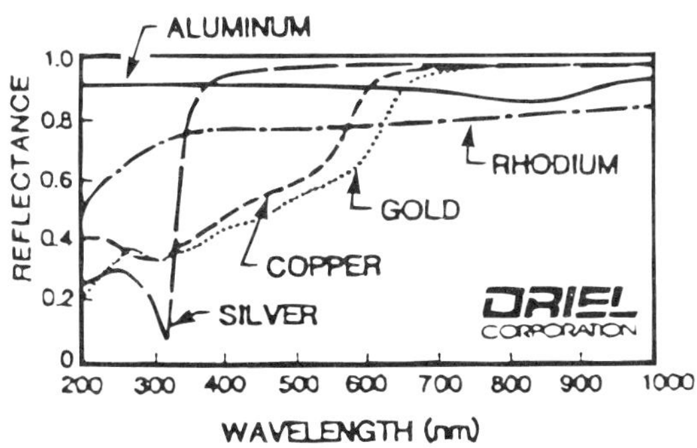

FIG. 8. Reflectance of freshly deposited metallic coatings (25).

product standards used to calibrate QC equipment. If a potential product standard is inhomogeneous, has a textured surface, or discolors on exposure, spectral measurements may not be reproducible. These specimens should not be used even as temporary transfer standards. Much care is taken to ensure stability, homogeneity, and critical geometric shapes of good standards. The NIST measurement assurance program (MAP), for example, examines these sample parameters and others critically before adopting specimens as round-robin test specimens (27, 28). Joint calibration round-robin programs such as MAP are just one part of a developing plan for society to advance toward more accurate spectral measurements. In Europe ISO 9000 standards are being adopted not only to calibrate instruments similarly but also to develop procedures for uniform sampling and data management.

B. Sample Types: Physical Considerations

From 0.3 to 2.5 μm refraction of light changes considerably under the influence of a prism, grating, lens, or specimen. Instrument components are designed to overcome refraction effects without specimens. It is important to realize that a specimen becomes part of the instrument optics during measurements. The physical shape, size, surface texture, and hardness of a specimen are important technical concerns impacting on instrument measurements of it after calibration with a dissimilar standard. Consider what effects the specimen types in Figs. 9–12 will have on the calibrated instrument. If the specimen is semitransparent, these concerns must be expanded to include specimen thickness, scattering, and absorption. Opaque specimen

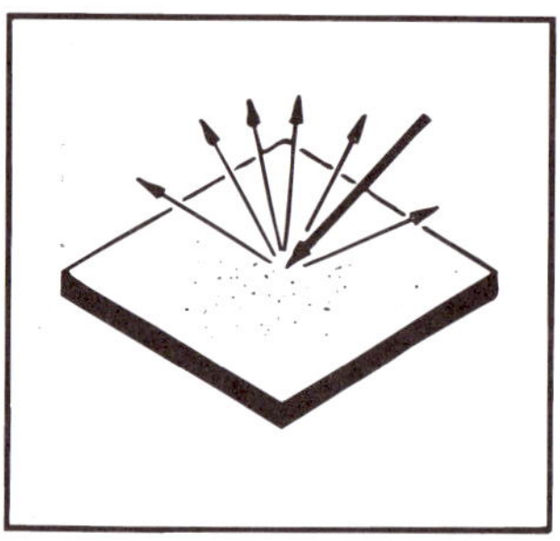

Fig. 9. Diffuse reflection is characteristic of light which is redirected over a range of angles from a surface on which it is incident. Diffuse reflection accounts for more of the color than any other type of distribution because most objects are opaque and reflect light diffusely (from HunterLab).

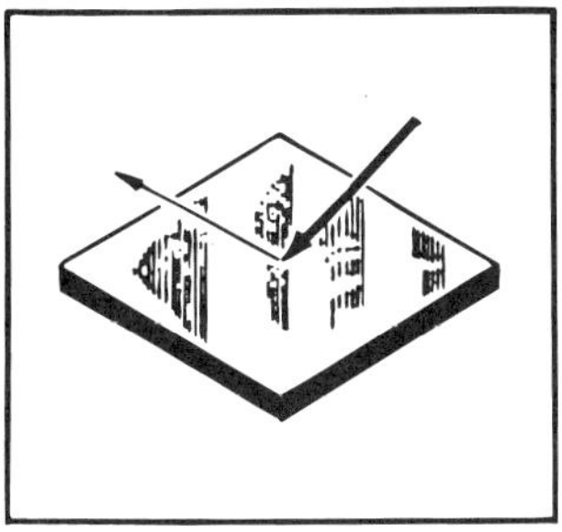

FIG. 10. Specular reflection is reflection as from a mirror. Specular reflection is thus highly directional instead of diffuse. Specular reflection is what gives objects a glossy or mirror-like appearance. There are a variety of ways of assessing or "seeing" this glossy appearance (from HunterLab).

spectral data must be examined for aborations caused by physical properties, such as surface flatness and smoothness, that influence flop angle, and specular and diffuse reflectivity. The effects may be in a local wavelength region, at a particular angle of measurement, or a combination outside the calibration parameters. Important information about the physical state of filmed specimens includes thicknesses, hardness, and durability of substrates and film. Soft films, which are used as front-surface calibration standards, are often overcoated to prevent oxidation and to improve durability for cleaning purposes (29). The overcoat material's type and thickness can considerably alter the spectral curves and data by interference (Fig. 13), or it

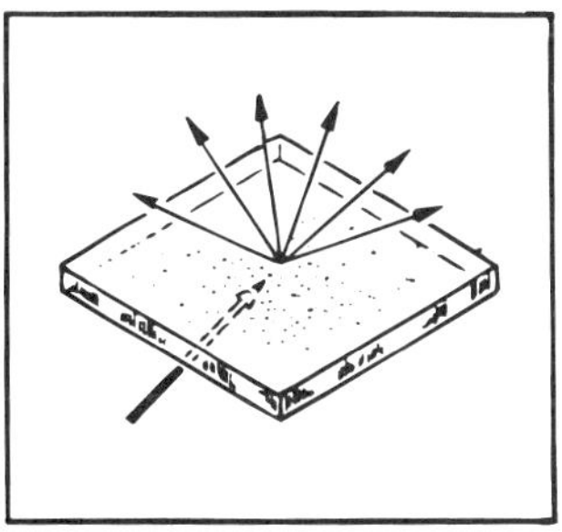

FIG. 11. Diffuse transmission occurs when light penetrates an object, scatters, and emerges diffusely on the other side. As with diffusely reflected light, the transmitted light leaves the object surface in all directions. Diffuse transmission is seen visually as cloudiness, haze, or translucency, each of which is of interest in appearance measurement (from HunterLab).

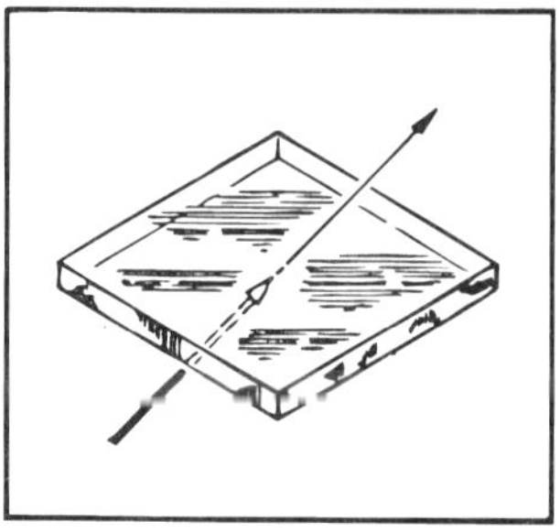

FIG. 12. Specular transmission refers to light passing through an object without diffusion. Specular transmission measurements are widely used in the chemical analyses and color measurements of liquids. Potentially, appearance attributes important for specular transmission should be roughly analogous with gloss attributes associated with specular reflection (from HunterLab).

may just enhance the reflectivity in a wavelength region of the solar spectrum that will require critical calibration for a particular specimen scan (Fig. 14).

Fluorescing samples must be measured in a monochromatic sample beam to minimize the exposure time at excitation wavelengths. Some instruments have reverse optics that allow specimens to be measured in either a polychromatic or monochromatic sample beam. If the specimen is measured both ways the spectral curves can be subtracted or simply overlayed to determine the flourescing influence the polychromatic beam exerted during the specimen run. Inhomogeneous specimens are the hardest to measure repeatably. Customized jigs and careful sample mounting information will minimize the effects but, in some cases, when high repeatability is required, the measurements should not be attempted.

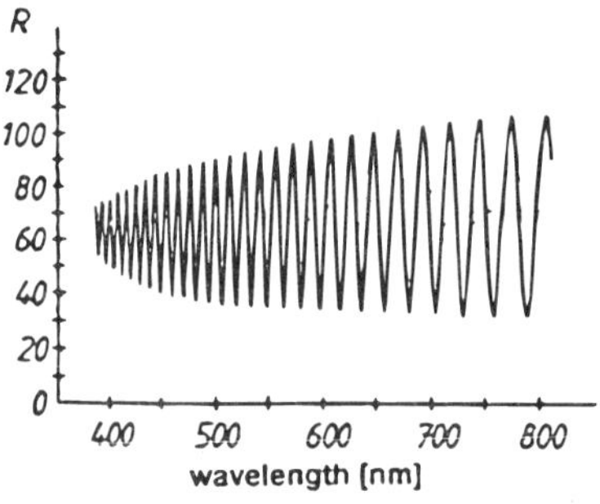

FIG. 13. Provided the geometrical thickness is known, the corresponding dispersion curve can be computed from the interference spectrum measured (from Zeiss Information).

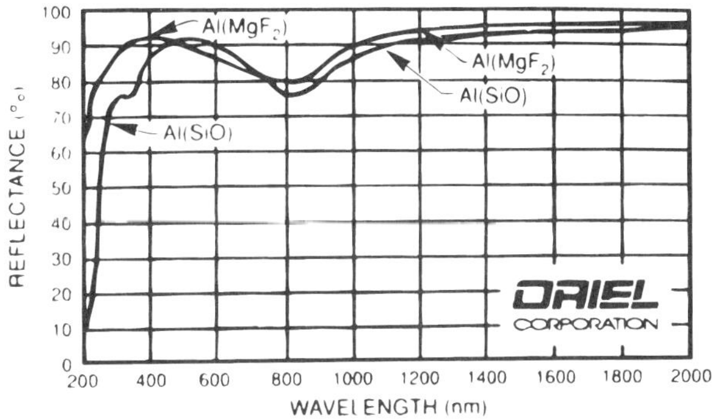

FIG. 14. Reflectance of AlSiO and $AlMgF_2$ coated reflectors for normally incident radiation (25).

C. Spectral Measurement Ranges: Purposes

The solar spectrum may be divided into three regions for discussion: the UV (280–400 nm), the "visible" (VIS, 400–760 nm), and the IR (760–2500 nm). There is no scientific consensus on the boundaries of these regions nor is there even agreement on whether they need to be defined for solar measurement purposes.

The UV region of the solar spectrum contains 5% or less of the total solar energy. Some feel that this is too small to make UV a separate solar issue. However, energy at these short wavelengths causes fading and cracking of many unprotected polymer materials as well as sunburn to humans. Companies that make products to protect materials from UV sunlight such as window glasses require numerical definitions of the UV. The region endpoint wavelengths, integration wavelength intervals (usually 5 or 10 nm), and spectral distributions (air mass 2.0 direct or 1.5 global) must be defined to enable comparisons of similar products. The official definitions of the UV regions are illustrated in Fig. 15. Quantitative calculations of potential UV protection are satisfactory for comparing materials but the actual protection will vary with the everchanging UV solar radiation due primarily to cloud cover and pollution (30).

The solar visible region is the wavelength band coincident with the eye's sensitivity. There is no photometric connection. The eye response is not involved in solar measurements. The color of average overcast daylight (illuminant C or D65) is used as a "no" spectral energy distribution for color matching but not for solar measurements. The visible region contains approximately 50% of the total solar energy at the earth's surface but the calculated visible solar transmittance or reflectance will not add any useful

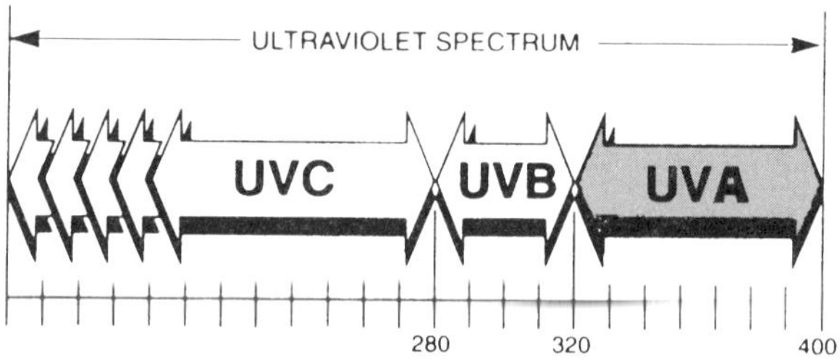

FIG. 15. The ultraviolet spectrum (25).

information about a specimen. Unfortunately, ASTM and others promote the eye–solar concept (31). There is considerable confusion about where the solar visible terminates. This is due to the mistaken conviction that the eye response has something to do with this judgment. Compounding the confusion is the uncertainty as to which wavelengths in the blue and red areas of the spectrum are the last to excite a visual response in the average observer. The division between the UV and visible is sometimes defined at 380 or at 400 nm or even counted twice between the two wavelengths—once for the visible and once for the UV (32).

The NIR region of the solar spectrum extends from the visible at approximately 760–2500 nm. Fortunately, there is a good IR detector for this range. It is a lead sulfide (PbS) detector that is about the size of one's little fingernail. Aligning all the light from the instrument source after it has been reflected from or transmitted through a specimen is a major instrument design problem. An indication that all the source energy has not reached the PbS detector can be seen as a disparity in curves from the specimen at the IR/VIS detector crossover point at approximately 800 nm. A technical note published by NIST (NBS) or its IR extension and check-out performed on its reference spectrophotometer will add much insight into NIR calibration and measurement tasks (33).

D. Static and Dynamic Measurements and Exposure Tests

The effectiveness of solar filters can be determined from deterioration rates of highly susceptible materials protected, to some degree, by the filters from intense (accelerated) solar radiation (34). Deterioration rates are determined by correlation with selective spectral changes in the materials over the wavelength solar range.

Spectral changes occur with changes of specimen thickness and/or absorber concentration. These static changes are the easiest to compare spectrally because the spectral response due to the absorber(s) is not changing during a sample scan. If the specimen contains unstable absorbers, the concentration may change during measurement from exposure to heat (thermochromic) or light (photochromic) from the instrument's source.

E. Sample "Fingerprints"

There is a difference between spectra of inorganic and organic materials. Generally, inorganic absorption bands are few but broad, and organic bands are many but narrow. Photometric accuracy is most important when measuring inorganic materials because the accuracy of the final integrated solar measurement depends more on this factor than the accuracy of the wavelength scale. Most spectrophotometers are built for extremely accurate wavelength determinations. This is necessary to identify organic materials from the wavelength positions of their narrow absorption bands. Photometric precision and linearity may not be as accurate. Therefore, if a spectrophotometer is to be used for solar data collection, one must run calibration standards and samples of the materials to be evaluated, calculate the solar properties, and compare them to values from reference spectrophotometers.

F. Quality and Quantity Measurements

Any spectrophotometer that scans the solar range will be expensive and very accurate. Diode array instruments are faster than conventional scanning spectrophotometers. They can collect total solar spectral data from a specimen in less than 2 min, whereas a scanning instrument must operate for approximately 20 min to capture the same data. Data quality from scanning spectrophotometers can be very high if the instrument parameters are properly adjusted before a scan. Diode array instruments also must have their operating parameters set properly by the operator in order to obtain consistently high-quality data, especially in the IR solar region. IR detectors such as PbS models have a slow response time that is compatible with the slow scan speeds of scanning instruments but may cause problems with faster data acquisition diode array systems. Most visible and UV detectors have fast response times and are fully compatible with diode array and scanning instruments. If operating parameters must be changed to improve a spectral run on a diode array instrument, adjustments can be made in real time while watching the affected output curves on the computer monitor. Scanning spectrophotometers must be preset from experience because the resultant curve and data are not available for analysis until the scan is completed. Diode array solar data acquisition repeatability and precision are improving as the technology develops along with experience. IR/VIS detector crossover speed changes should be monitored closely for both types of instruments.

An easy way to analyze changes in specimens that are undergoing chemical or exposure testing is to overlay the spectral curves from data during the experiment. Marked photometric changes in certain wavelength regions indicate how much influence the experiment had on what elements in the specimen. However, reaction times are not the only influences on spectral curves. Surface stains on solid samples from exposure testing can lower the spectral values because scattered light does not reach the detector or the surface deposits may absorb energy at selective wavelengths. Note all surface changes during exposure testing of solid samples to avoid misinterpretations of spectral data and curves at a later date.

Wavelength interval ranges are directly related to spectral data precision. Intervals shorter than 5 nm probably will not add significantly to solar spectral calculations in the UV, VIS or IR regions. Solar energy ramps sharply up from 300 to 400 nm. Good UV integrated solar calculations will require 5-nm data in this region. Note also that wavelength accuracy errors as small as 1 nm may cause large calculated errors (35). Five nanometers is also recommended in the visible region if the spectral data are also to be used to calculate color. Otherwise, 10- or 20-nm intervals are sufficient in the visible. In the IR region from approximately 750 to 2500 nm the interval usually can be extended to 50 or 100 nm without decreasing the accuracy of the solar integrations. These intervals vary in proportion to the amount of solar energy, and with the usual amount of selective absorption, in the spectral regions.

Long-term calibration stability is essential if an instrument is to be used continuously. More measurements yield better statistical QC averages if the instrument remains calibrated. Experiments should be made to determine how long an instrument will stay calibrated as it is typically used. Establish the sampling frequency based on these experiments.

IV. Measurement Techniques

Hands-on sample and instrument procedures will determine the value of a specimen's spectral data. Much of it comes from familiarity with the measuring instrument and its relation to the type of samples to be measured. The following are helpful hints and troubleshooting tips that may assist you in running spectrophotometers to collect solar spectral data from specimens:

- Choose the appropriate instrument for the specimen data required.
  - Does the instrument measure the spectral range required?
  - Is it in proper working condition (necessary attachments too)?
  - Does the operator have enough experience to recognize systematic and random errors (36)?
  - Is the sample beam small enough to cover small samples or large enough to integrate small specimen imperfections?
- Prepare specimens so that their introduction into the instrument will produce a minimum disturbance to the sample beam geometry.
  - Is the solid sample clean and free of scratches or labels?
  - Does the sample size allow it to cover the instrument's transmission or reflection port normal to the sample beam?
  - Can a paper, fabric, or paint reflectance specimen be prepared such that it will not leave a residue in a sphere attachment?
  - Are volatile liquid samples sealed so vapors will not affect instrument optics?
  - Are cuvets clean and are their surfaces clear?
  - Are gas holders clean and are their surfaces clear?
  - Are gas cell end plates parallel and sealed completely?
- Customize the instrument and accessories for the specimen run(s).
  - Is the specimen filmed, diffusing by transmission or reflection?
  - What spectral property is to be measured?
  - What wavelength range will be measured?
  - Does the specimen(s) fit the sample holder?
  - Will an integrating sphere attachment be used (37)?
  - Are the scan speeds and slit widths optimized for the specimen(s)?
  - Are many samples to be run routinely for QC?
  - Are specimens to be run before and after exposure tests?
  - Are specimens to be run as transfer standards?
  - Do specimens have gradient characteristics?
  - Are samples with nap or stress oriented properly in the sample beam?
  - Reflectance:
    - Is the reflectance attachment set up to calibrate and run samples in the specular, diffuse, or total reflectance mode (38)?
    - Is the appropriate software running to calibrate the instrument with the reflectance standards available?
  - Transmittance:
    - Are the sample and reference specimens geometrically equivalent?
    - Is polarization present in the specimen and sample beam?
- Examine run data and graphs for anomalies.
  - Do curves meet at the IR/VIS crossover?

- Are there noise spikes or sluggish absorption peaks in portions of the spectral curves? (Is the source signal adequate at all wavelengths at which the detectors have low sensitivity?)
- Are there too few data points to obtain a representative integrated solar value equal to the product of the area under the spectral curve by the spectral solar energy distribution?
- Do some of the calibration data points exceed 100%?
- Do some of the specimen data points fall below 0%?
- Are water bands present in the NIR spectra, i.e., system purged?
- Do the photometric levels roll off at lower wavelengths due to increasing scattered sample beam energy that is not detected?

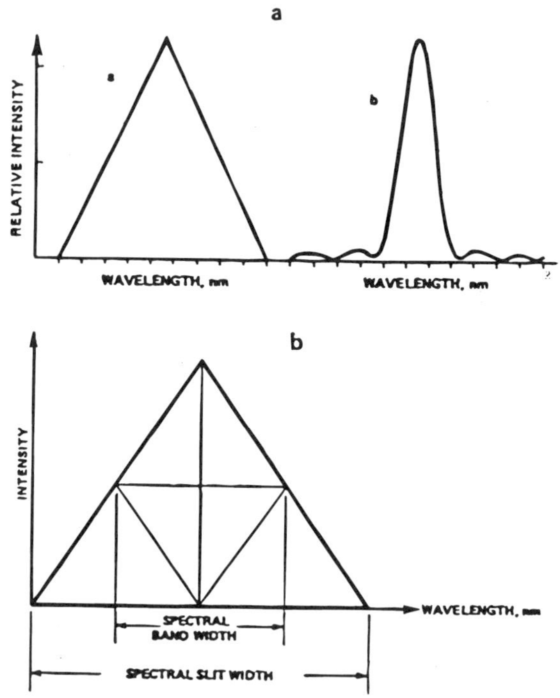

FIG. 16. (a) Light from the exit slit of the monochromator varies in intensity as a function of nearby wavelengths. The exit slit energy function is a measure of the relative intensities emerging from the monochromator. A triangular shape is obtained for normal slits (shown in a). Narrow slits cause a diffraction pattern similar to b [Coates and Hausdorff, *J. Opt. Soc. Amer.* **45**; 425 (1995)]. (b) Most slit widths can be assumed to have a triangular slit function. The total wavelength region that is isolated by the slit is the spectral slit width. Spectral bandwidth is the width of the function at one-half the peak intensity. For a triangular function, spectral bandwidth is three-fourths of the total radiant energy isolated by the slit (10).

Most of these questions are answered subconsciously by experienced operators but a new operator should consciously go through the checklist to avoid missing a crucial one. The questions about slit widths and scan widths do not have simple answers. The relationship between them is best understood visually from Figs. 16 and 17.

Operators must understand the 2:1 relationship between the spectral slit width and the spectral bandwidth. These terms are not always spoken or written about together and a 50% error is possible if the wrong "width" is used. Detailed discussions on all these issues are presented in ASTM standards (39, 40).

There are two ways to evaluate the differences in solar spectral properties of specimens. In single-beam instruments a reference specimen is run first, the data are stored, and then the sample specimen is run. The data from both are converted into curves that are presented together on a computer monitor's screen. This consecutive scan procedure can also be used on a double-beam instrument or the sample and reference specimens can be run simultaneously in their respective beams. This procedure eliminates any doubt about instrument parameters changing between runs. If both sample and reference specimens absorb strongly in a common wavelength region the data and curves may be wrong (drift) because there is not enough energy to activate the detector. Run the specimen against air first to deter-

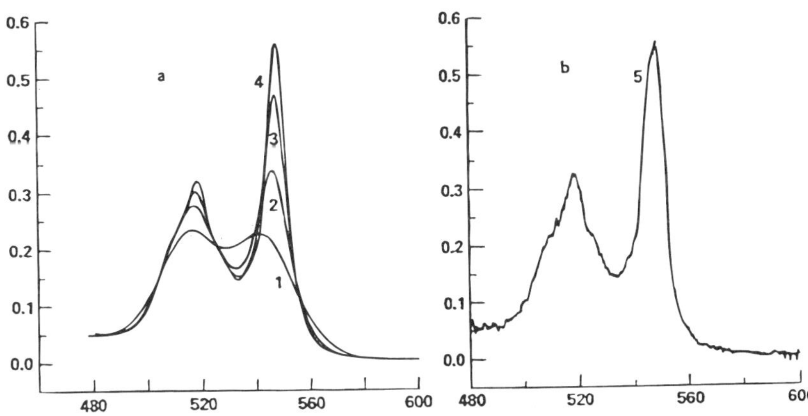

FIG. 17. An optimum spectral bandwidth will fully define an absorption band without introducing unnecessary noise. The α and β bands of reduced cytochrome c are scanned with several spectra band widths. A wide value of 20 nm (1) has completely distorted the separation and the peak heights of the two bands. Narrowing the spectral bandwidth to 10 nm (2), 5 nm (3) and 1 nm (4) shows a sequential improvement in band definition. However, adjustment of the spectral bandwidth to a narrow 0.08 nm (5) significantly increases the noise level with no noticeable improvement in peak height or band separation of the two bands (10).

mine "dead" wavelength areas of relative measurements in double-beam mode.

V. Mathematical Considerations

Calculating solar properties of materials from specimen spectral data is straightforward. Table 1, in wavelength abbreviated form, illustrates the total solar calculation of a tinted glass. The solar energy weight values are from Parry Moon (41). They represent the direct energy spectrum for air mass = 2.0. These weights have been referenced by International Commission on Illumination (42), American National Standards Institute (43), and ASTM (7).

Although Table 1 exemplifies the traditional "rectangular" way to integrate transmission data using the equation,

$$T_{(\lambda_1 \text{ to } \lambda_n)} = \frac{\sum_{\lambda=1}^{n} (T_\lambda \cdot E_\lambda)}{\sum_{\lambda=1}^{n} E_\lambda}, \tag{1}$$

other methods may yield different results. The next example compares the "rectangular" method to a commonly used "trapezoidal" method using the following equation:

$$T_{(\lambda_1 \text{ to } \lambda_n)} = \frac{\sum_{\lambda=1}^{n-1} \left(\frac{T_\lambda \cdot E_\lambda + T_{\lambda+1} \cdot E_{\lambda+1}}{2} \right)}{\sum_{\lambda=1}^{n-1} \left(\frac{E_\lambda + E_{\lambda+1}}{2} \right)}. \tag{2}$$

TABLE 1
TOTAL SOLAR CALCULATION OF A TINTED GLASS

Wavelength (nm)	% Transmittance (%T)	Solar energy (SE)	(%T) × (SE)
2000 Infrared	63.9	0.012	0.80
1600 Infrared	60.5	0.066	4.02
1300 Infrared	43.6	0.057	2.48
1060 Infrared	39.5	0.103	4.07
950 Infrared	41.2	0.063	2.61
850 Infrared	46.9	0.100	4.60
750 Dark red	58.2	0.121	7.06
650 Red	73.8	0.154	11.37
550 Yellow/green	84.0	0.160	13.42
450 Blue	79.4	0.134	10.65
370 Ultraviolet	62.4	0.024	1.48
330 Ultraviolet	0.4	0.006	0.00
Sums		1.000	62.6

The trapezoidal method sums one-half of each contiguous interval with one-half of the next. This is the equivalent to adding one-half of the first and last intervals to the sum of the intermediate ones:

$$T_{(\lambda_1 \text{ to } \lambda_n)} = \frac{\dfrac{T_1 \cdot E_1}{2} + \sum_{\lambda=2}^{n-1} (T_\lambda \cdot E_\lambda) + \dfrac{T_1 \cdot E_1}{2}}{\dfrac{E_1}{2} + \sum_{\lambda=2}^{n-1} (E_\lambda) + \dfrac{E_n}{2}}. \tag{3}$$

Table 2 calculates solar UV transmittance of a UV-absorbing glass specimen using Eqs. (1) and (3).

The 1.6% transmittance difference can appear to be a significant specimen difference if the computer printout does not specify the method of integration.

TABLE 2

SOLAR UV TRANSMITTANCE OF A UV-ABSORBING GLASS

Wavelength (nm)	Rectangular	Trapezoidal	%T	Rectangular %T	Trapezoidal %T
400	470.0	235.0	64.91	30,507.70	15,253.85
395	433.5	433.5	60.59	26,263.60	26,263.60
390	397.0	397.0	56.26	22,335.22	22,335.22
385	366.5	366.5	48.26	17,687.29	17,687.29
380	336.0	336.0	40.26	13,527.36	13,527.36
375	307.5	307.5	42.14	12,956.51	12,956.51
370	279.0	279.0	44.01	12,278.79	12,278.79
365	256.0	256.0	38.23	9,785.60	9,785.60
360	233.0	233.0	32.44	7,558.52	7,558.52
355	210.5	210.5	22.52	4,740.46	4,740.46
350	188.0	188.0	12.60	2,368.80	2,368.80
345	169.5	169.5	6.93	1,174.64	1,174.64
340	151.0	151.0	1.26	190.26	190.26
335	126.0	126.0	0.64	80.01	80.01
330	101.0	101.0	0.01	1.01	1.01
325	77.5	77.5	0.01	0.78	0.78
320	54.0	54.0	0.01	0.54	0.54
315	30.0	30.0	0.01	0.30	0.30
310	11.0	11.0	0.01	0.11	0.11
305	1.9	1.9	0.01	0.02	0.02
300	0.1	0.10	0.01	0.00	0.00
Sums	4199.0	3964.0		161,457.51	146,203.66
	P.M.-E2 dir.†	P.M.-E2 dir.	%T(UV) =	38.5	36.9

† P.M. = Parry Moon. E2 dir. = (solar) Energy at Air Mass 2 direct.

Computer spreadsheets are an ideal medium to calculate specimen properties at other thicknesses and concentrations. Expressing Beer's law in the following form can be useful in many ways:

$$\%T_{\text{calculated}} = \%T_{\text{maximum}} \left(\frac{\%T_{\text{original}}}{\%T_{\text{maximum}}} \right)^{\left(\frac{k_{\text{calculated}}}{k_{\text{original}}} \right)}, \tag{4}$$

where $\%T_{max}$ is the maximum transmittance of the glass specimen. The $\%T_{max}$ case occurs when the glass is infinitely thin and has no absorption. In other words, it is a measure of reflection losses from the air to glass surfaces. It is dependent on, and calculated from, the glass's index of refraction at each wavelength.

Table 3 is an abbreviated NIR solar calculation example of a specimen in its measured thickness (0.25 in.) and in a calculated thickness (0.10 in.). Glass thickness visible transmittance calculations can usually be simplified. If the index of refraction of a glass does not vary significantly over a spectral range, the maximum transmittance for the region may be averaged and a single preintegrated specimen transmittance value substituted in Eq. (4). For a window glass color calculation, for example, use its visible transmittance at 555 nm and a maximum transmittance (91.75) to calculate brightness at various glass thicknesses. Equations (4)–(6) must be solved one interval at a time and then integrated to calculate total solar, UV solar, and NIR solar spectral data. This is because the index or refraction varies too much over these ranges.

Another use of the versatile Eq. (4) is in the calculation of multiple components. For example, if plies of glass are in intimate contact using an index oil or thin clear plastic between each, then the visible transmittance of the

TABLE 3
NIR SOLAR CALCULATION

Wavelength	$\%T_{\text{Sample}(\lambda)}$	$\%T_{\text{Max}(\lambda)}$	E_{solar}	$\%T_s \cdot E$ ($k'' = 0.25$)	$\%T'_s \cdot E$ ($k'' = 0.10$)
2000	63.7	92.35	0.0239	1.5	1.9
1600	60.5	92.25	0.1272	7.7	9.9
1300	43.6	92.15	0.1091	4.8	7.5
1100	39.4	92.05	0.1974	7.8	12.9
950	41.2	91.95	0.1216	5.0	8.1
850	46.7	91.95	0.1883	8.8	13.2
750	58.1	91.85	0.2326	13.5	17.8
			$\%T$ (NIR)	49.1	71.3

combined laminate may be calculated as follows:

$$\%T_{\text{laminate}} = \%T_{\max} \cdot \left(\frac{\%T_1}{\%T_{\max}}\right) \cdot \left(\frac{\%T_2}{\%T_{\max}}\right) \cdot \left(\frac{\%T_3}{\%T_{\max}}\right) \cdots \tag{5}$$

A numerical example of a glass/plastic/glass laminate in which there is no internal reflection between the glass and plastic, due to matched indexes, could be stated and solved in the following way:

Problem: Find the transmittance of a ply in a three-layered laminate given the transmittance of the laminate (79.0), the other glass ply (89.4), and the internal transmittance of the plastic (99.3). $\%T_{\max}$ of the plastic is 100 because there is no reflection losses at its surfaces.

$$79.0 = \frac{(89.4) \cdot (99.3) \cdot (\%T_3)}{(100) \cdot (91.75)}, \quad \%T_3 = 81.7.$$

Spectral results will differ if the same three components are separated by air because of added internal reflectances. For the previous example, where the measured external transmittances of the components are known:

$$\%T_{\text{laminate}} = (\%T_1) \cdot (T_2) \cdot (T_3) \cdots \tag{6}$$

$$\%T_{\text{laminate}} = (89.4) \cdot (0.911) \cdot (0.817), \quad \%T_{\text{laminate}} = 66.5.$$

VI. Basic User Technique Pitfalls

Spectrophotometer operators need a bag of tricks for traps. Operator traps may be characterized as operator errors in procedures or data interpretation. Instrument traps are either calibration errors or innocuous computer program errors. Solutions to these traps are tricks the operator uses, or installs, to ensure correct final spectral outputs. The prime trick is experience. Prior knowledge sensitizes the operator to spot most operation errors and to correct them expeditiously. This section examines situations of, and possible remedies for, unpredicted spectral results caused by common instrument troubles or operator errors.

A. Operator Errors

New instruments pose a problem for inexperienced operators. The first word of advice from the manufacturer is to read the operator's manual. Obviously this is the best way to gain experience, but many of the words in the manual refer to concepts as yet unknown. Go slowly through the operations manual as you are learning to use a spectrophotometer. Most operator errors will be avoided in this way.

Most specimen measurements are made in conjunction with one or more instrument attachments and a sample holder. If the attachments or sample are incorrectly aligned with the optical system all of the sample beam may not reach the detector, resulting in low photometric values. Instruments that scan the NIR and visible wavelength ranges with two detectors will exhibit a larger than normal photometric disparity at the crossover wavelength during a scan when this condition occurs.

Many spectrophotometers must be calibrated with external standards. Operators may fail to use the correct standard or improperly clean the calibration standard. Front-surface mirrors, cleaned with tissue and cleaning fluid, may be ruined by the same procedure that is recommended to clean glass standards.

Samples should be mounted in the measuring beam, then examined to check that (i) the sample surface is clean (often more apparent in the beam than in room light), (ii) the beam is entirely incident on the sample where it is to be measured, and (iii) the beam does not graze the sample's edge.

Many similar calculations can transform instrument measurements into calculated results that are similar but different. Air mass = 1.5 direct (44) and others are published and are used. It is important to choose and state which air mass is to be used in the solar calculations. One must also choose a spectral energy distribution defined by direct or hemispherical (ASTM E892, and ISO/DIS 13837) reception, i.e., direct plus scattered components. The combinations are even more numerous if turbidity and irradiance at angles other than normal are to be specified (42). Distinctions are important where large amounts of solar energy must be assessed; for example, in a building's heat load calculation. Sun and specimen angles sometimes require calculations similar to celestial mechanics (45). The oversimplified cosine relationship between air mass and sun angle is skewed in the direction of increasing haze (46). The calculated results must be specified for one combination to compare results from different specimens.

Instruments are built around one of two principles. Either the instrument, often portable, is simple to use because it has many fixed parameters or it is a large research-grade instrument with many degrees of freedom. If the instrument is simple, the operator must not try to use it to obtain novel information from sample that is not prescribed in the operator's manual. If the instrument is "research grade," follow the manual explicitly when setting the scan rate, slit width, slit height, reference beam samples, photometric and wavelength scale factors, etc. Higher accuracy on both wavelength and photometric scales is attainable from research-grade instruments but they are more delicate. Most need a solid foundation but at the same time require isolation from floor vibrations. If the instrument is isolated, avoid irregular readings by not touching it during a calibration or specimen run.

Measurements in and around the water bands in the NIR require special attention. If the wavelengths of interest are outside the bands of approximately 1925 nm, and their presence is not objectionable on specimen spectral curves, the curves can simply be marked "water" in this region. Alternatively, the instrument must be purged with dry air or nitrogen and the spectral region must be monitored until the water bands are insignificantly small. Liquid samples, of course, must be in sealed cuvettes to avoid being purged.

Some instruments are automatically calibrated by computers. These instruments must be turned off and on over reasonable periods of time to ensure proper instrument calibration.

A systematic error will affect the instrument readings if volatile liquids are not sealed before they are inserted into the sample or reference beams. Many volatile chemicals will attack the lens coatings that are exposed to the sample and reference compartments.

Oversimplifications or overcomplications of procedures often result in bad specimen data. Standardized computer programs that control instrument operations can multiply mistakes quickly and sometimes imperceptibly (47). If a uniquely shaped specimen is measured according to a simplified "standard procedure," accuracy may be sacrificed for speed and run type uniformity. On the other hand, if an instrument is not set in an "auto" run mode, all the necessary parameters must be set correctly to obtain the correct specimen data from the instrument. Although better data can result from customized runs, higher accuracy is obtained at the expense of longer specimen and calibration runs.

B. Instrument Errors

Calibrations require physical standards. Most standards have associated spectral data files that usually are stored in a computer linked to the spectrophotometer. These files can become corrupted, sometimes in such a minor way that calibration errors, and associated specimen errors, will go undetected. A good practice is to periodically run an old sample whose spectral characteristics do not vary with time and compare previous data for changes outside of expected instrument variations.

Reflectance measurements are made relative to the perfect reflecting diffuser that cannot be realized. Sphere coatings, such as MgO, $BaSO_4$, and PTFE, only approximate the ideal diffuse reflector (48). Invariant correction factors are built into reflectance calculations that do not change as the reflectance of these coatings deteriorates with age.

Many spectrophotometers incorporate one or more moving parts that are subject to wear. A consequential slow loss of accuracy can go unnoticed unless the previous procedure is followed routinely. Another cause of slow loss of accuracy is instrument source aging. Variations from consecutive runs of a specimen or standard indicate that the source might be approaching the end of its useful life and should be changed. If this is the case, a new bulb will enable the instrument to output stable results.

C. Cleaning Tips

Lenses are coated with antireflecting films to increase the optical beam's energy through the instrument. These films are thickness and refractive index specific. If they are touched, residual skin oils will adversely affect both film thickness and refractive index. The lens will then reflect more and transmit less. Dirty lenses should be cleaned with a fresh cotton-tip swab saturated with a 3:1 mixture of denatured alcohol and ether. Touch the swab to the lens center and then wipe in a spirling motion to the edge. Rinse with denatured alcohol and allow to dry by evaporation. Do not attempt to clean coated lenses with silicon-coated lens tissues or cloths. The silicone will react with the antireflecting film to create a permanently clouded coating.

Cleaning integrating sphere attachments is necessary if the interior surface accumulates material from specimens or contaminates in the air. Fabric specimens, for example, may sluff some surface fibers through the sample port that will accumulate at the bottom of the sphere. This will lower sphere reflectance efficiency and change the calibration factors relative to instrument standards. Small flashlight-size, battery-powered vacuum cleaners fitted with a curved extension tube that is protected at the cleaning end by a soft brush are sold in many computer stores. These will remove the residual material without damaging the sphere's interior coating if the following procedure is followed. Turn on the source to light the sphere's interior. Slowly lower the brush through a sphere port almost to the bottom of the sphere. Take care not to rub the tube against the port edge or against an internal baffle. This operation must be performed slowly so that the operator's eye can adapt to distances between the brush and sphere wall without the aid of shadows. Recalibrate the instrument after cleaning and before running more specimens.

Slow losses of accuracy are particularly insidious when the instrument and specimens are changing. It is a good idea to calculate specimen color along with solar data to rule out instrument variations. For example, cloth samples measured before and after UV exposure testing will show a change of color data. How much of the post-exposure data difference is due to specimen exposure? Three precautionary procedures may help to resolve this question. First, rule out gross color data errors by a visual comparison between the sample and data after exposure testing. Second, run a control sample with the exposed specimens before and after the exposure tests. Third, compare the test data to expectations from the material manufacturer, from the exposure test equipment company, and from previous test results.

If changes in color data are greater than one would expect, check the condition of the integration sphere, that is generally employed to measure diffuse reflecting samples such as fabrics. Sometimes while measuring colored fabrics, for example, fibers will be sloughed off and fall into the integrating sphere. Look at the bottom of the sphere through the sample port when the interior is illuminated. Any fibers there must be removed. The following procedure is a relatively safe one to avoid damage to the coated sphere interior: Obtain a flashlight-size hand-held battery-powered vacuum cleaner for computers—They come with a bent extension tube and soft brush on the cleaning end. With the sphere lit, insert the tube slowly through the port toward the bottom of the sphere. Slowly is the operative word because one must avoid rubbing the tube against the port edge or against an interior baffle hard enough to damage the coating and also because there will not be any shadows inside of the sphere to help estimate the distance from the brush to the sphere bottom. Vacuum the lint and other loose material, remove the tube, replace and run the calibration standard, and report these data with the specimen results.

Even slower changes in color data will occur as the interior sphere coating ages. This usually is a yellowing process that occurs faster at the junction of the hemispheres. Yellowing is a reduction of blue reflection from the coating and may be detected in fluctuating blue data such as calibration transmittance or reflectance values at short wavelengths, integrated "Z" stimulus values, or "b*" CIELAB values.

Another cleaning caution concerns instrument lenses. Most lenses are coated with a quarter wave film to reduce reflections that decrease transmittance through them. These films are extremely sensitive to changes in thickness and refractive index. Fingerprints will change both for the worst. The best way to clean coated lenses is to touch a cotton swab, saturated with 1 part ether and 3 parts denatured alcohol solution, to the center of the lens and wipe with a spiraling motion to the lens edge. Rinse with alcohol and let dry by evaporation. Do not clean a coated lens with a lens tissue or cloth impregnated with silicone. Silicone reacts with lens coatings to produce a permanently clouded film that will destroy the lens.

VII. Limitations

A. Instruments

The optical design of all spectrophotometers is very limited. Many types of geometrical errors are involved in even the simplest designs (19). Whereas an observer constantly views objects under changing conditions, a spectrophotometer is carefully assembled with usually one view of specimens. The instrument's view is restricted to specified industry standards that describe conventional optical geometries, sources, detectors, data acquisition, and calculation algorithms. These restrictions are necessary to communicate solar parameters of a specimen without confusion.

The highest possible level of precision and accuracy an instrument can obtain is achievable only with the fewest operating limitations. Fewer limitations mean more variable controls. As the number of variable controls increases an operator must have more experience and knowledge of instrument theory to obtain the best results. The price of these more complicated instruments is generally also higher. The instrument operating manuals define parameters such as slit width, bandwidth, response time, scan rate, and data acquisition intervals over the wavelength ranges. These parameters must be programmed correctly if they are to have variable controls.

In addition to the intentional limitations of spectrophotometers, good data are obtainable only from limited types of specimens. Most specimens must be flat, homogenous, rigid, nonvolatile, nonphotosensitive to the sample beam, and be within a limited size and weight range. An integrating sphere attachment usually will solve specimen geometry problems but its inclusion into the system brings its own limitations. Integrating spheres have a fixed diameter, ratio of ports areas to total sphere surface area, a variable reflectance fact with wavelength, and internal baffle geometries that favor certain specimen types. The data from each specimen type, run either in reflectance or transmittance mode, should be compared to similar results obtained by others in round-robins using many instruments with different optical geometries.

For example, the following is a problem, cause, and solution to one geometry problem that arises using an integrating sphere to measure the reflectance from the unfilmed side of a mirror. An integrating sphere's diameter affects the ratio of mismatched sample and reference beam geometries. If the front surface of a mirror reflects the reference beam and the unfilmed side of a specimen mirror reflects the sample beam into an integrating sphere, the resulting specimen reflectance will measure low. There are two methods of solving this problem. First, one can use a second-surface mirror at the reference beam port to calibrate the instrument before measuring the second surface of the specimen. Second, one can run a series of tests to determine the correction function for the geometric asymetry between the sample and reference beams using the following procedure: Scan the entire instrumental wavelength range with graduated spacers between the sample port and an opaque front-surface mirror while a front-surface opaque mirror covers the reference port during all test scans. Both mirrors should be high reflectors such as freshly deposited aluminum films. A second series of test runs should be made with a lower reflecting but also spectrally "flat" mirror such as opaque chromium. The spacers should be machined to within a mil or two and have openings painted flat black that are considerably larger than the sample port. Attenuation factors can then be calculated and back corrected into second-surface specimen run data. This is probably the best method of the two, albeit more complicated at first, to solve this geometrical problem because first-surface mirrors, or very thin second-surface mirrors, are used to calibrate instruments for reflectance. The calibration will also be most accurate if the primary and reference standards present the same geometry to both the sample and reference beams.

B. Techniques

Technique limitations are often attributed to operator errors. Many researchers have become involved in spectrophotometry without much training (35). They may not be aware of the importance of proper positioning of samples in the sample beam or of proper specimen cleaning. Unless a jig is employed to position patterned specimens, for example, they cannot be repositioned and remeasured in exactly the sample place twice. Consequently, if spectral changes occur to the specimens during an exposure, test data changes cannot be attributed solely to the exposure test. Another example, the failure to define the initial measurement parameters, can lead to different solar calculations following exposure tests if the second set of tests are made by another uninformed operator. Some specimens cannot be cleaned properly before testing, e.g., front-surface mirrors filmed with soft films or hygroscopic films on glass. Operators must know which cleaning techniques to avoid as well as which ones to use in such cases.

VIII. Advances Needed

The weak points of spectrophotometer systems surface from users as questions or gripes. Most of these weaknesses can be classified into three arbitrary categories; confusing operating instructions and procedures, mechanical failures, and technological lag behind spectral needs. This section examines examples of each and predicts future solutions.

A. Instructions and Procedures

The first requisite of a good spectrophotometer operator is to know basic concepts of the instrument. Operating manuals are usually written to describe procedures particular to an instrument but not to inform the reader about the concepts and operating precautions of spectrophotometry. Today these instruments are commanded by increasingly sophisticated software and less by operator skill and choice. It is increasingly each to produce incorrect data because the basics of operation were overlooked by managers and technicians alike and the assumption is made that the software has built-in safeguards against errors of procedure or operation. This is not the case. Program error routines usually focus on calibration and calculation operations. Warning "help" windows should be displayed to inform instrument users about the system limitations and where the operator's responsibilities begin. A basic theory pamphlet should accompany all new instruments as well as the operating manual.

B. Mechanical Failures

As instruments become easier to operate, they invariably get more use (wear) in a shorter period of time. Repair and replacement of expendable supplies should be made easy and reliable for operators with little experience in instrument maintenance to perform. Technicians with limited backgrounds are now expected to run and maintain very sophisticated equipment such as spectrophotometers (35). Owners also expect their instruments to be almost independent from the manufacturer's service section. As instrument manufacturers begin to realize that those people who are enlisted to run their products do not expect to need much outside instrument service, primer manuals and instruction courses will become part of the instrument purchase. More prealigned rugged replacement parts with more instructional step-by-step illustrations will appear on the market.

C. Technological Lag

We all have grown accustomed to expecting higher quality in all the products we buy. As providers we must reciprocate by supplying our customers with better products. Familiar spectral parameters required for quality control of a products's color are expected to be higher precision, acquired in less time, in the production plant or out in the field rather than in the lab. To meet all these demands the instrument manufacturer must design a smaller but more rugged spectrophotometer with faster responding detectors over a larger spectral range that is also higher in sensitivity to improve the signal-to-noise ratio.

Intensive investigations are under way in many companies to solve all these problems at once by using the right combination of detectors and sources with a diode array technology and fiber optics elements. Currently, these technologies are being used to satisfy all the design conditions except photometric precision and extended spectral range beyond 1100 nm. The breakthrough will come when a source and detector combination is discovered that will allow data acquisition to 2000 nm with the same speed and precision one can obtain only in the ultraviolet, visible, and near-infrared.

References

1. Thekaekara, M. P., Evaluating the light from the sun, in *Optical Spectra*, p. 32–34 (1972).
2. Gates, D. M., Spectral distribution of solar radiation at the earth's surface. *Science* **151**(3710); 523–529 (1966).
3. Drummond, A. J., and Wade, H. A., *Instrumentation for the Measurement of Solar Ultraviolet Radiation*, The Eppley Laboratory (1969).
4. Jones, L. A., and Condit, H. R., Sunlight and skylight as determinants of photographic exposure. I. Luminous density as determined by solar altitude and atmospheric conditions. *J. Optical Soc. Am.* **38**(2), 123–178 (1948).
5. Laurin, T. C., *et al.*, The photonics dictionary, in *The International Handbook and Reference to Photonics* (T. C. Laurin, Ed.). Laurin, Pittsfield, MA (1992).
6. Bernstein, N. S., *Light Measurement Terminology and Techniques*, EG&G Electro-Optical Division (1973).
7. American Society for Testing and Materials *E424 Standard Methods of Test for Solar Energy Transmittance and Reflectance (Terrestrial) of Sheet Materials*, Discontinued ed., Appearance of Materials, E-12. Author, Philadelphia (1971).
8. International Standards Organization, *Glass in Building—Determination of Light Transmittance, Solar Direct Transmittance, Total Solar Energy Transmittance and Ultraviolet Transmittance, and Related Glazing Factors*, Author (1990).
9. ASTM, *E131 Standard Definitions of Terms and Symbols Relating to Molecular Spectroscopy*, Molecular Spectroscopy, E-13, *Annual Book of Standards*. Author, Philadelphia (1990).
10. Varian Instrument Division, *Optimum Parameters for Spectrophotometry* (1973).
11. Mavrodineanu, R., An accurate spectrophotometer for measuring the transmittance of solid and liquid materials. *J. Res. Nat. Bureau Standards* **76A**(5) (1972).
12. Perkin–Elmer, Ltd., *Guide to UV*/VIS/NIR Reflectance Spectroscopy (1992).
13. Spectra-Tech, Inc., The complete guide to FT-IR accessories, supplies, microscopes, in *Catalogue* (1993).
14. ASTM, *E179 Standard Guide for Selection of Geometric Conditions for Measurement of Reflection and Transmission Properties of Materials*, 1994 ed., Appearance of Materials, E-12, Vol. 06.01, *Annual Book of Standards*. Author, Philadelphia 1991.

15. ASTM, *E167 Standard Practice for Goniophotometry of Objects and Materials*, 1994 ed., Appearance of Materials, E-12, Vol. 06.01, *Annual Book of Standards*. Author, Philadelphia (1991).
16. Boero, F. J., Absolute reflectometry measurements using digital ratio recording spectrophotometers, in *Instrumentation-Research*, pp. 40–44 (1986).
17. Hawes, R. C., Technique for measuring photometric accuracy. *App. Optics* **10**(6), 1246–1253 (1971).
18. Gibson, K. S., and Belknap, M. A., Permanance of glass standards of spectral transmittance. *J. Nat. Bureau Standards* **44**, 463–473 (1950).
19. Weidner, V. R., Mavrodineanu, R., and Mielenz, K. D., Spectral transmittance characteristics of holium oxide in perchloric acid solution. *J. Nat. Bureau Standards* **90**(2), 115 (1985).
20. Weidner, V. R., Barnes, P. Y., and Eckerle, K. L., A wavelength standard for the near infrared based on the reflectance of rare-earth oxides. *J. Res. Nat. Bureau Standards* **91**(5), 243–253 (1986).
21. Mavrodineanu, R., and Baldwin, J. R., *Glass Filters as a Standard Reference Material for Spectrophotometry—Selection, Preparation, Certification, Use SRM 930*, National Bureau of Standards (1975).
22. NPL SRM, *Infrared Spectral Regular Reflectance and Emissivity and Total Emissivity of Two Aluminized Reference Mirrors*, Author, Teddington, UK (1989).
23. Uriano, G. A., *et al.* (Ed.), *NBS Calibration Services Users Guide 1986–88*/Revised, NBS Special Publication No. 250. U.S. Government Printing Office, Washington, DC (1987).
24. Furler, R., and Rubin, M., *Standard Reference Materials*, Lawrence Berkeley Laboratory (1992).
25. Oriel Corp., *ORIEL Corp. Light Sources, Monochromators & Spectrographs, Detectors & Detection Systems, Fiber Optics*, 1994 ed., The Oriel Three Volume Catalog, (E. Esposito and N. Fernandes, Eds.), Vol. 2. Author, Stratford, CT (1994).
26. Hass, G., Specular reflectance of unprotected and protected evaporated metallic front-surface mirrors at various angles of incidence. *J. Vac. Sci. Technol.* **16**(2), 113–116 (1979).
27. Eckerle, K. L., Hsia, J. J., and Weider, V. R., *Transmittance MAP Service*, National Bureau of Standard (1985).
28. Eckerle, K. L., *et al.*, Measurement assurance program transmittance standards for spectrophotometric linearity testing: Preparation and calibration. *J. Res. Nat. Bureau Standards* **88**(1), 25–36 (1983).
29. Technical Staff of Laser Optics, Inc., *Thin Films Optics*. Optical Spectra (1970).
30. Brennan, P., and Fedor, C., Sunlight, UV, & accelerated weathering, in *Society of Plastics Engineers Automotive RETEC* (1987).
31. ASTM, *E971 Standard Practice for Calculation of Photometric Transmittance and Reflectance of Materials to Solar Radiation*, Appearance of Materials, E-12, Vol. 06.01, *Annual Book of Standards*. Author, Philadelphia (1992).
32. CIE, *Recommendations for the Integrated Irradiance and the Spectral Distribution of Simulated Solar Radiation for Testing Purposes*, Publ. No. 20 (TC-2.2). Author (1972).
33. Eckerle, K. L., *et al.*, *Extension of a Reference Spectrophotometer into the near Infrared*, National Bureau of Standards (1983).
34. Heraeus DSET Laboratories, Inc., Weathering services and equipment catalogue, in *Catalogue*, p. 36 (1992).
35. Optronic Laboratories, Inc., *The Measurement of Solar Ultraviolet Spectral Irradiance Problems and Solutions* (1990).
36. Berns, R. S., Empirical modeling of systematic spectrophotometric errors. *Color Res. Appl.* **13**(4), 243–256 (1988).

37. ASTM, *E903 Standard Test Method for Solar Absorptance, Reflectance, and Transmittance of Materials Using Integrating Spheres*, 1994 ed., Appearance of Materials, E-12, Vol. 06.01, *Annual Book of Standards*. Author, Philadelphia (1988).
38. ASTM, *E429 Standard Method for Measurement and Calculation of Reflecting Characteristics of Metallic Surfaces Using Integrating Sphere Instruments*, 1994 ed., Appearance of Materials, E-12, Vol. 06.01, *Annual Book of Standards*. Author, Philadelphia (1978).
39. ASTM, *E932 Standard Practice for Describing and Measuring Performance of Dispersive Infrared Spectrometers*, 1994 ed., Appearance of Materials, E-12, Vol. 06.01, *Annual Book of Standards*. Author, Philadelphia (1989).
40. ASTM, *E275 Standard Practice for Describing and Measuring Performance of Ultraviolet, Visible, and Near Infrared Spectrophotometers*, 1989 ed., Molecular Spectroscopy, E-13, Vol. 14.01, *Annual Book of Standards*. Author, Philadelphia (1989).
41. Moon, P., Proposed standard solar-radiation curves for engineering use. *J. Franklin Inst.* **230**, 583–617 (1940).
42. CIE, *Solar Spectral Irradiance*, Publ. No. 85 (TC2-17). Author (1989).
43. ANSI, *American National Standard for Opthalmics—Nonprescription Sunglasses and Fashion Eyewear—Requirements*. Author (1986).
44. ASTM, *E891 Standard Terrestrial Direct Normal Solar Spectral Irradiance Tables for Air Mass 1.5*, 1994 ed., Appearance of Materials, E-12, Vol. 06.01, *Annual Book of Standards*. Author, Philadelphia (1982).
45. Nettleton, J., and J. Murdoch, Determination of total external illuminance on a sloped surface from sunlight and skylight. *J. Illuminating Eng. Soc.* **12**(4), 260–274 (1983).
46. Halldane, J. F., A designer–scientist approach to describing the light from the sun, sky, and cloud, in *Lighting Design* + Application, pp. 44–56 (1985).
47. Paralusz, C. M., Internal reflection spectroscopy, in *ASTM Standardization News*, pp. 42–46 (1985).
48. Erb, W., Requirements for reflection standards and the measurement of their reflection values. *Appl. Optics* **14**(2), 493–499 (1975).
49. Oriel Corp., *Light Sources, Monochromators, Detection Systems*, 1989 ed., The Oriel Catalog, (E. G. Arthurs, Ed.), Vol. 2. Author, Stratford, CT (1989).

APPENDIX A

SOURCES, DETECTORS, AND WINDOW MATERIALS FOR UV–VIS, NIR, AND IR SPECTROSCOPY

UV–VIS, NIR, AND IR EMISSION SOURCES (USEFUL WAVELENGTH AND FREQUENCY WORKING RANGES ARE GIVEN)

Source type	Useful emission range	
	μm	cm^{-1})
Quartz tungsten halogen monofilament lamp	0.22–2.7	45,455–3,704
DC deuterium lamp for UV	0.185–3.75	54,054–2,667
Pulsed Xenon arc Lamp	0.18–2.5	55,556–4,000
DC arc lamp	0.20–2.5	50,000–4,000
Globar	1.0–100	10,000–100
Nernst Glower	0.30–35	33,333–286
Carbon arc	0.50–100	20,000–100
Mercury lamp	0.30–100	33,333–100
Visible and NIR lasers	Helium : neon (He : Ne) at 632.8 nm	15,802.8
	Neodymium : yttrium aluminum garnet (Nd : YAG) at 1064 nm, generally 0–3 W output power	9,398.5

UV–VIS, NIR, and IR Detectors (Useful Wavelength and Frequency Working Ranges Are Given)

Detector type	Useful detection range	
	μm	cm^{-1}
Silicon	0.30–1.1	33,333–9,091
PbS (lead sulfide)	1.1–3.0	9,091–3,333
InAs (indium arsenide)	1.7–5.7	5,882–1,754
InGaAs (indium gallium arsenide)	0.90–1.7	11,111–5,882
Ge:X (germanium)	2–40	5,000–250
Ge:Au (germanium gold)	2–9	5,000–1,111
Ge:Cd (germanium cadmium)	2–24	5,000–417
PbSe (lead selenide)	1.7–5.5	5,882–1,818
Ge:Zn (germanium zinc)	2–40	6,667–250
InSb (indium antimonide)	1.8–6.8	5,556–1,471
PbTe (lead telluride)	1.5–4.5	6,667–2,222
DTGS/KBr (deuterated triglycine sulfate)	0.83–25	12,050–400
DTGS/PE (deuterated triglycine sulfate)	10–120	1,000–83
MCT (mercury cadmium telluride) or HgCdTe	1–11	10,000–909
MCT (photovoltaic)	1–17	10,000–588
TGS (triglycine sulfate)	10–120	1,000–83
HgCdTe (mercury cadmium telluride	1.5–16	6,667–625
PLT (pyroelectric lithium tantalate; $LiTaO_3$)	1.5–30	6,667–333

UV–VIS, NIR, and IR Window Materials (Useful Wavelength and Frequency Working Ranges Are Given)

Window material	Useful transmittance range	
	μm	cm^{-1}
CsI (cesium iodide)	0.3–50	33,333–200
PbS (lead sulfide)	1.1–3.0	9,091–3,333
KBr (potassium bromide)	0.25–26	40,000–385
KCl (potassium chloride)	0.25–20	40,000–500
NaCl (sodium chloride)	0.25–16	40,000–625
KRS-5 (thallium bromide iodide)	0.6–40	16,667–250
Ge (germanium)	1.1–30	9,091–333
As_2S_3 (arsenic sulfide)	0.6–15	16,667–667
MgF_2 (magnesium fluoride; IRTRAN-1)	0.6–9.5	16,667–1,053
ZnSe (zinc selenide; IRTRAN-4)	0.6–26	16,667–385
BaF_2 (barium fluoride)	0.2–13	50,000–769
ZnS (zinc sulfide; Cleartran or IRTRAN-2)	0.6–15	16,667—667
CaF_2 (calcium fluoride; IRTRAN-3)	0.2–9	50,000–1,111
Al_2O_3 (aluminum oxide; sapphire)	0.2–7	50,000–1,429
SiO_2 (fused silica or quartz)	0.2–4	50,000–2,500
AgBr (silver bromide)	0.5–35	20,000–286
Polyethylene (high density)	16–333	625–30

FIBER OPTIC MATERIALS (USEFUL WAVELENGTH AND FREQUENCY REGIONS ARE GIVEN)

Fiber optic material	Useful transmittance range	
	μm	cm^{-1}
SiO_2 (fused silica or quartz)	0.2–1.25	50,000–8,000
SiO_2 (anhydrous quartz)	0.4–2.64	25,000–3,788
ZrF (zirconium fluoride)	0.9–4.76	11,111–2,100
Chalcogenide	2.22–11.1	4,505–901

SELECTION OF USEFUL MEASUREMENT PATHLENGTHS (BASED ON PURE HYDROCARBONS)

SW-NIR, 800–1,100 nm, 5–10 cm
LW-NIR, 1,050–3,000 nm, 0.1–2 cm
MIR, 2,500–20,000 nm, 0.1–4 mm
Raman, 2,500–20,000 nm, N/A

MATRIX/MEASUREMENT TECHNIQUES OF CHOICE

Gases: long-path MIR (0.5–20 m)
Solids: DR or DT
Liquids: all techniques
Organics: all techniques
Wastewater: sparging MIR
Caustics: MIR-ATR

SOLVENTS FOR UV SPECTROSCOPIC MEASUREMENTS

UV–VIS solvent cutoff wavelengths (nm):
Chloroform (235–240)
Hexanes (195–202)
Methanol and ethanol (205)
Water (190)
1,4-Dioxane (205)
Acetonitrile (190)

I. Solvents for Near-Infrared Spectroscopy

For measurement in the NIR, solvents most nearly transparent from 1.0 to 3.0 μm include carbon tetrachloride, tetrachloroethylene, carbon disulfide, and chlorofluorocarbons. None of these materials can be considered safe from an environmental or health perspective and thus are not recommended. The only signal for any of these solvents occurs at 2.21 μm for carbon disulfide (which should not be used at pathlengths above 1 cm). All of the solvents are transparent even to 10-cm pathlengths over the NIR spectral region, with the only exception so noted. Solvents such as chloroform, methylene chloride, dioxane, di(*n*-butyl) ether, triethylene glycol (dimethyl ether), heptane, benzene, acetonitrile, dimethylformamide, and dimethyl sulfoxide have all been used for solvents for a portion of the NIR region, generally from 1.0 to 1.6 and 1.8 to 2.2 μm. Silicone lubricants and Nujol (liquid paraffin) have also been suggested as solvents, meeting the requirements of nontoxicity, low cost, and availability. These materials, however, are far from ideal spectroscopically. For solid samples, oven-dried (105°C for 2 hr) sodium chloride, potassium bromide, or potassium chloride can be used to dilute samples across the entire 1.0–3.0 μm range.

II. Solvents for Infrared Spectroscopy

Common solvents used for infrared analysis include chloroform, acetone, methanol, and hexane. Cast films are often made by dissolving or extracting a soluble solid or liquid sample in one of the previously mentioned solvents, filtering the extract, evaporating most of the solvent in a suitable hood, and casting a thin film (by evaporation) onto an infrared transparent window material. Chloroform has IR bands at 3020 (weak), 1215, and 755 cm^{-1} (strong). Acetone has several sharp IR bands near 3000 (weak), 1715 (strong), 1420 (weak), 1365 (strong), 1220 (strong), 1090, and 535 cm^{-1}. Methanol has bands at 3650–2750 (three broad and strong bands), 1450 (medium-broad), 1120, 1030 (strong), and 640 cm^{-1} (broad). Hexanes have bands at 2750–3000 (strong and sharp), 1445 (strong), 1250 (weak), 900 (weak), and 850 cm^{-1} (weak).

For solid samples, oven-dried (105°C for 2 hr) potassium chloride can be used to dilute samples across the entire range of 40,000 to 385 cm^{-1}.

APPENDIX B

PRACTICES OF DATA PREPROCESSING FOR OPTICAL SPECTROPHOTOMETRY

I. Scaling Methods

A. Mean Centering

The mean spectrum is subtracted from all the spectra in the teaching set prior to calibration (Fig. 1).

B. Autoscaling

The mean spectrum is subtracted from all the spectra in a teaching set (mean centering), with the additional step of dividing all the spectra in the teaching set by the standard deviation spectrum prior to calibration (Fig. 2).

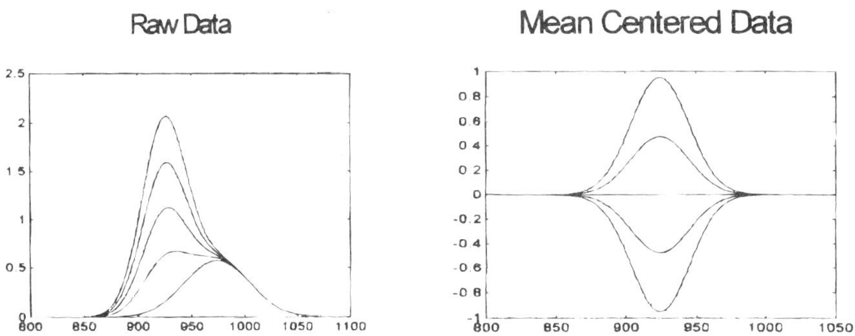

FIG. 1

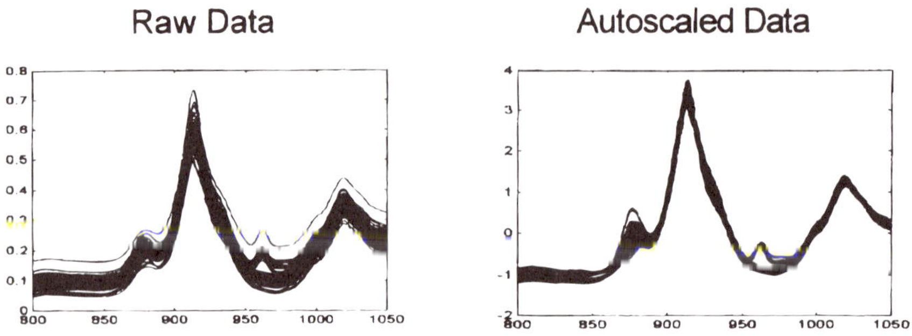
Raw Data
Autoscaled Data
0.8
0.7
0.6
0.5
0.4
0.3
0.2
0.1
0
800
850
900
950
1000
1050
4
3
2
1
0
-1
-2

FIG. 2

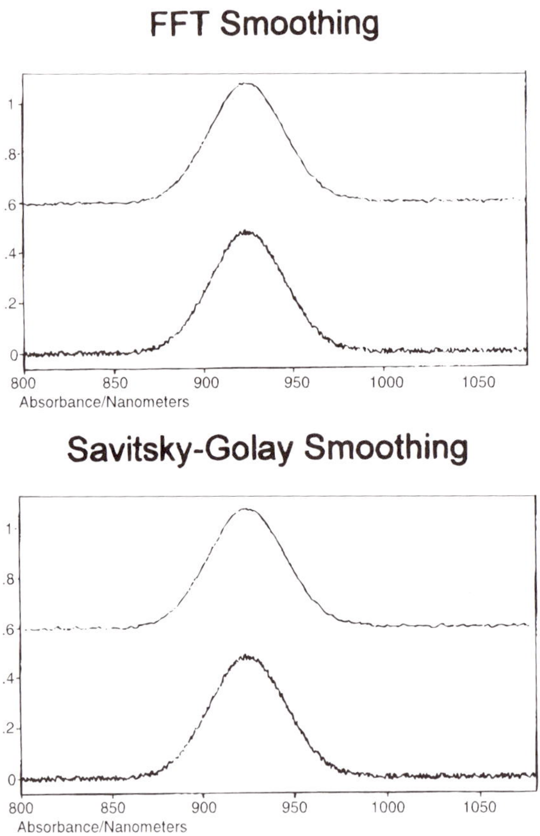
FFT Smoothing
Savitsky-Golay Smoothing
1
.8
.6
.4
.2
0
800
850
900
950
1000
1050
Absorbance/Nanometers

FIG. 3

II. Smoothing

A. Boxcar Smoothing

Boxcar smoothing attempts to improve the signal-to-noise ratio in data by averaging successive data points to remove random variation. This process also broadens bands and removes some fine structure in the data; over-smoothing can cause x-axis shifts in the signal data.

B. Fourier Domain Smoothing

Fourier domain smoothing (FFT) involves a Fourier transformation of the signal, the application of a filter function (with a set filter factor), and then performing a reverse Fourier transform to the data. A triangular filter function is generally applied. The FTT smooth can remove high-frequency noise from the signal but can also add artifacts that appear as structure when overfiltering is applied (Fig. 3, top).

C. Savitsky–Golay Smoothing

This smoothing procedure performs a best fit quadratic polynomial through successive data points. This smoothing technique determines the best fit center point for the polynomial fit as constrained by the data point segment (or window). The signal-to-noise ratio is improved, bands are broadened, and fine structure can be lost (Fig. 3, bottom).

III. Normalization

A. Normalize by Pathlength Correction

Individual pathlength data is used as a scalar multiplier term for correction of individual or groups of spectra.

B. Normalize by Area

All band areas are set to a single value for the purpose of signal comparison and/or correlation techniques (Fig. 4).

C. Normalize by Reference Band

Divide set of spectra for calibration by peak height or peak area of a reference band of consistent height or area, respectively. This ratio will "correct" the signal for quantitative analysis when the baseline is poorly resolved.

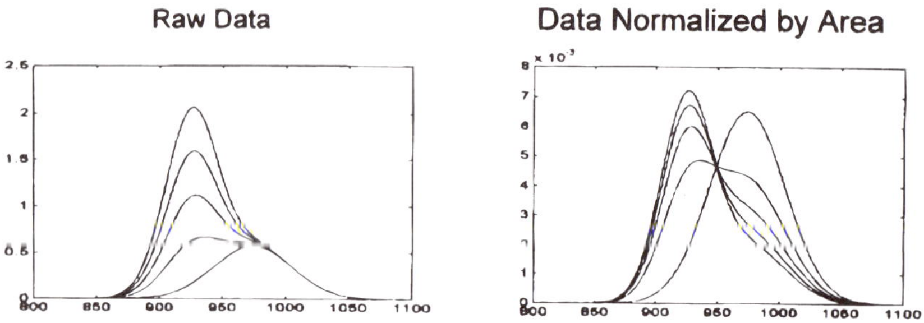

FIG. 4

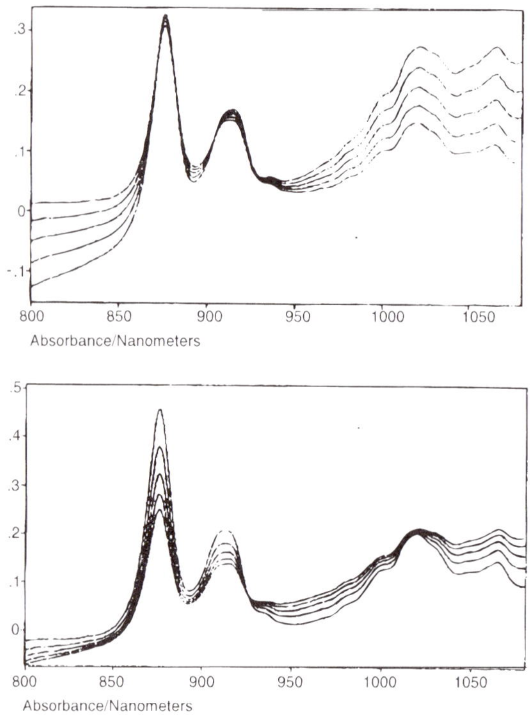

FIG. 5. (Top) Five equal increments of slope added to a single spectrum (with increasing multiplicative effect dependent on the degree of slope). This effect is typical with reflectance measurements of crystalline materials. (Bottom) MSC applied to five equal increments of slope added to a single spectrum (with increasing multiplicative effect dependent on the degree of slope). The multiplicative effect still remains following the MSC treatment (From *Multivariate calibration*, p. 345, Wiley, New York, 1989).

D. Normalize Using Multiplicative Signal Correction

Multiplicative signal correction, also known as multiplicative scatter correction (Martens and Naes), was developed to correct for light scattering in reflectance measurements (or measurements containing a strong multiplicative component). Offset and scaling are performed for this algorithm (Fig. 5).

E. Normalize Using Kubelka–Munk Transform

The Kubelka–Munk (K–M) model is applied as a linearization function to signals with scattering and absorptive characteristics as often encountered in diffuse reflectance. This relationship is given as follows (from V. P. Kubelka and F. Munk, *Z. Tech. Physik* **12**, 593, 1931):

$$\text{Signal}_{\text{K-M}} = \frac{k}{s} = \frac{(1 - R_{\lambda_i})^2}{2R_{\lambda_i}}$$

IV. Baseline Correction

A. Baseline Offset Correction

Offset correction is performed by selecting a single point of multiple points on a spectrum and adding or subtracting a y value (intensity value) from the point or points to correct the baseline offset. This preprocessing step is used to align the baseline of two or more spectra causing them to overlap, or it is used to bring the minimum point to zero (Fig. 6).

B. Derivatives

Derivatives are used to remove offset and slope due to background differences (Fig. 7).

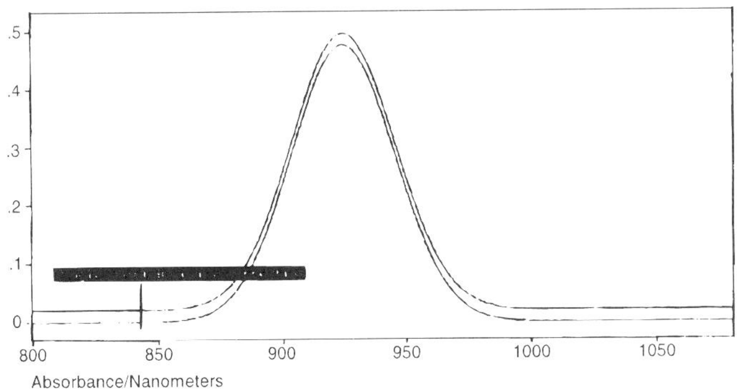

Fig. 6. Offset (single-point) baseline correction.

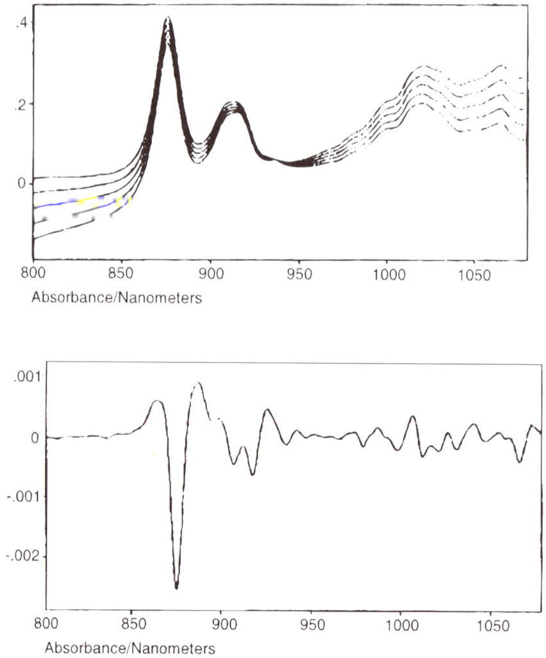

FIG. 7. (Top) Five equal increments of slope added to a single spectrum. Second derivative (Savitsky–Golay, 10 convolution points) applied to five equal increments of slope added to a single spectrum. (Bottom) The derivative treatment removes slope for this case.

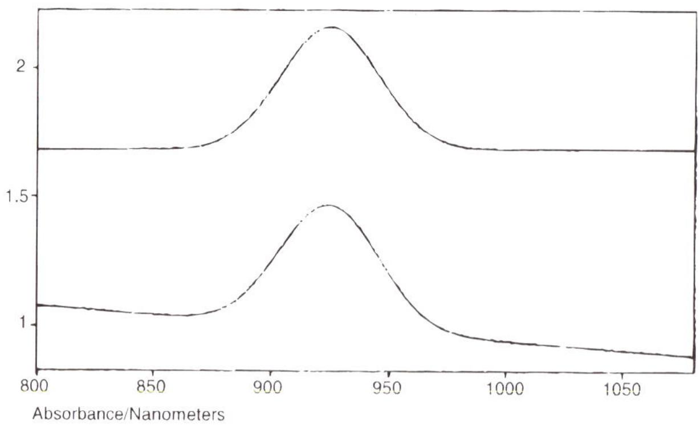

FIG. 8. Flat correction.

C. Flat Baseline Correction

Two data points are selected on the signal, and a line connecting these points is subtracted from the signal. This procedure removes first order uniform offset and slope characteristics (Fig. 8).

D. Kramers–Kronig Correction

Specular reflectance of infrared energy from a sample surface typically interacts with approximately 3 to 5 μm of the sample surface when the angle of incidence for the radiation is 30° or more. Spectra measured using specular reflectance are useful for observations of surface degradation or composition for coated surfaces. The specular reflectance spectrum consists of two major spectral components convoluted into a single spectrum. These components are (1) the molecular absorption spectrum and (2) the refractive index spectrum. The unusual appearance of a specular reflectance spectrum is due to the principle that some frequencies of light reflect more than others for any particular sample (i.e., the refractive index for any sample changes as a function of the frequency of the incident energy). The Kramers–Kronig correction is also termed the dispersion correction, and is used to mathematically correct for the refractive index spectrum. After applying the Kramers–Kronig correction to the specular reflectance spectrum, the molecular absorption spectrum results.

V. Traditional Ultraviolet/Visible and Near-Infrared Spectroscopic Data Preprocessing Methods

The common linearization of reflectance (R) or transmission (T) data to absorbance is given by

$$A = -\log R_\lambda$$

$$A = -\log T_\lambda,$$

where A is absorbance expressed as reflectance or transmission when R and $T = I/I_0$.

Absorbance ratio has been used for near-infrared (NIR) data by K. Norris and B. Hrushka [In *Near Infrared Technology* (P. Williams and K. Norris, Eds.), AACC, St. Paul, MN, 1987] and is the ratio of absorbance at one wavelength divided by the absorbance at a second wavelength. This ratio method is common to infrared spectroscopists because using it results in a term that compensates for baseline offset. The ratio of transmission data is also useful for normalizing or reducing baseline offset in spectroscopic measurements. A typical absorbance ratio is given as

$$\frac{A_{\lambda_1}}{A_{\lambda_2}}.$$

The use of the moving-averaged-segment convolution (MASC) method for computing derivatives brings about the following expression that describes a first derivative term as

$$A_{\lambda_2} - A_{\lambda_1} = A_{\lambda_{1+\delta}} - A_{\lambda_{1-\delta}},$$

where the derivative is given as the difference in absorbance values at two different wavelengths, with the position of each wavelength being determined as the gap (δ) distance $\pm$ a center wavelength (λ). Ratios of derivative terms have been applied to NIR data and have been empirically shown to produce lower prediction errors in special cases. The following expression is the first derivative ratio where two different center wavelengths have been used:

$$\frac{A_{\lambda_2} - A_{\lambda_1}}{A_{\lambda_4} - A_{\lambda_3}} = \frac{A_{\lambda_{1+\delta}} - A_{\lambda_{1-\delta}}}{A_{\lambda_{2+\delta}} - A_{\lambda_{2-\delta}}}.$$

First derivative ratio
(with two different center wavelengths)

The MASC form of second derivative is shown below [from W. F. McClure, *NIRNews* **4**(6), 12, 1993; and **5**(1), 12–13, 1994]. In this case, the second derivative term is defined by the sum of absorbances at two wavelengths ($\lambda_{1+\delta}$, and $\lambda_{1-\delta}$) minus two times the absorbance at a center wavelength (λ_1). In this case the second derivative gap size is designated as

$$A_{\lambda_1} + A_{\lambda_3} - 2A_{\lambda_2} = A_{\lambda_{1-\delta}} + A_{\lambda_{1+\delta}} - 2A_{\lambda_1}.$$

The second derivative preprocessing step is quite effective in removing slope and offset variations in spectral measurement baselines. It also "assists" the calibration mathematics in defining spectral regions where small response changes can be useful in calibration modeling. Without the use of derivatives, these regions would not be beneficial for use in calibration.

Second derivative ratio preprocessing has also been demonstrated in the Norris regression method. Empirically, this data preprocessing technique has proved useful for some applications. The following relationship designates a second derivative ratio with two different center wavelengths:

$$\frac{A_{\lambda_1} + A_{\lambda_3} - 2A_{\lambda_2}}{A_{\lambda_4} + A_{\lambda_6} - 2A_{\lambda_5}} = \frac{A_{\lambda_{1-\delta}} + A_{\lambda_{1+\delta}} - 2A_{\lambda_1}}{A_{\lambda_{2-\delta}} + A_{\lambda_{2+\delta}} - 2A_{\lambda_2}}$$

Second derivative ratio
(with two different center wavelengths)

The traditional Kubelka–Munk expression is shown below. This expression is used for samples exhibiting scatter and has proven beneficial for some measurements.

$$\frac{(1 - R_{\lambda_1})^2}{2R_{\lambda_1}}$$

Kubelka–Munk expression

The log Kubelka–Munk expression has been demonstrated and may empirically prove useful for some scattering sample data sets.

$$\log\left[\frac{(1 - R_{\lambda_1})^2}{2R_{\lambda_1}}\right]$$

Log Kubelka–Munk expression

APPENDIX C

INFRARED MICROSPECTROSCOPY

The infrared (IR) microscope can be used for many samples typically measured using an IR macro bench. It can be faster and easier than using the bench top system for routine samples due to less requirements for sample preparation. The only restrictions for microscope use are the size and hardness of the sample. Most samples are compatible for measurement using the microscope. A variety of objectives are available for IR measurements using transmittance, diffuse and specular reflectance, reflectance absorption, attenuated total reflectance (ATR), and grazing angle. Visible objectives are used for finding and aligning the sample into the proper position for measurement. A description of microscope terms and procedures is outlined below. Microscopic measurements are more optimized by purging the measurement area using dry nitrogen to eliminate absorption bands due to carbon dioxide and water vapor. A plastic ring is often employed between the microscope objective and sample stage for purging. Purge gases can also blow the sample out of the field of view.

Alignment: The procedure used to maximize energy throughput and thus increase the signal-to-noise ratio of the microscope. The noise is measured and recorded over time to ensure optimal microscope performance.

Alignment procedure for microscope optics for use in reflectance measurements: A gold-coated mirror is generally used for microscope alignment in the reflectance mode. Use upper source (reflectance configuration on single-illumination systems) to focus a bright image of the gold surface. Perform standard noise performance measurement as outlined later.

Alignment procedure for microscope optics for use in transmission measurements: The alignment procedure for IR microscopes follows three main steps and uses the optical survey mode of the microscope:

Step 1: Focus the objective onto the sample (generally a 100-μm pinhole) by moving the stage in the X, Y, and Z directions to adjust the sample to the focal spot of the objective. This step is performed with upper illumination (reflectance configuration on single-illumination systems) and with the upper aperture removed. With proper alignment the pinhole should be exactly on center and appear as a dark spot.

Step 2: Focus the condenser to the focal point of the sample. This step is performed using the lower source (transmittance configuration on single-illumination systems) and with the lower aperture present. No sample or pinhole should be present for this step. Move the condenser focus knob in the *Z* direction until the brightest possible spot is observed. The condenser thumbscrews are then used to complete the *X* and *Y* adjustment to exactly center the bright spot.

Step 3: Position both the upper and lower apertures in place and align the condenser image on the center focal point of the objective. This step is performed using the 100-μm pinhole in exact center position with both the upper and lower light sources (transmittance configuration on single-illumination systems). Once completed, this configuration should provide the maximum energy throughput for the microscope optics and thus provide the highest signal-to-noise ratio.

Aperture: The opening within the microscope optics that is responsible for producing an optical image and for specifying the area of the sample to be measured using transmittance or reflectance. The aperture is generally made of aluminum or steel coated with carbon black to absorb extraneous IR light. Coated glass apertures are available to allow visual observation of a sample without allowing IR energy to pass the area around the opening. The size of the aperture should be compatible with the image size generated by the specific objective and condenser optics. The following table presents the objective or condenser magnification and the corresponding compatible aperture.

Objective/condenser magnification	Compatible aperture
10 Times	1000 μm (1.0 mm)
15 Times	1500 μm (1.5 mm)
20 Times	2000 μm (2.0 mm)
25 Times (some ATR)	2500 μm (2.5 mm)
30 Times (some grazing angle)	3000 μm (3.0 mm)
32 Times	3200 μm (3.2 mm)

Compensation ring: A focusing mechanism for both the objective and the condenser optics that is used to a adjust for the refraction of light caused by a specific window thickness over or under the sample. If a window is used over the sample the compensation ring of the objective is used to adjust for the refraction of light. If a window is placed under the sample the compensation ring of the condenser is used to adjust for the refraction of light. To calculate the compensation setting required for each window thickness and material use the following relationship:

Compensation setting

$$= \text{thickness of window (in mm)} \times \frac{\text{refractive index of window}}{1.5}.$$

Compression cell: Allows two sample windows with flat surfaces to be squeezed together for the flattening of samples to make them transparent or to flatten them for easier transmittance measurement. KBr and diamond are common window materials.

Dichroism: Dichroism is the use of infrared polarized light (as *p*- and *s*-polarized light) to measure the molecular orientation of crystalline polymers and other highly oriented molecules. For microscope work, the requirements are (1) the capability to generate *p*- and *s*-polarized infrared energy, and (2) the capability of minimizing the stray radiant energy (stray light) by using redundant apertures (see discussion of this term). The difference between a spectrum of a material taken with *p*-polarized light and the spectrum of a material illuminated with *s*-polarized light can reveal molecular orientation sensitivities of the measured molecules. When the electronic field vector of the sample molecules is perpendicular to the field direction vector of the infrared energy there is very little interaction and slight or no absorption of the light. When the electronic field vector of the sample molecules is parallel to the field direction vector of the infrared energy there is a large interaction and significant absorption of the light. One can measure the affect of stretching on molecular orientation by making dichroic measurements before and after stretching the sample. The dichroic ratio (in arbitrary units) can be measured from +1.0 to −1.0 by using the ratio (see Appendix E):

$$\text{Dichroic ratio} = A_{\parallel} - A_{\perp}/A_{\parallel} + A_{\perp}.$$

Gain setting of detector: The gain is the voltage amplification of a detector (above a nominal voltage setting) that has the effect of increasing the signal strength of the detector in a linear fashion over the linear range of a detector. For example, if the signal-to-noise ratio of a measurement A is 50% of a previous measurement B, moving the gain setting to 2 for this weaker measurement A will yield the same voltage at the detector as that of measurement B. This gain enhancement will provide increased detector sensitivity for measurements over a limited linear detection range.

Light sources: Microscopes are generally provided with an upper light source and a lower light source. The upper source is used to view the objective image for alignment or for the reflectance measurement mode. The lower source is used to view the condenser image for alignment or for the transmittance measurement mode. Some microscopes have a single visible light source. For aligning this microscope configuration, the reflectance mode is used for objective focusing, and the transmittance mode is used for condenser focusing.

Motorized stage: Allows the microscope to "memorize" the point position (or positions) on a sample and is able to reproduce a precise location on the stage automatically.

Noise performance measurement. The optical performance of the microscope is evaluated by completing the alignment procedure to optimize energy throughput and signal-to-noise ratio and then recording the noise of the microscope system. The noise and spectrum are recorded over time to verify continuous optical performance of the system. The measurement parameters should be measured in exactly the same manner for each measurement. The number of scans, resolution, apodization function, data format, and frequency range should always be adjusted to the same parameter setup each time a noise measurement is made. The noise should be recorded as the peak-to-peak (maximum percentage transmittance minus minimum percentage transmittance) over a prespecified frequency region of the spectrum. Generally, peak-to-peak noise should be less than 0.1% *T* for most IR microscopes. Note: Peak-to-peak noise should approximately double when moving from bench-top IR transmittance measurements to microscope IR transmittance measurements. The peak-to-peak noise factor should increase approximately four times when moving from the bench-top IR transmittance measurements to microscope IR reflectance measurements.

Objective: The optical device taking the light from the microscope and focusing it onto the sample. The higher the magnification the lower the throughput. A 15× objective is a general use objective. Higher magnification allows more specific sample position measurements but reduces the energy throughput and signal-to-noise ratio. Types of objectives include standard, specular reflectance, ATR, and grazing angle.

ATR objective: ATR is used for surface analysis using physical contact with a sample surface. The objective utilizes a crystal of material for the actual physical contact with the sample. Typical crystal materials include diamond (Di), germanium (Ge), silicon (Si), and zinc selenide (ZnSe). ATR measurements can yield excellent quality spectra provided that the contact pressures of the ATR crystal and the sample are held constant; reproducible data require reproducible contact pressure. Commercial contact gauges are available from suppliers of microscope accessories. One-percent reproducibility is typically achieved when keeping the contact pressure constant. ATR crystals are generally ZnSe with a refractive index of 2.42 and an angle of incidence of 45°. Sample penetration is approximately 40 μm using this ATR crystal. An air background is taken for ATR measurements. The depth of penetration is calculated for ATR crystals using the relationship:

$$\text{Depth of penetration for ATR} = \frac{\text{Wavelength of incident light}}{2\pi \times n_1(\sin^2 \alpha - n_2^2)^{1/2}},$$

where n_1 is the refractive index of the ATR crystal—refractive indices for typical ATR crystals are ZnSe (2.42), Di (2.2), Si (3.6), Ge (4.0); n_2 is the refractive index of the sample; and α is the angle of incidence (and reflection) of the ATR crystal—typically 45°.

Grazing angle objective: A measurement configuration in which light from the objective strikes the sample at a high angle of incidence for measuring the coating or surface characteristics of a sample. This measurement technique has a sample penetration of typically less than 1 μm. This technique can be used to measure very thin surface characteristics of reflective samples. The technique is qualitative and is not particularly reproducible and thus not useful for quantitative analysis. The grazing angle objective typically is configured for a 65–80° angle of incidence. Apertures are not generally used when using grazing angle measurements; there is a fivefold pathlength increase when using grazing angle compared to standard transmittance measurement geometry.

Visible objective: Microscope objectives for use in visual survey of the sample for the purpose of inspection, sample positioning, and alignment. Cross-hair features are often available for measuring sample distances, sample size, and sample alignment.

Redundant apertures: A pair of apertures can reduce the diffraction of light caused when the light passing through and aperture is of a wavelength close to the size of the aperture. A second aperture matched to the first aperture can reduce or eliminate the spurious energy due to diffraction. Use of a second aperture (termed the redundant aperture) allows a single sample layer less than 50 μm to be measured on a routine basis.

Refraction of light: Refraction is defined as the change in direction of light due to a change in the velocity of the light while passing through two or more materials of differing refractive indices. In microscope systems this phenomenon causes the offset of the light based on the refractive index and thickness of the sample and sample accessory window material(s). The refracted light must be adjusted for by using a compensation ring, which is a feature of both the objective and the condenser optics.

Spurious energy: Spurious energy in a microscope system is unfocused light caused by the diffraction of IR energy passing through an aperture that is nearly the size of the wavelength of the IR energy. If the wavelength of light is longer than the size of an aperture the incident light will be reflected back. The closer the wavelength is to the size of the aperture the greater is the percentage of light diffracted. Typically for microscopes, apertures less than 50–60 μm are too small and create diffraction problems. The use of multiple apertures can reduce the diffraction of light passing through the microscope optics.

Stray light: Stray light is dealt with in other sections of this book; however, the main source of stray light (also termed stray radiant energy) within the IR microscope is the spurious energy caused by diffraction of infrared energy through small apertures.

Window materials for microscopy: Typical window materials for microscopy include barium fluoride (BaF_2) for use with polar solvents (including water), potassium bromide (KBr) for solids and nonpolar solvent use, zinc selenide (ZnSe) with its high refractive index for use in diamond cell filler (background measurement), and diamond for compression cell work in which higher pressures are required.

APPENDIX D

I. Diffuse Transmittance and Optical Geometry

A. Introduction

The following discussion illustrates diffuse transmittance measurement and the design advantages of employing diffuse illumination for measuring the ultraviolet blocking offered by sun-protective fabrics.

As with many methodology standards, the measuring apparatus details are generalized and subject to interpretation. AS/NZ 4399 (1) does not strictly mandate the use of collimated illumination as some have interpreted. It does mandate the use of an integrating sphere that in practice can be used as either a diffuse light collector or a diffuse light source, utilizing the principle of optical reciprocity. The AS/NZ standard implies optical reciprocity without being completely explicit. In comparison, methodology standards prepared for the colorimetry of materials are clearly explicit in their endorsement of reciprocal geometries (CIE, ISO, and ASTM).

B. Diffuse Transmittance Measurements of Textiles

Many textiles are translucent materials that diffuse incident light. A ray of light incident onto a textile sample will be scattered. Light that is not transmitted is reflected or absorbed.

For translucent samples, the radiation intensity is strongest in proximity to the regular transmitted direction as shown in Fig. 1. More opaque

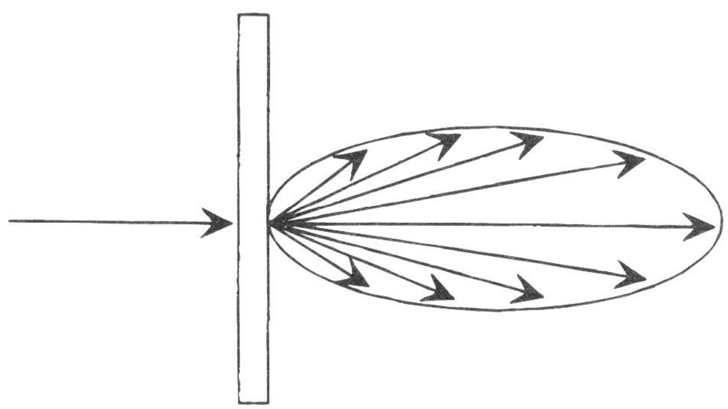

FIG. 1

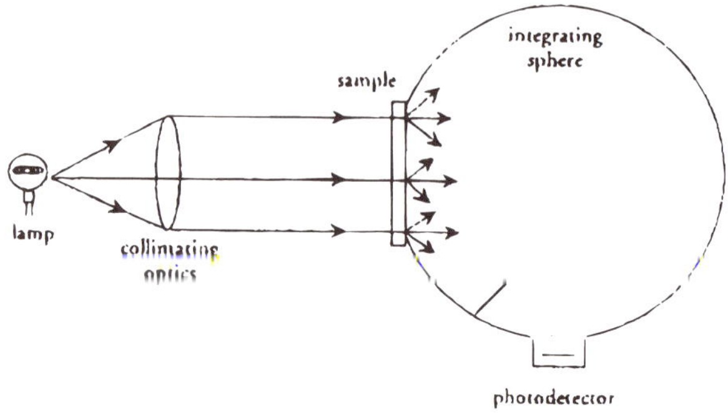

FIG. 2

samples will produce an intensity pattern that approaches a uniform, hemispherical distribution. The ratio of the total transmitted light to the total incident light is known as the transmittance, a measurable quantity.

Total hemispherical transmittance is measured by the use of an integrating sphere to collect the light scattered at all angles.

The interior walls of the integrating sphere are coated with a white, highly reflective material. In Fig. 2, light from an external source is collimated and strikes the sample surface at normal incidence. A photodetector responds proportionally to the internal illumination produced on the sphere wall. A baffle prevents direct illumination of the detector after scattering from the sample. The incident beam flux is recorded initially without the sample in place to determine the measurement baseline.

C. Reciprocity of Illuminating and Viewing Geometry

The geometry depicted in Fig. 2 is known as normal/hemispherical, which refers to the illuminating/viewing conditions. It is often abbreviated as 0°/h or more commonly 0°/d (diffuse) geometry.

The Helmholtz reciprocity principle (2, 3) states that the loss of light flux within a ray bundle will not be changed if the direction of travel is reversed. As applied to measuring instruments, the results will not change if the geometry of the illuminating and viewing beams are interchanged (4). Optical engineers often make full use of this principle in ray tracing calculations.

Consider the light scattering depicted in Fig. 1. Instead of an incident ray, consider an observer's field of view (FOV). The sample scattering

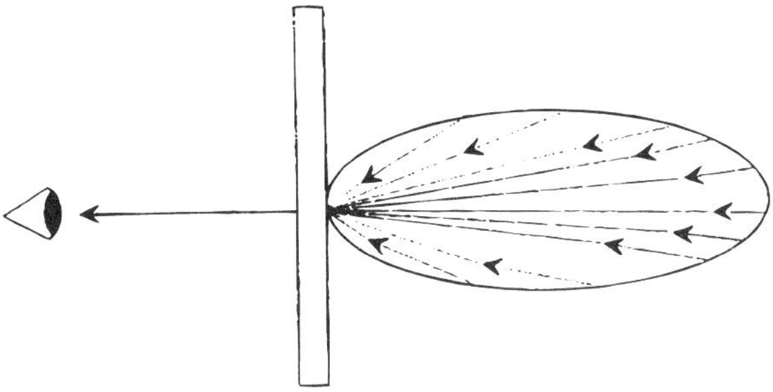

FIG. 3

extends the FOV to multiple points in space beyond the sample. The scattered FOV is specific to the sample's scattering profile (Fig. 3).

In practice, illuminating material samples over the entire hemisphere is the reciprocal method for determining transmittance for a given FOV and any scattering profile. Therefore, the light source and photodetector in Fig. 2 can be reversed in order to construct a diffuse/normal (d/0°) measuring instrument (Fig. 4).

It can appear to those not completely familiar with geometrical optics that diffuse illumination will produce a different measurement result than collimated illumination because so many more incident rays and angles are transmitted through the sample. This is not true. The collimated viewing system with hemispherical illumination only accepts the incident rays that are reciprocal to the radiation pattern produced by an identically collimated incident beam.

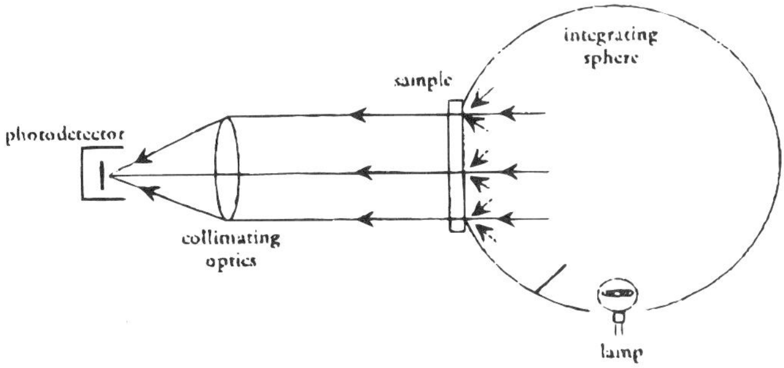

FIG. 4

D. Radiometric Advantage of d/0° Geometry

The major advantage of the d/0° system compared to the 0°/d is in radiometric system efficiency. In Fig. 2, the amount of light that is collected from the lamp is a function of the *f*-number (*f*/#) of the collimating optics. Those familiar with photography understand that *f*/# expresses the light gathering power of a camera lens. Lower values of *f*/# are required for slower speed film.

The total flux collected from the lamp in Fig. 2 can be expressed as

$$\Phi_{0°/d} = I \cdot \frac{\pi}{4f/\#^2} \quad (\mathrm{W}), \tag{1}$$

where I is the intensity of the lamp in units of W/sr. The second half of the equation expresses the collected solid angle as a function of *f*/# in units of steradians.

If the same lamp is placed inside the integrating sphere, the total flux from the lamp is collected:

$$\Phi_{0°/d} = I \cdot 4\pi \quad (\mathrm{W}), \tag{2}$$

where 4π expresses the solid angle subtended by a sphere.

The ratio of Eq. (2) to Eq. (1) quantifies the efficiency improvement of the d/0° design. This is equal to

$$\frac{\Phi d/0°}{\Phi 0°/d} = 16 \cdot f/\#^2. \tag{3}$$

Therefore, even for a highly efficient *f*/1 lens, the d/0° geometry would be 16 times more efficient. When one applies this analysis to diffraction grating spectrophotometers, which are generally *f*/3 and larger, the d/0° system offers greater than two orders of magnitude improvement in radiometric efficiency.

II. Spectral Measurements and Sample Fluorescence

A. Introduction

The measurement of the ultraviolet protection factor (UPF) of clothing can be performed by measuring the material's diffuse transmittance in the ultraviolet (UV) spectrum.

A common textile additive known as an optical brightening agent (OBA) enhances a fabric's appearance through the optical effect of fluorescence. A brief overview of the mechanism of fluorescence, as it relates to OBAs, is presented. Ultraviolet spectrophotometers, used to measure spectral transmittance, can exhibit systematic errors when fluorescent samples are introduced.

In addition to geometric requirements, AS/NZ 4399 describes the effect of fluorescent samples on diffuse transmittance measurements for two types of spectrophotometer designs. The discussion is specific to whether the instrument monochromator is placed before or after the sample: in other words, whether the sample illumination is monochromatic or polychromatic.

B. Measurement of Ultraviolet Protection Factor

AS/NZ 4399 determines the UPF of a clothing material as an *in vitro* test method based on measuring its diffuse spectral transmittance. UPF is analogous to the SPF for a topical sunscreen. However, SPF by definition is determined *in vivo* as the increase in exposure time required to induce erythema, i.e., SPF 4 means four times longer to induce erythema. UPF is calculated as follows:

$$\mathrm{UPF} = \frac{\sum_{290\text{ nm}}^{400\text{ nm}} E_\lambda \times S_\lambda \times \Delta\lambda}{\sum_{290\text{ nm}}^{400\text{ nm}} E_\lambda \times S_\lambda \times T_\lambda \times \Delta\lambda}, \tag{4}$$

where E_λ is the CIE erythemal spectral effectiveness, S_λ is the solar spectral irradiance, and T_λ is the spectral transmittance of the sample. The two standardized functions, E_λ and S_λ, are illustrated in Fig. 5.

The two functions describe the relative sensitivity of erythema to individual wavelengths and the spectral distribution of sunlight as it reaches the earth's surface.

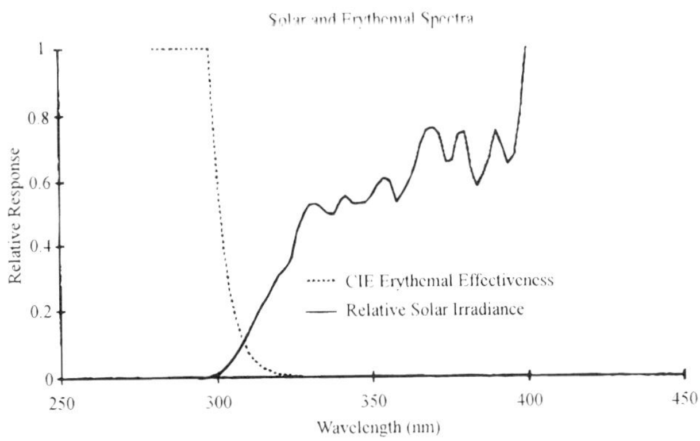

Fig. 5

A more revealing plot is the product $E_\lambda \times S_\lambda$, which appears in both the numerator and denominator of Eq. (4). In Fig. 6, the determination of UPF is weighted most heavily by a sample transmittance in the UVB portion of the spectrum, with maximum weighting near 305 nm.

C. Fluorescence and Optical Brightening Agents

Fluorescence is one type of photoluminescence, the physiochemical phenomenon in which the absorption of a photon within a material species induces the emission of a photon of a longer wavelength, being in excess of that due to thermal emission. It is accurately defined as "photoluminescence in which the emitted optical radiation results from direct transitions from the photoexcited energy level to a lower level, these transitions taking place generally within 10 nanoseconds after the excitation" (5). From chemistry, we know that these are generally *p′p** transitions, involving *p* orbitals from either conjugated aromatic compounds (organic) or inorganic transition metals or rare earth metal compounds and complexes.

In the case of fabrics, the typical fluorescent material is an optical brightening agent. An OBA is used to give an enhanced appearance in the blue region of the visible spectrum. A decrease in the blue reflectance of a natural white material will make it appear more yellow and less appealing; therefore, many new white fabrics are treated. OBAs are typically highly conjugated derivatives of stilbene or benzimidazole, which have been derivatized to give them water solubility or polar characteristics to make them attracted to the fabric substrate on which they will be used (Fig. 7). Different OBAs are used for substrates, such as cotton, polyester, and nylon.

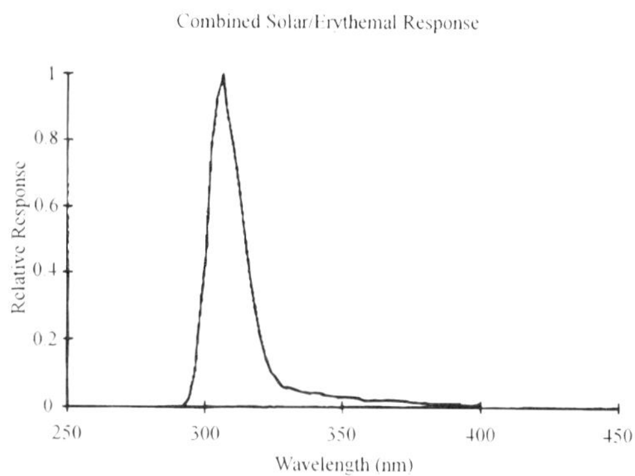

Fig. 6

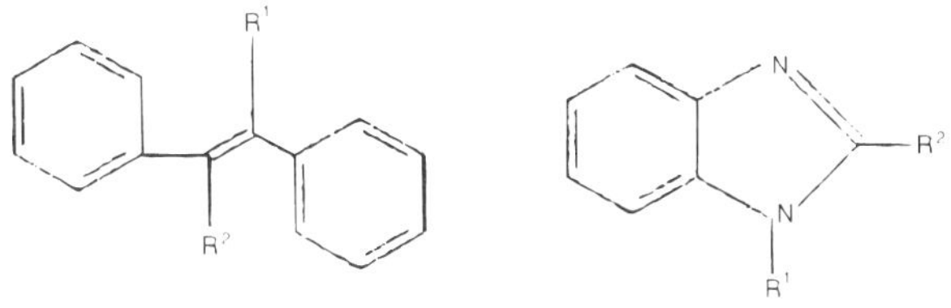

FIG. 7

Optical brighteners and whiteners are also used in laundry detergents to suppress the yellowing that can accompany the repeated laundering of white garments. Most OBAs have excitation maxima in the very near UV or at the lower end of the visible scale—wavelengths in the range from 340 to 400 nm. Emissions typically peak at blue wavelengths at approximately 435–440 nm. Typical excitation and emission spectra are illustrated in Fig. 8.

D. SPECTROPHOTOMETERS AND FLUORESCENT MATERIALS

A spectrophotometer is an analytical instrument used to determine the optical transmittance, reflectance, or absorbance of a material as a function of wavelength. The essential components of a spectrophotometer are a light source, a monochromator to separate the wavelengths of light, and a photodetector. As discussed in Section 1, an integrating sphere is required for

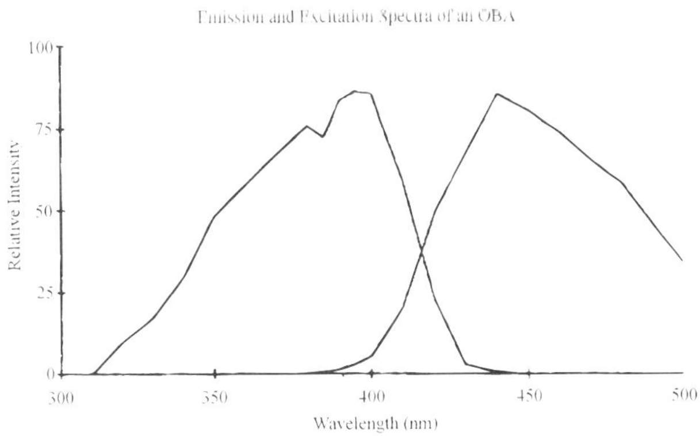

FIG. 8

diffuse transmittance measurement. It can be used in conjunction with either the illumination or the collection subsystem.

There are basically two types of spectrophotometers that differ by position of the monochromator with respect to the sample. Sample illumination is either monochromatic or polychromatic. Each has its own characteristic response to fluorescent materials. Measurement data from both types of spectrophotometers are presented here. The ideal type of instrument for fluorescent samples would use both monochromatic illumination and monochromatic detection. The design is difficult to realize in practice, especially with respect to diffuse transmittance measurements and integrating spheres.

E. Monochromatic Illumination

In most commercial spectrophotometers, the monochromator precedes the sample, illuminating sequentially with single wavelengths. This type of instrument is intended for nonfluorescent materials. A distinct systematic error occurs when a fluorescent sample is introduced as depicted in Fig. 9.

The instrument is set to measure the transmittance at wavelength λ_t. However, if λ_t also corresponds to an excitation wavelength for that material, it will emit an associated fluorescent wavelength λ_f. The photodetector in this instrument is capable of responding to both λ_t and λ_f but incapable of distinguishing between the two. The instrument would display an erroneously high transmittance value for λ_t because it adds the signal generated by photons emitted with wavelength λ_f.

The systematic error can be eliminated by using an optical filter that effectively removes the fluorescent component before it reaches the photodetector as depicted in Fig. 10.

Practical application of this method and proper filter selection presumes a foreknowledge of the excitation and emission spectra. Fortunately, this can be applied to measuring the ultraviolet transmittance of fabrics containing OBAs.

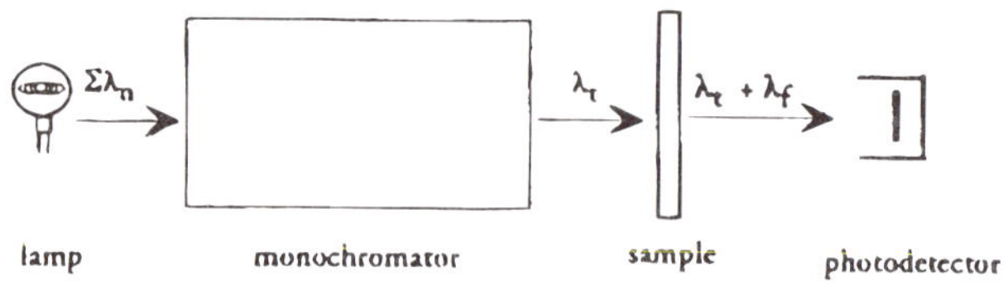

Fig. 9

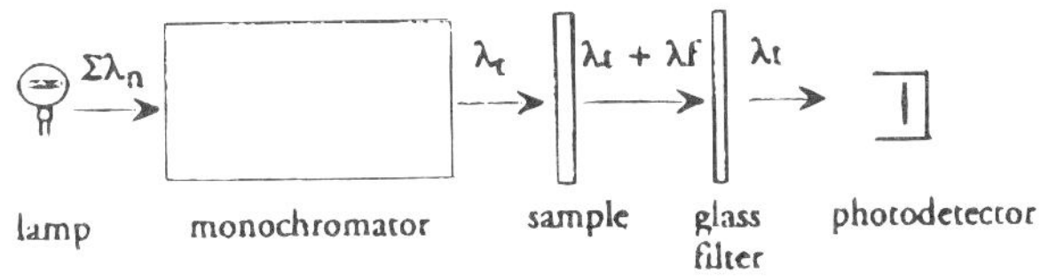

FIG. 10

Because OBAs excite below 400 nm and emit above 400 nm, AS/NZ 4399 recommends the use of a Schott UG-11 filter glass for monochromatic illumination. Its spectral transmittance is shown in Fig. 11.

The measurement error associated with monochromatic illumination can be quite substantial in determining UPF if the fluorescent emission is not removed. A white fabric containing an OBA was measured on this type of spectrophotometer with and without a UG-11 filter. The results are shown in Fig. 12.

The data without the filter reveal that the OBA is excited by virtually all the incident wavelengths. The dramatic error most adversely affects the calculation of UPF. Without correcting for the systematic error due to fluorescence, this fabric would seem to offer a UPF of only 7. However, the true effect of the OBA absorption provides a UPF of 40, revealed when the UG-11 filter is utilized. Incidentally, the instrument fitted with the UG-11 filter did not produce useful data for wavelengths above 380 nm due to the filter's sharp cutoff in transmittance.

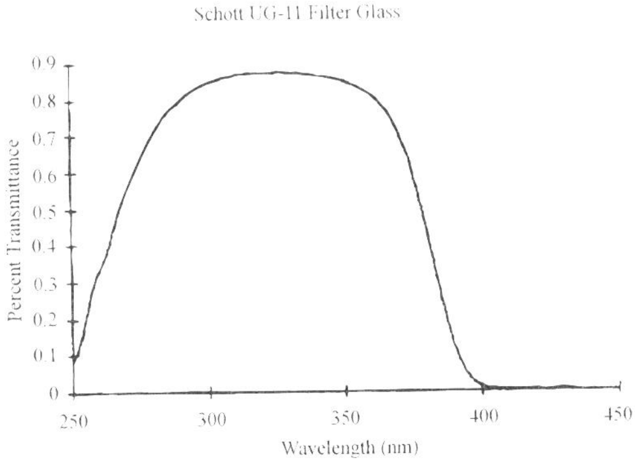

FIG. 11

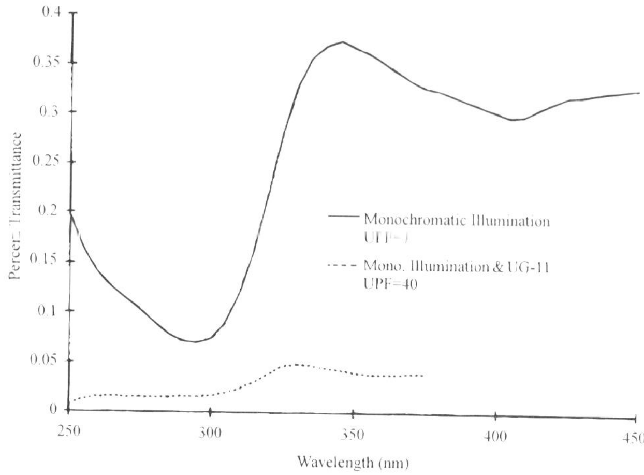

FIG. 12

F. POLYCHROMATIC ILLUMINATION

The second type of spectrophotometer design uses polychromatic illumination of the sample and monochromatic detection. Placing the monochromator after the sample separates both λ_t and λ_f before they are detected. The division of wavelengths can be distributed spatially as depicted in Fig. 13.

The measured spectrum correctly displays the transmittance for wavelength λ_t. However, the result for λ_f combines the material's transmittance and fluorescent emission at that wavelength. The combined result at λ_f also depends on the UV content of the instrument's light source., i.e., the measurement result will change for a different lamp type.

Spectrophotometers that use polychromatic illumination reveal the effect of OBAs on a measured spectrum. The use of OBAs is to enhance the blue reflectance of white fabrics. This type of instrument in a reflectance

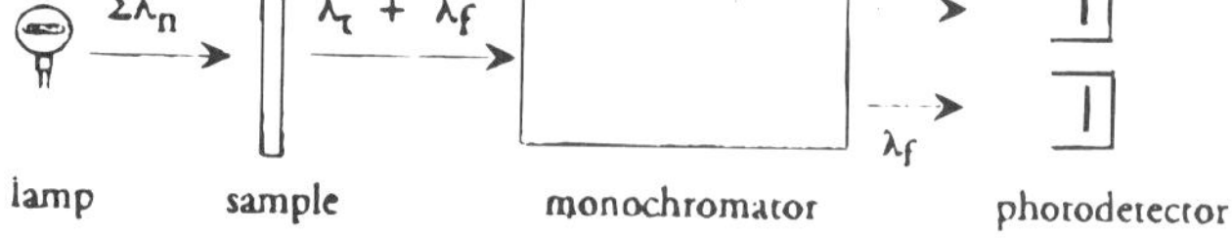

FIG. 13

geometry is used for measuring the whiteness of textiles containing OBAs (6). Commercial instruments that characterize fluorescent whiteness must also use light sources filtered to provide a standardized UV content (7).

Because fluorescence is diffuse by nature, the emitted component is both reflected and transmitted. White cotton fabrics measured in the transmittance geometry of the UV-1000F analyzer will also display the effect of OBA additives as illustrated in Fig. 14.

The xenon flashlamp in the UV-1000F contains sufficient UV content to excite the fluorophores. The effect of the OBA is clear. The emission spectrum appears at visible blue wavelengths, beyond 400 nm with a peak at 435 nm.

AS/NZ 4399 recommends that the polychromatic illumination should conform to the requirements of a solar simulator, i.e., in UV spectral distribution. However, as revealed by Fig. 14, this would only be relevant to quantifying the emission spectrum or whiteness of the sample. The source spectrum does not impose on the UV absorption measurement.

G. The Effect of OBAs on Sun Protective Clothing

As seen in Fig. 14, the OBA has increased the UV absorption of the fabric. This is a desired effect in the manufacture of sun protective clothing. The OBA fluorescence emission measured by the UV-1000F has no bearing on evaluating a fabric's UV protection.

The measured transmittance data for the cotton samples from Fig. 14 can also be converted to UPF values. The increase in UPF offered by the OBA additive is apparent from Table 1. The same advantage applies to the

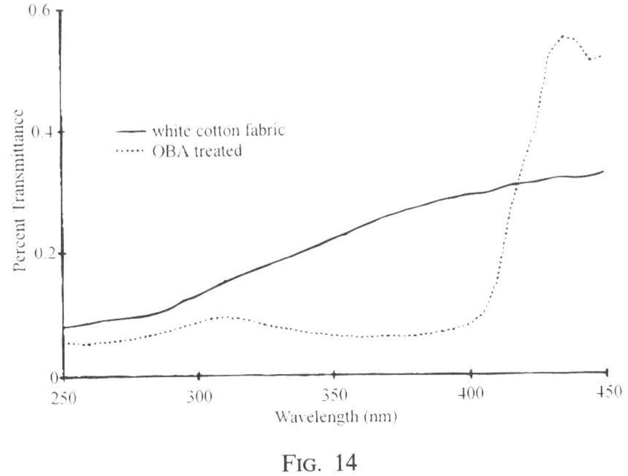

Fig. 14

TABLE 1

Sample	UPF	UVA blocking (%)	UVB blocking (%)
White cotton	6.5	77	88
OBA treated	11.4	93	92

total blocking (100%-T%) for either UVA (315–400 nm) or UVB (280–315 nm). The OBA-treated fabric offers improved sun protection.

H. Measuring OBA Concentration

The emission spectrum in Fig. 14 was isolated by the measured difference between the treated and untreated fabrics. The increased absorption at UV wavelengths is also evident. The relative strengths of both the absorption and the emission spectra are proportional to the concentration of the OBA. Quantitative analysis is possible by devising a calibration curve using a sample set of known OBA concentration.

I. Summary: Part Two

Optical brightening agents are often applied to enhance the whiteness of textiles by inducing fluorescence by UV excitation and visible blue emission. OBAs have the added benefit of increasing the UV absorption and hence a textile's sun protective ability.

References

1. Australian/New Zealand Standard, *Sun Protective Clothing*—Evaluation and Classification. AS/NZ NO. 4399 (1966).
2. Helmholtz, H. von, *Handbuch der Physiologischen Optik*, 3rd ed., pp. 198–199 (1909).
3. ASTM, *Selection of Geometric Conditions for Measurement of Reflection and Transmission Properties of Materials*, E 179-91a.
4. *The Helmholtz relationship also considers the state of polarization for incident and emergent fluxes.*
5. Gundlach, D., and Terstiege, H., Problems in measurement of fluorescent materials, *Color Res. Appl.* **19**(6) 427–436 (1994).
6. AATCC 10, *Whiteness of Textiles.*
7. CIE, *A Method of Assessing the Quality of Daylight Simulators for Colorimetry*, Publ. No. 51.

APPENDIX E

DICHROIC MEASUREMENTS OF POLYMER FILMS USING INFRARED SPECTROMETRY

I. Introduction

Infrared Dichroism is the use of infrared polarized light as *p*-(90°) and *s*-(0°) polarized light to measure the molecular orientation of polymer films and other highly-oriented molecules. The capability to generate *p*- and *s*-polarized infrared energy is required to complete these measurements. The difference between a spectrum of a polymer film taken with *s*-polarized light ($A_{\parallel}$) and the spectrum of a film illuminated with *p*-polarized light ($A_{\perp}$) will reveal molecular orientation of the polymer backbone and attached side groups. When the electric field vector of the infrared active molecule (dipole) is perpendicular to the field direction vector of the infrared energy there is no interaction of the incident infrared energy with the dipole and thus no infrared absorption. When the electric field vector of the infrared active molecule (dipole) is parallel to the field direction vector of the infrared energy there is a large interaction and significant infrared absorption. The molecular orientation of the various molecules in a polymer film can be measured by making dichroic (*s*- and *p*-polarized light) measurements before and after stretching of the film sample.

II. Theory

The intensity of an infrared absorption band is proportional to the square of the transition moment (or infrared active dipole moment). The absolute intensity of an infrared band also depends upon the direction of the transition moment (dipole electric field vector) and the field direction vector (electric field vector) of the incident infrared radiation. The proportion of the transition moment (TM_P) in the direction of the infrared electric field direction vector (E) is given as

$$TM_P = tm \times \cos \beta$$

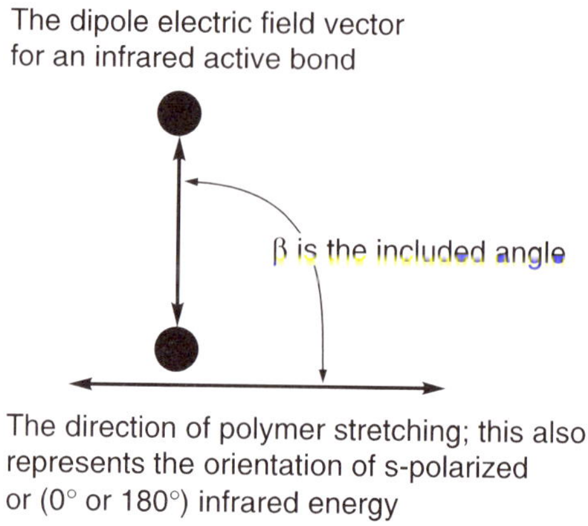

FIG. 1. An illustration of the bond angle between the dipole electric field vector and the direction of polymer stretching.

Noting that $(tm \times \cos \beta)^2$ is proportional to the infrared absorbance, and that *tm* is the transition moment of the molecule (dipole) of interest and β is the included angle. An illustration of the dipole electric field vector and the electric field vector of the *s*-polarized infrared radiation (with respect to the angle of polymer stretching) is illustrated in Figs 1 and 2.

Polymer films may be stretched resulting in uniaxial elongation (orientation along the stretching axis). During the process of stretching the polymer backbone will align in a parallel manner along the direction of stretching. However under the stretching conditions the attached groups may assume no preferred orientation with respect to the direction of stretching. For polymer films the direction of stretch is generally chosen in the horizontal (↔) direction. When this stretching direction is chosen, the *s*-polarized or (0° or 180°) infrared energy is designated as the infrared polarizer orientation for measuring $A_{\parallel}$ (Absorbance spectrum representing parallel polarized light). Likewise, the *p*-polarized or (90° or 270°) is the orientation for measuring $A_{\perp}$ (Absorbance spectrum representing perpendicularly polarized light) (1).

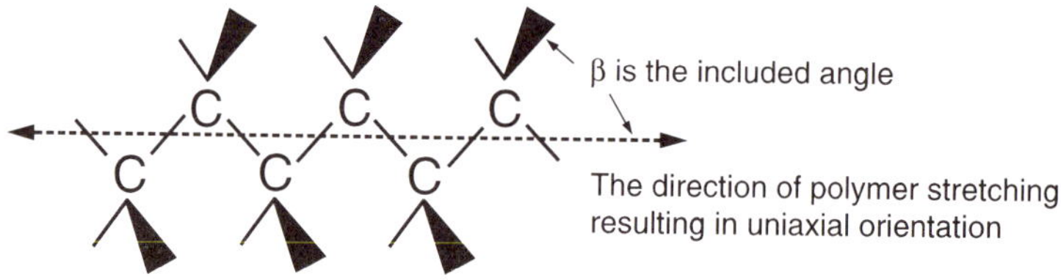

FIG. 2. An illustration of Fig. 1 for a molecular example.

III. Methods of Calculating Dichroic Parameters

The dichroic calculations (in arbitrary units) can be reported using the following relationships for polymer films:

$$\text{Dichroic ratio } (R) = \frac{A_{\parallel}}{A_{\perp}} \qquad (2)$$

$$\text{Dichroic Difference } (\Delta A) = A_{\parallel} - A_{\perp} \qquad (2)$$

Dichroic Difference Ratio (ΔAr)

$$= A_{\parallel} - A_{\perp}/A_{\parallel} + A_{\perp} \quad (\text{from } +1.0 \text{ to } -1.0) \qquad (3)$$

$$\text{Polymer Orientation Ratio } (Pr) = K_1(R - 1/R + 2) \qquad (4)$$

where $K_1 = 2/(3\cos^2\beta^{-1})$, β = the bond moment angle (in degrees) between the absorption band of interest and the polymer backbone. Note: If β is unknown, replace the constant K_1 with the value 1.0 for relative calculation of Pr.

A. Calculation of Bond Moment Angles (β) (1)

$$\text{Dichroic ratio } (R) = \frac{A_{\parallel}}{A_{\perp}} = 2\left(\frac{\cos^2\beta}{\sin^2\beta}\right) = 2\cot^2\beta \qquad (5)$$

Thus by using Eq. (5) the values for R at any angle β can be calculated where β is the included angle, or the infrared dipole angle relative to the direction of stretch (i.e., *s*-polarized orientation). The angle β and the corresponding R values are shown in Table 1.

TABLE 1

β (The infrared dipole angle relative to the direction of stretch)	R (Dichroic ratio)
90°	0.000
80°	0.062
70°	0.265
60°	0.667
54°44′	1.000
50°	1.407
40°	2.843
30°	3.000
20°	15.094
10°	64.667
$\beta \to 0°$	$R \to \infty$

IV. Graphical Representation

The polymer stretching ratio (α) is plotted on the abscissa versus the (Pr) Polymer orientation ratio as the ordinate. The polymer stretching ratio (α) is given by $\alpha = Ls/Lo$ = Stretched Length/Initial Length. The initial length should be set to approximately 1.0 cm by making two marks 1.0 cm apart directly on the polymer area to be stretched. The value for the polymer stretching ratio (α) for no stretching should be 1.0.

V. Method

A. Infrared Polarizers

Two basic types of infrared polarizers are commercially produced. These include the Brewster Angle cross-plate polarizers and the Wire Grid polarizers. The Brewster Angle cross-plate polarizers use a high refractive index material (such as germanium) and have a useful transmission range of 5000–500 cm^{-1}. The Wire Grid polarizers consist of 0.2 micron wide aluminum strips separated by a space of approximately 0.4 microns. The matrix for placing the aluminum strips is often KRS-5 with a useful working range of 5000–285 cm^{-1} (2).

All polarizers are placed within the sample compartment directly in the infrared beam. They are mounted for complete 360° rotation. At 0° or 180° rotational settings the transmitted electric field vector (E) of the infrared energy is horizontal (s-polarized). This represents the $A_{\parallel}$ or the absorbance spectrum for parallel polarized light. Likewise at 90° or 270° rotational settings the transmitted electric field vector (E) of the infrared energy is vertical (p-polarized). This represents the $A_{\perp}$ or the absorbance spectrum for perpendicularly polarized light.

B. Polymer Stretching

If an automated polymer stretcher is available, this is used as per manufacturer's instructions. If no special apparatus is available the following manual technique is used. A manual devise is constructed for stretching the polymer film over a small area for measurement. The polymer films are cut in approximately 4 cm (width) by 6 cm (length) strips. Each side of the strip is mounted using standard adhesive tape to 2 metal pieces (approximately 4 cm by 7 cm). The polymer is measured at initial length (without stretching) using both 0° ($A_{\parallel}$) and 90° ($A_{\perp}$) polarizer settings. Then by gently stretching the polymer film in approximately 2–3 mm increments the sample is repeatedly measured for each length change using both 0° ($A_{\parallel}$) and 90° ($A_{\perp}$) polarizer settings. The stretch length change is determined by marking the film at initial length with 2 marks approximately 1 cm apart. These marks are best made using a fine point permanent marker and are manually measured as each 2–3 mm stretch is completed. The spectral measurements are best made using a standard infrared film holder during measurement.

C. Infrared Measurements and Reporting

Measurements are made as 0° ($A_{\parallel}$) and 90° ($A_{\perp}$) starting at $\alpha = 1.0$ (initial) and at 5 or more increasing stretch lengths (*Ls*). Select the functional group bands of interest on the infrared spectrum (Note: The absorbance of the band should be less than ≤ 0.2 A and the bandwidth at half-height should be ≤ 30 cm^{-1}. This assures the highest linearity and photometric accuracy of the infrared instrument). The absolute absorbance values are measured for these selected infrared bands for both the 0° ($A_{\parallel}$) and 90° ($A_{\perp}$) spectra. When these results are tabulated, calculate and plot Pr (ordinate) versus α (abscissa). Report final dichroic results in tabular form as R, ΔA, ΔAr, and Pr for all 6 measurements for two bands per sample following the equations given in Section C of this Appendix E.

References

1. N. B. Colthup, L. H. Daly, and S. E. Wiberley, "Introduction to Infrared and Raman Spectroscopy," Academic Press, San Diego, CA, 1990, p. 99.
2. J. Harrick, Harrick Scientific Corp., Ossining, NY, 1995.

Index

D

T